Paleontology and Paleoenvironments

Edited by Brian J. Skinner

A volume in the series EARTH AND ITS INHABITANTS
Readings from *American Scientist*

Readings from
American Scientist

Paleontology and Paleoenvironments

Edited by

Brian J. Skinner

Yale University

Member, Board of Editors
American Scientist

WILLIAM KAUFMANN, INC. LOS ALTOS, CALIFORNIA

ISBN 0-913232-93-9

Contents

Introducing

Earth and Its Inhabitants

A new series of books containing readings originally published in *American Scientist*.

The 20th century has been a period of extraordinary activity for all of the sciences. During the first third of the century the greatest advances tended to be in physics; the second third was a period during which biology, and particularly molecular biology, seized the limelight; the closing third of the century is increasingly focused on the earth sciences. A sense of challenge and a growing excitement is everywhere evident in the earth sciences—especially in the papers published in *American Scientist*. With dramatic discoveries in space and the chance to compare Earth to other rocky planets, with the excitement of plate tectonics, of drifting continents and new discoveries about the evolution of environments, with a growing human population and ever increasing pressures on resources and living space, the problems facing earth sciences are growing in complexity rather than declining. We can be sure the current surge of exciting discoveries and challenges will continue to swell.

Written as a means of communicating with colleagues and students in the scientific community at large, papers in *American Scientist* are authoritative statements by specialists. Because they are meant to be read beyond the bounds of the author's special discipline, the papers do not assume a detailed knowledge on the part of the reader, are relatively free from jargon, and are generously illustrated. The papers can be read and enjoyed by any educated person. For these reasons the editors of *American Scientist* have selected a number of especially interesting papers published in recent years and grouped them into this series of topical books for use by nonspecialists and beginning students.

Each book contains ten or more articles bearing on a general theme and, though each book stands alone, it is related to and can be read in conjunction with others in the series. Traditionally the physical world has been considered to be different and separate from the biological. Physical geology, climatology, and mineral resources seemed remote from anthropology and paleontology. But a growing world population is producing anthropogenic effects that are starting to rival nature's effects in their magnitude, and as we study these phenomena it becomes increasingly apparent that the environment we now influence has shaped us to be what we are. There is no clear boundary between the physical and the biological realms where the Earth is concerned and so the volumes in this series range from geology and geophysics to paleontology and environmental studies; taken together, they offer an authoritative modern introduction to the earth sciences.

Volumes in this Series

The Solar System and Its Strange Objects Papers discussing the origin of chemical elements, the development of planets, comets, and other objects in space; the Earth viewed from space and the place of man in the Universe.

Earth's History, Structure and Materials Readings about Earth's evolution and the way geological time is measured; also papers on plate tectonics, drifting continents and special features such as chains of volcanoes.

Climates Past and Present The record of climatic variations as read from the geological record and the factors that control the climate today, including human influence.

Earth's Energy and Mineral Resources The varieties, magnitudes, distributions, origins and exploitation of mineral and energy resources.

Paleontology and Paleoenvironments Vertebrate and invertebrate paleontology, including papers on evolutionary changes as deduced from paleontological evidence.

Evolution of Man and His Communities Hominid paleontology, paleoanthropology, and archaeology of resources in the Old World and New.

Use and Misuse of Earth's Surface Readings about the way we change and influence the environment in which we live.

Introduction: Paleontology and Paleoenvironments

What preceded us and what were past environments like? Patterns of the present are the result of a multitude of past events, and thus it is not surprising that nothing seems to stir more interest than the past. Where life itself is concerned we start with a statement of cold fact: life is a chemical system and, like all chemical systems, reflects the kind and abundance of its chemical components. Our chemical constituents are what they are because of the still imperfectly understood chemistry of the primitive ocean and atmosphere whence our earliest ancestors drew their formative components.

But life is also a chemical system that interacts with its surroundings; after life originated it had a rapid, widespread, and irreversible effect on the atmosphere, the surface waters, the soils, and the sediments. Preston Cloud's paper makes a most appropriate beginning for this volume because it deals with the evolution of the global ecosystem in which life appeared and in which it played a shaping role thereafter. But mysterious and unique though life may be, it must still be viewed as a chemical system; and in the second paper Calvin discusses some of the most probable chemical steps along the way to the earliest life forms and then the subsequent molecular steps that led to complex life with its genetic code.

But how is it that a living world now occupied by five different kingdoms of organisms—bacteria, single-celled protozoa and algae, fungi, plants, and animals—apparently descended from one kind of ancestor? Lynn Margulis presents an intriguing case to suggest that an evolutionary process called serial symbiosis may have operated on common, primitive, bacteria-like ancestors. The Margulis suggestion is challenging and exciting; many aspects cannot yet be evaluated. One intriguing step along the way, as discussed by Cohen, is the possibility that certain organelles, such as photosynthesizing chloroplasts found within the cells of plants, actually originated as primitive, bacteria-like organisms living symbiotically within larger host cells. This symbiotic relationship, which is one major feature distinguishing the relatively complex eukaryotic cells from the simpler cells of prokaryotes, seems to record one of the great steps of evolution. With the appearance of eukaryotes, paleontological evidence accumulated rapidly and the number of species blossomed; but interpretation of the evidence as to the actual mechanisms regulating species growth remains controversial. The papers by Valentine and Campbell and by Raup address the question.

Some life forms of great antiquity still remain on Earth and still produce the same kinds of structures found in the fossil record. Most ancient fossils are microscopic and very poorly preserved in the rock record. The great diversity of animal life first becomes apparent in rocks little more than 600 million years old. Prior to that time, as discussed in the example by Cloud, Wright, and Glover, all complex animals were soft-bodied and left only tantalizing imprints of their existence—if they left any evidence at all—when they happened to die and be buried under very special circumstances. From 600 million years ago to the present day, however, hard coverings and skeletons developed in a number of animal groups; the fossil record proliferates and evolutionary developments can be studied in great detail. Pojeta and Runnegar discuss the early history of the mollusks, Robert Linsley the extraordinary development of gastropods, while Carpenter discusses the evolution of that most sizable, intriguing, and diverse of all animal groups, the insects.

Fossil specimens considered alone are important, but equally interesting is a re-creation of the environment and communities in which long-dead plants and animals lived, a topic analyzed by Léo Laporte. Environment is, of course, influenced by climate, which in turn is influenced by latitude and the disposition of land masses. Once we can decipher paleoenvironments we can, in theory, understand past climates and geographies. Indeed, our understanding of paleoenvironments and the great advances of paleoecology have been the key to one of the most exciting advances of modern geology. Palmer discusses the Cambrian world, while Bambach, Scotese, and Ziegler, in a paper that demonstrates one of the triumphs of modern geology, recreate the changing geography of the world throughout the Paleozoic. The subsequent two papers, by Wolfe and by Raven and Axelrod, analyze the geographies, floras, and faunas of North and South America during the Tertiary.

We humans are vertebrates, so it is not surprising that the fossils that hold the greatest fascination for many of us are vertebrate fossils. The first paper on vertebrates, by Ostrom, focuses on those once-prominent reptiles, the dinosaurs. Flight arose among one branch of the dinosaurs—probably as a food-gathering aid. Therefore, suggests John Ostrom, birds are present-day descendants of the dinosaurs. Fascinating though dinosaurs are, the mammals, and particularly the hominids, are the greatest attraction for many non-zoologists. Among the best-preserved mammalian remains are teeth and jaws, and therefore much of what we know of mammalian evolution is drawn from this record. The evolution of dentitions is discussed by Osborn, the entire chewing apparatus by Crompton and Parker.

The final papers in the volume focus attention on the extraordinary advances in primate and hominid paleontology over the past 30 years. Radinsky considers the evolution of the primate brain; Tattersall and Eldredge weed out some of the fantasy and myth that have grown up around the recent discoveries; and Richard Leakey, a member of a prominent family among the revolutionaries of hominid paleontology, discusses the remarkable richness and diversity of fossil discoveries made in Africa.

Suggestions for Further Reading

Banks, H.P., *Evolution and Plants of the Past* (Belmont, California: Wadsworth, 1970).

Cloud, Preston, *Cosmos, Earth and Man: A Short History of the Universe* (New Haven, Connecticut: Yale University Press, 1978).

Colbert, E.H., *Evolution of the Vertebrates: A History of the Backboned Animals Through Time*, 3rd ed. (New York: John Wiley & Sons, 1980).

Dott, R.H. and R.L. Batten, *Evolution of the Earth,* 3rd ed. (New York: McGraw-Hill, 1980).

Hallam, A., editor, *Patterns of Evolution; As Illustrated by the Fossil Record* (Amsterdam, Holland: Elsevier, 1977).

Laporte, L.F., *Ancient Environment,* 2nd ed. (Englewood Cliffs, New Jersey: Prentice-Hall, 1979).

McAlester, A.L., *The History of Life,* 2nd ed. (Englewood Cliffs, New Jersey: Prentice-Hall, 1977).

Stanley, S.M., *Macroevolution: Pattern and Process* (San Francisco: W.H. Freeman and Co., 1979).

Vermeij, G.J., *Biogeography and Adaptation: Patterns of Marine Life* (Cambridge, Massachusetts: Harvard University Press, 1978).

Authoritative and up-to-date reviews, summaries, and analyses of many of the topics discussed in this volume can be found in the volumes published by Annual Reviews, Inc., Palo Alto, California 94306. Articles of special interest will be found in the annual volumes of *Annual Review of Earth and Planetary Sciences,* starting with Volume 1, 1973, and *Annual Review of Ecology and Systematics,* starting with Volume 1, 1970.

Particularly appropriate articles from the *Annual Review of Earth and Planetary Sciences* are:

Volume 1, 1973

Crompton, A.W. and F.A. Jenkins Jr., "Mammals from Reptiles: A Review of Mammalian Origins," p. 131–156.

Volume 2, 1974

Clark, George R. II, "Growth Lines in Invertebrate Skeletons," p. 77–100.

Volume 3, 1975

Ostrom, John, "The Origin of Birds," p. 55–78.

Ubaghs, George, "Early Paleozoic Echinoderms," p. 79–98.

Schopf, J.W., "Precambrian Paleobiology: Problems and Perspectives," p. 213–250.

Stanley, S.M., "Adaptive Themes in the Evolution of Bivalvia (Mollusca)," p. 361–386.

Volume 4, 1976

Simpson, G.G., "The Compleat Paleontologist," p. 1–14.

Volume 5, 1977

Palmer, A.R., "Biostratigraphy of the Cambrian System," p. 13–34.

Wright, H.E. Jr., "Quaternary Vegetation History: Some Comparisons between Europe and America," p. 123–158.

Volume 6, 1978

Durham, J.W., "The Probable Metazoan Biota of the Precambrian as Indicated by the Subsequent Record," p. 21–42.

Volume 7, 1979

Russell, D.A., "The Enigma of the Extinction of the Dinosaurs," p. 163–182.

Volume 8, 1980

Gingerich, P.D., "Evolution of Early Tertiary Mammals," p. 407–424.

PART 1 *Origin and Diversification of Life*

Preston Cloud

Evolution of Ecosystems

Analysis of the sedimentary record and its contained nannofossils enables the biogeologist to organize the last 3.4 aeons of Earth's history in a skeletal phylogeny of ecosystems

How did we get where we are and where do we go from here? These absorbing questions, so much in the public consciousness nowadays, are also central concerns of ecology, including paleoecology and environmental biogeochemistry. To deal with the environment is to discover the importance of historical perspective (e.g. Whittaker and Woodwell 1970). And historical perspective illuminates the questions and focuses the search for solutions.

The relevant basic principle was articulated by Henry Louis LeChâtelier (1850–1936): when a stress is brought to bear on a system in equilibrium, the system tends to react in such a way as to achieve a new state of equilibrium. In biological systems we call this homeostasis. The principle becomes highly general if we equate equilibrium with steady state and allow that the new condition need not be identical with that disturbed, so that systems may evolve. In this sense, LeChâtelier's prediction is germane to all systems.

The author is in the Department of Geological Sciences at the University of California, Santa Barbara, where he teaches and directs the work of the Biogeology Clean Laboratory. Before that he spent 16 years with the U.S. Geological Survey and taught briefly at several other universities. He is a member of various professional societies and has been honored with several medals and awards. *The research on which this paper is based was supported by NSF Grant No. GB-23809 and NASA Grant No. NGR-05-010-035. It is Contribution No. 46 of the Biogeology Clean Laboratory, and it was originally presented as the keynote speech at a symposium on environmental biogeochemistry, held at Logan, Utah, in late March 1973 under the joint auspices of the National Science Foundation, the Utah State University, and the Organic Geochemistry Division of the Geochemical Society. It is here published with the kind permission of the organizing committee of that symposium.* *Address: Department of Geological Sciences, University of California, Santa Barbara, CA 93106.*

I have myself long been occupied with the reconstruction of early ecosystems, bemused by the complexity of modern ones, conscious of the connections between them, and acutely aware of the range of expertise that must be called upon in seeking to reconstruct an integrated history of ecosystem evolution. It is evident that our perceptions of the present, the past, and the future states of the world are in some sense functions of the degree to which we grasp the connections between them, and that these connections become more general as we are able to reduce them to ever smaller numbers of independent elements.

Thus I will not attempt here to summarize the wealth of previously published analytical detail about the biogeochemical composition of sediments, most of which is available in published compendia by Breger (1963), Kuznetsov et al. (1963), Manskaya and Drozdova (1968), Eglinton and Murphy (1969), and Ponnamperuma (1972). In fact, the compounds reported are of as-yet-uncertain origin and obscure evolutionary significance, although Scott et al. (1970) and others are probably on the right track in tackling the kerogen (insoluble polymeric substances) as at least presumptively coeval with sedimentation.

Nor will I attempt to treat comprehensively of pre-Phanerozoic paleomicrobiology (summarized from the yet meager published record by Schopf, in nearly identical papers in 1970 and 1972a), microbial evolution (e.g. Hall 1971), geochemical evolution, or the diverse records of Phanerozoic evolution—all of which have been done elsewhere. Rather, my aim here is to suggest some of the points of contact that appear to me to be significant in creating a sense of ecosystem evolution and of its connections with paleomicrobiology and biogeochemistry.

In contrast to biological communities, which consist of assemblages of interacting and coevolving species, the ecosystem is the spatial complex (including biological and physical sectors and interactions) within which community evolution takes place. With ecosystems there is an intimate functional interchange of energy and components between living and nonliving sectors. Within the global ecosystem at any given time are many lesser ecosystems, each inhabited by distinctive, usually evolving communities. All of these are interconnected in some sense with all other components of the global ecosystem, and with all other ecosystems that have existed on earth back to the origin of life itself, perhaps 3.4 aeons (years times 10^9) ago. They are the antecedents and in significant part the determinants of those ecosystems that exist today and that will exist in the future.

I seek the elements of a phylogeny of ecosystems from earliest times to the present. Ecosystems have a central biochemical-biogeochemical component and are the product of an evolutionary past. They would not be ecosystems without organisms which interact with and modify their physical and chemical environments in profound and pervasive ways. An unending succession of plants and animals has both molded the directions of environmental evolution and left within the sedimentary record a palimpsest of its own morphogenetic, biogeochemical, and paleoecological evolution.

Reconstructing the past

It is self-evident that, once life appeared on Earth, the geochemical record could never be the same again. Since life is a chemical system that involves constant interchange of components with the surrounding media, its appearance necessarily had an immediate and irreversible effect on the composition of atmospheric gases, water, and sediments deposited. Similarly, life could not have arisen without an atmosphere and hydrosphere to provide the indispensable fluid media within which the essential components of living organisms could be suspended and transferred.

The only concrete evidence we have as to the time of origin of an atmosphere and hydrosphere is the appearance in the geological record of normal sedimentary rocks showing evidence of chemical weathering and subaqueous deposition. The oldest surely sedimentary rocks we know on Earth of this or any other kind—in eastern Transvaal, South Africa (Anhaeusser 1972; Hurley et al. 1971)—appear to be slightly less than 3.4 aeons old. [While this was in press, metasediments 3.76 aeons old or older were reported from West Greenland by Moorbank, O'Nions, and Pankhurst 1973.] Evidence of chemical weathering and subaqueous deposition preserved in these rocks is important above all for its demonstration that a substantial atmosphere and hydrosphere already existed on Earth at that time. Chemical evolution leading to the origin of life, with the diversity of biogeochemical interactions implied by that term, was now possible if not inevitable. Indeed, experience with later developments in evolution suggests that whenever environment (including its biological components) became conducive to new biological directions, such directions were likely to ensue. Thus, we may hypothesize that there would be little lag between the onset of water-deposited sediments and the appearance of life on Earth.

That is, of course, provided there was little or no free oxygen in atmosphere and hydrosphere. Free oxygen in more than evanescent amounts would preclude the origin of life in several ways. Essential precursor molecules for living systems as we know them would not form, first because they are so readily oxidized, and second because high-energy ultraviolet radiation (UV), the most likely energy source, would not penetrate the ozone screen that results from the presence of much free oxygen (>1 percent present atmospheric level).

If such molecules did somehow form, rapid oxidation would probably keep them from accumulating in sufficient quantities over enough time for more complex molecules to evolve, accumulate, and combine to form a bounded, negentropic, self-replicating, and mutable system of the type we would call living. If a living cell did, nevertheless, manage to evolve in some local reducing environment, it could not survive transfer to the larger oxidizing environment in the absence of suitable oxygen-shielding systems, O_2 being lethal to all forms of life in the absence of such systems. Additional evidence that life originated under essentially anoxygenous conditions is that the most basic metabolic processes of all organisms are anaerobic and that biochemical evolution has gone to great lengths to carry out most of its oxidations by removal of hydrogen rather than by addition of oxygen (e.g. Wald 1964).

Thus we may surmise with some confidence that life on Earth arose beneath an essentially anoxygenous atmosphere not much later than about 3.4 aeons ago when the oldest known sediments appeared. Although formaldehyde and other polyatomic organic molecules have now been found to be widespread in interstellar space and may have been present in the dust cloud from which our planet condensed, that roughly 3.4-aeon-old terrestrial event was the beginning of the only ecosystem evolution and environmental biogeochemistry we yet know.

The geological framework within which this ecosystem evolution took place consists of five great divisions, each hundreds of millions of years long. The divisions reflect the major modalities of biogeochemical and physical evolution, including the broader aspects of the separate but related and interacting evolutions of atmosphere, hydrosphere, biosphere, and lithosphere. These five major divisions of earth history, summarized in Figure 1, from oldest to youngest, are:

1. The *Hadean*. From the origin of the earth to just before the appearance of the oldest sedimentary rocks—about 4.6 to 3.4 aeons ago. Records of Hadean events are largely obscure or missing except for the last 300 million years or so, when there seems to have been extensive production of granitic igneous rocks and gneissic metamorphism. During this interval, the aggregation of the earth was completed, Earth's core and mantle were differentiated, the moon came into orbit, and a major terminal thermal event widely reset older radiometric clocks—apparently on both Earth and its moon. Whatever primary atmosphere may have survived the aggregation of Earth was probably lost at that time or swamped by the voluminous escape of juvenile volatiles associated with pervasive igneous intrusion (plutonism), volcanism, and resetting of radiometric clocks. The Hadean came to a close when atmospheric weathering and hydrospheric condensation became sufficient for the generation and preservation of aquatic sediments.

2. The *Archean*. From the appearance of the oldest sedimentary rocks to the onset of widespread but generally unoxidized platform-type sediments—about 3.4 to 2.6 aeons ago. The Archean was a time of continued widespread, mainly basic volcanism and granitic intrusion in which both volcanic and plutonic rocks were, as a rule, relatively low in potassium and high in sodium (geochemically primitive). To these were added immature, unoxidized sediments consisting primarily of volcanic source materials. Volcanics

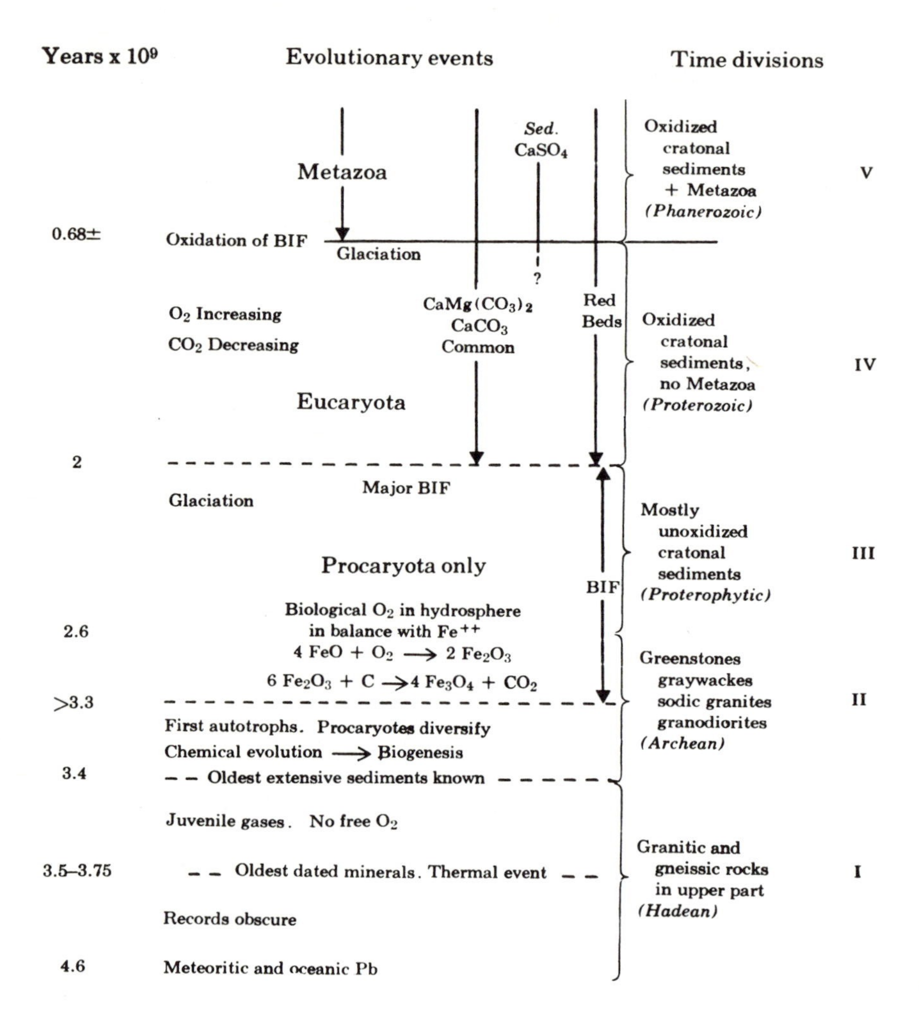

Figure 1. Main events on the evolving earth. (Modified after Cloud 1972.)

of types we would now call oceanic, together with volcanigenic and local immature quartzose sediments, were deposited in subsiding belts between rising, lighter, in part contemporaneous and in part older plutonic rocks. Sediments from granitic sources are found in the upper parts of some such sequences, and local hematitic banded iron deposits indicate the presence of restricted sources of oxygen in a generally anoxygenous system. The Archean was terminated by the appearance of widespread, but still mainly unoxidized, shelf sediments of the kind that are deposited in generally shallow waters on the submerged surfaces of continents or cratons. Archean types of rocks are found in later times but only locally, to the exclusion of conspicuously cratonal sediments.

3. The *Proterophytic*. From the first widespread appearance of typically cratonal sediments until subaerially oxidized sediments (e.g. red beds) and carbonate rocks became important and widespread—a range from about 2.6 to 2 aeons ago. Proterophytic deposits are the oldest we know that are at many places relatively undeformed. Deposited extensively in the oldest truly epicontinental seas and on the oldest truly continental surfaces, they have been widely (but by no means universally) shielded from the compressive forces that disrupt and fold the deposits of continental margins and intercratonal belts.

The granitic and most other rocks of this interval generally have higher (less primitive) potassium/sodium ratios than those in the Archean, except where the latter have undergone potassium metasomatism. The oldest extensive glacial deposits known are of upper Proterophytic age (e.g. Gowganda and related formations), although older ones are reported locally. The Proterophytic was also a time of exclusively procaryotic (essentially asexual) evolution and of hematitic and siliceous banded iron formations. It closed with the near-disappearance of banded iron formation and the onset of important continental and marginal marine detrital hematitic sediments—the so-called red beds.

4. The *Proterozoic* (or Paleophytic, if a new term is needed). From the onset of conspicuously oxidized cratonal sediments to the origin of differentiated multicellular animal life (Metazoa)—about 2 to 0.68 aeons ago. Along with the absence of Metazoa, Proterozoic deposits are characterized by the frequency of the detrital hematite-rich sediments called red beds, the prevalence of carbonate rocks (dolomite and limestone), a variety and abundance in the latter of the laminated blue-green algal-bound (also bacterial) deposits called stromatolites, the relative scarcity of banded iron formation, and the rarity of sedimentary sulfate deposits except locally in the upper part. Following, or toward the end of, a long and widespread interval of glaciation, the Proterozoic was terminated with the appearance of the Ediacarian fauna—a conspicuous assemblage of soft-bodied Metazoa of floating and nonburrowing benthic types that marks the beginning of Phanerozoic time.

5. The *Phanerozoic*. From the appearance of Metazoa until the present—the last 680 million years. This is above all the time of elaboration of metazoan and metaphytic variety during which the modern world took shape. Sediments are similar to those of the Proterozoic except for the common and locally voluminous occurrence of sedimentary sulfates and other evaporites and the common preponderance of limestone over dolomite among carbonate rocks. The Phanerozoic will end when multicellular plant and animal life becomes extinct.

Hadean

We do not know whether there were any Hadean ecosystems. What we suspect, from the extreme depletion in noble gases of the modern terrestrial atmosphere as compared with cosmic abundances (Russell and Menzel 1933), is that either very little primary atmosphere ever existed, or that it was mostly lost in some subsequent thermal event that raised surface temperatures to above 5,000°K—otherwise, the terrestrial noble gases should approach cosmic abundances. [Fanale 1971, a paper that came to my attention while this was in press, argues forcefully for selective primary outgassing.] If there was Hadean life, it was probably wiped out during the same late Hadean global and lunar thermal event that obscured older records, if any, of sedimentary deposits and of atmospheric and hydrospheric history.

This same thermal event, however, set the stage for the beginning of Earth's recorded biogeochemical and ecosystem evolution through the voluminous outgassing that accompanied its plutonism and volcanism. That, I suggest, was the beginning of the atmosphere and hydrosphere from which ours has evolved: an atmosphere primarily of gases once chemically bound in Earth's interior—of CO, H_2O, N_2, H_2, and probably CO_2, HCl, H_2S, and some but probably not much NH_3 and CH_4 (Rubey 1951; Holland 1962)—and a hydrosphere from condensation of the H_2O. This outgassing set the stage for the chemical evolution that would lead to the origin of life and thereby of ecosystems.

In the discussion that follows, I speak primarily of very broadly defined and long-lasting global ecosystems. That is the best we can do at this juncture for the first nearly four aeons of geologic time, and it would warp balance and unduly prolong discussion to attempt here to probe the decipherable complexity of Phanerozoic ecosystems.

Archean

The oldest definite evidence of a substantial atmosphere and hydrosphere consists of the sedimentary rocks of the Swaziland System, eastern Transvaal, Republic of South Africa, of earliest Archean age (summarized by Anhaeusser 1972). The general sequence and nature of these rocks is outlined in Figure 2. The oldest have been

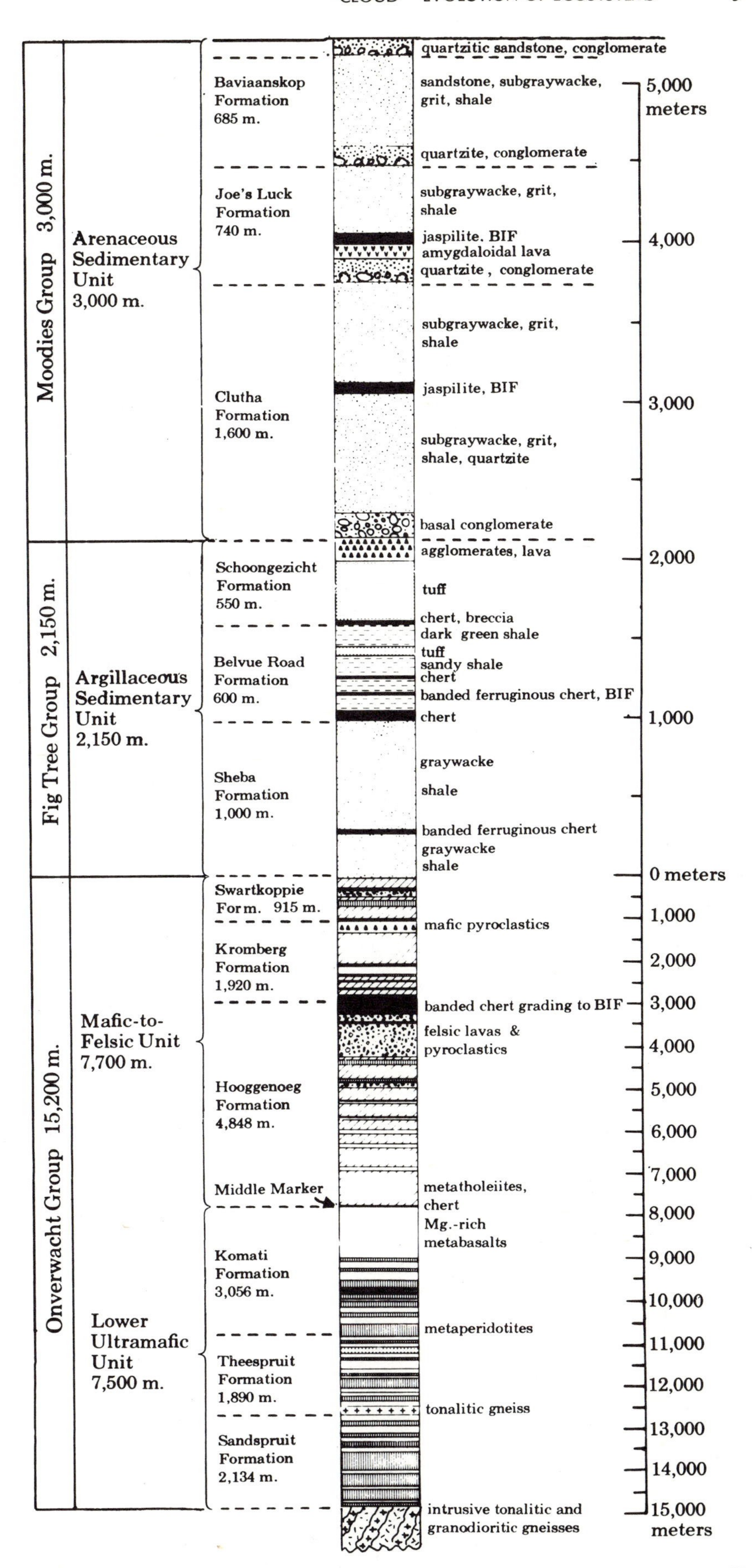

Figure 2. Swaziland System, Barberton Mountain Land, South Africa. Composite section Onverwacht Anticline, Kromberg Syncline, Stolzberg Syncline, Eureka Syncline. (Modified after Anhaeusser 1973, Fig. 1. Basic data from work of R. and M. Viljoen on Onverwacht, T. Reimer on Fig Tree, and C. Anhaeusser on Moodies.)

dated by Hurley et al. (1971) from a whole-rock Rb-Sr isochron in a cherty claystone near the middle of the basal Onverwacht Group as having an age of 3,375 ± 20 m.y.—call it 3.4 aeons. The atmosphere was probably dominated by outgassed compounds. Hoering and Abelson (Abelson 1966) have produced HCN, CH_4, and a variety of amino acids from UV irradiation of mixtures of CO, N_2, and H_2—presumably common products of the outgassing visualized.

Such reactions, plus probably the formation of formaldehyde, would have been continuous in an atmosphere of juvenile gases in the absence of free O_2; and despite the abundance of combined oxygen, no juvenile source of O_2 is known. A number of investigators, in many well-known experiments, have now produced amino acids from various plausible primitive atmospheres using UV and other plausible energy sources, linked them into polypeptides, made nucleotides and polynucleotides, and taken most of the essential steps in biogenesis up to, but still short of, the point of showing how the polynucleotides could be linked together with sugar-phosphate chains and packaged with other components within lipid and lipoprotein membranes to form self-replicating, mutating, negentropic units we would all recognize as living cells.

Can we set a time for the origin of life? One interesting response to this question is that of Scott et al. (1970) who thought they found a change from mainly aromatic kerogen deposition products in the Onverwacht beds to aromatics plus an abundance of *N*-alkanes in the overlying Fig Tree Group, which they cautiously hinted could coincide with the appearance of life. Yet unpublished results obtained in my laboratory by Togwell Jackson suggest, however, that the aromatic component of the Onverwacht kerogen may be simply less highly condensed, and not more plentiful, than that of the Fig Tree, negating its significance for the timing of biogenesis.

Another approach to the question is that of Oehler et al. (1972), who found anomalously heavy $\delta^{13}C$ values of −16.5 per mil in the kerogen and total organic fraction of lower Onverwacht sediments (Theespruit Formation), as compared with values mainly less than −25 per mil for the middle and upper Onverwacht (Hooggenoeg and Kromberg formations). Oehler and her associates point out that the relatively large depletion of ^{13}C with reference to ^{12}C observed in the upper formations, in contrast to the lesser depletion in the lower part of the Onverwacht, would be consistent with the interpretation that photosynthesis had not evolved at the time the lower Onverwacht was being deposited, but that it did evolve soon thereafter. That attractive line of reasoning seems not only to push back the possible time of origin of photosynthesis to a very early date but also implies a still older time for the origin of life itself, simply because of the large improbability that the first organisms would have been photosynthesizers.

Flimsy as it still is, the evidence implies that the first ecosystem already existed in early Onverwacht time, and that, in the absence of oxygen and the opportunity for more advanced biochemical evolutionary development, it involved only the most basic interactions with simple procaryotic communities of anaerobic heterotrophs that would probably be called bacteria if they were alive today. These presumably obtained their vital energy from elementary substrate-level phosphorylation combined with assimilation of organic compounds of nonvital origin and their own cellular degradation products, as is common among *Clostridium* (a bacterium) and the Lactobacillaceae today (Hall 1971). An expectable early advance would have been the coupling of phosphorylation to anaerobic electron transport, as in other clostridia (Hall 1971).

The addition of the dead cells and metabolic products of such organisms to sediments being deposited, and their gaseous interactions with the watery milieu essential to survival and thus with the atmosphere, began a mutual interaction that has continued to the present. The impounded hydrosphere in which the primary interactions occurred was, on grounds advanced and effectively supported by Rubey (1951), presumably salty from the beginning; but the absence of bedded sulfate *deposits* (individual gypsum pseudomorphs are not uncommon) anterior to the last aeon or less of earth history can be taken to imply that sulfate was not a common ion in older seas, and that the composition of their dissolved salts may have been different from and the salts less than now. (The salinity may have continued to increase until about 680 m.y. ago, by which time the sea was probably approaching its present salinity and volume.) Such was the simple and, by present standards, highly polluted initial ecosystem. We may call it Archean 1.

Archean 2 is visualized as a relatively brief and rather uncertain interval that began with the appearance of $\delta^{13}C$ ratios indicative of photosynthesis in early upper Onverwacht time, perhaps 3.35 aeons ago, implying that autotrophs probably arose rather soon after life itself. The selective advantages of autotrophy in the presence of a fixed food supply, or one to which only limited additions were being made, make this predictable. The chemoautotrophs, however, are all biochemically advanced aerobes, while the presence of abundant, easily oxidized, detrital pyrite (FeS_2) and uraninite (UO_2) in stream sediments as young as about 2.1 to 2.3 aeons reinforces other evidence that there could have been no significant amount of free O_2 around in Archean time. In addition, banded ferruginous cherts that grade locally into banded iron formation (BIF) are known as far back as upper Onverwacht (top of Hooggenoeg Fm., R. and M. Viljoen, letter of 12 July 1973), while BIF is well developed in the middle and upper Swaziland, Fig Tree and Moodies groups (Fig. 2). Thus, although there is no indication that any free oxygen was being generated (biologically or otherwise) within the hydrosphere of lower Onverwacht time, there is evidence of its at least fleeting presence as early as upper Onverwacht.

We may infer from this that the initial photoautotrophs whose fixation of carbon caused relative enrichment of ^{12}C in middle Onverwacht sediments may have employed an anaerobic system, as in the photosynthetic bacteria. If inorganic acceptors could be used for electrons generated in the oxidation of organic substrates, more of the potential energy from these substrates could be used. Common among such systems, and perhaps primitive, is the reaction $2H_2S + CO_2 \rightarrow (CH_2O) + H_2O + 2S$ (the sulfate- and nitrate-reducing bacteria would have to await adequate substrates). As Yčas (1972) has pointed out, variations of this reaction lead to the production of gaseous H_2 and CH_4, which in turn would provide sinks for photosynthetically (and photolytically) generated O_2, thereby enhancing the production of CO_2 and H_2O and inhibiting accumulation of free O_2.

A condition for the functioning of this ecosystem is that heme-containing electron-carrier proteins, cytochromes or their antecedents, as well as a chlorophyll, had already made their appearance. As T. Jackson points out elsewhere (1973; in prep.), a prerequisite for such evolution may have been the prior evolution of UV-absorbing aromatic secondary metabolites, especially condensed phenolic compounds that could have offered shielding against UV irradiation and which, his data suggest, increased in amount until about 2 aeons ago, after which they declined again. Such an adaptation would have increased the range of living sites available to early microbial procaryotes and could have set the stage for more advanced forms of photosynthesis utilizing similar absorptive pigments. This second postulated global and Archean ecosystem was a little advanced over the first wholly heterotrophic one, but not much.

I suggest, using the oldest BIF as my benchmark, that the third discernible ecosystem, characterized by oxygen-releasing photoautotrophy—with chlorophyll *a* or an antecedent, ferredoxin or an antecedent, and a primitive cytochrome system—began somewhat before 3.3 aeons ago in latest Onverwacht time (early but not earliest Swaziland time). Free oxygen, however, did not accumulate in the hydrosphere or atmosphere until over a billion years later. In addition to the previously mentioned detrital FeS_2 and UO_2, the basis for that conclusion is the abundant presence of cherty BIF in the geological column during the interval specified, combined with the apparent absence of red beds or other oxidized terrestrial sediments over the same time.

BIF is a chemically deposited sediment of open water bodies that consists of laminae and bands of iron-rich and iron-poor silica (chert) whose episodic banding suggests cyclical systems in which divalent iron in solution alternately became widely dispersed and then precipitated in ferric combinations during limited time and in fluctuating balance with internal sources of oxygen. A biological source of oxygen would serve nicely. Hence, I hypothesize that ferrous iron in solution was the oxygen-acceptor that kept the ambient O_2 level of the aquatic ecosystem sufficiently depressed to be tolerated by the first oxygen-releasing photosynthesizers in the absence of suitably *advanced* internal oxygen-mediating systems, which presumably were not fully developed when the first protoalga learned to split the water molecule, releasing free oxygen. Just as even now (Stewart and Pearson 1970) some blue-green algae depend on H_2S to hold the oxygen level down (while others seem capable of heterotrophic existence: Saunders 1972), so in Archean and Proterophytic times their early predecessors may have found ferrous iron useful for the same purpose.

The appearance for the first time of BIF in the Hooggenoeg Formation of upper Onverwacht age, as well as in the Fig Tree and Moodies groups, signals the beginning of the third, last, and longest-lasting Archean ecosystem. This ecosystem was characterized, indirect evidence suggests, by assemblages of minute and morphologically simple, but probably biochemically diverse, bacteria and procaryotic protoalgae whose gaseous interchanges with their environment affected the sedimentation of BIF and whose remains account for the carbonaceous shales and graphitic schists of this interval. (I discuss the paleoecological implications of BIF in detail elsewhere: Cloud 1973b.)

In the general absence or very low level of oxygen (<1 percent of the present atmospheric level) and hence no ozone screen, these microorganisms would have been limited to sites sheltered from UV radiation, as at depths of water below about 10 meters, except to the extent that UV-absorbing (and intracellular oxygen-shielding) aromatic chromophores allowed them to range higher. Locally, as in stromatolites and other sediment-binding algal-mat communities, they may also have been sheltered within intertidal to shallow subtidal or even terrestrial sediments, but evidence of such communities is rare in rocks of this age. Some may even have lived in permanently shaded areas below or above water. At one site, in young Proterophytic rocks perhaps 2.1 aeons old, in northeastern Minnesota, we know a teeming and exquisitely preserved procaryotic microbiota that lived embedded in a silica-gel deposit in deep irregular cracks in the wave zone of a rocky beach (Pokegama Formation, Cloud and Licari 1972).

The prevalence of DNA repair mechanisms among the bacteria and blue-green algae (BGA) is a further UV-countering device that may have originated at this time. It is also likely that biological nitrogen fixation was an Archean invention. Essential nitrates would not readily form abiologically in an atmosphere devoid of free O_2, just as the presence of free O_2 inhibits biological nitrogen fixation. Nitrogen fixation, so characteristic of the procaryotes, thus would have been essential to the Archean 3 ecosystem but unlikely to have originated after free O_2 became common. Unfortunately, the Archean is as yet poor in authentic records of demonstrably indigenous organic remains. Apart from rare stromatolites (implying, where well-laminated, the activity of filamentous microbes; C. D. Gebelein, letter of 10 June 1973), the only reasonably well-documented record is of rather characterless microspheroids in the Fig Tree Group (Barghoorn and Schopf 1966). All others need verification either as to biologic origin or source.

Archean 3 ended with the onset of extensive cratonal sedimentation, beginning about 2.6 aeons ago. The continued differentiation of Earth's mantle, with increase in the granitic fraction of the crust, led to the emergence and agglomeration of protocontinents into one or a few large continents, compressing and eliminating many of the subsiding depositional basins in which the primitive and largely volcanigenic Archean sediments had accumulated. This presented extensive continental areas to weathering, created the first known epicontinental seas, and gave rise to extensive cratonal sediments such as sandstones, arkoses, and, in local depressions, shales and mudstones. Continental types of sediments are uncommon in the Archean, but in post-Archean time they became conspicuous.

Proterophytic

Thus began the ecosystem we may call Proterophytic 1—or just Proterophytic, as no others are yet recognized in this roughly 600 million-year-long interval from 2.6 to 2 aeons ago. It is differentiated from Archean 3 mainly in the kinds and sites of sediments deposited and by the appearance late in Proterophytic times of filamentous and spheroidal procaryotes showing a cellular differentiation and revealing ultramicroscopical details of the cell envelope that are characteristic of modern nostocacean and chroococcalean BGA (Licari and Cloud 1968b; Cloud and Licari 1972) as well as a diversity of other procaryotic types (Barghoorn and Tyler 1965). The sole Proterophytic ecosystem recognized in the broad sense of this discussion was a precursor blue-green algal and bacterial ecosystem in which there were doubtless a number of variants. They include the classic and important Gunflint microflora from mainly stromatolitic cherts (Barghoorn and Tyler 1965; Cloud 1965), and the somewhat older and less varied benthic but nonstromatolitic, and well-preserved Pokegama microflora (Cloud and Licari 1972).

By Gunflint time, we not only have a variety of filamentous and coccoidal blue-green algae, probably bacteria, and perhaps actinomycetes, for the most part morphologically and perhaps biochemically similar to common modern forms, but also an array of unusual morphologies, some almost identical to forms which today occupy very restricted habitats. One is the peculiar umbrellaform *Kakabekia*, resembling though not identical to a living soil bacterium of ammonia-rich environments. Another is a morphologically diverse assemblage showing specialized reproductive stages that are morphologically identical to those of the living aerophilic but oxygen-transferring genus *Metallogenium*, which Perfil'ev et al. (1964) have described from Fe- and Mn-rich environments in modern Karelian lakes. Identical but younger and even more diverse forms of this genus are still more abundant in the 1.6-aeon-old Paradise Creek Formation of northwest Queensland, now being studied in my laboratory by Gary Kline.

For the first time also in the Proterophytic, we have evidence of extensive continental, as well as marine, sediments, although the continental sediments (e.g. the Huronian Supergroup of southern Ontario) are of odd sorts. As in Archean time, the expected flux of UV radiation makes an extensive Proterophytic land vegetation improbable (except perhaps within soils or in permanently shaded places), and there were certainly no rooted plants that could effectively hold overbank flood sediments in place. Thus, wide braided rivers occupying the entire alluvial valley appear to have been the norm in terrestrial environments. Meandering streams, which do not develop without a substantial floodplain vegetation, were presumably absent (Schumm 1968). In such a situation, finer sediments would be swept downstream to local depressions or the sea, the characteristic later fluviatile associations of interlensing coarse and fine sediments would be absent, and the product of fluviatile regimes would be cross-bedded blanket sands, arkoses, and gravels, with a conspicuous downstream directional component.

Regrettably, demonstrable older Proterophytic sediments are uncommon. (Although it is sometimes suggested that frequency of exposure is a more-or-less symmetrically decreasing exponential function of age, that is not borne out by the record. On the whole, Archean rocks are relatively common at the cratonal cores, while lower Proterophytic rocks seem to be relatively uncommon even for their great age —perhaps because older cratonal sedimentary blankets were generally stripped away.) BIF is one of the most distinctive Proterophytic sediment types. It is prominent in Proterophytic successions, especially in the upper part, apparently converging on an age of around 2 to 2.1 aeons. Deposits of this age on most continents contain individual reserves of tens of billions of tons of minable iron ore associated with even larger volumes of rock that is somewhat poorer in iron but still geologically BIF. Unlike the more limited and lensy Archean BIF and the rare post-Proterophytic ones, these late Proterophytic deposits, as well as individual beds and fine laminae within them, can be traced laterally for hundreds of kilometers (Trendall 1972, 1973). The big iron formations appear to be somewhat younger than the oldest records of clearly large-scale glaciation (Gowganda, Ramsy Lake, and Bruce tillites of Huronian Supergroup and equivalent deposits).

I have suggested elsewhere (Cloud 1973b) that the latitudinal thermal gradient associated with and following this glaciation may have upset the equilibrium of a previously stratified and semistagnant ocean in which the bottom waters were rich in ferrous iron that had accumulated over a long period of time and which was also being replenished from the leaching of low-lying continents. Overturning of the formerly stratified ocean would have brought waters rich in dissolved ferrous iron into the photic zone and across the continental shelves, where planktonic and sediment-binding BGA were living a precarious existence, dependent on oxygen-depressing ions or molecules in the ambient waters to maintain populations of any substantial size.

We know that real BGA of essentially modern types were involved because some very common filamentous forms show cellular differentiation into convincing heterocysts and akinites (Licari and Cloud 1968b). That, in the absence of tapering and false-branching, implies close affinity with the living family Nostocaceae and seems to confirm that biological nitrogen fixation had previously been achieved (e.g. Fay et al. 1968). In addition, common associated unicellular forms were probably chroococcaceans. We postulate that they were still dependent on external ambient oxygen depressors because otherwise it becomes difficult to account for the distribution in time and space and the anomalous geochemistry of the BIF. The very widespread dispersal of chemically deposited iron oxides in layers a few microns to a few millimeters thick is hard to explain unless it was dispersed in solution in the ferrous state in an oxygen-poor aquatic system, while its precipitation as ferric or ferro-ferric oxides requires combination with free oxygen.

I visualize a mutually dependent relationship—biological generation of O_2 in the hydrosphere in balance with ferrous iron, such that $4FeO +$ biol. $O_2 \rightarrow 2Fe_2O_3$. In turn, the Fe_2O_3 would react with sedimented carbonaceous matter such that $6Fe_2O_3 + C \rightarrow 4Fe_3O_4 + CO_2$ (Perry et al. 1973), accounting for the prevalence of magnetite and the rarity of carbon in the unaltered BIF. Associated iron- and manganese-oxidizing bacteria, such as *Metallogenium* and *Siderococcus* (Perfil'ev et al. 1964), could also have helped to maintain low oxygen levels while deriving energy from the oxidation of ferrous iron to the ferric state. Enrichment in iron is episodic within apparently continuously deposited silica on several scales—in laminae from a few microns to a millimeter or so thick, in mesobands a few millimeters to a centimeter or so thick, and, in some areas, in macrobands a meter or more thick alternating with shale. Excellent descriptions are given by Trendall (1968, 1972). The alternation observed implies an episodic balance consistent with the postulated seasonal climatic regime. Some iron was being added most of the time, with larger amounts at seemingly regular intervals. Microbial oxygen-releasing photosynthesizers could fluorish as long as oxygen-depressing ferrous iron and essential mineral nutrients were conveniently available, but would die back whenever ambient ferrous iron or mineral nutrients were temporarily depleted and would not recover to precipitate another iron-rich lamina or band until iron or nutrients were again plentiful.

An episodicity, either of ferrous iron supply or of microbial growth or activity in the presence of a constant supply of iron, could account for the banding observed. Or perhaps both supply of iron and microbial growth varied independently, accounting for different aspects of the banding. The microbanding is thought by Trendall (1972, 1973) and Cloud (1973b) to be probably an annual banding. Trendall (1972) also considered but discarded the possibility of a daily banding, while Walter (1972) has not unreasonably suggested that some very fine banding in the micron range may be daily on the basis of rather striking analogy with modern diurnally banded (and nonbiogenic) siliceous hot spring deposits in Yellowstone Park. Trendall (1972) has further observed that deposits of the Hamersley Group average around 25 microbands per mesoband, which would be consistent with interpreting the mesobands as reflecting a double sunspot cycle, as Bradley (1929) postulated for the Eocene Green River varves. More work will be needed to see whether some of the macrobanding suggests the solar variation cycles noted by Johnsen et al. (1970) in cores of Greenland ice. For instance, might the 600- to 1,750-year periodicity suggested by Trendall (1972) reflect an older variation of one of these cycles?

The source and mode of precipitation of the silica which forms the matrix of the BIF and the rarity or absence of carbonate rocks are puzzles that may have a reverse biogeochemical twist. Inasmuch as there are no known silica-secreting procaryotes and none that directly induce silica precipitation, and since no eucaryotes are certainly known before about 1.3 aeons ago (Cloud et al. 1969; Licari in press), a biological source for the silica seems unlikely. In the absence of biological fixation, the contemporaneous hydrosphere was probably saturated with monosilicic acid (H_4SiO_4)—perhaps solubilized in part from igneous precursor materials by bacteria (Webley et al. 1963) and BGA (Jacks 1953). The polymerization and precipitation of monosilicic acid to form SiO_2 are both favored with decreasing acidity of solutions to a neutral or slightly alkaline state (Krauskopf 1956; White et al. 1956; Siever 1962; summarized in Walter 1972).

Conversely, all alkalinity components at such a low pH are bicarbonate—no carbonate ion is observed at a pH less than 7 and very little carbonate is present below pH 7.5 (Harvey 1955, p. 153). Thus, the polymerization and precipitation of silica would be favored at a pH of around 6.8 to 7.5, while the potential components of carbonate sedimentation would tend to remain in solution as ions of HCO_3^- or come out as ferrocarbonate in some microbands. Such a range of pH seems a not improbable consequence of the expected relatively high partial pressure of CO_2 of the primitive atmosphere, combined with the bicarbonate and silicate buffering systems. We may visualize a continuing polymerization and precipitation of silica gel in the same places to which ferric oxides were being intermittently added to make the BIF. In such a model, it is relatively easy to explain the abundant carbonates younger than the last great BIF as a consequence of pH and carbonate ion increase related to reduction of CO_2 levels caused by the photosynthetic generation and segregation of carbohydrates and O_2. The uncommon but locally thick, extensive, and even stromatolitic pre-BIF carbonates, such as those of the Transvaal System in South Africa, are more difficult. Although we can invent special explanations for these older carbonates, they stand as a warning that the model proposed is not without difficulties.

Consider now the biological, evolutionary, and sedimentological consequences of introducing an abundant supply of oxygen depressor, namely ferrous ion, to an environment where its rarity had previously been limiting. Larger phytoplanktonic and sediment-binding precursor blue-green algal populations would be able to survive than previously. In procaryotic evolution, where genetic recombination is unusual and biological variation arises wholly or mainly from mutation, rapidity of evolution is directly related to number of mutations, hence of individuals. In mainly asexual procaryotic reproduction, any surviving mutant begins a new strain. Diversification, mainly biochemical, would be accelerated with an increase of procaryotic populations. An increase in O_2-releasing photosynthesis would lead to biological depression of dissolved CO_2, increase in pH, increase in carbonate ion, precipitation of $CaCO_3$ from ambient waters, saturation in O_2 of the aquatic environment, evasion of free O_2 to the atmosphere, and oxidation of accumulated reduced substances below and above water.

Because the oldest oxidized terrestrial sediments (red beds) known follow or barely overlap the great late-Proterophytic episode of BIF, we can logically hypothesize that it was during the postulated expansion of procaryotic photosynthesizers and the ensuing burst of biochemical diversification that the oxygen-mediating enzymes first reached something like their present state of effectiveness. Such an advance, freeing its possessors from dependence on external oxygen-depressors and able to spread rapidly because of the brief asexual generations involved, would seem almost instantaneous in the geologic record. The generation of oxygen could now proceed apace, sweeping the oceans free of iron forever, except along the beach, in local interstitial waters, and in euxinic depressions. This in no sense contradicts the probability of a transition interval of still relatively low oxygen tolerance such as among some living blue-green algae (Stewart and Pearson 1970), during which the ferrous-ferric acceptor system kept the oxygen level low in the sea but still permitted the accumulation of atmospheric concentrations sufficient to account for the limited local formation of oxidized terrestrial sediments before they became an important general phenomenon.

Molecular oxygen, being primarily a product of photosynthesis, must of course be in approximate geochemical balance with reduced carbon, with which it persistently seeks to recombine.

Free O_2, whether or not later combined with iron or other noncarbon elements, implies a geochemically equivalent amount of sedimentarily segregated (reduced) carbon. The carbon content of sediments needs to be much more extensively, critically, and quantitatively analyzed than it has been for its bearing on the question of swings in the amount and rate of oxygen storage and accumulation. As previously mentioned, Perry et al. (1973) offer an elegant explanation for the simultaneous rarity of carbon and abundance of magnetite in BIF in the equation $6Fe_2O_3 + C \rightarrow 4Fe_3O_4 + CO_2$; and some of the carbon may be buried with the black pyritic shales of basin centers where it may reach 10 to 12 percent of the deposits (verbal communication from H. L. James, November 1962).

The pyrite itself may have had its source in a reaction such as $2H_2S + FeO \rightarrow FeS_2 + H_2O + H_2\uparrow$, or in the reactions $H_2S + 2CO_2 + 4H_2 \rightarrow H_2SO_4 + 2CH_4$, followed by $2H_2SO_4 + Fe^{++}$ + bacterial action $\rightarrow FeS_2 + 2H_2O$ + 2 bacterial O_2 (combined with an alcohol or sulfide), the H_2S being of volcanic origin or produced by sulfate-reducing bacteria from sulfates generated in a previous cycle of bacterial oxidation. The increase of free oxygen here postulated as beginning about 2 aeons ago should also have its counterpart in carbon-rich deposits somewhere—either the carbon-rich basin sediments mentioned above or something like the perhaps 1.9 to 1.8-aeon-old (early Proterozoic) shungites of Karelia and Siberia. Quantitative comparison, however, remains to be done.

The upper Proterophytic microbial ecosystem with its sediment-binding and -shielding stromatolitic variant in inter- and supratidal to shallow subtidal reaches and its deeper planktonic variant in more open waters came to a close with the evolution of essentially modern oxygen-mediating systems, the systematic and widespread evasion of free biogenic oxygen to the atmosphere, and the onset of terrestrial red-bed sedimentation about 2 aeons ago.

One point worth mentioning before going on to Proterozoic ecosystems (younger than about 2 aeons) has to do with the source of free O_2. The discussion so far has gone along as if it were settled that it was practically all of biologic origin. I have elsewhere discussed (Cloud and Gibor 1970; Cloud 1972, 1973a) why I think that is the case, but, as Brinkman (1969) has made a strong argument for a major input of O_2 from photolytic disassociation of H_2O, I will briefly recapitulate the arguments for a predominantly photosynthetic origin. The main evidence rests on Brinkman's omission of the major oxygen sinks, plus an assessment of the geochemical balance between C and O_2. If the primary source of both C and O_2 is green-plant photosynthesis, they should show an approximate balance as chemically equivalent fractions of CO_2 ($CO_2 + H_2O \rightarrow CH_2O + O_2 \rightarrow C + H_2O + O_2$). It turns out that when all O_2, C, and CO_2 sinks are counted (e.g. Cloud 1972), a fair balance is seen, with enough C left over to account for about 73 geograms (73×10^{20} grams) or 2.6 geomoles of CO having been converted to CO_2 over geologic time. Of course a good deal of this excess carbon could also have been the product of bacterial photosynthesis; but this is countered by the probability that a substantial amount of conversion of CO to CO_2 utilized photolytic O_2 and that much is obscured by the above described reaction of carbon with hematite to give magnetite and new carbon dioxide.

It is difficult to draw a balance sheet for a photolytic source of O_2 because we do not know how much hydrogen has escaped Earth's gravity field or been implanted in it by the solar wind; nor do we know the magnitude of juvenile or bacterial hydrogen added. All variables considered, we have good reason to believe the rough carbon-oxygen balance observed and may conclude from it that photosynthesis was the overwhelmingly important source of free oxygen and the process that, in combination with sequestration of carbon, was and is responsible for most of the oxygen that is now or ever has been in Earth's atmosphere.

Proterozoic

The next four ecosystems in our evolutionary sequence are reconstructed from the Proterozoic record, covering the interval from 2 billion to 680 million years ago. This interval is much more extensively represented by dated rocks and shows a greater variety of sedimentary types and associated microorganisms than does the Proterophytic. We could, if space permitted, discuss a number of delimitable ecosystems, global and otherwise. But space permits the consideration of only a few salient aspects.

If we take seriously the rule I suggested early on that biologic innovations tend to appear soon after environmental conditions become favorable to them, we might expect that the aerobic, organellular, mitosing, eucaryotic cell would have made its debut with or soon after the first appearance of free oxygen. And there are ambiguous suggestions that it may have done so (e.g. Licari and Cloud 1968b). On the other hand, the roughly 1.6-aeon-old Paradise Creek microflora of northwest Queensland (Licari and Cloud 1972) has revealed no eucaryotes, while the 1.3-aeon-old Beck Spring microflora of eastern California has (Cloud et al. 1969; Licari in press). Thus, eucaryotes may have first appeared between 1.6 and 1.3 aeons ago. The dominant elements of the Paradise Creek (ca. 1.6 aeons), Gunflint (2 aeons), and Pokegama (2.1? aeons) microfloras are quite similar filamentous BGA, and the main difference may be that the Paradise Creek microflora had learned to cope with free oxygen without the assistance of oxygen-depressing ferrous ion. It appears to have lived in an exposed position, for some of the associated stromatolites encrust ancient sea stacks that were probably in the intertidal zone or perhaps even emergent. We may infer that freely oxygen-tolerating procaryotes became common and diversified during this interval from 2 to 1.6 aeons ago. We might call this ecosystem Proterozoic 1.

The next big step in ecosystem evolution—Proterozoic 2—came with the origin of the eucaryotic cell and its special organellular and biochemical peculiarities sometime before about 1.3 aeons ago. Among other biochemical requirements, the appearance of eucaryotes demands the origin of the oxygenases, essential for the oxidation of cell membrane steroids and implying a minimum oxygen pressure of 10^{-5} present atmospheric level (M. Yčas, verbal communication, 24 March 1973). Well before that time, indeed before the beginning of the Proterozoic in the sense here employed, crustal evolution had advanced to the point where continents and oceans were well defined and commonly shared abrupt, mobile margins. Plate tectonic mechanisms (e.g. McKenzie 1972) were in effect; continental margins were likely to be marked by asymmetrically paired eu- and mio-geosynclines; and potential marine habitats were more diversified than in older times.

If the Proterophytic or Archean lands were colonized, they were locally colonized within soils or permanently shaded places by BGA, bacteria, or actinomycetes. Fungi and hence lichens could not have taken hold until the eucaryotic cell appeared. That event—marked by advances noted, plus the evolution of cytochrome *c* (Dickerson 1971) and initiating Proterozoic 2—was fraught with significance however it happened. But the biogeological record on that problem is, so far, silent. In this interval lay the seeds of advanced oxidative metabolism, metaphyte and metazoan evolution, and colonization of the lands. Colonization may not have been long in coming. Although we see no direct evidence of it, we begin to find fluviatile deposits about 1.3 aeons old (Pioneer Shale, Apache Group of southern Arizona; McConnell in press) that have lenses of coarse channel sands in more fine-grained floodplainlike deposits, suggesting that something was at least locally and weakly holding overbank sediments in place to permit the evolution of more modern-looking alluvial sediments than we know in older rocks.

The known marine microbiota of Proterozoic 2, in both stromatolitic and nonstromatolitic black cherts of the Pahrump Group of eastern California (and possibly the Apache Group of southern Arizona), includes benthic nostocacean and planktonic chroococcacean BGA, large (diameter up to >50μm) unicellular and branching siphonaceous chlorophytes, and perhaps chrysophytes (Cloud et al. 1969; Licari in press). They have been found in near-shore slope, intertidal, and perhaps in part supratidal environments, both as free-living and probably phytoplanktonic forms and embedded in calcareous and diagenetically silicified algal stromatolites. The large size of the spheroids and the systematic presence in them of regularly located organellelike bodies, the large diameter and regular branching of the siphonaceous filaments, and the presence of objects resembling the statocysts of chrysophycean algae all imply a eucaryotic level of development. Slightly younger (and somewhat problematical) microbiotas described from south Australia and Montana offer no refinements to the general scheme and call for no elaboration here.

Proterozoic deposits younger than about an aeon old in a number of parts of the world, and notably in the USSR, contain, among other things, an abundance of relatively large unicellular forms assignable to the Sphaeromorphitidae of Timofeev (1966). Many of these are well upwards of 50μm in diameter and larger than forms known from older rocks, which may permit a rough stratigraphy. Many also look disturbingly like large living bryophyte spores, which calls for further investigation. Their presumed indigenousness seems to be reinforced by the fact that we do not find in the upper Proterozoic spiny acritarchs of the sort common in basal Phanerozoic deposits. Similar occurrences in apparently much older rocks of unlikely lithology may, however, be contaminants from the modern surface (e.g. that described by Gowda and Sreenivasa 1969). These structures and their containing deposits also need further study. Although now generally regarded as some kind of phytoplankton, they could be the records of a very primitive terrestrial flora. We may tentatively call this Proterozoic 3.

Rocks between the ages of about 800 to 900 and 680 m.y. ago constitute Proterozoic 4. This interval is notable for extensive glacial deposits and for some remarkably well-preserved eucaryotic microfloras. No evidence of Metazoa is found in these or older rocks, although plenty of pseudometazoa have been described (summarized in Cloud 1968b). The stunning Bitter Springs microflora of central Australia, found by Barghoorn and described by Schopf (1968, 1972) and by Schopf and Blacic (1971) occurs in both stromatolitic and nonstromatolitic black cherts of this age. It includes a variety of unicellular higher algae as well as BGA—a total of 50 species, of which 9 are demonstrably eucaryotic, and one of which *(Eotetrahedrion)* produces sporelike tetrads. The latter are reasonably interpreted by Schopf as evidence that the previously evolved mitotic cell-division had by then advanced to the special and fully sexual variation known as meiosis, wherein, during the reproductive phase, a reduction or haploid division of chromosomes is followed by recombination to the normal diploid number. Unfortunately, very similar tetrads are also observed in some of the nonmeiotic chlorosarcinacean green algae (e.g. *Chlorosarcinopsis*, Bold 1970), so we cannot be sure. It does seem likely, however, that a fully sexual reproductive cell-division and biogeochemistry had probably been achieved among the eucaryotic algae by about 800 to 900 m.y. ago (whether they be primitive rhodophytic or chlorophytic algae, or both).

Of the 19 species of BGA described from the Bitter Springs microbiota, 14 have been referred to living families, and, in fact, are virtually indistinguishable on morphological grounds from living genera and species, although not occupying identical habitats. As is obvious from inspection, and as Schopf has repeatedly stated, they constitute a striking example of phenotypical evolutionary conservatism. The Bitter Springs nannofossils, like the older Beck Spring microflora, include both planktonic coccoid species and benthic filamentous species, single and in clusters, embedded in the deposits of a shallow but quiet segment of an ancient epicontinental sea. Their relation to the glacial regime of apparently mostly younger age in central and south Australia is not known, but as they are associated with a thick and extensive carbonate unit, they presumably lived in a relatively warm area or interval.

In my laboratory, we have been finding an abundant, morphologically diverse, apparently monospecific, and probably planktonic procaryotic microflora in uppermost Proterozoic deposits from Alberta (Licari and Cloud 1968; Moorman 1974), as well as in east central Alaska and the Wasatch Mountains of Utah (Cloud, Moorman, and Pierce in press). These tiny pelagic-seeming creatures apparently settled into fine black muds in euxinic shelf-basin deposits of the late Proterozoic Cordilleran geosyncline. They represent an endosporangiate, probably chroococcalean species with close affinities to living entophysalidacean forms, and they show a fair amount of ultrastructural detail under the electron microscope. They may be part of the global algal bloom that Berkner and Marshall (1965) predicted for the late Proterozoic on rather shaky theoretical grounds, calling for a stepwise increase in the amount of free O_2 generated at this time. Had such an increase occurred, we should see evidence of it in the sequestration of carbon and in oxidized sediments of that age.

Interestingly enough, the Soviet geologists Ronov and Migdisov (1971) have found a pronounced increase in the ratios of carbon in sediment beginning about 680 m.y. ago. In addition, sedimentary sulfate, with its large oxygen requirements, first became important in the geologic column shortly before this. And the oxidative enrichment of the BIF, at least in North America, dates from the same time. This was a time of glaciation and probably diversified climate. The piling-up of continental ice lowered water tables and allowed deep oxidation of surface ores by presumably oxygen-rich meteoritic waters. Vigorous turnover of the oceans enriched the surface waters with nutrients so that phytoplankton could flourish. And flooding of continental margins upon melting of glacial ice provided a range of new habitats for benthic algae, with or without planktonic growth stages.

We may hypothesize, since Metazoa appear at about the same time (or slightly later) as this diversity of potential habitats, that oxygen may then have reached a level conducive to metazoan metabolism—to perhaps 3 to 10 percent of the present atmospheric level. We are justified in doing this because eucaryotes had already been extant for around 600 m.y., meiosis had probably been invented at least 100 to 200 m.y. earlier, and the only previously unsatisfied condition for metazoan origins was a sufficient level of free oxygen to support a metazoan level of oxidative metabolism and type of biochemistry. Finally, as Schopf et al. (1973) have recently emphasized, the oldest records of metaphytes only slightly precede the oldest Metazoa and may have helped to set the stage for the wave of multicellularization that led to their emergence.

Phanerozoic

Thus it is not surprising that the oldest authentic Metazoa we know are just about 680 m.y. old (Evans et al. 1968). This, the Ediacarian fauna (Glaessner 1958, 1961), comprises a suitably archaic initial metazoan assemblage. The appearance of that mainly (but not exclusively; see Germs 1972) soft-bodied assemblage marked the end of Proterozoic ecosystem 4 and the beginning of the Phanerozoic ecosystems. A key biochemical whose evolution must have marked or preceded the appearance of Metazoa is hydroxyproline, needed for collagen formation.

Even in the earliest Phanerozoic, we have a much greater variety of interacting organisms and environments than previously (Glaessner 1958, 1961; Pflug 1970, 1972; Ford 1958; Sokolov 1972; Misra 1969). Although fascinating environmentally connected major biologic and accompanying biochemical innovations such as the diversification of the invertebrates, the acquisition of hard parts, and the origin of chordates, vertebrates, tracheophytic plants, quadrupeds, reptiles, birds, mammals, flowering plants, social insects, bipeds, and man still lay ahead, they are, in a sense, elaborations on the metazoan and metaphytic theme. In the long range of time and of morphogenetic and biochemical evolution, the early Phanerozoic aspect was essentially modern. Although later environmental and biogeochemical interactions were complex and pervasive, they were probably less profound than those of pre-Phanerozoic time. The main events of biological, biogeochemical, and ecosystem evolution had already taken place by the beginning of Phanerozoic time. But, of course, new complexities are added, as the Phanerozoic elaboration leads to man himself, and to an enormous flowering of biological and ecosystem variety along the way.

Still, some significant aspects of Phanerozoic ecosystem evolution are poorly enough appreciated to warrant emphasis here. An abundance and diversity of algal stromatolites is a feature of Proterozoic time. They were less common and diversified before and after—before probably because of the generally inhibitive effects of excessive UV on shallow benthic algae (and the rarity of environments favoring carbonate precipitation), afterward probably because of predatory elimination. Although Monty (1972) ingeniously argues competitive exclusion of the BGA by more advanced algae, Garrett's suggestion (1970) that they are simply trimmed back by metazoan browsers in post-Proterozoic time is certainly consistent with the observed time of onset of the Metazoa and observations on living stromatolites in western Australia and the Persian Gulf (Paul Hofmann, personal communication, spring 1972).

Interesting and subtle contrasts are found, too, between pre-Phanerozoic and Phanerozoic cherts and carbonate sediments that may be related to silica- and $CaCO_3$-precipitating organisms. Phanerozoic cherts, commonly associated with carbonates, are often rich in siliceous tests or spicules of organisms, whereas silica in most natural waters is kept in a state of great undersaturation by biological precipitation to form such tests and spicules. The silica of the BIF, as concluded earlier, is probably a chemical precipitate from waters of pH generally too low for carbonate precipitation. Later pre-Phanerozoic carbonates were probably mainly a consequence of a water chemistry that was strongly influenced by biological assimilation of CO_2, while Phanerozoic carbonates include, in addition, some large skeletal fractions.

These younger carbonates nevertheless remain conspicuous among shelf sediments until after the appearance of the calcareous plankton in Cretaceous time, when the bulk of $CaCO_3$ sedimentation is shifted seaward.

Epilogue

We have seen that pre-Phanerozoic ecosystem evolution is rich in great but widely separated events, such as the origin of life, early biological-biochemical-environmental interactions, the crossing of that major biological-biochemical gap between procaryote and eucaryote, the origin of meiosis and sex, the oxygenation of the atmosphere, and others that still lie concealed within the immense remaining biologically and biogeochemically unexplored reaches of early geologic time. Phanerozoic evolution, on the other hand, displays a bewildering wealth of lesser but still richly significant details.

Now that we are beginning to see the main elements of primitive Earth ecosystems, some of us are eager to pursue their evolution toward more comprehensive and satisfying reconstructions. At the same time, our outlooks toward events closer to the origin, present, and future of man are enhanced by the perspective that already emerges from a consideration of the first 4.6 aeons of earth history—from "natural pollution" to man-made pollution! Torn as I feel between the pressing present and the pullulating past, I am reminded of Robert Benchley's *My Ten Years in a Quandary and How They Grew.* Maybe the quandary carries a lesson. The present, we are fond of saying, is the key to the past. But isn't the past also a deeply significant key to the present and perhaps the future? And don't we have to apportion our thinking time somehow between both in order to understand either well?

Finally, a word about the future. We have seen the profound effects on atmospheric, hydrospheric, and lithospheric evolution of simple but biochemically potent microorganisms. Can we believe that man is a less pregnant force? If the lowly blue-green algae could convert the primitive anoxygenous atmosphere into an oxygenous one, who is to say what an agent for biological and ecological change as powerful and prevalent as *Homo sapiens* might not eventually bring about? And who will monitor his activities to anticipate and attempt to counteract potentially adverse changes? Anyone who calls himself an environmental biogeochemist will find a challenge here to match that of the primitive earth itself.

References

Abelson, P. H. 1966. Chemical events on the primitive earth. *Proc. Natl. Acad. Sci. USA* 55:1365–72.

Allison, C. W., and M. A. Moorman. 1973. Microbiota from the late Proterozoic Tindir Group, Alaska. *Geology* 1(2):65–68.

Anhaeusser, C. R. 1972. The evolution of the early Precambrian crust of southern Africa. Univ. Witwatersrand, Econ. Geol. Rsch. Unit, Inf. Circ. No. 70. 31 pp.

Barghoorn, E. S., and S. A. Tyler. 1965. Microorganisms from the Gunflint chert. *Science* 147:563, 577.

Barghoorn, E. S., and J. W. Schopf. 1966. Microorganisms 3 billion years old from the Precambrian of South Africa. *Science* 152:758–63.

Berkner, L. V., and L. C. Marshall. 1965. History of major atmospheric components. *Proc. Natl. Acad. Sci. USA* 53:1215–25.

Bold, H. C. 1970. Some aspects of the taxonomy of soil algae. *Ann. N.Y. Acad. Sci.* 175:601–16.

Bradley, W. H. 1929. The varves and climate of the Green River epoch. U.S. Geol. Survey, Prof. Paper 158, pp. 87–110.

Breger, I. A., ed. 1963. *Organic Geochemistry.* Pergamon Press. 658 pp.

Brinkmann, R. T. 1969. Dissociation of water vapor and the evolution of oxygen in the terrestrial atmosphere. *J. Geophys. Res.* 74:5355–68.

Church, W. R., and G. M. Young. 1972. Precambrian geology of the southern Canadian Shield with emphasis on the Lower Proterozoic (Huronian) of the north shore of Lake Huron. 24th Internatl. Geol. Cong. (Montreal), Guidebook A36-C36. 65 pp.

Cloud, Preston. 1965. Significance of the Gunflint (Precambrian) microflora. *Science* 148:27–35.

———. 1968a. Atmospheric and hydrospheric evolution on the primitive earth. *Science* 160:729–36.

———. 1968b. Pre-metazoan evolution and the origins of the Metazoa. In E. T. Drake, ed., *Evolution and Environment.* Yale Univ. Press, pp. 1–72.

———. 1972. A working model of the primitive earth. *Am. J. Sci.* 272:537–48.

———. 1973a. Atmosphere, development of. *Encyclopedia Brittanica* 2:313–19.

———. 1973b. Paleoecological significance of the banded iron formation. *Econ. Geol.* 68:1135–43.

———, and Hannelore Hagen. 1965. Electron microscopy of the Gunflint microflora: Preliminary results. *Proc. Natl. Acad. Sci. USA* 54:1–8.

———, and G. R. Licari. 1968a. Microbiotas of the banded iron formations. *Proc. Natl. Acad. Sci. USA* 61:779–86.

———, and G. R. Licari. 1968b. Morphological criteria for biogeochemical processes. GSA Ann. Mtgs. (Mexico City), 11–13 Nov. *Abstracts,* p. 57.

———, G. R. Licari, L. A. Wright, and B. W. Troxel. 1969. Proterozoic eucaryotes from eastern California. *Proc. Natl. Acad. Sci. USA* 62:623–31.

———, and Aharon Gibor. 1970. The oxygen cycle of the biosphere. *Sci. Am.* 223(3)110–23.

———, and G. R. Licari. 1972. Ultrastructure of some two-aeon-old Nostocacean algae from northeastern Minnesota. *Am. J. Sci.* 272:138–49.

———, Mary Moorman and David Pierce. In press. Sporulation and ultrastructure in late Proterozoic Cyanophyta—some implications for cyanophyte taxonomy and plant phylogeny.

Dickerson, R. E. 1971. The structure of cytochrome *c* and the rates of molecular evolution. *J. Molec. Evolution* 1:26–45.

Eglinton, G., and M. T. J. Murphy, eds. 1969. *Organic Geochemistry.* Springer-Verlag. 828 pp.

Evans, A. M., T. D. Ford, and J. R. L. Allen. 1968. Precambrian rocks. In P. C. Sylvester-Bradley and T. D. Ford, eds., *The Geology of the East Midlands.* Leicester Univ. Press, pp. 1–19.

Fanale, F. P. 1971. *Chem. Geol.* 8:79–105.

Fay, P., W. D. P. Stewart, A. E. Walsby, and G. E. Fogg. 1968. Is the heterocyst the site of nitrogen fixation in blue-green algae? *Nature* 220:810–12.

Ford, T. D. 1958. Precambrian fossils from Charnwood Forest. *Proc. Yorkshire Geol. Soc.* 31, pt. 3(8), pp. 211–17.

Frarey, M. J., and S. M. Roscoe. 1970. The Huronian Supergroup north of Lake Huron. Geol. Surv. Canada, Paper 70-40, pp. 143–58.

Garrett, Peter. 1970. Phanerozoic stromatolites: Non-competitive ecologic restriction by grazing and burrowing animals. *Science* 169:171-73.

Germs, G. J. B. 1972. New shelly fossils from the Nama Group, South West Africa. *Am. J. Sci.* 272:752-61.

Glaessner, M. F. 1958. New fossils from the base of the Cambrian in South Australia. *Trans. Roy. Soc. S. Aust.* 81:185-88.

———. 1961. Pre-Cambrian animals. *Sci. Am.* 204(2)72-78.

Gowda, S. S., and T. N. Sreenivasa. 1969. Microfossils from the Archean complex of Mysore. *J. Geol. Soc. India* 10:201-08.

Hall, J. B. 1971. Evolution of the procaryotes. *J. Theoret. Biol.* 30:429-54.

Harvey, H. W. 1955. *The Chemistry and Fertility of Sea Waters.* Cambridge Univ. Press. 224 pp.

Holland, H. D. 1962. Model for evolution of the earth's atmosphere. Geol. Soc. America, *Buddington Volume in Petrologic Studies*, pp. 447-77.

Hurley, P. M., W. H. Pinson, Jr., Bartholomew Nagy, and T. M. Teska. 1971. Ancient age of the Middle Marker Horizon, Onverwacht Group, Swaziland Sequence, South Africa. Mass. Instit. Tech., *19th Ann. Prog. Rept. for 1971 to U.S. Atomic Energy Commission*, MIT-1381-19, pp. 1-4.

Jacks, G. V. 1953. Pedology. *Sci. Prog.* 41:301-05.

Jackson, T. A. 1973. "Humic" matter in the bitumen of ancient sediments: Variations through geologic time—a new approach to the study of pre-Paleozoic (Precambrian) life. *Geology* 1(4):163-66.

Johnsen, S. J., W. Dansgaard, H. B. Clausen, and Ċ. C. Lanquay. 1970. Climatic oscillations 1200-2000 A.D. *Nature* 227:482-83.

Krauskopf, K. B. 1956. Dissolution and precipitation of silica at low temperature. *Geochim. et Cosmochim. Acta* 10:1-26.

Kuznetsov, S. I., M. V. Ivanov, and N. N. Lyalikova. 1963. *Introduction to Geological Microbiology.* McGraw-Hill Book Co. 252 pp. (Trans. from the Russian by P. T. Broneer).

Licari, G. R. In press. Paleontology and paleoecology of the Proterozoic Beck Spring Dolomite in eastern California: *J. Paleont.*

———, and Preston Cloud. 1968a. Eucaryotic nannofossils in kerogen from the pre-Paleozoic Windermere Series of Alberta. GSA Ann. Mtgs. (Mexico City), 11-13 Nov. *Abstracts*, pp. 174-75.

———, and Preston Cloud. 1968b. Reproductive structures and taxonomic affinities of some nannofossils from the Gunflint Iron Formation. *Proc. Natl. Acad. Sci. USA* 59:1053-60.

———, and Preston Cloud. 1972. Prokaryotic algae associated with Australian Proterozoic stromatolites. *Proc. Natl. Acad. Sci. USA* 69:2500-04.

McConnell, R. L. In press. *Stratigraphy, paleoecology, and paleontology of the 1.2 to 1.4 aeon-old Apache Group, Arizona, and its relation to equivalent rocks in the southwestern United States.* Museum of Northern Arizona, Bull. 51.

McKenzie, D. P. 1972. Plate tectonics and sea-floor spreading. *Am. Sci.* 60(4)425-35.

Manskaya, S. M., and T. V. Drozdova. 1968. *Geochemistry of Organic Substances.* Ed. and trans. L. Shapiro and I. A. Breger. Pergamon Press. 345 pp.

Misra, S. B. 1969. Late Precambrian(?) fossils from southeastern Newfoundland. *Bull. Geol. Soc. Am.* 80:2133-40.

Monty, C. L. V. 1972. Recent algal stromatolite deposits, Andros Island, Bahamas—preliminary report. *Geol. Rundschau* 61:742-83.

Moorbath, S., R. K. O'Nions, and R. J. Pankhurst. 1973. *Nature* 245:138:39.

Moorman, M. 1974. Microbiota of the late Proterozoic Hector Formation, southwestern Alberta, Canada. *J. Paleont.*

Oehler, D. Z., J. W. Schopf, and K. A. Kvenvolden. 1972. Carbon isotope studies of organic matter in Precambrian rocks. *Science* 175:1246-48.

Perfil'ev, B. V., D. R. Gabe, A. M. Gal'penna, V. A. Rabinovich, A. A. Sapotnitskii, E. E. Sherman, and E. P. Troshanov. 1964. *Applied Capillary Microscopy—The Role of Microorganisms in the Formation of Iron-Manganese Deposits.* Akad. Nauk USSR, F. P. Savorenskii Laboratory of Hydrogeological Problems. (Trans. from the Russian by Consultants Bureau, New York, 1965). 122 pp.

Perry, E. C., Jr., F. C. Tan, and G. B. Morey. 1973. Stable isotope geochemistry of the Biwabik Iron Formation. *Econ. Geol.* 68:1110-25.

Pflug, H. D. 1970. Zur Fauna der Nama-Schichten in Südwest-Afrika I. II. *Palaeontogr. Abst. A* 134(4-6)226-62; 135(3-6)198-230.

———, 1972. The Phanerozoic-Cryptozoic boundary and the origin of Metazoa. 24th Internatl. Geol. Cong. (Montreal), Sec. 1, *Precambrian Geology*, pp. 78-84.

Ponnamperuma, Cyril, ed. 1972. *Exobiology.* North Holland Publ. Co. 485 pp.

Ronov, A. B., and A. A. Migdisov. 1971. Geochemical history of the crystalline basement and the sedimentary cover of the Russian and the North American Platforms. *Sedimentology* 16:137-85.

Rubey, W. W. 1951. Geologic history of sea water: *Bull. Geol. Soc. Am.* 62:1111-48.

Russell, H. N., and D. H. Menzell. 1933. The terrestrial abundance of the permanent gases. *Natl. Acad. Sci. USA* 19:997-1001.

Saunders, George W. 1972. Potential heterotrophy in a natural population of *Oscillatoria agardhii* var. *isothrix* Skuja. *Limn. and Oceanog.* 17:704-11.

Schopf, J. W. 1968. Microflora of the Bitter Springs Formation, Late Precambrian, central Australia. *J. Paleont.* 42:651-88.

———. 1970. Precambrian micro-organisms and evolutionary events prior to the origin of vascular plants. *Biol. Reviews* 45:319-52.

———. 1972a. Precambrian paleobiology. In C. Ponnamperuma, ed., *Exobiology.* North Holland Publ. Co., pp. 16-61.

———. 1972b. Evolutionary significance of the Bitter Springs (Late Precambrian) microflora. 24th Internatl. Geol. Cong. (Montreal), Sec. 1, *Precambrian Geology*, pp. 68-77.

———, and J. M. Blacic. 1971. New microorganisms from the Bitter Springs Formation (Late Precambrian) of the north-central Amadeus Basin, Australia. *J. Paleont.* 45:925-60.

———, B. N. Haugh, R. E. Molnar, and D. F. Satterthwait. 1973. On the development of metaphytes and metazoans. *J. Paleont.* 47:1-9.

Schumm, S. A. 1968. Speculations concerning paleohydrologic controls of terrestrial sedimentation. *Bull. Geol. Soc. Am.* 79:1573-88.

Scott, W. M., V. E. Modzeleski, and Bartholomew Nagy. 1970. Pyrolysis of early pre-Cambrian Onverwacht organic matter (>3 × 10⁹ years old). *Nature* 225:1129-30.

Siever, Raymond. 1962. Silica solubility 0-200°C and the diagenesis of siliceous sediments. *J. Geol.* 70:127-50.

Sokolov, B. S. 1972. The Vendian stage in earth history. 24th Internatl. Geol. Cong. (Montreal), Sec. 1, *Precambrian Geology*, pp. 78-84.

Stewart, W. D. P., and H. W. Pearson. 1970. Effects of aerobic and anaerobic conditions of growth and metabolism of blue-green algae. *Proc. Royal Soc.* (London) ser. b, 175:293-311.

Timofeev, B. V. 1966. *Microphytological investigations of the ancient strata.* Akad. Nauk. USSR, Laboratory of Precambrian Geol. 147 pp., 89 pls. (in Russian).

Trendall, A. F. 1968. Three great basins of Precambrian banded iron formation deposition—a systematic comparison. *Bull. Geol. Soc. Am.* 79:1527-44.

———. 1972. Revolution in earth history. *J. Geol. Soc. Australia* 19:287-311.

———. 1973. Varve cycles in the Weeli-Wolli Formation of the Hamersley Group, Western Australia. *Econ. Geol.* 68:1089-97.

Wald, George. 1964. The origins of life. *Proc. Natl. Acad. Sci. USA* 52:595-611.

Walter, Malcolm. 1972. A hot spring analog for the depositional environment of Precambrian iron formations of the Lake Superior region. *Econ. Geol.* 67:965-80.

Webley, D. M., M. E. K. Henderson, and I. F. Taylor. 1963. The microbiology of rocks and weathered stones. *J. Soil Sci.* 14:102-12.

White, D. E., W. W. Brannock, and K. Murata. 1956. Silica in hot spring waters. *Geochim. et Cosmochim. Acta* 10:27-59.

Whittaker, R. H., and G. M. Woodwell. 1970. Evolution of natural communities. In J. A. Wiens, ed., *Ecosystem Structure and Function.* Univ. of Oregon Press, pp. 137-59.

Yčas, M. 1972. Biological effects on the early atmosphere. *Nature* 238:163.

"There goes the ecology."

Melvin Calvin

Chemical Evolution

Life is a logical consequence of known chemical principles operating on the atomic composition of the universe

My interest in chemical evolution began in the fall of 1949 when I read *The Meaning of Evolution*, by George Gaylord Simpson. I devised an experiment to test some of the ideas arising from that reading —namely, to determine whether or not an energy input into the collection of primitive molecules believed to exist on the surface of the original earth could lead to molecules of biological consequence. In 1951 such an experiment was performed, and it was indeed found that the beginnings of molecular growth could be demonstrated under prebiotic conditions.

The first external publication from our laboratory of a comprehensive view of the problem appeared in *American Scientist* (44:248–63, summer 1956) under the title "Chemical Evolution and the Origin of Life." At that time only a few laboratories throughout the world were engaged in experimental efforts to demonstrate the possible origins of life. These included experiments in the laboratory of Harold Urey, the preliminary results of which were published in 1955, and in the laboratory of A. I. Oparin in Moscow, who had published a comprehensive discussion of his point of view in English in 1936, as well as a few scattered photochemical experiments in England and India.

Melvin Calvin is known for his scientific achievements in fields ranging from metal-organic chemistry to the chemical origin of life and for his contributions to the understanding of photosynthesis in green plants and, more recently, of chemical oncogenesis. Dr. Calvin obtained the Ph.D. in chemistry from the University of Minnesota in 1935. After two postdoctoral years at the University of Manchester, in England, in 1937 he joined the Department of Chemistry at the University of California, Berkeley, becoming a professor in 1947. He has served on the President's Science Advisory Committee, as chairman of the committee on Science and Public Policy of the National Academy of Sciences, and as president of the American Chemical Society. He is a member of the National Academy of Sciences, the Royal Society of London, the Japan Academy, and other distinguished societies. In 1961 he was awarded the Nobel Prize for Chemistry, in 1964 he received the Davy Medal of the Royal Society, and in 1975 the Virtanen Medal in Finland. The preparation of this paper was supported by the U.S. Atomic Energy Commission. It was originally presented, in slightly different form, as an address to the Mitsubishi Kasei Institute of Life Sciences, Asahi-Kodo Hall, Tokyo, on 18 May 1974. Address: Laboratory of Chemical Biodynamics, University of California, Berkeley, CA 94720.

In the two decades since the original publication, the subject has evolved into a many-sided discipline. The demonstration of the formation of biologically important molecules under all sorts of abiogenic conditions has been repeated many times in many places all over the world. Four international conferences on the origin of life have taken place, and the International Society for the Study of the Origin of Life was formed in 1971. Whereas the initial problem was simply whether or not organic molecules of biological consequence could be formed in a prebiotic environment, the questions now before us are far more sophisticated and difficult. Such things as the origin and evolution of the mechanisms of directed energy transfer and information storage and transfer now concern us.

The question of the origin of life on earth and the nature of chemical evolutionary processes that could have given rise to it has engaged the minds of men since they first contemplated the nature of their place on the earth and in the universe (*1*). The most acceptable view in scientific terms today stems primarily from the concepts first carefully and clearly enunciated by Charles Darwin in his early writings. The basis for his discussion was the great variety of the morphological and functional forms of present living things, as well as knowledge of the vast array of extinct forms whose morphological features were preserved in the paleontological record in the rocks.

On that basis, Darwin was able to formulate his general hypothesis of biological evolution, which is perhaps best expressed by the title of a paper proposed by Wallace but never used—"On the Tendency of Varieties to Depart Indefinitely from Original Types." These words seem to me to express best the fundamental idea of Darwinian evolution—namely, that two species which exist today as independent species, if followed back in time, were originally two varieties of the same species. Thus, if we go forward in time, individual variations would gradually separate to become new species.

If we follow the reverse-time process sufficiently far back, we must arrive at a time, and a condition, when there was only one ancestral species—one type of organism —which included varieties that became today's separate species. If we go still farther back, reaching that point when the individual living thing was only one variety of many different kinds of molecular aggregations, we can see that the transition from molecules to a living thing is a continuous one. Thus, we can reach back into the history of the earth to a period when the earth had no living things, only molecules. And we can go even farther back to that time when there were no molecules, only atoms, bringing us to the period of the evolution of the elements themselves.

As Darwin and Wallace realized—and expressed in the unpublished title—there was a continuity in evolution, ultimately arriving at a successful starting point whose descendants survived. That starting point is what we tend to mean by the "origin" of living matter. Darwin wrote a very interesting letter (*2*) about this idea in 1882 in response to a query, stating:

> You expressed quite correctly my views where you said that I had intentionally left the question of the Origin of Life uncanvassed as being altogether *ultra vires* in the present state of our knowledge, and that I dealt only with the matter of succession. I have met with no evidence that seems in the least trustworthy, in favour of so-called Spontaneous Generation. I believe that I have somewhere said (but cannot find the passage) that the principle of continuity renders it probable that the principle of life will hereafter be shown to be a part, or consequence, of some general law.

The statement to which Darwin refers, and which he had forgotten, was written earlier, in 1871:

> It is often said that all the conditions for the first production of a living organism are present, which could ever have been present. But if (and oh what a big if) we could conceive in some warm little pond with all sorts of ammonia and phosphoric salts—light, heat, electricity, etc.—present, that a proteine compound was chemically formed, ready to undergo still more complex changes, at the present day such matter would be instantly devoured, or absorbed, which would not have been the case before living creatures were formed (*3*).

I propose to discuss a certain period in this time sequence—the period in which the molecules themselves were being formed and transformed and built up to reach, eventually, a size and complexity that could contain and sustain the living process leading to life as we now know it (*4, 5, 6*). We will begin by looking at the molecular nature of living things as we understand them today, to see, first, what it is we must arrive at by chemical means. It is not yet possible for us (and I am not sure it ever will be) to find a record in the rocks of the molecular events that may have taken place prior to the appearance of what we call a living thing. Therefore, we must try to reconstruct those possible processes from what we know about present-day chemistry and see how far we can carry them out experimentally in the laboratory.

Figure 1 is a diagrammatic representation of the essential principles of present living organisms and their construction. It contains all the elements that we must eventually describe in molecular terms—the appearance of small molecules (amino acids) and of polymers of those amino acids and nucleic acids which form large mole-

Figure 1. The mechanism of protein biosynthesis is illustrated by the reproductive process in a living cell. The genetic code of the cell—DNA—is located within the nucleus (1). In order to construct a new cell, the cell must first transcribe the information contained in the DNA to the messenger RNA (mRNA) templated on pieces of the DNA (2). The mRNA combines with ribosomes (3) in the cytoplasm to form a polyribosome (polysome) (4). The simple amino acids present in the cytoplasm (5), after being activated by a chemical catalyst (6), become attached to transfer RNA (tRNA) (9). The polysome then unites the amino acids from tRNA, in an order designated by mRNA (10), thus creating a protein (11). The process is a combination of information transfer and energy transfer. Steps 2, 3, 4, and 10 represent the cell-specific process of information transfer, while steps 5, 6, 7, and 9 represent the general biological information processing common to all cells. (After Calvin, ref. 7.)

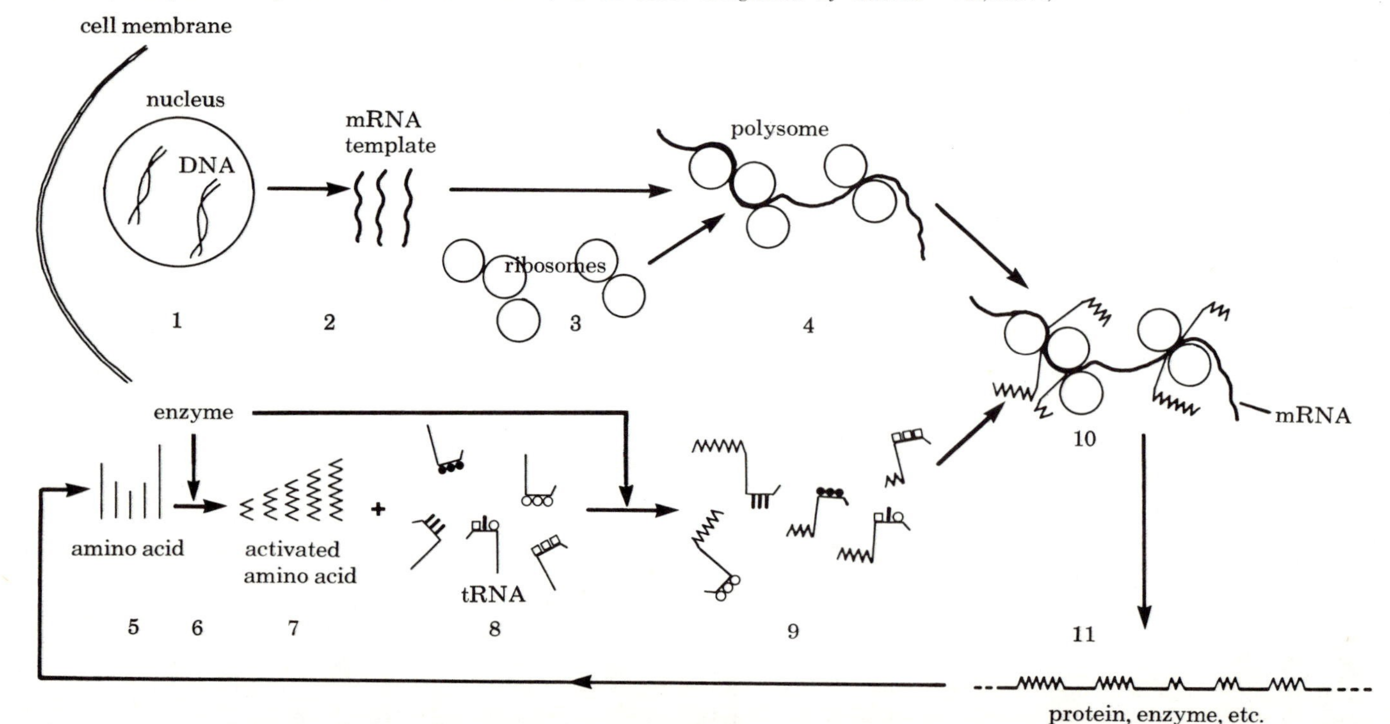

cules that ultimately have specific structure, giving rise to specific shapes and sizes. Thus, there are four elements of molecular evolution that we must try to understand: the evolution of molecules themselves, the simple molecules of which living things are made; the evolution of macromolecules (polymers) and structure; the evolution of catalysis—the ability to make specific reactions occur with a high degree of efficiency; and the evolution of information-storage and information-transfer processes that allow the two systems to coalesce.

Evolution of molecules

Figure 2 outlines the time sequence in which results of evolution were and are being achieved. In the beginning, most of the elements of the universe were in the form of hydrogen, which eventually had to undergo fusion reactions, giving rise to the higher elements in the periodic table, particularly those important to living things: carbon, nitrogen, oxygen, sulfur, phosphorus, halides, and certain metals, particularly iron, which are important for catalytic functions in living organisms.

Then, the primitive (prebiotic, primeval) molecules were formed from the organogenic elements with which the earth was initially coated: methane, ammonia, carbon monoxide, water, carbon dioxide, hydrogen sulfide, and, of course, hydrogen. These first three stages present no chemical problem, since the first two are nuclear and the third is simply the result of the presence of carbon, hydrogen, nitrogen, and oxygen at a low enough temperature to produce the small, primitive molecules.

The next stage of chemical evolution—from the organogenic molecules to the biomonomers—does present a chemical problem, and it has been an area of major progress in the last twenty years (*6, 7*). The conversion of organogenic molecules into amino acids, sugars, nucleic acid bases, and other carboxylic acids (acetic acid and citric acid, for example) has been achieved in the laboratory under the influence of a wide variety of energy sources, ranging from the ultraviolet light of the sun to radioactive energy (in the form of ionizing radiation) to mechanical energy (in the form of meteoritic shock waves) (*8*). All these energy sources give rise to the transformation of the organogenic molecules to biomonomers.

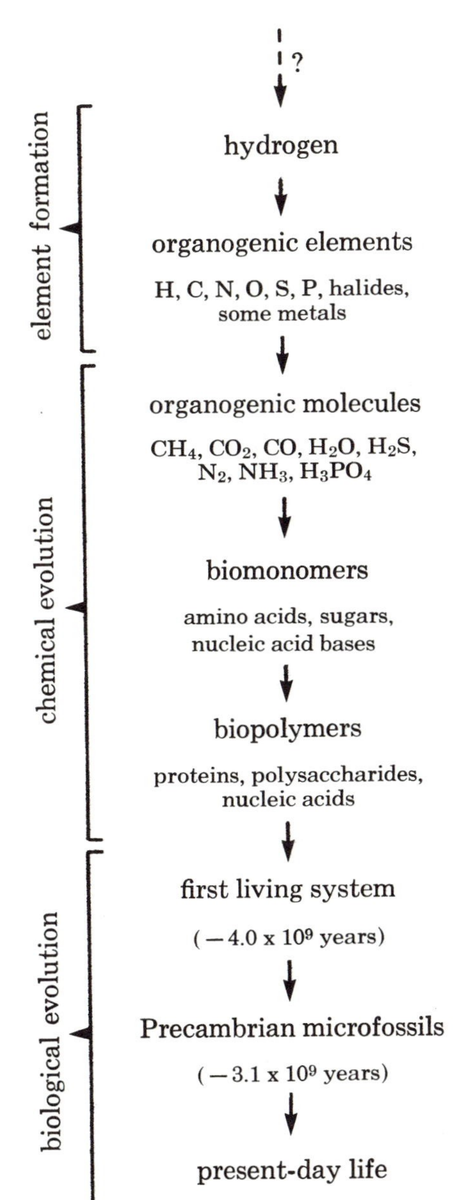

Figure 2. Time sequence of evolution from the formation of the elements to the present.

The next stage—the transition from biomonomers to biopolymers—is more difficult to achieve in terms of chemical evolution (*9*), and most of the rest of the discussion will be concentrated in this area and on the formation of structure and function in the biopolymer region, which eventually gave rise to the first living organisms about four billion years ago.

Evolution of polymers

The transition from biomonomers to biopolymers is best illustrated in the formation of polypeptides and nucleic acids (Fig. 3). In each of these cases, in order to make the biopolymer from the biomonomer, it is necessary to remove a water molecule between the two monomers. The removal of the water molecule, the essential chemical reaction that must be achieved in the presence of water itself (*10, 11*), is sometimes very difficult. We know that it can be done if the conditions are correct, because that is what occurs today in every living organism. Proteins, polysaccharides, and lipids are all made by such a water-removal process, in the presence of water.

It was necessary to devise a special kind of chemical reaction that would allow the condensation polymerization with polypeptide or the nucleic acid formation to take place. We used a variety of chemical reagents that store the energy of ionizing radiation or ultraviolet radiation, agents which are formed very readily from methane and ammonia. These are reagents in which the carbon-nitrogen multiple bond is contained (either a double bond as in the tautomer of cyanamide in which one of the hydrogens has moved, or a triple bond such as in cyanide ion, HCN). These multiple carbon-nitrogen bonds are relatively stable high-energy sources for absorption of water, and they do not react very rapidly with water themselves; they react preferably with the sources of water. By mixing dicyandiamide with glycine in water solution we were able to make glycine polymers at least up to the tetrapeptide (*12*). The polymer thus formed is one in which the glycine loses a water molecule between the carboxyl group of one molecule and the amino group of another to form a peptide. This occurs, of course, in water, and the products are diglycine, triglycine, and tetraglycine.

We have thus demonstrated that it is possible to unite two amino acids to form a peptide link, even in water. In fact, that process takes place with some degree of specificity. When one mixes several amino acids in the same solution and in the same reaction, it is possible to see a certain selectivity of amino acids for each other, as shown in Table 1. The experimental value for the coupling of glycine-glycine is taken as the standard, and the relative rates of coupling for each of the other peptides are compared to diglycine. The close similarity between the experimental value and the calculated value (the value obtained from known proteins) suggests that the original polypeptide was similar to the ones we have today. By suitable modification, that is, elongation, insertion, and other transformations, we get the variety of proteins that we now have—about a billion different ones.

Table 1. Dipeptide yields, as determined experimentally and as calculated from known protein sequences

Dipeptide*	Frequencies (relative to Gly-Gly)	
	Experimental	*Calculated*
Gly-Gly	1.0	1.0
Gly-Ala	0.8	0.7
Ala-Gly	0.8	0.6
Ala-Ala	0.7	0.6
Gly-Val	0.5	0.2
Val-Gly	0.5	0.3
Gly-Leu	0.5	0.3
Leu-Gly	0.5	0.2
Gly-Ile	0.3	0.1
Ile-Gly	0.3	0.1
Gly-Phe	0.1	0.1
Phe-Gly	0.1	0.1

* The dipeptides are listed in order of increasing volume of the side chains of the constituent residues. Gly, glycine; Ala, alanine; Val, valine; Leu, leucine; Ile, isoleucine; Phe, phenylalanine. Example: Gly-Ala = glycylalanine.
Source: Ref. *13*.

Another method of hooking amino acids together in water solution was described by Katchalsky (*14*), who made use of the fact that polypeptides are formed today via the formation of an amino acyl adenylate. He used these amino acyl adenylates with montmorillonite, a naturally occurring clay, functioning as a catalyst to demonstrate not only the formation of peptides but polypeptides containing 20–40 amino acid units (Fig. 4).

Here, also, it was possible to examine the reaction for any specificity that might occur with respect to the combination of one amino acid with another. Paecht-Horowitz, one of Katchalsky's students, performed this same type of reaction with mixed amino acyl adenylates to determine whether there was any selectivity (*16*). Table 2 shows the results of that study. When the reaction is done with a mixed amino acyl adenylate, for example glycine and alanine, the homopolymer is preferred over the heteropolymer. This is true for almost all

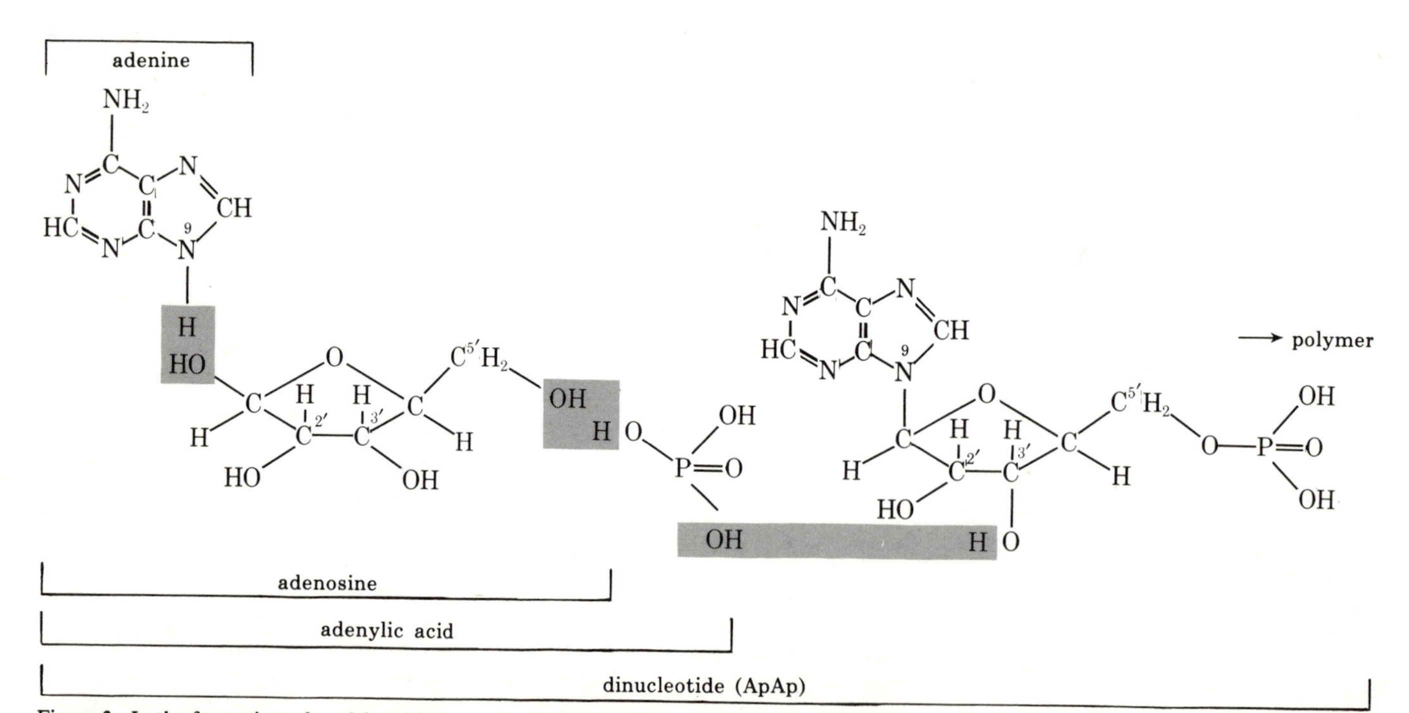

Figure 3. In the formation of nucleic acid, the removal of water is required at three different sites *(grey)* to produce the polynucleotide: between the No. 9 hydrogen of the adenine and the hydroxyl of the ribose to form adenosine; between the No. 5′ carbon atom of adenosine and a hydroxyl group of the phosphoric acid; and, finally, between the second phosphoric acid hydroxyl group and the No. 3′ carbon atom of another adenosine to form the polymer. (The DNA lacks OH on 2′ position.)

cases studied by Paecht-Horowitz. If we examine the heteropolymers themselves, we see that in some cases there is no distinction as to which way the heteropolymer will go, and in other cases there is a large distinction. This observation is important because it demonstrates that even in a simple polymerization, in which there is no nucleotide template to guide the sequence of amino acids, there is already a selectivity in the order in which the polyamino acids will be formed. This, I believe, indicates that there is some possibility that the earliest protein catalysts, which were formed around trace metals such as iron, were formed prior to the appearance of the information-storing and -transferring process represented in Figure 1 (*17*).

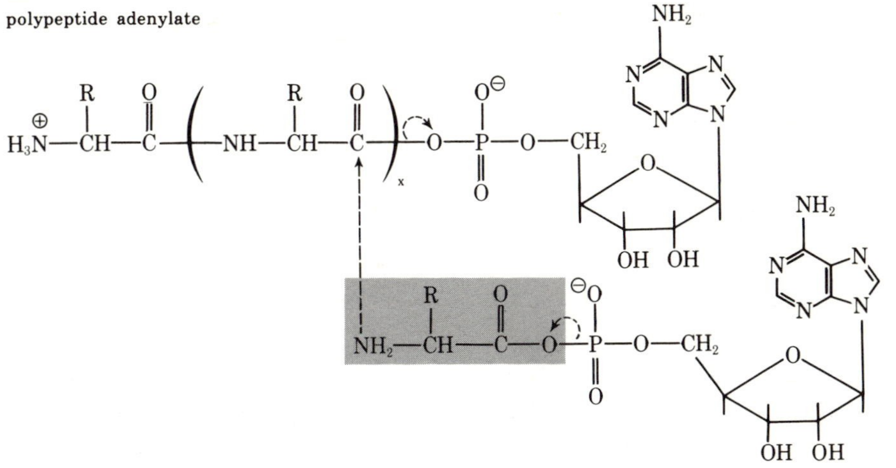

polypeptide adenylate

Figure 4. The polymerization of amino acyl adenylate illustrates the process of polypeptide formation. The molecule containing an amino acyl group on the phosphate group of adenylic acid *(grey)* is catalytically transferred by montmorillonite clay. the result of the reaction (dashed line) is a polypeptide adenylate. (After Katchalsky, published in ref. 15.)

Evolution of structure

Once the polymers have evolved under the conditions of the prebiotic earth, we know they will have a secondary and tertiary structure that is intrinsic to the primary sequence of amino acids or nucleotides. The secondary structure is helical, determined primarily, in the case of polypeptides, by the hydrogen bonding between the amide hydrogen and the amide carbonyl, three peptides removed—the familiar alphahelical structure. In addition, there is a tertiary structure into which the coils are folded in a specific manner, and this factor is important for the catalytic and structural function of proteins (*18*). Higher orders of structure may also arise. Figure 5 shows the secondary and tertiary structures of myoglobin. The highly convoluted structure of a protein is contained in the primary sequence of the protein itself. An excellent example of the degree to which structure is contained in the sequence is provided by the tobacco mosaic virus (TMV) (Fig. 6).

I was seeking a reaggregated cell membrane, but as yet a complete reassembled cell membrane was unavailable. However, a partial synthetic reaggregation of lipid and protein producing a membranelike structure could be achieved in the laboratory (Fig. 7) (*20*). Again, the characteristics of the membrane structure are contained in the structure of the molecules of which

Table 2. Relative yields of bonds in the copolymerization reactions of adenylates of pairs of amino acids

Interacting substances	*Bonds*	*Relative yields of bonds (%)*
alanine-adenylate glycine-adenylate	Al-Al	40
	Gly-Gly	32
	Al-Gly	15
	Gly-Al	13
alanine-adenylate valine-adenylate	Al-Al	23
	Val-Val	52
	Al-Val	12
	Val-Al	13
alanine-adenylate aspartyl-adenylate	Al-Al	47
	Asp-Asp	49
	Al-Asp	2
	Asp-Al	2
alanine-adenylate serine-adenylate	Al-Al	37
	Ser-Ser	37
	Al-Ser	12
	Ser-Al	14
aspartyl-adenylate glycine-adenylate	Asp-Asp	55
	Gly-Gly	21
	Asp-Gly	9
	Gly-Asp	15
aspartyl-adenylate serine-adenylate	Asp-Asp	59
	Ser-Ser	22
	Asp-Ser	10
	Ser-Asp	9
aspartyl-adenylate histidyl-adenylate	Asp-Asp	36
	Hist-Hist	44
	Asp-Hist	8
	Hist-Asp	12

Source: Ref. *16*.

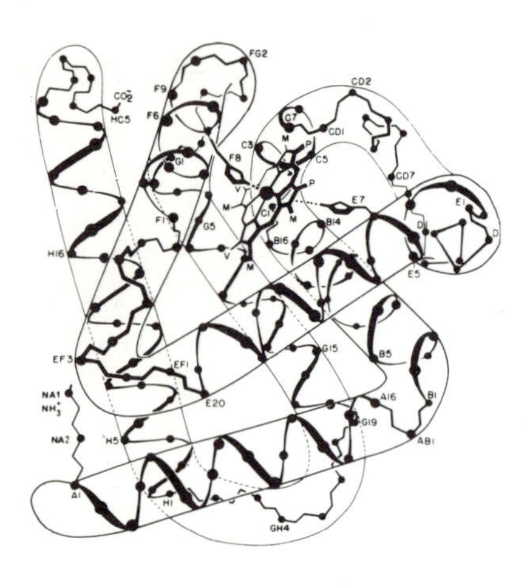

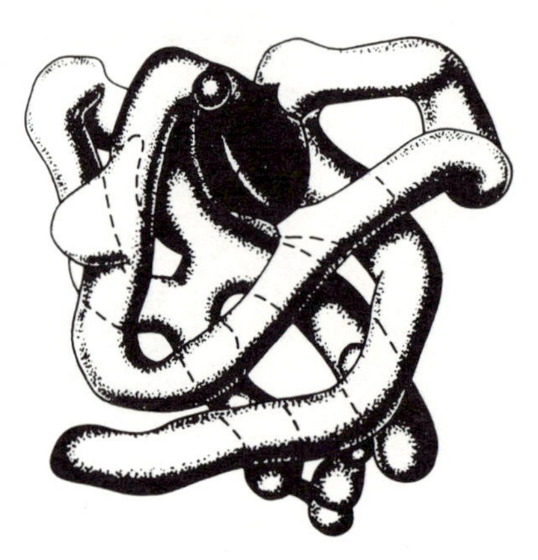

Figure 5. The two kinds of structure inherent in the primary sequence of amino acids and nucleotides are illustrated using a molecule of myoglobin. The secondary structure (*left*) is visible as a helical pattern of connected dots, with the sequence of amino acids clearly apparent. The tertiary structure is seen as the individual, specific way in which the helix is bent and folded (*right*).

it is composed. Since the time of the experiment shown in Figure 7 many new experiments have been reported in which phospholipids and proteins are reassembled in a manner bearing structural similarity to a cell. Even more important, some of the membrane functions have also been recovered—for example, permeability and ion pumping mechanisms. At present the mechanisms by which selective permeability is achieved are under close investigation. The structure of the membrane is commonly believed to be a combination of protein and lipid molecules, in which the protein molecules having specific functions are embedded in the bilipid membrane (*21*). Considerable progress has been made toward reconstructing biomembranes that are active in function as well as in structure.

Thus, even the highly selective membrane functions of living cells are contained in the structure of the phospholipids and proteins of which they are made. The reassembled components produce spontaneously formed membranes having at least partly the structure and function of natural living membranes. We have, thus, passed through the second stage of evolution mentioned above: we have evolved the biomonomers, the biopolymers, and the structure.

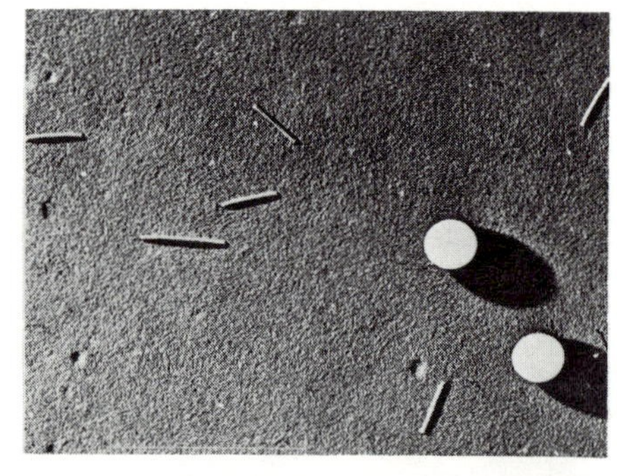

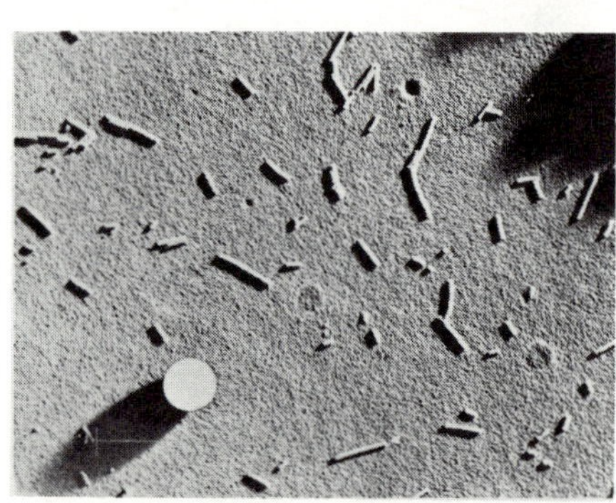

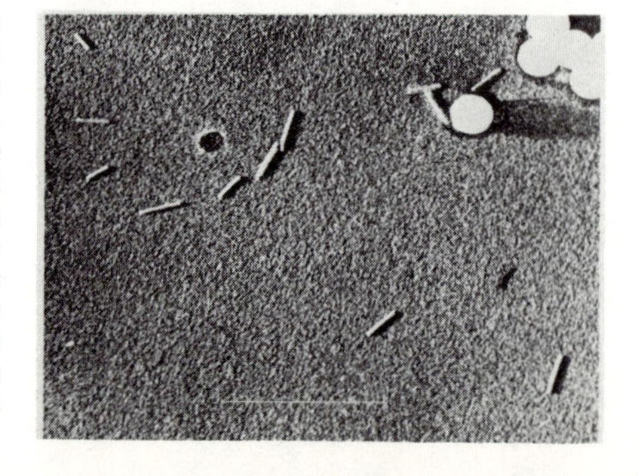

Figure 6. The isolated native tobacco mosaic virus (TMV) (*top*) has a specific diameter and length and is made up of a single strand of nucleic acid surrounded by a tightly packed group of a specific protein molecule. The virus particle can be decomposed to produce one solution containing the protein and another containing the genetic material. When the proteins are allowed to reaggregate in the solution (*center*), they form small, rodlike particles of indeterminate length. If the TMV RNA is added to the protein solution, the intact virus particle reappears (*bottom*). Polystyrene spheres (markers) are about 60 nm (600 Å).

Evolution of catalysis

In the evolution of the catalytic function there is an interesting development that can be traced in the laboratory as well as in the nature of existing catalysts in the animal and plant worlds. Basic to the evolution of catalysis is the role that reflexive catalysis, or autocatalysis, plays in all biological systems (*22*). Autocatalysis is a process in which a chemical reaction produces a product which is a catalyst for its own formation. In the sense that one living organism catalyzes the organization of organic matter into another living organism similar to itself, each living organism is simply a system of reflexive catalysis for manufacturing itself. Autocatalysis as the chemist normally knows it is simply a single reaction of a complex process that goes on in all living organisms.

The evolution of a catalyst for the reaction of hydrogen peroxide to water and molecular oxygen is a well-known example of autocatalysis (*23*). This example illustrates how the catalytic function can be improved with time and selection by showing what happens to the ability of iron ion to catalyze the decomposition of hydrogen peroxide to water and oxygen. The bare aqueous ferric ion has a catalytic ability represented by 10^{-5}. If, however, that iron ion is incorporated into a porphyrin molecule, such as a heme, the catalytic capability of the iron is increased by a thousandfold, to 10^{-3}. If, further, one builds that heme into a protein such as catalase, which has a rather specific structure holding the two imidazole groups on either side of the iron, the catalytic function increases once more, to 10^{5}. Thus, the primitive catalytic function of the

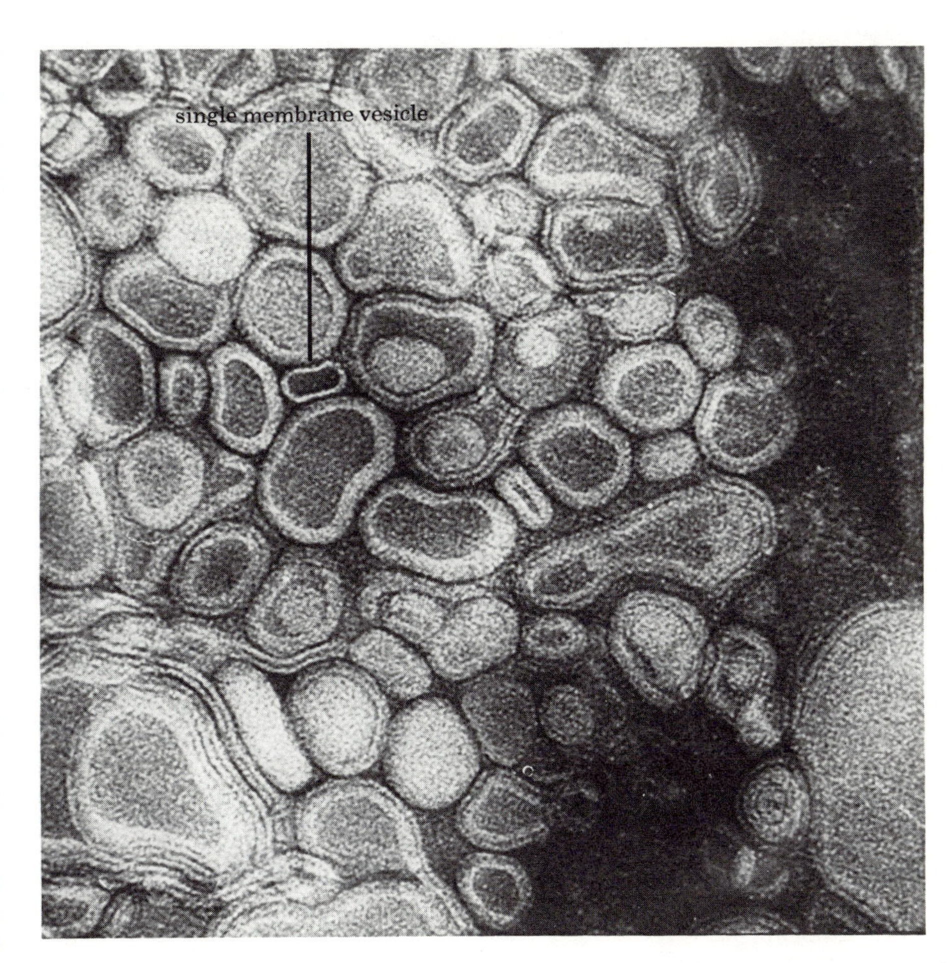

Figure 7. Membranelike structures are created in the laboratory by combining cytochrome and phospholipid molecules and submitting the mixture to sonication. Near the center of the photograph is a single membrane surrounding a protein solution (*arrow*); the other vesicles are surrounded by multiple membranes. (From ref. *19*.)

iron present in seawater as ferric ion evolves as it is first incorporated into heme and then into a protein more complex in structure, developing the highly efficient catalytic function of catalase.

In addition to producing O_2, the photodecomposition of water that takes place in photosynthetic organisms enables some of these organisms to use or evolve hydrogen. We are just beginning to learn how we can organize iron into a catalyst that might be capable of reacting with or generating hydrogen (*24, 25, 26*). In a number of primitive photosynthetic organisms, as well as in some bacteria, there exists a catalyst known as hydrogenase, an enzyme that will catalyze the reaction of hydrogen with a variety of other materials. In some cases, the hydrogenases can be used to evolve hydrogen from the organism when it is receiving its energy from the light (*27, 28, 29*).

The structure of this type of compound, in which the hydrogen is activated (and in which in some cases the nitrogen may be reduced), is very different from the structure of catalase. The one known structure of this class that has been published is that of bacterial ferredoxin (Fig. 8), a complex protein containing two distorted cubes of iron and sulfur atoms which have exactly the same structure and which occur in two different places in the protein molecule. These distorted cubes (and structures related to them) have been found in both single and double units in a variety of bacteria and algae.

Hydrogenase structure seems to be simpler, containing only two iron atoms instead of four (Fig. 9). Although it is very similar in its properties to the known bacterial ferredoxin cube, the structure of hydrogenase and green plant ferredoxin has not yet been demonstrated unequivocally. However, there is enough circumstantial evidence to enable us to say with some confidence that, when those materials are analyzed by X-ray crystallography, their structure will be found to be similar to the structure of bacterial ferredoxin.

The question is: How could such a structure evolve? Let us examine the structure of the commonest iron-sulfur mineral—iron pyrite—to see if it can give rise to a dimeric iron-sulfide structure by interaction with hydrogen sulfide, which we know to have been present in the primitive atmosphere of the earth. Such an interaction seems to be able to give rise to a dimeric structure which, upon further electron and hydrogen transport followed by ligand exchange (*32*), would give rise to a dimeric structure exactly analogous to that of hydrogenase and ferredoxin. We now need experiments to determine whether a system such as this can indeed operate.

The two examples discussed here show that the nature of the catalytic function is sharply affected by the nature of the surround of the iron atom: in one case, porphyrin and protein are used to produce an iron atom that is involved in oxygen reactions, and in the other, mercaptide and sulfide are used to produce a set of iron atoms involved in hydrogen reactions.

Evolution of information transfer

Figure 1 depicted the coupling of the information-carrying polynucleotide and the catalytic-function-carrying polypeptide. The last major question we will address here is: How could this coupling have occurred? The code for the coupling of these functions is shown in Figure 10. It is clear that the code had some redundancy, and, in many cases, it looks as though it started out as a two-letter rather than a three-letter code. For example, two A's in sequence are enough to code for lysine, whether the third base is an A or a G. However, two A's in sequence are also the code word for asparagine, and a third letter is thus necessary to dis-

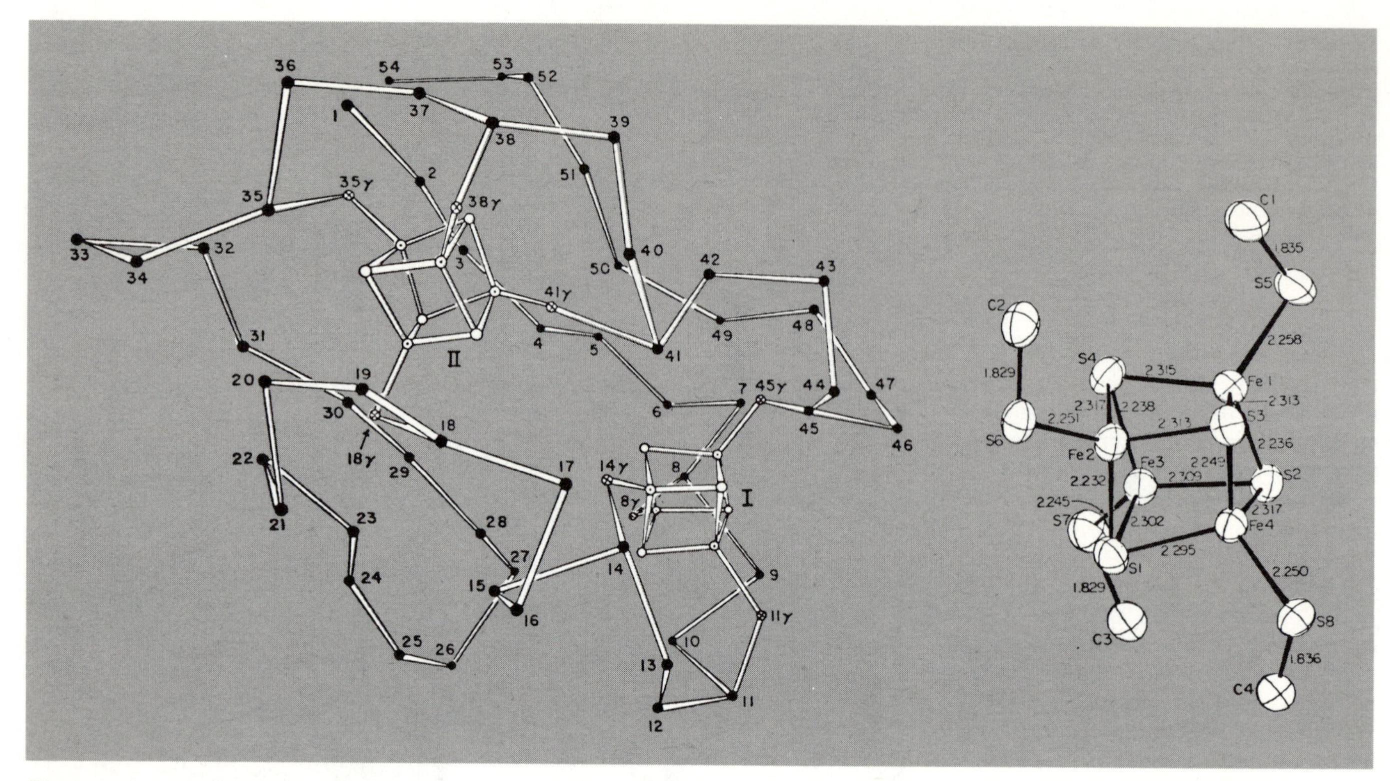

Figure 8. A marked feature in the structure of bacterial ferredoxin (*left*), the only iron-sulfide protein yet elucidated, is the two identical distorted cubes of iron and sulfur atoms. (After Adman et al., ref. *30*.) A synthetic material (*right*) very similar to the distorted cube in bacterial ferredoxin has been produced in the laboratory. This synthetic molecule is composed of four iron atoms, four sulfur atoms, and four mercaptides. (After Herskovitz et al., ref. *31*.)

tinguish between them. The code probably started as a single-base code, became a doublet, and is now a triplet.

I think this code arose not by accident (*33, 34, 35*) but because of the peculiar chemistries of the various bases and amino acids. Two sets of experiments give some idea of how the coupling of the information-bearing molecules with the catalyst-bearing molecules might have occurred. In the laboratory we have tried to discover specific relations between some amino acids and some specific bases. In the first experiment a particular base—either adenine or cytosine—was attached to a synthetic polymer, and the relative efficiency, or rate, at which various amino acids reacted with that base was measured. This was, in a sense, a "model" of a very primitive tRNA (*36*). This experiment was performed for two bases and two amino acids (*37*). Adenine and cytosine were separately coupled to polystyrene, and we then measured the efficiency with which phenylalanine is coupled to adenine or cytosine and the efficiency with which glycine is bound to the adenine or cytosine. The relative efficiencies of the four possible coupling reactions are: Phe-A, 6.7; Phe-C, 2.9; Gly-A, 10.0; Gly-C, 6.5. We thus have the beginnings of evidence that, even with one amino acid and one base, there is a kind of selectivity intrinsic in the structures. We must now explore this idea with larger groups—two (and more) bases and different amino acids.

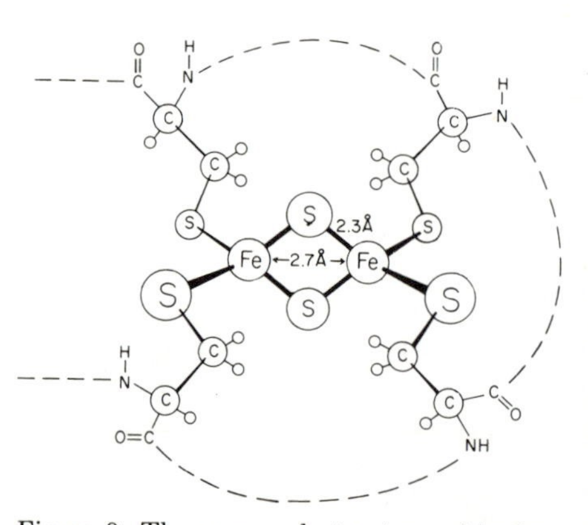

Figure 9. The proposed structure of hydrogenase and green plant ferredoxin, which has not yet been completely elucidated, is much simpler than that of bacterial ferredoxin, containing only two iron atoms, two sulfur atoms, and four mercaptides. However, the structure has similarities to that of bacterial ferredoxin. (After Calvin, ref. *27*.)

A different experiment of this type was devised by C. A. Ponnamperuma (*38*). Instead of hanging the base on the polymer and measuring the coupling of the free amino acid to the base, he did the reverse. Ponnamperuma put the amino acid on the polymer and examined the ability of that polymer to hold various nucleotides noncovalently. The results of this experiment (Table 3) are given for di- and tri-nucleotides and several different amino acids. It is interesting to observe the different strengths with which the glycine-bearing polymer adsorbs the various bases, and that the polymer has a greater selectivity for GAU than for AGU, the same three bases in different sequence. The same distinction appears with the tryptophan but in the opposite direction.

Thus two kinds of evidence and even some theoretical background (*39, 40*) exist to support my belief that the code is not an accident but

Second letter \ First letter	A	C	G	U
A	AAA AAG Lys AAC AAU Asn	CAA CAG Gln CAC CAU His	GAA GAG Glu GAC GAU Asp	UAA UAG terminate UAC UAU Tyr
C	ACA ACG Thr ACC ACU	CCA CCG Pro CCC CCU	GCA GCG Ala GCC GCU	UCA UCG Ser UCC UCU
G	AGA AGG Arg AGC AGU Ser	CGA CGG Arg CGC CGU	GGA GGG Gly GGC GGU	UGA terminate UGG Trp UGC UGU Cys
U	AUA Ilu AUG Met AUC AUU Ilu	CUA CUG Leu CUC CUU	GUA GUG Val GUC GUU	UUA UUG Leu UUC UUU Phe

Figure 10. The code for the coupling of polynucleotide and polypeptide in a living cell (represented in Fig. 1) consists of base triplets corresponding to each amino acid. For example, three adenine bases in a single sequence code for lysine (*upper left*).

has a chemical base (*41*). The high degree of selectivity that exists today has yet to be demonstrated, but the rudiments of such selectivity have been shown in at least these two preliminary types of experiments.

Our search for answers to questions about the origin of life has had, over the past two decades, a number of practical consequences in new and usable reactions. We may expect similar progress as we continue to look for answers to the more difficult problems before us, particularly in our ability to control and manipulate the intimate molecular events of living things.

Table 3. Selectivity coefficients for the binding of oligonucleotides to immobilized amino acids

	UpGp	*GpUp*	*ApUp*	*ApApUp*	*GpApUp*	*ApGpUp*
Gly	10.32	14.05	23.6	63.9	16.6	10.9
	13.44	14.39	27.5	60.0	19.4	13.0
Trp	95.1	42.1	187.5	2,045	60.4	173
	101.3	56.2	177.6	1,817	65.4	197

Source: Ref. *38*.

References

1. There have been many reviews and books written on the subject of chemical evolution. The following contain overall summaries of the work: (a) Melvin Calvin, 1969. *Chemical Evolution: Molecular Evolution Towards the Origins of Living Systems on Earth and Elsewhere.* Oxford: The Clarendon Press. (b) Sidney W. Fox and Klaus Dose, eds. 1973. *Molecular Evolution and the Origin of Life.* San Francisco: W. H. Freeman. (c) L. E. Orgel. 1973. *The Origins of Life: Molecules and Natural Selection.* N.Y.: Wiley.
2. *Notes and Records of the Royal Society, London.* 1959. No. 14, Vol. 1.
3. Letter from Darwin to Hooker, 1 February 1871.
4. Ref. *1a*, chaps. 5 and 6.
5. R. M. Lemmon. 1970. *Chem. Rev.* 70:95.
6. R. M. Lemmon, 1973. *Environ., Biol., and Med.* 2:1.
7. M. Calvin. 1974. *La Recherche* 41:44–57.
8. A. R. Hochstim. 1963. *Proc. Nat. Acad. Sci.* 50:200.
9. M. Calvin. 1974. *Angew. Chem. Internat. Ed.* 13:121.
10. D. H. Kenyon and Gary Steinman. 1969. *Biochemical Predestination.* N.Y.: McGraw-Hill.
11. Ref. *1a*, pp. 160–70.
12. D. H. Kenyon, G. Steinman, and M. Calvin. 1966. *Biochim. Biophys. Acta* 124:339.
13. G. Steinman and M. N. Cole. 1967. *Proc. Nat. Acad. Sci.* 58:735.
14. M. Paecht-Horowitz, J. D. Breger, and A. Katchalsky. 1970. *Nature* 228:636.
15. M. Paecht-Horowitz. 1972. *Angew. Chem. Internat. Ed.* 11:798.
16. M. Paecht-Horowitz. 1974. *Origins of Life* 5:173.
17. Roger Acher. 1974. *Angew. Chem. Internat. Ed.* 13:186.
18. Ref. *1a*, pp. 187–93.
19. D. Papahadjopoulos and N. Miller. 1967. *Biochim. Biophys. Acta* 135:624; D. Papahadjopoulos and W. Watkins. 1967. *Biochim. Biophys. Acta* 135:639.
20. A. D. Bangham, J. DeGier, and G. D. Grevitch. 1967. *Chem. Phys. Lipids* 1:225.
21. S. J. Singer and G. L. Nicholson. 1972. *Science* 175:721.
22. Ref. *1a*, pp. 145–52.
23. Ibid., p. 146.
24. R. H. Wickramsinghe. 1973. *Space Life Sciences* 4:341.
25. D. O. Hall, R. Cammack, and K. K. Rao. 1973. *Space Life Sciences* 4:455.
26. W. H. Orme-Johnson. 1973. *Ann. Rev. Biochem.* 42:159.
27. M. Calvin. 1974. *Science* 184:375.
28. G. N. Schrauzer, G. W. Kiefer, K. Tano, and P. A. Doemeny. 1974. *J. Amer. Chem. Soc.* 96:641.
29. R. R. Eady and J. R. Postgate. 1974. *Nature* 249:805.
30. E. T. Adman, L. C. Silkin, and L. H. Jensen. 1973. *J. Biol. Chem.* 248:4987.
31. H. Herskovitz, B. A. Averaill, R. H. Holm, J. A. Ibers, W. D. Phillips, and F. J. Weiher. 1972. *Proc. Nat. Acad. Sci.* 69:2437.
32. L. Que, Jr., M. A. Bobrik, J. A. Ibers, and R. H. Holm. 1974. *J. Amer. Chem. Soc.* 96:4168.
33. M. Eigen. 1971. *Naturwissenschaften* 58:465.
34. F. H. C. Crick. 1968. *J. Mol. Biol.* 38:267.
35. L. E. Orgel. 1968. *J. Mol. Biol.* 38:381.
36. M. A. Harpold and M. Calvin. 1968. *Nature* 219:486.
37. M. A. Harpold and M. Calvin. 1973. *Biochim. Biophys. Acta* 308:117.
38. C. Saxinger and C. A. Ponnamperuma. 1974. *Origins of Life* 5:189.
39. G. Melcher. 1974. *J. Mol. Evol.* 3:121.
40. C. W. Carter and J. Kraut. 1974. *Proc. Nat. Acad. Sci.* 71:283.
41. M. Calvin. 1969. *Proc. Roy. Soc. Edinburgh* 70:273.

Lynn Margulis

The Origin of Plant and Animal Cells

The serial symbiosis view of the origin of higher cells suggests that the customary division of living things into two kingdoms should be reconsidered

Both primitive peoples and modern men divide the living world into two vast groups, plants and animals. This dichotomous view is not only consistent with intuition but, until recently, has been quite universally accepted by biologists. Animals, in general, are motile heterotrophs (other-nourished) and plants are immotile autotrophs (self-nourished). Certain organisms such as fungi and bacteria seem to be dependent plants, that is, they neither contain chlorophyll nor make their own organic food compounds; yet they do not have the characteristics of animals. By default, botanists have traditionally claimed them and placed them within the Thallophytes, or lower plants. Although many modifications and criticisms of this dichotomous view of life have been advanced at various times by botanists and systematists, it was not until recently that the entire concept underlying the kingdom level of taxonomy has come under severe fire (*1*).

Dr. Lynn Margulis, associate professor of biology, Boston University, received her bachelor's degree from the University of Chicago, the M.A. in zoology from the University of Wisconsin, and the Ph.D. in genetics from the University of California, Berkeley. Prior to her appointment at Boston University, she worked for several years developing science curricula for U.S. and African schools and training Peace Corps biology teachers for Latin America. Her continuing work on the symbiotic theory of the origin of eukaryotic cells was first published in the Journal of Theoretical Biology *(14:225). Recently the Yale University Press brought out her monograph,* Origin of Eukaryotic Cells, *which expands on her controversial theory. In 1967, Dr. Margulis received the Shell Award, the Boston University faculty publication merit award. Her present research is focused on a study of the metabolism of basal body and cilia regeneration in Stentor as it relates to her theory.* *The author makes grateful acknowledgment to T. N. Margulis, E. S. Barghoorn, G. E. Hutchinson, and V. Tartar for their aid, and the NASA and the National Science Foundation for support.* *Address: Department of Biology, Boston University, Boston, MA 02215.*

In general, scientists have believed that the only evolutionary mechanism for producing new populations of organisms is the progressive differentiation of descendants via mutations of many kinds and their natural selection. The evolutionary sequence usually envisioned for lower organisms is as follows: primitive heterotrophic bacteria led to photosynthetic bacteria and eventually to photosynthetic algae. A primitive phytoflagellate alga (sometimes called the "uralga") is considered to be the common ancestor to higher green plants on the one hand and, by loss of photosynthetic capabilities, the ancestor to fungi and animals on the other. This classical view of the phylogeny of animals and green plants is shown in Figure 1.

The alternative "serial symbiosis" point of view, while not denying the paramount importance of these processes, adds another relevant evolutionary mechanism: the acquisition of intracellular organelles by symbiosis and the subsequent joint evolution of the symbiotic partners as a unique entity (Fig. 1, right side). The unsuspicious reader should be warned that the progressive differentiation idea is still very widely accepted and that the alternative view—that the eukaryote cell arose by a series of symbioses—described in this article still represents a minority opinion, only very recently even discussed in "polite biological society" (*2*).

The symbiotic theory is actually based on several ideas that have a long classical tradition. The concept that cellular organelles, such as the chloroplasts in algae and green plants and the mitochondria of plants and animals, originated as endosymbionts has been present in the classical cytological literature since the discovery of these organelles (*3*, *4*). The notions of genetic autonomy of such organelles derive mainly from observations of the growth and division of the organelles inside the cytoplasm. While the sizes, staining properties, and correlation of the presence of the organelles with certain inherited traits always reinforced these views, critics of the endosymbiont origin idea have been correct in their assertion that the concept of the independent origin of organelles, although attractive, has been untestable. Now that biologists have developed a clearer picture of the actual workings of a minimal self-replicating system—the prokaryote cell (consisting basically of DNA, messenger RNA coded off that DNA, protein synthesis on ribosomes coded by messenger RNA surrounded by lipoprotein membrane containing an energy source and so forth)—the concept of a cell can be applied to the workings of an organelle within a cell.

In comparing the classical and symbiotic views (Fig. 1), it is clear that there is much agreement and overlap between them. Both the classical view and the symbiotic view agree that bacteria are primitive and that photosynthetic bacteria are ancestral to photosynthetic blue-green algae. They also agree that green algae are ancestral to higher green plants. The area of disagreement lies in the relationship between the admittedly primitive prokaryotic blue-green algae and the eukaryotic green algae and protozoans, that is, in the middle section of Figure 1. Whereas the classical view holds that blue-green algae evolved into photosynthetic phytoflagellates that later lost autotrophy to evolve animals and fungi, the symbiotic view insists that the blue-green algae were ancestral only to the plastid of eukaryotic algae.

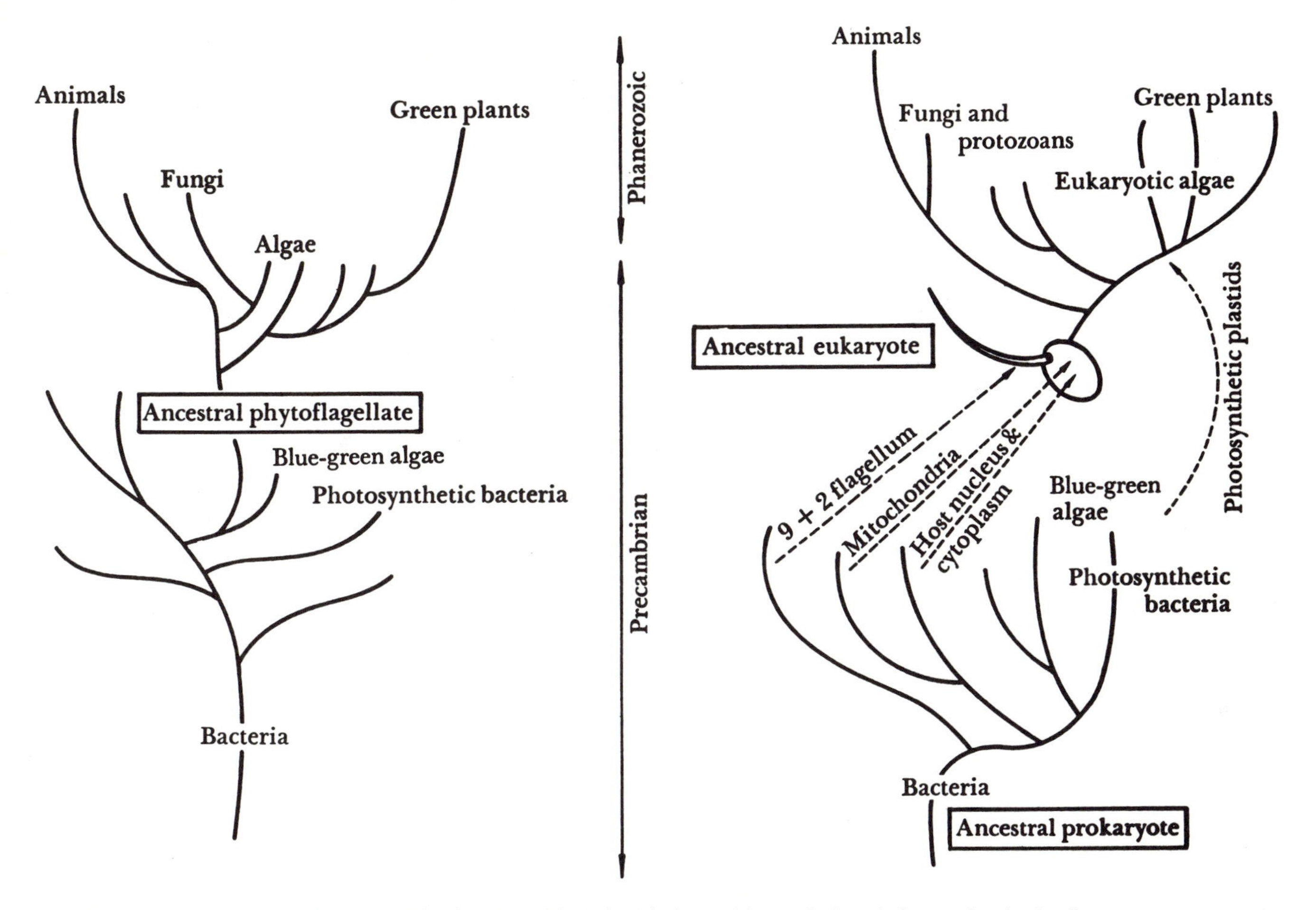

Figure 1. Comparison between the classical (*left*) and symbiotic (*right*) views of the evolution of plant and animal cells.

The cell symbiosis theory holds that a primitive amoeboflagellate, a heterotrophic cell, is ancestral to all eukaryotes: fungi, animals, nucleated algae, and higher plants. Some descendants of the amoeboflagellate became photosynthetic when they acquired intracellular blue-green algae-like symbionts. The photosynthetic symbiont was retained, selection acted on the entire complex, and with time the symbiont eventually differentiated into the membrane-bound photosynthetic plastid of algae and plants. Thus according to this view, asking What photosynthetic organism is the ancestor of the green algae? is quite analogous to asking What plant is ancestral to the lichen? Just as lichens have two immediate ancestors, a photosynthetic algal ancestor and a heterotrophic fungal ancestor (*5*), the serial symbiotic argument claims that all nucleated plants have at least two immediate ancestors, a photosynthetic ancestor to the plastid and a heterotrophic ancestor to the rest of the cell. The theory has really even been developed much further (*6*), claiming that the ancestral heterotroph itself formed as a product of intracellular symbiosis. That is, protozoans, fungi, animals, nucleated algae, and green plants had common heterotrophic ancestors—mitochondria-containing amoeboflagellates. These ancestral amoeboflagellates were cells in which mitosis and eventually meiosis evolved. Before the sequence of steps suggested for the origin of the eukaryote cell is described, some explanation is in order of the recently available evidence that makes such views defensible.

First, there is the fundamental recognition of the enormous difference in cell structure between the prokaryotes and the eukaryotes. Table 1, which summarizes some of these differences, is based on a large accumulation of twentieth-century cytological and microbiological work culminating in the recognition of this great evolutionary discontinuity in cell type first made by C. B. van Niel, R. Y. Stanier, and their colleagues (*7*). Unlike the distinction between animals and plants which becomes more and more blurred as one studies the flagellate algae and protozoans, the distinction between eukaryote and prokaryote cells has become progressively sharper and more valid with new microbiological investigation. Any given population of microbes may unequivocally be assigned to one or the other of these non-overlapping groups. The phylogeny and the symbiotic theory presented here is entirely dependent on the prior recognition of the eukaryote-prokaryote dichotomy of cell type.

Furthermore, the symbiotic theory rests upon recent discoveries concerning the metabolic capabilities of three types of eukaryotic organelles: mitochondria, photosynthetic plastids, and flagellar basal bodies. At least the first two of these are known definitely to contain their own nucleic acids and basic components of the protein synthesizing system that characterizes free-living prokaryotic cells. Although, no doubt, they have been modified by perhaps a billion years' association with nucleated cytoplasm, the case that mitochondria and plastids fulfill criteria for organelles originating inside cells as symbionts (Table 2) can be argued now on the basis of enor-

Table 1

Major Differences between Prokaryotes and Eukaryotes

Prokaryotes	Eukaryotes
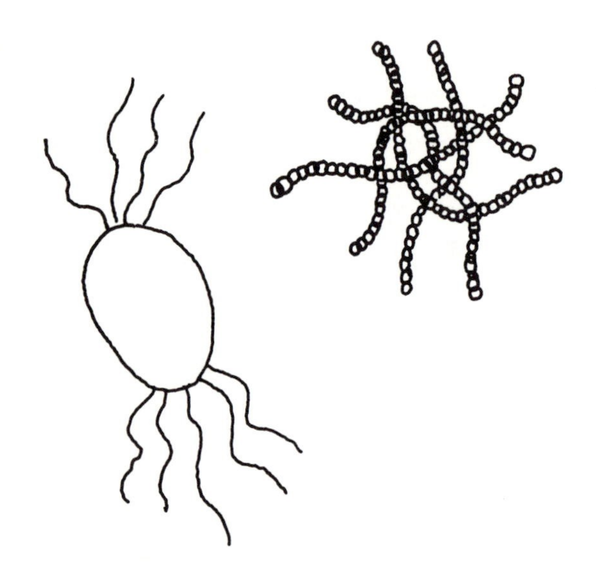	
Mostly small cells (1–10 μ); all microbes; the most morphologically complex are filamentous or mycelial with fruiting bodies	Mostly large cells (10–10^2 μ); some are microbes, most are large organisms; the most morphologically complex are the vertebrates and the flowering plants
Nucleoid, not membrane-bound	Membrane-bounded nucleus
Cell division direct, mostly by "binary fission," chromatin body which contains DNA and polyamines; does not stain with the Feulgen technique. No centrioles or mitotic spindle	Cell division by classical mitosis; many chromosomes containing DNA, RNA, and proteins; stains bright red with Feulgen technique; centrioles, mitotic spindle present
Sexual systems absent in most forms; when present, unidirectional transfer of genetic material from donor to host	Sexual systems present in most forms; participation of both partners (male and female) in meiotic production of gametes
Multicellular organisms never develop from diploid zygotes, show no tissue differentiation	Multicellular organisms develop from diploid zygotes, show extensive tissue differentiation
Includes strict anaerobes (killed by O_2), and facultatively anaerobic, microaerophyllic, and aerobic forms	All forms aerobic (need O_2 to live; exceptions clearly secondary modifications)
Enormous variations in the metabolic patterns of the group as a whole; mitochondria absent; enzymes for oxidation of organic molecules bound to cell membrane, i.e. not "packeted"	Same metabolic patterns of oxidation within the group (i.e. Embden-Meyerhof glucose metabolism, Krebs cycle oxidations, molecular oxygen combines with hydrogens from foodstuffs, catalyzed by cytochromes, water produced); enzymes for oxidation of 3-carbon organic acids within "packeted" membrane-bounded sac; mitochondria present
Simple bacterial flagella, if flagellated	Complex "(9+2)" flagella or cilia, if flagellated or ciliated
If photosynthetic, enzymes for photosynthesis bound to cell membrane (chromatophores); not "packeted" in chloroplasts; anaerobic and aerobic photosynthesis—sulfur deposition and O_2 elimination	If photosynthetic, enzymes for photosynthesis "packeted" in membrane-bounded chloroplasts; O_2 eliminating photosynthesis.

Table 2. Criteria for organelles originating as endosymbionts

1. Symbiont originally had its own DNA, messenger RNA complimentary to that DNA, a source of ATP and other nucleotides, a functioning protein synthesizing system on ribosomes, and cell membrane synthesizing system. With progressive evolution of the symbiosis, intracellular symbiont may lose from none to all its independent synthetic capabilities except ability to replicate its own DNA. The symbiont may dedifferentiate down to the level of DNA.

2. If an organelle has been acquired symbiotically it will be retained only if there exists some mechanism insuring that at each cell division each daughter receives at least one copy of the symbiont genome.

3. If an organelle was acquired by symbiosis there should be no species that contain intermediate intracellular stages of that organelle.

4. If the symbiont is lost, all metabolic characteristics coded for on the symbiont genome must be lost together. Once lost the symbiont can only be regained by reingestion.

5. Since any intracellular symbiont must have its own genes, a correlation can be made between the genetic traits conferred on the host by the symbiont and the morphological presence of the symbiont (*3*). (For example "cytoplasmic" or "uniparental" inheritance.)

6. If an organelle originated as a free-living cell, it is possible that naturally occurring counterparts may still be found among extant organisms. Even if precise co-descendants can not be found, the organelle must have genetic and physiological characteristics known to be consistent with those generally present in terrestrial cells. (Possibilities of organelles and free-living counterparts might be: chloroplasts—blue-green algae, mitochondria—gram negative Krebs cycle containing rod shaped eubacteria, flagella—spirochaetes, and so forth.)

mous quantities of data from many different fields and many different organisms (*6*). The case for the basal body (the same entity recognized as the centriole in mitotic cell division) and its product the eukaryote cilium originating symbiotically is much less closed. Evidence has recently been presented that these highly characteristic eukaryotic organelles have DNA and RNA (*6, 8*). Even more recent evidence for nucleic acids in basal bodies is negative, but this may not preclude symbiont origin (*9*).

The other profound contribution to new concepts of cellular evolution has come from an entirely different and hitherto unrelated field, Precambrian paleontology. Even this juxtaposition of words may seem unfamiliar to those who believe that the vast stretches of Precambrian time are unfossiliferous. Now this belief must be reconsidered in view of the recent demonstrations by E. S. Barghoorn and his colleagues J. W. Schopf and P. E. Cloud, Jr. that the Precambrian is full of fossils (Table 3) (*10–15*). Although some remains, such as algal stromatolites, are several feet across (Figure 2), the morphologically recognizable fossils of the era are mainly microbial; the algae that are primarily responsible for the initial deposition of the sediments are measured in microns. Be-

Table 3. Summary of evidence for Precambrian fossils

$\times 10^9$ years	*Location and sediment*	*Type of fossil*	*Investigators*
3.0–3.3	South Africa Fig tree	Bacteria, blue-green algae	Barghoorn and Schopf (*9*)
2.7–2.0	North America Great Slave Lake, Canada	Algal stromatolites	Hoffman (*10*)
2.1–1.9	North America Gunflint	Blue-green algae, bacteria, green algae?	Cloud (*12*), Barghoorn and Tyler (*11*)
1.0–0.8	Australia Bitter Springs	Blue-green algae, green algae	Schopf (*13*)
0.7–present	Worldwide	Metazoans Beginning of continuous fossil record	Cloud (*14*)

Figure 2. *Above:* Rocks between 1.8 and 2.5 billion years old from the Great Slave Lake area in the Northwest Territory of Canada. These are called "stromatolites" and were formed in the intertidal environment by blue-green algae. *Below:* Such stromatolites are being formed today, for example, in Shark Bay, Western Australia. Photographs courtesy of Dr. Paul Hoffman, Geological Survey of Canada, from his unpublished work. See (*6*) pp. 137–41, and (*11*).

sides the impressive finds of intact Precambrian microbes, organic chemists have determined that ancient rocks contain "chemical fossils" (*15*), which are organic materials corresponding to geological derivatives of common biotic molecules. These are found in even the oldest sediments.

Taken together, I believe all this evidence can be summarized most usefully in terms of the concepts of the symbiotic theory diagrammed on the right side of Figure 1. The Precambrian can be thought of as the "Age of Prokaryotes," the time during which the atmosphere changed from reducing or neutral to oxidizing as a result of blue-green algal photosynthetic activity. The Phanerozoic, on the other hand, is the well-known "Age of Eukaryotes." During this time, the aerobic metabolism and elegant genetic systems of these "higher cells" led to the eventual dominance of the most complex eukaryotes, the metazoans, and the green plants. These ideas have a significant effect upon the choice of classification system at the level of the highest taxa (Table 4).

Summary of the serial symbiosis view

This journal is not the place in which to elaborate on the recently published (*6*) serial symbiosis view. Rather, it can only be stated in simplified terms so that interested readers can consult the references and evaluate it for themselves. Essentially, the symbiotic view can be outlined as follows: all living organisms have a common ancestry; all were ultimately derived from a prokaryote heterotrophic bacteria-like cell that contained the present genetic code and the present protein synthesizing system (*17*). By mutation and selection the primitive heterotroph gave rise to many metabolically diverse populations of prokaryote organisms; among these were microbes capable of synthesizing porphyrins and isoprenoid derivatives, microbes able to fix atmospheric carbon dioxide and atmospheric nitrogen, microbes capable of reducing sulphur, and so forth. In early Precambrian times all of these microbes were anaerobic.

Eventually, from a population of microbes capable of fixing CO_2 and synthesizing porphyrins, there arose a class of bacteria that could utilize visible radiation from the sun to produce the key biological intermediate in energy transactions, adenosine triphosphate (ATP). Atoms from hydrogen gas, hydrogen sulfide, or from small organic compounds were used to reduce CO_2 to form the specific organic compounds requisite for cellular reproduction. These bacteria were ancestral to the present anaerobic photosynthesizers. Mutations occurred in this population of organisms that eventually led to the use of H_2O as hydrogen donor in CO_2 reduction. Such microbes eliminated unused oxygen from water into the atmosphere as a waste product. These photosynthetic microbes were the first green-plant photosynthesizers—ancestors to our present blue-green algae—and, according to the view presented here, ancestors to the photosynthetic plastids of nucleated algae and higher plants.

The elimination of highly reactive oxygen gas into the atmosphere led to a crisis among prokaryotes—a crisis that has left an indelible mark on the metabolism of these cells. They either had to adjust to the increasing presence of oxygen in the atmosphere by evolving metabolic mechanisms to cope with it or they had to find some anaerobic niche. Even a superficial perusal of prokaryotes will convince anyone that many different responses to oxygen—from obligate anaerobiosis to obligate aerobiosis—exist in the group as a whole. Thus, it was only during or after the transition to the oxidizing atmosphere that any eukaryote cell arose.

According to this theory, the ancestor to all eukaryotes became an aerobe when a large anaerobic heterotrophic microbe capable of catabolizing glucose to pyruvate established an intracellular symbiosis with a smaller aerobic bacterium. The aerobic endosymbiont metabolized 3-carbon organic compounds completely to CO_2 and water via the Krebs cycle oxidations. The increasing presence of atmospheric oxygen selected for the symbiotic complex eventually gave rise to all mitochondria-containing cells. This primitive mitochondria-containing heterotroph, now an amoeboid, acquired a further population of symbionts, motile organisms that stuck to the surface of the amoeboid much as spirochaetes are known to stick to the surfaces of certain protozoans. As in the modern case of *Myxotricha* (a hypermastigote flagellate found in the guts of termites and containing cortical spirochaetes responsible for the movement of their host), these motile symbionts were initially selected because they conferred motility on their amoeboid host, thus helping it to procure food more efficiently. By hypothesis, the motile symbionts were themselves ancestral to what later evolved into flagella, cilia, and all the other "9 + 2" homologues so universally characteristic of eukaryotes and so lacking in any prokaryote. In fact, again by far-out but testable hypotheses, the motile symbionts eventually differentiated into the "achromatic apparatus" of mitosis, that is the spindle, centrioles, and other nonchromatin portions of the mitotic figure.

This differentiation must have taken millions of years and, by hypothesis, produced along the way the many and fascinating variations upon the mitotic theme found in protozoans, certain fungi, nucleated algae, and other lower eukaryotes. Once mitosis and meiosis were perfected, the advanced tissue differentiation which is based upon the Mendelian genetic foundation and which is characteristic of higher plants and animals could evolve and eventually take over the earth, producing myriads of large organisms. In essential outline then, this is the serial symbiotic theory of the origin of eukaryotic cells. Even if eventually proved invalid, it provides a unified framework upon which to evaluate both the rapid influx of data from Precambrian sediments and from research on the genetics and biochemistry of eukaryote organelles (*18*).

Table 4. Comparison of largest taxa derived from alternative views of the origin of plant and animal cells

Classical view		Serial symbiosis view	
Two kingdoms	Major members	Five kingdoms (*1*)	Major members
Plantae	Bacteria Fungi Algae Green plants	Monera (prokaryotes)	Bacteria Blue-green algae
Animalia	Protozoans Metazoans	Protista (lower eukaryotes)	Protozoans Nucleate algae
		Fungi	Mushrooms, molds, yeasts
		Plantae	Green plants (bryophytes, tracheophytes)
		Animalia	Metazoans

References

1. R. H. Whittaker. 1969. New concepts of kingdoms of organisms. *Science* 163:150–59.
2. "More recently Wallin (1922) has maintained that chondriosomes [mitochondria] may be regarded as symbiotic bacteria whose associations with other cytoplasmic components may have arisen in the earliest stages of evolution... to many, no doubt, such speculations may appear too fantastic for present mention in polite biological society; nevertheless, it is in the range of possibility that they may some day call for more serious consideration." E. B. Wilson. 1925. *The Cell in Development and Heredity.* New York: Macmillan.
3. D. B. Roodyn and D. Wilkie. 1968. *The Biogenesis of Mitochondria.* London: Methuen and Co., Ltd.
4. J. T. O. Kirk and R. A. E. Tilney-Bassett. 1967. *The Plastids.* London and New York: W. H. Freeman.
5. V. Ahmadjian. 1963. The fungi of lichens. *Scientific American* 208:122–32.
6. L. Margulis. 1970. *Origin of Eukaryotic Cells.* New Haven: Yale University Press.
7. R. Y. Stanier, E. Adelberg, and M. Douderoff. 1970. *The Microbial World.* 3rd ed. Englewood Cliffs, N.J.: Prentice Hall, Inc.
8. J. Randall and C. Disbrey. 1965. Evidence for DNA at basal body sites in *Tetrahymena pyriformis. Proc. Roy. Soc.* Series B., 162:473–91.
9. L. Margulis. 1971. Cytoplasmic Genes: Our Precambrian Legacy, in *Stadler Symposia,* Vols. 1 & 2. G. P. Redei, ed. Columbia, Mo.: Univ. of Missouri Press.
10. E. S. Barghoorn and J. W. Schopf. 1967. Alga-like fossils from the Early Precambrian of South Africa. *Science* 156: 508–12.
11. P. Hoffman. 1968. See A. L. McAlester. *The History of Life.* Englewood Cliffs, N.J.: Prentice-Hall, p. 14. See also (*6*).
12. E. S. Barghoorn and S. A. Tyler. 1965. Microorganisms from the Gunflint Chert. *Science* 147:563–77.
13. P. E. Cloud, Jr. 1965. Significance of the Gunflint (Precambrian) microflora. *Science* 148:27–35.
14. J. W. Schopf. 1969. Microorganisms from the Late Precambrian of South Australia. *J. Paleontology* 43:111–18.
15. P. E. Cloud, Jr. 1968. Pre-Metazoan evolution and the origins of the Metazoa. In *Evolution and Environment,* E. T. Drake, ed. New Haven and London: Yale University Press, pp. 1–72.
16. G. Eglinton and M. Calvin. 1967. Chemical fossils. *Scientific American* 216: 32–43.
17. J. D. Watson. 1970. *Molecular Biology of the Gene.* New York: Benjamin Publishers.
18. S. S. Cohen. 1970. Are/Were Mitochondria and Chloroplasts Microorganisms? *American Scientist* 58:281–89.

Seymour S. Cohen

Are/Were Mitochondria and Chloroplasts Microorganisms?

A biochemist's inquiry into the origins of animal and plant cells

"*Upon analysis all Insects generally yield some acid. In that they differ from other animals, and can be considered the transition from the* règne animal *to the* règne vêgêtal *which abounds in acid; just as cruciferous plants which give a great deal of volatile alkali, or gramineous plants which contain a very large quantity of mucous matter and can be considered the transition from the* règne vêgêtal *to the* règne animal."—Guillaume-François Rouelle, 1762

In attempting to understand evolutionary relationships today, we have many more data than were available to M. Rouelle. New data have raised new questions, and the solutions to the basic questions of the transitions among kingdoms are far from complete, even when pursued at the level of cells, a form of life with which M. Rouelle was not familiar. Nevertheless the scientific discoveries and data collection of the last decade have permitted us to explore the problem of these transitions at a new level of sophistication. It is the purpose of this paper to summarize our present position in relation to one of these complex questions.

We are asking if certain organelles which are found in the cytoplasm of all higher cells may not have been derived from bacteria or bacteria-like microorganisms. Mitochondria are particles concerned with energy production which are found in all higher cells, including the cells of plants and animals. Chloroplasts are the photosynthesizing particles of plant cells. These latter bodies contain even more elaborate membrane systems, and these membranes contain the light-absorbing pigments and the numerous compounds involved in converting light energy to chemical energy. A cleavage of water results eventually in the generation of molecular oxygen. This chemical energy is used to generate a reductant essential to the fixation of carbon dioxide.

Almost thirty years ago, Stanier and van Niel (*1, 2*) distinguished procaryotic and eucaryotic cells. The Procaryota, comprising bacteria and blue-green algae, may be readily differentiated from animal or plant cells. Examining nuclear and genetic features, we note that procaryotes lack a nuclear membrane and multiple chromosomes, and do not divide by mitosis or exhibit reciprocal recombination. Very few of their genes are redundant. These cells lack membranous organelles such as mitochondria and chloroplasts. Procaryotic cells have characteristic cell wall structures, as well as several unique wall constituents, such as the carbohydrate derivative, muramic acid, and the amino acid, diaminopimelic acid. When such cells are motile they lack the characteristic flagellar structure, comprised of 9 fibrils surrounding two central fibrils, found in eucaryotic cells. We are asking, then, if mitochondria and chloroplasts have the numerous cytological features of procaryotic cells, and indeed if the eucaryotic cells in which they are found are harboring symbiotic procaryotes.

The fitness of the scientific environment

The question whether mitochondria and chloroplasts are symbionts within the cells they inhabit is old,

Seymour S. Cohen received his Ph.D. in biochemistry from Columbia University in 1941. He began his postdoctoral studies on plant viruses at the Rockefeller Institute and in 1945 began the first studies on the biochemistry of phage multiplication at the University of Pennsylvania, where he is now Professor of Biochemistry and Chairman of the Department of Therapeutic Research. As Jesup Lecturer at Columbia University, he wrote Virus-Induced Enzymes, *which was published in 1968. In May 1970 he was Visiting Professor at the Collège de France.*
Having touched on the nucleic acids in his dissertation and expanded this interest in an exposure to urology, he has continued to work on the biochemistry of the nucleic acids. His current research interests include the much neglected polyamines which appear to serve as organic cations on the nucleic acids, as well as aspects of the application of nucleic acid biochemistry to clinical problems. His scholarly interests also relate to problems of comparative biochemistry, of which this article is an outgrowth. When he has fulfilled his administrative and editorial commitments, he plays tennis and browses in Western Civilization 1780–1840.
This paper is adapted from an article which will appear in Essays in Biology, *to be published by Columbia University Press. Address: The School of Medicine, University of Pennsylvania, Philadelphia, PA 19104.*

having been posed even in the last century. The great advances in biology and biochemistry during the past quarter of a century have revealed numerous new approaches to this problem. Hundreds of significant papers on the biogenesis of these organelles have appeared in the last five years. The specific events which led to this activity were the discoveries of DNA in chloroplasts by Ris and Plaut in 1962 (*3*) and in mitochondria by Nass and Nass in 1963 (*4*).

Genetic evidence that inheritance of the organelles is non-Mendelian has been acquired over many years; this evidence is still being collected. The existence of cytoplasmic units endowed with genetic continuity has been actively discussed throughout the past twenty-five years; Lwoff's analysis of the role of kinetosomes in ciliate development is one leading example (*5*). Important experiments such as that of Luck with choline-labeled mitochondria in 1963 (*6*) strongly suggested organelle division. He concluded that "existing mitochondrial material is directly and randomly transmitted to progeny in formation of new mitochondria." It may be noted that Luck's paper was written before the existence of mitochondrial DNA was well known. Perhaps less familiar to biochemists and molecular biologists, a well-developed literature existed on the intracellular symbiosis and parasitism (these terms frequently may be difficult to distinguish by experimental or ecological criteria, as pointed out by Theobald Smith) of bacteria and blue-green algae. Blue-green algae frequently inhabit eucaryotic cells (*7*) and chloroplasts may function within animal cells in nature (*8*).[1]

The actual demonstrations that these organelles contained the genetic material, DNA, appear to have aroused a scientific public for whom these magic letters suddenly assumed the role of an addictive mind-expander. At a time when the role of DNA in the cell was acquiring unchallenged preeminence, the problems of polymer synthesis dominated by DNA naturally drew first attention, pointing to even more specific biological questions. These questions were "Does the DNA of the organelle act as the genetic substance of these organelles and is this substance different from the nuclear DNA of the eucaryotic cells in which they operate?" and "To what extent do these genetic systems of the organelle determine their structure and function?"

As one might expect, these specific questions have not yet been entirely answered. However, the general problem may now be posed as a serious scientific challenge, warranting the most systematic and penetrating exploration. Just as the Copernican revolution demonstrated that man is not the center of the Universe, so the investigation of this problem may show that a man (and indeed any higher organism) is merely a social entity, combining within his cells the shared genetic equipment and cooperative metabolic systems of several evolutionary paths. We suspect that governments should be interested in such a possibility, although their responses may not be readily predictable.

On the biochemistry of genetic independence

The questions posed above have been tested by all the modern and growing armamentarium of molecular biology. The DNAs of mitochondria and chloroplasts have been isolated and characterized with respect to size, shape, density, and base composition. A mitochondrion appears to contain multiple (2 to 6) copies of its DNA, each in individual compartments, as in some filamentous bacteria which have managed to replicate but not to divide into separate cells. The DNA of animal mitochondria is double-stranded and circular as in bacteria, but it is small (only about 5 μ long or about 10×10^6 molecular weight [*10*]) and therefore unable to code for all the proteins and enzymatic activities one finds in a mitochondrion. The mitochondrial DNA differs from the nuclear DNA in base composition, and the two DNAs do not interact in hybridization experiments. The DNAs of plant and fungal mitochondria also appear clearly different in polynucleotide sequence from those of their respective nuclear DNAs. Plant mitochondrial DNA, although not as well studied as the animal material, shows some interesting differences from that in animal mitochondria. This DNA has been isolated as significantly longer linear pieces, more nearly approaching those isolated from chloroplasts. Indeed chloroplast DNA does appear to exist in a size comparable to that of a bacterium or a blue-green algae, 10^{-15}–10^{-14} g per chloroplast depending on the source (*11*, *12*), and therefore chloroplast DNA probably codes for many more proteins than does mitochondrial DNA.

The organelles are equipped with organelle-specific enzymes active on DNA, both DNA polymerase and DNA-dependent RNA polymerase, which are presumably active in DNA replication and transcription *in situ*. Recently it has been found that the DNA-dependent RNA polymerases of mitochondria and chloroplasts are sensitive to the antibiotic rifampicin, unlike the nuclear polymerases of the same eucaryotic cells. This antibiotic is known to inhibit bacterial RNA polymerases selectively.[2] The organelles also appear to contain the entire panoply of metabolic equipment that is necessary to make specific proteins.

Several classes of RNA are found in the organelles, including two characteristic ribosomal RNA (rRNA) molecules and at least several species of transfer RNA (tRNA). Little is known of other classses of RNA, such as messenger RNA (mRNA), in the organelle. The mitochondrial ribosomal RNA fractions consist of two nucleic acids thought to derive from the mitochondrial ribosomal subunits. These nucleic acids, as well as the mitochondrial ribosomes and their subunits, possess slightly but significantly smaller sedimentation rate constants than those of their cytoplasmic counterparts. Similar differences have been found between these components of chloroplasts in spinach and *Euglena* and the cytoplasmic constituents. In these respects, the mitochondrial and chloroplast structures resemble bacteria, whose ribosomes and rRNA are also significantly smaller than the comparable units of eucaryotic cytoplasm. However, these components of the organelles and of bacteria do differ somewhat among themselves and as yet the data do not permit any designation of phylogenetic relationships. Furthermore the hybridization data in some instances have indicated a small cross-reaction of mitochondrial and chloroplast RNA with their respective nuclear DNA, suggesting either that some organelle RNA may be made in the nucleus, or that the RNA isolated from the organelles may be slightly contaminated with cytoplasmic components.

The study of other structures involved in protein syntheses has produced clearer evidence of intramitochondrial biogenesis. Mitochondrial amino acid acyl synthetases were shown to differ from the cytoplasmic enzymes. Transfer factors derived from either *E. coli* or *Neurospora* mitochondria were found to be relatively active for bacterial and mitochondrial ribosomes and far less active with the ribosomes of *Neurospora* cytoplasm. Thus bacterial and mitochondrial factors showed marked functional homology (*13*). Completing this experiment it was shown that transfer factors from *Neurospora* cytoplasm were inactive with *E. coli* ribosomes and much more active with cytoplasmic than mitochondrial ribosomes. Even more definitively, mitochondrial tRNAs have been found not only to differ from their cytoplasmic counterparts, but also to hybridize specifically with mitochondrial DNA (*14*, *15*).

The initiation of protein synthesis in bacteria, mitochondria and chloroplasts seems to have common features, involving the formation of N-formylmethionyl-tRNA and the beginning of the polypeptide chain with N-formylmethionine. This process common to bacteria and these organelles clearly differs from the inception of protein synthesis in the cytoplasm of eucaryotic cells.

Thus in only a few years, molecular biology has revealed astonishingly close similarities among procaryotic cells and the DNA-containing cytoplasmic organelles of eucaryotic cells in the systems active in the synthesis of RNA and proteins. Of course numerous major points as well as fine details have still to be clarified. For example it will be important to compare the elements of transcription and translation among the nucleus and the several cytoplasmic organelles of a single type of plant cell to see if the three genetic systems are structurally different. As will be seen below, a perfectly reasonable question is whether the ribosomal proteins of cytoplasm, mitochondria, and chloroplasts have some units in common. It would interest me to know if polynucleotide phosphorylase, an enzyme possibly involved in mRNA degradation, which is unequivocally present in bacteria and blue-green algae (*16*) and has been reported to be present at a low concentration in spinach (*17*), is actually present in chloroplasts or mitochondria or in both.

Mitochondriacs and organelle composition

The long pursuit of the mechanisms of respiration and of oxidative phosphorylation in eucaryotic cells has led to a chemical and structural dissection of the mitochondrion. After some years of fractionation and structural analysis, we may speak of chemically distinguishable and physically separable outer and inner membranes. From the latter the internal invaginated fingers of membrane, the cristae, as well as DNA circles, extend into a central matrix. The outer and inner membranes are rich in lipid; nevertheless removal of all lipid does not appear to destroy the double-layered structure of the inner mitochondrial membranes and cristae.

The lipids of mitochondria include some substances not commonly found in bacteria, such as the choline-containing lecithin and sphingomyelin; these are predominantly in the outer structure. Although the organelle does not contain all the enzymes for assembly of complex phospholipids, the outer membrane appears to possess transferases which permit numerous steps in fatty acid exchange, and choline incorporation into lecithin. Acetate is also incorporated into fatty acid in isolated mitochondria, but such a process appears to involve mainly an extension of preformed fatty acid, as in the conversion of palmitate to stearate. This mitochondrial activity does not involve the formation of malonyl derivatives. Nor can the mitochondria desaturate the long chain fatty acid stearate to oleate. This reaction, characteristic of cytoplasmic synthesis, is similarly missing in bacteria.

The organelle is rich in the phosphatidyl glycerol derivative cardiolipin, which is not known in the remainder of the cell; this substance is found rather widely in bacteria. Mitochondria are quite low in sterols, which are usually absent in bacteria, except for *Mycoplasma* grown in the presence of sterols. It is also of interest that polyunsaturated fatty acids (linoleate and γ-linolenate), also absent from bacteria, are reported to be inessential to mitochondrial development, structure, and function in some special fibroblast strains after 10 years of subculture in lipid-free media (*18*). Thus the so-called essential fatty acids do not appear essential to respiration and oxidative phosphorylation, at least in these cells.

A similar situation was found in respect to chloroplast structure and photosynthesis. Organisms capable of photosynthetic O_2 production, from the blue-green algae on up, were found in initial analyses to possess characteristic lipids, including such substances as galactosyl diglycerides and an unusual sulfolipid, both of which contained large amounts of the polyunsaturated acid, α-linolenate.[3] The anaerobic photosynthetic bacteria, like most bacteria, lacked this fatty acid and contained as unsaturated fatty acid mainly monoenoic acid, such as cis-vaccenic acid. This distribution seemed to suggest a role for α-linolenate in O_2 evolution. However, the O_2-evolving blue-green alga *Anacystis nidulans* was found to lack this fatty acid (*19*). In hindsight it is difficult to understand how α-linolenate, presumed to be derived from oleate, which itself is derived from stearate via an O_2-requiring reaction, could have been imagined to be essential to O_2 production. Nevertheless I imagine that eventually we shall learn that α-linolenate does help in some way in the development of a conformation or environment favorable to O_2 production. Similarly, γ-linolenate and other essential fatty acids requiring O_2 for their biosynthesis may have facilitated in a later evolutionary step the production of complexes in which certain respiratory complexes and the machinery of oxidative phosphorylation can operate stably and optimally.

Mitochondrial membranes contain numerous proteins, among which are found many enzyme activities characteristic of the organelle. The integrated elements of the respiratory chain are found mainly in the inner membrane. From this basal structure arises the characteristic stalked spheres which contain the oligomycin-sensitive ATPase, apparently essential to oxidative phosphorylation. These structures and sequences provide the crucial steps by which the small discrete steps of the free energy change of electron transfer, terminating in the acceptor, O_2, are transduced to form ATP.

A eucaryotic cell deficient in this mitochondrial activity will behave in most aspects like an anaerobe, and will grow slowly in the presence or absence of O_2. However, such a cell defective in mitochondrial function may still possess many O_2-utilizing reactions. With only a few exceptions these are centered in the lipid-rich membranes of the endoplasmic reticulum, and control such essential functions as sterol biosynthesis, formation of oleate, conversion of gulonolactone to ascorbate, the hydroxylation of phenylalanine to tyrosine, and the oxidation of tryptophan to N-formyl kynurenine (and through this to nicotinic acid). These O_2-requiring reactions are characteristic of the eucaryotic cell and to a significant extent are essential to the profusion of membranous structures found in such cells. Thus anaerobic yeast requires an exogenous sterol for growth.

In almost every instance such functions are determined by nuclear genes. It appears therefore that although the division of labor in a eucaryotic cell has centered the respiratory function in bodies which bear some resemblance to bacteria, numerous enzymatic functions crucial to the structure of eucaryotic cells, such as lipophilic oxygenases and other O_2-utilizing systems, are actually determined genetically and constructed outside of the mitochondrion.

These "new" activities are almost totally absent from all but a very few bacteria, such as some *Pseudomonads*. The aerobic bacteria are thought to have arrived relatively recently on the phylogenetic scene. Some of these rare bacterial oxidative enzymes, such as phenylalanine hydroxylase, tryptophan pyrrolase, cytochrome oxidase, have actually been purified, even crystallized from *Pseudomonads*. Although the mechanisms of the bacterial and mammalian enzymes such as phenylalanine hydroxylase may have numerous similarities, they do have many differences. Examination has not proceeded in these instances to the level of polypeptide sequence, that is, to the search for homology, to determine whether there is in fact any genetic relation between these procaryotic and eucaryotic enzymes, as has been found among the cytochrome c's of eucaryotic cells.

When we return to mitochondrial enzymes, many other significant opportunities for the detection of homology become evident. For example we might ask about the possible homology among components in α-keto acid (pyruvate or α-ketoglutarate) dehydrogenase complexes. The classical studies of Reed and his collaborators (*20*) have revealed remarkable similarities between the complexes derived from *E. coli* and mitochondria with respect to details of size, structural features and mechanism. Do these similarities represent homology or an extraordinary evolutionary convergence, that is, an independent discovery of optimally efficient structure? One can hardly wait for a chemically clear answer.

It would be unfair to leave this sketchy summary without recording a curious and potentially disturbing anomaly. It is well known that bacteria, a few fungi, and almost all plants (other than *Euglenids*) make lysine exclusively via α,ε-diaminopimelic acid (DAP). DAP is a characteristic component of many bacterial cell walls (as well as those of blue-green algae[4]). The formation of certain intermediates in the biosynthesis of DAP appears also to be essential to sporulation, another unique bacterial activity. On the other hand, fungi such as yeast make lysine via a series of reactions in which homocitrate is converted to α-amino adipate (AAA), eventually to saccharopine [ε-N(L-glutaryl-2)-L-lysine], and finally lysine. Many of these reactions, such as formation of homocitrate, occur in the mitochondria. Of course an animal is incapable of synthesizing lysine, but it is not quite clear why, since lysine is metabolized by animal mitochondria by an apparent reversal of the biosynthetic steps to saccharopine and other components of the AAA pathway. Thus animal mitochondria appear to contain remnants (?) of a path of lysine biosynthesis absent from bacteria and found in some fungal mitochondria.

Where do the proteins come from?

The complete machinery for protein synthesis, from DNA template on, seems to be present in the organelle but this DNA is too limited in size to code for all the proteins and enzymes found in the mitochondria. Assuming the homogeneity of mitochondrial DNA, it has been estimated that at least 30% of the mitochondrial DNA of 10×10^6 codes for species of mitochondrial rRNA and tRNA. Thus we are left with information for at most about 18 mitochondrial proteins of molecular weight 20000. Which of the very many proteins and enzymes found in the organelle are actually coded for and made in the mitochondria?

The "petite" mutants of yeast are characterized by damaged mitochondria and are unable to respire. These mutants, induced by dyes such as acriflavin or ethidium bromide, display the now classical pattern of cytoplasmic inheritance. The loss of a protein component from "petite" mitochondria can be seen in polyacrylamide gel electrophoresis (*21*). More specifically the mutant mitochondria have been shown to lack an inner membrane component essential for the attachment of otherwise apparently normal ATPase to the inner mitochondrial structure (*22*). It is of interest that the loss of this component does not merely affect oxidative phosphorylation, some components of which are controlled by nuclear genes (*23*). "Petites" contain cytochrome c but lack a number of heme components, such as cytochromes b and a(a_3), and do not possess a functional cytochrome oxidase. It has been found that the apoprotein of cytochrome oxidase is present in the mitochondria of these mutants and that a functional cytochrome oxidase can be reconstituted by incubation of mutant submitochondrial particles with heme a (*24*).[5] The conclusion was thus reached that in this important class of mutants the mitochondria had lost an important step in the biosynthesis of heme a which was coded for and synthesized by the organelle. It may be noted that the cytoplasmic mutant of *Neurospora* (poky), which also has respiratory-deficient mitochondria, possesses a defective cytochrome oxidase in which the protein itself is altered (*25*).

Protein synthesis in isolated mitochondria as in bacteria is quite sensitive to chloramphenicol, unlike that of cytoplasmic ribosomes. On the other hand protein synthesis in cytoplasm is very sensitive to cycloheximide, unlike the bacterial and mitochondrial syntheses. The chloroplast pattern of sensitivity also resembles the mitochondrial or bacterial pattern. Amino acid incorporation into isolated mitochondria normally labels some components of the inner membrane and it appears that under such conditions one major fraction of these components is inhibited by low levels of chloramphenicol (*26*). In whole cells the antibiotic prevents the synthesis of several internal mitochondrial enzymes, including cytochrome oxidase, without inhibition of synthesis of cytochrome c (*27*). However, at high concentrations the antibiotic rapidly inhibits respiration and cell growth (*28*), as well as mitochondriogenesis (*27*).

Inhibitor techniques have been used to show not only that enzymes such as cytochrome oxidase or those involved in its biogenesis are mitochondrial in origin but that numerous others such as the proteins of the mitochondrial ribosomes are actually synthesized on cytoplasmic ribosomes (*29*). Kinetic techniques which had been used to show that cytochrome c of the mitochondria are derived from cytoplasmic synthesis (*30*) have also been interpreted to indicate that the protein of cytochrome oxidase is probably not constructed in the mitochondria. This result seems to be at variance with the result obtained with the *Neurospora* poky mutant.

Anaerobically grown yeast contain poorly formed mitochondria (promitochondria) which contain DNA and incomplete inner membranes (*31*). Sterols are not essential to such anaerobic syntheses. The membranes of the promitochondria lack cyanide-sensitive respiration and numerous cytochrome continuents but can adapt to form these components on aeration in low glucose media (*32*). The nature of this oxygen-induced adaptation has been quite puzzling for many years. Recently it has been found that a burst of mitochondrial DNA synthesis precedes the formation of respiratory activity (*33*). Both cycloheximide and chloramphenicol block the O_2-induced adaptation although cycloheximide does not inhibit synthesis of mitochondrial DNA (*34*). Thus analysis of this remarkable response to the gaseous environment has similarly revealed the dependence of mitochondrial biogenesis upon both its own genetic and metabolic apparatus and other nuclearly determined metabolic systems.

Careful electrophoretic analysis of the membranes of the mitochondria has revealed at least 12 and 23 components in the outer and inner membranes respectively. These membranes appear to have only a single protein in common (*35*). Of the 15 proteins detected by this technique in the microsomal membrane, only about three seem to be present among components of the outer membrane. Despite the apparent segregation of these proteins in the mitochondrial compartment, there appears to be a growing agreement that most of the mitochondrial proteins are of cytoplasmic origin. The problems of the mechanisms by which nuclear proteins made in the cytoplasm reach their specific sites in the nucleus can now be extended to cytoplasmic organelles as well.

On the comparative biochemistry of O_2-producing organisms

At first sight it might seem a bit easier to relate chloroplasts to blue-green algae. Cytoplasmic inheritance of the organelle is well known. Chloroplasts have a larger complement of DNA presumably containing a greater number of polynucleotide sequences presumably translatable into a larger number of characteristic proteins. The presence of polymerases, ribosomes, tRNA, etc., has been affirmed, and indeed chloroplast systems *in vitro* do very well in making complete proteins, such as tobacco mosaic virus subunits. Blue-green algae fulfill the role of photosynthetic symbionts in many systems and ingested plants can furnish symbiotic photosynthetic chloroplasts even in animal tissues. The photosynthetic membrane structures from vesicles and lamellae of photosynthetic bacteria, structures in blue-green algae, and the fused membranes of spinach chloroplasts form a continuous sequence of steps in the evolution of invaginations from the inner cytoplasmic membrane of the bacterium, algae, and organelle.

The blue-green algae and eucaryotic chloroplasts both function with chlorophyll a instead of bacteriochlorophyll. The former pigment is metabolically more primitive, by the law of Granick. This, in addition to the peculiarities and distribution of photoassimilation and photosynthesis by the blue-green algae, has suggested that the common ancestor of both groups carried out photoassimilation with the aid of chlorophyll a. The evolution of bacteriochlorophyll systems in facilitating light gathering reduced the possible free energy change and prevented the possible development of O_2 evolution, for which the presence of a pigment such as chlorophyll a is essential. Blue-green algae (and chloroplasts) unlike photosynthetic bacteria thus have been able to evolve the linked elaborate photosystems I and II which culminate in the Hill reaction and O_2 evolution on one side (of the potential diagram) and reduced ferredoxin and NADPH at the other.

The O_2 thus evolved has permitted the biosynthesis of sterols, oleate, ascorbate, etc., as well as formation of the "bile pigment" prosthetic group of the phycobilins which help the blue-green algae (and the eucaryotic red algae as well) to gather more light to effect more photosynthesis.[6] Blue-green algae and chloroplasts have also evolved similar enzyme systems crucial to their specialized photosynthetic CO_2 fixation. These include the characteristic ribulose diphosphate carboxylase which generates phosphoglyceric acid and is a larger and more highly organized enzyme in blue-green algae and in chloroplasts than in most photosynthetic bacteria. In photosynthetic bacteria, such as *Chromatium*, both the photosynthetic cycle and glycolysis use an NAD-utilizing triose phosphate dehydrogenase. A triose phosphate dehydrogenase specific for NADP is found only in eucaryotic chloroplasts and in O_2-evolving photosynthetic microorganisms. The fact that the appearance of this enzyme parallels the rate of chlorophyll synthesis in regreening Chlorella has suggested that the NADP enzyme is active specifically in the photosynthetic carbon cycle while the well-known NAD enzyme in these cells acts primarily in glycolysis and oxidative reactions.[7]

Despite these differences between photosynthetic bacteria and chloroplasts, it is exciting to record the apparent existence of structural homology with respect to amino acid sequences among ferredoxins of bacteria and chloroplasts (*37*). Such studies may eventually serve to demonstrate an evolutionary sequence from procaryote to eucaryote organelle. However no data are yet available on the amino acid sequences in the ferredoxins of the presumed intermediate, the blue-green algae.

On the biogenesis of chloroplasts

An examination of the biosynthesis of chloroplast components has revealed a complex division of labor every bit as puzzling as that we have seen in our examination of mitochondrial independence. Despite the existence of non-Mendelian inheritance governing chloroplast properties, the genetic analysis of steps in the synthesis of chloroplast constituents has demonstrated that many controlling genes are nuclear. These determine such functions as the synthesis of carotenoids, chlorophyll a, and plastoquinone. In studies of *Chlamydomonas*, it has been found that all mutations affecting photosynthetic electron transport are in nuclear genes (*38*). In selected mutants of this organism, growth in the dark leads to cells which lack chlorophyll and membranes and discs containing this pigment. Greening occurs in such cells in the light, as does photoreductive power. Chloramphenicol does not significantly block greening and membrane formation but does block the fusion of membranes to form grana, and the development of photoreductive activities characteristic of the two photosystems. On the other hand, cycloheximide blocks greening functions without affecting the proportionate development of photoreduction (*39*). Other studies (*40*) have shown that chloramphenicol also blocks the synthesis of ribulose diphosphate carboxylase and the NADP-specific triose phosphate dehydrogenase, while cycloheximide does not. The synthesis of these enzymes is thus thought to center in the chloroplast in contrast to those relating to electron transport.

The totality of evidence, of which that presented above is but a small part, clearly indicates that as in the case of mitochondrial biosynthesis, the synthesis of chloroplast structures and functional entities involves the products of both nuclear and chloroplast genes, and the activity of both cytoplasmic and chloroplast ribosomes. Thus for both of these organelles, inquiries about their evolutionary origins and their degrees of genetic, biosynthetic, metabolic and functional independence have now revealed a previously unsuspected degree of complexity. The organelles have established an elaborate interplay with the nucleus and cytoplasm of their cells such that both sets of genomes and ribosomes are essential to organelle biosynthesis and function. In all eucaryotic cells the close integration of both sets of these structures and their metabolic activities is crucial to the development of a successful life cycle of the cell as a whole.

The possibility that chloroplasts may have stemmed more or less directly from blue-green algae symbionts is still very real. Detailed searches for homologies among both their polynucleotides and polypeptides are required, searches which have barely begun. Is the evidence for such descent real, such as the apparent homology of bacterial and plant ferredoxin or that of 5 S RNA of *E. coli* ribosomes and of KB cell ribosomes (*41*), or is it illusory? This question will certainly be enlightened in this coming decade. Furthermore how shall we explore the possibility, for which we have no example at the present time, that the symbiotic procaryote, ancestral to the organelles, may have lost a large portion of its genome to become part of the nuclear genetic complement? Will some DNA viruses prove to represent terminal events of such an evolutionary process?

On the origin of eucaryotic cells

Even if a mitochondrion is the direct descendant of some ancestral *Pseudomonad*, *Rickettsia*, or *Mycoplasma*, or a chloroplast is the direct descendant of this or that blue-green alga, a far more difficult question remains. What is the nature of the cell with which the presumed symbiont established its early working relation? It is possible that the inherent difficulty of this question forestalled serious inquiry until the discovery of DNA in the organelles made it inescapable. Until very recently the theories of cellular evolution have presupposed a more or less continuous development of eucaryotic cells from procaryotes. Indeed some biologists seem to prefer this type of monophyletic pattern (*42*, *43*). Such evolutionary and phylogenetic schema have tended to minimize the sharp discontinuities in the geological, cytological, compositional, and metabolic records which relate specifically to the differences between such cells. These differences have appeared mainly with the production of atmospheric O_2 and of blue-green algae. If such discontinuities had been widely recognized the classical phylogenists might have constructed hypotheses relating the development of the cytoplasm of a eucaryotic cell to the structural and functional differentiation of a procaryotic membrane.

In the context of such a hypothesis how can we account for the evolution of mitochondria and chloroplasts? We may imagine the formation and separation of a cytoplasmic nodule made of fatty acid and keto acid dehydrogenases, electron transport and ATP-generating systems plastered together with lipids and other membrane glues. The addition of a copper-containing cytochrome oxidase to this functionally integrated unit may have been a later acquisition of function. All other membrane-forming O_2-utilizing systems would have been left to help form the endoplasmic reticulum outside of the mitochondria. We must formulate comparable and even more elaborate notions in imagining the evolution of the organelles of a primitive O_2-evolving eucaryote. It is of interest that no one has attempted as yet to come to grips seriously with this type of hypothesis, although some views of the organization of the mesosome in gram-positive bacteria or of the differentiation of the membrane in some photosynthetic procaryotes are perhaps a beginning in this direction.

These classical phylogenies have already begun to be rejected by serious biologists (*44*). However, the alternative concept of the existence of our organelles as once free-living symbionts creates other major problems of the possible nature of the eucaryotic host. Bluntly stated, can we imagine a viable eucaryote devoid of all the enzymatic apparatus, such as cyanide-sensitive respiration, present in the mitochondria?

The most daring and serious effort in pursuit of the symbiont hypothesis to date has been that of Margulis (*45*, *46*), who has attempted to construct a comprehensive picture of the evolution of all eucaryotic cells as the product of the symbiotic fusion of several procaryotic cells. She has explored the geochemical and paleontological record, as well as the data of cytology, microbiology, and biochemistry to seek the relevant data. It may be noted that much relevant cytological and embryological evidence which was so assiduously collected in the first third of this century has tended to be neglected by many modern biologists. Margulis has also attempted to place this evidence in the context of the symbiotic theory. Briefly, it is postulated that eucaryotic cells arose between 0.5 and 1.0 billion years ago, perhaps a billion years after the evolution of O_2-forming blue-green algae. The presence of atmospheric O_2 presumably permitted the evolution of aerobic bacteria. The ancestral eucaryote then represents the fusion of an aerobic bacterium and another procaryote to give rise to a mitochondrion-containing amoeboid cell. The subsequent addition of a spirochaete-like organism provided the motile ancestral amoeba form, containing the crucial 9 + 2 flagellar symbiont, subsequently active in the differentiation of centrioles and mitotic apparatus. This primitive cell may have evolved to form protozoa or fungi or eventually multicellular animals. Another later possibility was "infection" of this cell by a blue-green alga to generate the ancestral eucaryotic alga on the way to the formation of higher plants.

This hypothesis seems no less likely to me than does any other. Any existing symbiotic hypothesis on these questions is oversimplified, as I have detailed in the body of this essay. We must account not only for the multiplicity of discrete genomes in nucleus and organelles but also for the mechanism of their fragmentation and assembly. We must also account for the division of labor and mechanisms by which a nuclear genome provides many gene products essential to the structures, function, and activity of cytoplasmic organelles that contain their own systems of genetic determination and polymer synthesis. However, even without pressing to this unexpected level of complexity we may need years or decades to evaluate Margulis' scheme, which is just finding its way to a publisher. Will the details or the broad outline of this hypothesis hold up under the close scrutiny which is clearly on its way in this decade?

I would like to believe that each new relevant piece of information will fit within the framework presented above. This article has had its own evolution, initially as an amusing intellectual exercise, then as a serious teaching effort of five symbiotic colleagues confronting a somewhat dubious graduate class, and finally as a serious approach to understanding the literature to come.

Footnotes

1. My colleague M. Nass has recently fed chloroplasts to mouse fibroblasts. The resulting enlarged green cells, each containing many chloroplasts, can still grow and divide. The chloroplasts may be reisolated after several days and have been shown to retain their DNA and photosynthetic activity (*7*).
2. The antibiotic is highly active against many infectious agents, such as tubercle bacillus. A therapeutic trial will have to watch closely for possible effects against the mitochondria. If such long term toxicities are actually found, it may be asked if we can select for rifampicin-resistant mitochondria and if we can seed tissues with such forms.
3. It is of interest that blue-green algae and leaf chloroplasts do not contain phosphatidyl choline, -ethanolamine, -inositol, and cardiolipin, although these are present in leaf extracts (perhaps in leaf mitochondria and other structures) (*17*).
4. Do chloroplasts contain DAP and is all lysine production in higher plants a chloroplast function? No one has yet tested this question, nor has the apparent anomaly of *Euglenid* lysine synthesis via α-aminoadipate been explored in appropriate depth. Have *Euglena* chloroplasts lost the presumed DAP path and is lysine synthesized from AAA in *Euglena* mitochondria?

5. The cytochrome oxidase of eucaryotic cells contains copper, which is believed to be essential to the enzymatic activity. However the appropriate fractions of the "petite" mitochondrial extract contained normal amounts of copper (*21*) consistent with the block occurring specifically in heme a biosynthesis. Substances that chelate copper, such as cuprizone, produce giant deformed mitochondria, and a deficiency in Cu transport, as in Wilson's disease in the mammal, leads to a deficiency of cytochrome oxidase. Copper is also present in other enzymes of eucaryotic cells, such as the monoamine oxidase of mitochondria or in plastocyanin and ribulose diphosphate carboxylase of chloroplasts. However, the cytochrome oxidase (a_3) of *Pseudomonas* lacks Cu. Indeed it has not been proven that any procaryotic cell has functional copper. Although *Pseudomonas* possess a copper protein capable of reducing cytochrome oxidase *in vitro*, it has not been demonstrated that this protein has any role in the respiration of the organism. The presence of Cu in the plastocyanin and ribulose diphosphate carboxylase of chloroplasts poses the problem of the possible existence of this metal in blue-green algae and in some of their presumed colorless descendants, such as the flexibacteria and some autotrophs.
6. It would be of interest to know if O_2 is also essential in blue-green algae and chloroplasts in the conversion of coproporphyrinogen to protoporphyrin IX as it apparently is in mitochondria. Isolated chloroplasts are capable of converting δ-aminolevulinate to heme but it is not yet known whether the formation of protoporphyrin in the chloroplast may not be an evolutionary remnant comparable to protoporphyrin synthesis in an obligate anaerobe, such as *Desulfavibrio*.
7. Some recent findings in the field of nitrogen fixation are so interesting in an evolutionary sense that I feel compelled to include them, even though they may seem a bit off the main subject of our discussion. In the last two decades nitrogen fixation has been revealed as a reductive capability of many organisms possessing a bewildering assortment of oxidative activities. These range from aerobes such as *Azotobacter* and the *Rhizobia* which symbiotically infect legume root nodules to anaerobes such as the *Clostridia* and the photosynthetic bacteria. Some filamentous blue-green algae also fix N_2 but green leaves do not and this is one more significant difference between blue-green algae and chloroplasts. The study of N_2 fixation in extracts has revealed the importance of exclusion of O_2 from the reductive enzymes, even in such systems as the "aerobic" symbiotic *Rhizobia* or the blue-green algae. N_2 fixation in filamentous blue-green algae has recently been revealed as the activity of the unusual and previously mysterious heterocyst cells. Such cells have reorganized their lamellae for this reductive function and have lost the capacities of CO_2 fixation and presumably of O_2 evolution (*36*). It apparently has been demonstrated that the symbiotic N_2-fixing blue-green algae in the root nodules of *Cycadaceae* perform this function with a poor rate of CO_2 fixation. Thus the specialized N_2-fixing cells of the blue-green algae seem to have taken a step back to life as a bacterium. O_2-evolving chloroplasts have learned to live with the consequences of this evolution of their membranes and enzymes, which have included a greater dependence on other organisms for the reduction of N_2.

References

1. Stanier, R. Y., and C. B. van Niel. 1941. *J. Bact.* 42:437.
2. Stanier, R. Y., and C. B. van Niel. 1962. *Arch. Mikrobiol* 42:17.
3. Ris, H. and W. Plaut. 1962. *J. Cell. Biol.* 13:383.
4. Nass, M. M. K. and S. Nass. 1963. *J. Cell. Biol.* 19:593.
5. Lwoff, A. 1950. *Problems of Morphogenesis in Ciliates.* New York: John Wiley and Sons.
6. Luck, D. J. L. 1963. *Proc. Natl. Acad. Sci.* 49:233.
7. Hall, W. T., and G. Clause. 1967. *J. Phycol.* 3:37.
8. Taylor, D. L. 1968. *J. mar. biol. Assn. U. K.* 48:1.
9. Nass, M. M. K. 1969. *Science* 165:1128.
10. Nass, M. M. K. 1969. *Science* 165:25.
11. Kirk, J. T. O. 1966. *Biochemistry of Chloroplasts I.* New York: Academic Press.
12. Edelman, M., D. Swinton, J. A. Schiff, H. T. Epstein and B. Zeldin. 1967. *Bact. Rev.* 31:315.
13. Küntzel, H., and H. Noll. 1969. *FEBS Letters* 4:140.
14. Buck, C. A., and M. M. K. Nass. 1969. *J. Mol. Biol.* 41:67.
15. Nass, M. M. K., and C. A. Buck. 1969. *Proc. Natl. Acad. Sci.* 62:506.
16. Capesius, I., and G. Richter. 1967. *Z. fur Naturforsch* 22B:204.
17. Brummond, D. O., M. Staehelin, and S. Ochoa. 1957. *J. Biol. Chem.* 225:835.
18. Kagawa, Y., T. Takaoka, and H. Katsuta. 1969. *J. Biochem. (Japan)* 65:799.
19. James, A. T., and B. W. Nichols. 1966. *Nature* 210:372.
20. Reed, L. J., and R. M. Oliver. 1968. *Brookhaven Symposia*, p. 397.
21. Tuppy, H., R. Swetly, and I. Wolff. 1968. *European J. Biochem.* 5:339.
22. Schatz, G. 1968. *J. Biol. Chem.* 243:2192.
23. Kovac, L., T. M. Lachowicz, and P. P. Slonimski. 1967. *Science* 158:1564.
24. Tuppy, H., and G. D. Berkmayer. 1969. *European J. Biochem.* 8:237.
25. Edwards, D. L., and D. O. Woodward. 1969. *FEBS Letters* 4:193.
26. Yang, S., and R. S. Criddle. 1969. *Biochem. Biophys. Res. Comm.* 35:429.
27. Kroon, A. M., and H. DeVries. 1969. *FEBS Letters* 3:208.
28. Firken, F. C., and A. W. Linnane. 1968. *Biochem. Biophys. Res. Comm.* 52:398.
29. Küntzel, H., and H. Noll. 1967. *Nature* 215:1340.
30. Kadenbach, B. 1969. *European J. Biochem.* 10:312.
31. Criddle, R. S., and G. Schatz. 1969. *Biochemistry* 8:322.
32. Ephrussi, B., and P. P. Slonimski. 1950. *Compt. rend.* 230:685.
33. Mounolou, J. C., G. Perrodin, and P. P. Slonimski. 1968. *Biochemical Aspects of the Biogenesis of Mitochondria*, p. 133, Adriatica Editrica.
34. Rabinowitz, M., G. S. Getz, J. Casey, and H. Swift. 1969. *J. Mol. Biol.* 41:381.
35. Schnaitman, C. A. 1969. *Proc. Natl. Acad. Sci.* 63:412.
36. Stewart, W. D. P., A. Haystead, and H. W. Pearson. 1969. *Nature* 224:226.
37. Weinstein, B. 1969. *Biochem. Biophys. Res. Comm.* 35:109.
38. Levine, R. P. 1968. *Science* 162:768.
39. Hoober, J. K., P. Siekevitz, and G. E. Palade. 1969. *J. Biol. Chem.* 244:262.
40. Smillie, R. M., D. Graham, M. R. Dwyer, A. Grieve, and N. F. Tobin. 1967. *Biochem. Biophys. Res. Comm.* 28:604.
41. Forget, B. G., and S. M. Weissman. 1969. *J. Biol. Chem.* 244:3148.
42. Klein, R. M., and A. Cronquist. 1967. *Q. Rev. Biol.* 42:105.
43. Allsopp, A. 1969. *New Phytol.* 68:591.
44. Whittaker, R. H. 1969. *Science* 163:150.
45. Sagan (Margulis), L. 1967. *J. Theor. Biol.* 14:225.
46. Margulis, L. 1969. *J. Geol.* 77:606.

James W. Valentine
Cathryn A. Campbell

Genetic Regulation and the Fossil Record

Evolution of the regulatory genome may underlie the rapid development of major animal groups

The abrupt appearance of higher taxa in the fossil record has been a perennial puzzle. Not only do characteristic and distinctive remains of phyla appear suddenly, without known ancestors, but several classes of a phylum, orders of a class, and so on commonly appear at approximately the same time without known intermediates. Darwin recognized that such gaps presented a major obstacle to demonstrating that evolution proceeded by the slow accumulation of change within lineages; he attributed the lack of antecedents to the incompleteness of the fossil record. Over a hundred years later, we still face the problem of missing ancestors of many higher taxa. Indeed, our present knowledge of the fossil record demonstrates even more clearly the episodic nature of the origin of new higher taxa. If we read the record rather literally, it implies that organisms of new grades of complexity arose and radiated relatively rapidly.

Generally, these bursts of evolutionary advances have been treated as problems of evolutionary ecology, and Simpson (1944, 1953) in particular has developed a model that regards them as resulting from the broaching of an ecological barrier. Upon crossing the barrier into a new adaptive zone, a lineage is usually without close competitors and may undergo rapid diversification to realize the many ecological opportunities that have become available to it. While it is actually passing the barrier, the lineage may evolve very rapidly and be small in population size, thus having only a slight chance of being preserved in the fossil record. Once it truly enters the new adaptive zone, however, the lineage radiates into a number of diverging and expanding populations, and there is commonly a sudden appearance of fossils of these new types.

Such ecological models have a great deal of explanatory power, but major problems still remain concerning the genetic mechanisms involved. Evolutionary genetics has been concerned primarily with changes in the frequencies of *structural* genes—those genes that are transcribed into RNA and then translated, causing the synthesis of proteins. Incorporation of structural gene mutations into the gene pool of a species leads to a gradual change in the enzymes and other proteins. However, organisms belonging to different phyla and classes are organized so differently from each other as to represent different levels of complexity or, at least, qualitatively different approaches to the ways of life within a given level of complexity. Probably these differences are based upon genetic changes beyond the mere substitution of amino acid sequences in proteins. Major genetic differences between taxa in higher categories may involve the pattern of genetic regulation (Mayr 1963; Dobzhansky 1970).

The effects of genetic regulation are illustrated by the cells of the various tissues of an organism. All possess identical genetic material, or genomes, yet they may have highly distinctive morphologies and perform diverse functions. The differences result from the presence and functioning of different proteins and enzymes within the various types of cells. Clearly, the genes in the different cells are switched on and off in different combinations and in different sequences; genes that are important in one tissue may be dormant in another. The activity of these protein-specifying genes is controlled by other genes which form the regulatory apparatus of the genome, that is, of all hereditary material in a cell.

While the types of proteins in the cells of animals belonging to different phyla are certainly not identical, their similarities nevertheless far outweigh their differences. Yet the morphologies of the phyla are strikingly diverse. This morphological heterogeneity may be primarily a matter of the way in which genomes are organized—of differences in the batteries of genes and in the sequences of gene activities that occur during the development of individuals. The major morphological differences, then, may be due to differences in gene regulation.

In this article, we will examine a

James W. Valentine is Professor of Geology and of Environmental Studies at the University of California, Davis. His primary interest is in the evolution of such ecological units as populations, communities, and provinces, especially in relation to the sweeping environmental changes associated with continental drift. He is Past President of the Paleontological Society. Cathryn A. Campbell, a graduate student in ecology at the University of California, Davis, received her B.A. from Wellesley College and her M.S. from Davis, with an emphasis on paleoecology. Her major interests center on the genetic aspects of adaptive strategies in marine invertebrates. *The authors are grateful for critical reading of this paper by John C. Avise and Francisco J. Ayala, University of California, Davis, and Eric H. Davidson, California Institute of Technology.* *Address: Department of Geology, University of California, Davis, CA 95616.*

model of gene regulation in order to get some idea of the sorts of changes in the complexity and form of animals that may be achieved by the evolution of the regulatory apparatus. Turning to the fossil record, we shall see that the first appearances of organisms with radically novel anatomical structure, and often of increased complexity, are neither regular nor random but rather are clumped together in time so as to form episodes of broad evolutionary advance. These episodes can be interpreted as indicating that at some few times in the past conditions were particularly propitious for significant changes in the regulatory apparatus of animal genomes. Animals responded to major environmental opportunities or challenges by enlarging or repatterning the regulatory portions of their genomes rather than by simply employing novel mutations or gene combinations within their structural gene complements. The conditions under which such important regulatory changes occur are not completely understood and indeed form a fundamental question in evolutionary biology.

Genome regulation

Most evolutionary models deal with the selection, mutation, recombination, migration, and drift of structural genes. The products of these genes are proteins which function as enzymes or form the physical structures of the organisms. The operation of structural genes must be closely coordinated and regulated to ensure harmonious development and metabolism, and some regulatory system is therefore necessary. Even in the simplest bacterial cell, different proteins are synthesized at different times, chiefly in response to changes in the environment. For example, the bacterium *E. coli* may be grown in environments with or without lactose (milk sugar). The enzyme that breaks down lactose can be manufactured by bacterial cells, and when lactose is present, over 1,000 times as many molecules of this enzyme may be synthesized as when it is absent. Thus there is a system to activate and deactivate structural genes according to the needs of the cell and the environmental conditions.

Some of the regulatory mechanisms in prokaryotic organisms are fairly well understood. Jacob and Monod (1961) proposed a model of gene regulation that has since been verified in many particulars. The transcription of some structural genes is controlled by *operator* genes—segments of DNA that are adjacent to the structural genes. The operator genes may be inhibited by a class of proteins called *repressors*. Each repressor is specific for one operator; by combining with this operator, the repressor prevents transcription of messenger RNA molecules from the adjoining structural gene (or genes). Like all proteins, repressors are themselves coded by segments of DNA.

Repressors exist in two forms: an active one which is free to bind to the operator and an inactive one which is not. In some systems—*inducible* systems—the repressor is normally in the active form except in the presence of certain deactivating substances called *inducers*, to which it preferentially binds. In other systems—*repressible* systems—the repressor is normally inactive and can only bind to the operator when it is complexed with a substance called a *corepressor*. The inducer will often be the substrate of the enzyme or protein coded by the structural gene involved; the corepressor is often the very enzyme or protein itself. Thus, the substrates and end-products provide positive and negative feedback to ensure that adequate but appropriate amounts of protein are maintained. Still other regulatory mechanisms are now known in prokaryotes.

Gene regulation is less well known in eukaryotes. The complex development of higher organisms requires that various sets of many genes be activated in a coordinated manner—either contemporaneously or in a controlled sequence—during the life of the organism. Further, the cells of multicellular eukaryotes are differentiated as a result of the activity of different suites of structural genes in different cells. Eukaryotic cells contain much greater amounts of DNA than do prokaryotes, and this material is compartmentalized into chromosomes. Because genes whose activities must be coordinated are often located on separate chromosomes, the regulatory system that controls these carefully coordinated and yet greatly diversified processes must be relatively elaborate. Several models of eukaryotic gene regulation have been suggested (e.g. Kauffman 1971; Holliday and Pugh 1975); all have certain common attributes, and for simplicity we will confine our attention to a single model.

The Britten-Davidson model

Britten and Davidson (1969, 1971) have formulated a general regulatory model for higher organisms that is as simple as possible and yet consistent with the requirements of the genetic system and with our present knowledge of cell biology. As they emphasize, their particular model may well require modification as new knowledge comes to light, but some such regulatory system is called for by the evidence. We will review their model briefly before examining the implications of genome evolution from the standpoint of the fossil record.

Figure 1 depicts the main elements of the Britten-Davidson model. *Sensor* genes receive a stimulus that will result in the switching on of a battery of structural genes, which have products that function along a biosynthetic or metabolic pathway. The substance that is responsible for inducing activity in the sensor gene might be a hormone, perhaps acting through an intermediate agent. When the proper inducing agent is present, the sensor gene activates the adjacent *integrator* gene, which produces a specific *activator RNA* molecule which suffuses throughout the genome and complexes with certain specific *receptor* genes. The complexing is possible because the sequence of bases in the activator RNA molecule is complementary to that of the receptor gene; therefore, the integrator gene from which the RNA was transcribed and the receptor gene must be similar or identical in sequence. When complexed with activator RNA, each receptor gene causes a structural gene, which lies in close proximity, to transcribe, resulting in the elaboration of polypeptides at appropriate sites in the cell. Thus, in short, the need for a suite of gene products is detected by a sensor which causes an integrator to transcribe; the product of this transcription—activator RNA—is detected by a number of receptors which cause structural genes to transcribe.

If many structural genes (which may be scattered widely through the genome) are associated with receptors that have identical (or similar) base sequences, then a single integrator can cause them all to transcribe (Fig. 1). Such coordination would be salutory if the products of such a battery of genes were all required for integrated activity in a biochemical cycle or pathway. If the product of a given structural gene is required in more than one activity, that gene would be associated with more than one receptor and would belong to more than one battery. The polygene batteries are thus not mutually exclusive in this model. Furthermore, if the products that result from genes controlled by a given integrator could stimulate one or more sensors, then additional gene batteries would be activated and sets of integrated gene batteries would be brought into play. A whole hierarchy of such sets could be controlled by a series of master sensors.

A possible variation on this model is that each structural gene could have a unique receptor, and batteries of structural genes could be organized by multiple integrators, all (in any given battery) activated by a given sensor. The evolutionary consequences of genome regulation are somewhat different in this case; however, as details of the organization of regulatory systems are not yet demonstrated in any event, we shall simply employ the earlier model in our discussion. For the evidence of regulator systems of these general types, see Britten and Davidson (1969, 1971) and Davidson and Britten (1973).

It is possible that some roles in the regulatory apparatus are played by proteins, although direct evidence on this point is still lacking. For example, integrators may produce messenger RNA rather than activator RNA; the messenger RNA would cause the elaboration of a protein that would form the activating substance, complexing with receptors or with substances which repress the activity of the receptors (Davidson and Britten 1973). A regulating mechanism utilizing histone and certain nonhistone proteins has been proposed (see Stein et al. 1975). In this model, histones repress the activity of structural genes but are neutralized by nonhistone proteins that recognize specific genes and complex with the histones that bind them. The genes are than free to transcribe. Whatever the complications may be, the general notion of a hierarchical regulatory system that controls the operations of batteries of genes seems required. The Britten-Davidson model permits us to examine the

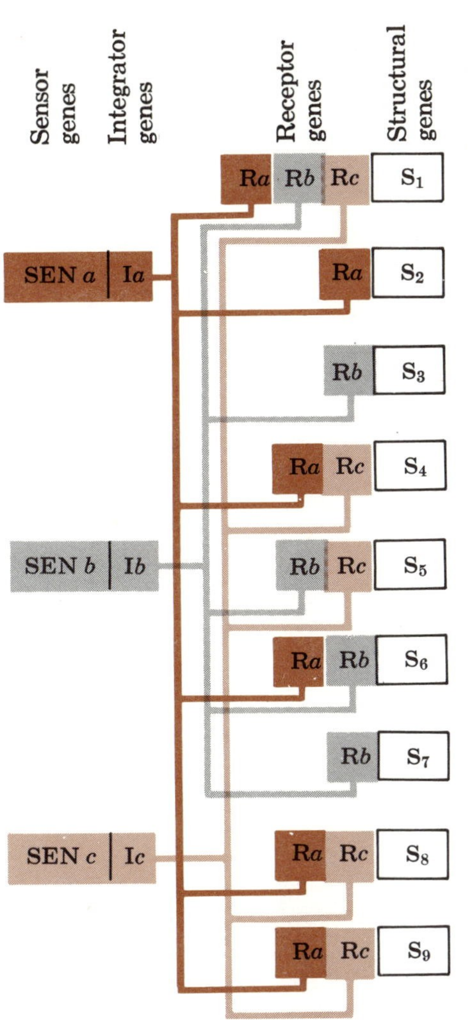

Figure 1. The Britten-Davidson model of gene regulation in eukaryotes. Sensor genes (SEN) detect a substance that causes them to switch on integrator genes (I), which form activator RNA (*colored lines*). This RNA diffuses through the genome and complexes with receptor genes (R), wherever they may be located, owing to a complementarity in base sequences. Receptors then cause or permit transcription of messenger RNA by associated structural genes (S), and the polypeptide products are employed in development or metabolism.

evolutionary results that may accrue from changes within the regulatory apparatus of the genome.

The importance of new regulatory patterns in explaining evolutionary novelties was well appreciated by Britten and Davidson (especially 1971), and we paraphrase some of their points. Higher and lower organisms have similar enzymes; perhaps over 90 percent of enzyme activities are common to mammals and prokaryotes, and thus significant differences between them must be primarily regulatory. Higher metazoa require the operation of different gene batteries and battery sequences in different cell types, and thus numerous cell ontogenies must be programmed in a single genome. As this sort of differentiation becomes more complicated during phylogenesis, the regulatory apparatus must be modified and elaborated, especially when new levels of organization are being evolved.

One of the major evolutionary consequences of the existence of a regulatory system in the genome is that changes may occur not only among the structural genes but among the regulatory genes as well. Most mutations to structural genes are deleterious, and presumably most regulatory gene mutations are deleterious as well, but occasionally a mutation may enhance regulatory activity. Assuming that the sequence specificities need not be perfect at the various steps depicted in Figure 1, then a point mutation at, say, an integrator site might still permit activation of most receptors, although some might no longer accept the activating substance from the mutant integrator (Britten and Davidson 1969).

For a given gene battery, effects of a mutation at a sensor site could range from deactivation of the battery to switching it to an entirely new activating substance so that the battery would function in new biochemical circumstances. Mutation of the integrator might also deactivate the battery, or it might change the composition of the battery, eliminating some and adding other structural genes. And mutation of the receptor could cause a particular producer to become part of one or more new batteries and eliminate it from some or all of its

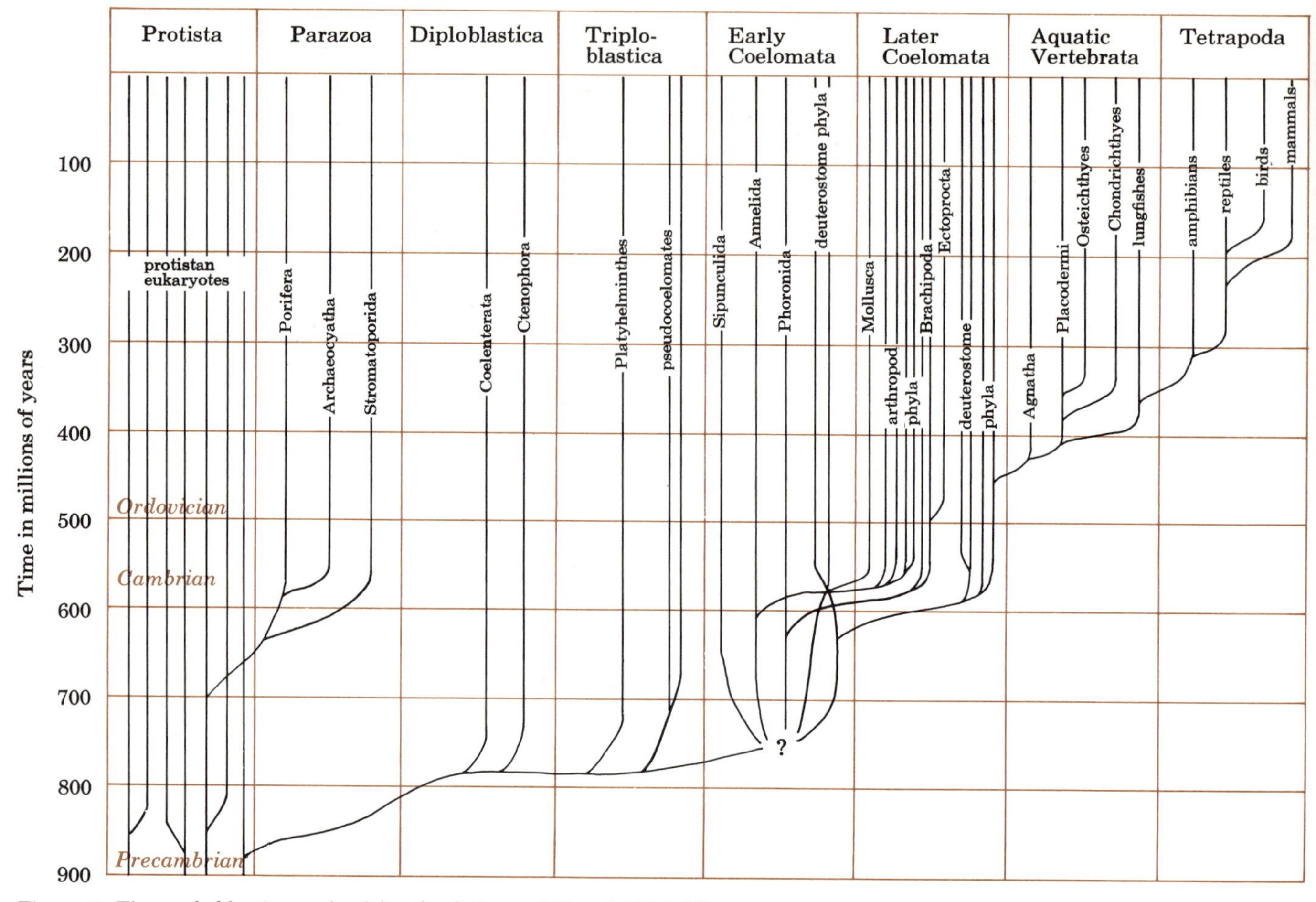

Figure 2. The probable times of origin of major groups of animals. Most phyla appear to have originated in the late Precambrian, between 800 and 700 million years ago, or about 570 million years ago, the beginning of the Cambrian.

former gene batteries. Thus, gene activity could be extensively repatterned through mutations, whole batteries or sets of batteries being activated under new circumstances, or reconstituted, or individual genes moved from one set of batteries to another. Even without any structural gene mutations, major changes in biochemical pathways can arise in a variety of ways, and if mutations occurred to more than one type of regulatory gene, the resulting changes could create a wide variety of novel patterns involving large numbers of structural genes. We assume that the novel patterns arise in small steps, since mutations with large phenotypic effects are very unlikely to increase fitness.

In addition to these mutations within the genes, chromosomal alterations could also create major changes in the regulatory patterns. The Britten-Davidson model implies that position is a major determinant of the function of different segments of DNA. Thus, translocations, transpositions, or other chromosomal rearrangements that cause the relocation of certain segments of DNA may create new regulatory gene complexes (Britten and Davidson 1971). For example, if a novel gene is emplaced next to a sensor, it will replace or supplement the former integrator and produce a novel activator substance. The effect of this new integrator would depend on the extent to which appropriate receptors were present in the genome.

Sources of new DNA are available to the genome which might be used to expand or repattern the regulatory apparatus. The presence of numerous highly repetitive DNA sequences is well documented (Britten and Kohne 1968), although the mechanisms of their replication are not understood. However they arise, repetitive sequences are eventually scattered throughout the genome, presumably through translocations, cross-overs, and other chromosomal rearrangements. New genetic material can also be added to the genome by chromosomal duplication (Ohno 1970).

The fossil record of novelty

The first fossils appear in rocks of very great age, well over 3 billion years old. These early forms somewhat resemble living bacteria and blue-green algae. Animals do not appear until much later, approximately 700 million years ago. During this span of perhaps 2½ billion years, the complex eukaryotic type of cell evolved, with sophisticated metabolic pathways and genetic machinery; such cells form the basic units of which multicellular organisms are constructed. The transition between prokaryotic and eukaryotic cell types is not well documented from the fossil record, but it may have occurred between about 1.7 and 1.4 billion years ago (Cloud et al. 1969; Schopf and Blacic 1971). We shall restrict our attention to the animal kingdom, for which the fossil record is best. The probable times of origin and the first appearances of many of the striking evolutionary novelties in animals are diagrammed in Figure 2.

The earliest fossils that represent animals are probably burrows and similar traces that have been preserved in old sea-bottom deposits. They become widespread though rare between 700 and 600 million years ago; during this interval, body fossils of nonskeletonized marine animals also appear (Glaessner 1969, 1971). The first animal remains include coelenterates (jellyfish and their allies) and several types of wormlike forms of uncertain affinities. Claims of earlier animal fossils have been disproved or are of doubtful validity (Cloud 1968, 1973).

About 570 million years ago, at the beginning of the Cambrian Period, mineralized skeletons made their first appearance (Fig. 2), and thereafter the fossil record improves greatly. All but two of the living phyla that are now well skeletonized had appeared by the end of the Cambrian, and it seems likely that primitive members of even those two (the Ectoprocta, or bryozoan "moss animals," and the Chordata) had appeared but were not then represented by well-skeletonized groups and thus were not preserved. Both these phyla appear in rocks of the next geological period, the Ordovician.

A number of inferences as to the major steps in the origin of animal phyla can be drawn from their fossil records and from the extensive knowledge of their functional anatomy that is now available. The earliest metazoan animals must have been multicellular descendants of single-celled protozoans, but no fossil record is known for either of these groups in Precambrian times. The advantages of multicellularity probably include increased size and homeostasis, enhanced feeding ability, enlarged reproductive capacity, and longevity. The primitive metazoans had to regulate the growth patterns of collections of cells rather than of single cells, and thus required a more extensive regulatory system than their ancestors, one which was sensitive to signals from other cells.

Multicellularity led to the development of differential functions among cells—some of which become chiefly reproductive, others digestive, and so forth—with the advantages of efficiency that accompany functional specialization. This required further regulatory elaboration in order to permit genes to be associated with different batteries in different cells. Sponges exhibit this stage of complexity but appear to have descended independently of other living animal phyla, although also from protozoan ancestors. It is possible that an extinct group of early Cambrian fossils, the Archaeocyatha, were representatives of this level of organization on the main metazoan line. Archaeocyathans resemble sponges in some but not all skeletal features.

The first group we actually find that is, in our opinion, associated with the main line of descent is the coelenterates. These forms have advanced further than sponges, with cells differentiated to form organs such as gonads. Again, the coordination of the differentiated cells that this requires must involve another important step in the evolution of the regulatory portion of the genome. Another advance involved the addition of a middle tissue layer to produce the three-layered "triploblastic" plan exemplified by flatworms. Perhaps the new layer permitted more efficient locomotion on the substrate by providing a more rigid body than that of jellyfish. This simple architectural change alone need not have involved a major regulatory step, yet the flatworm anatomy is so much advanced over that of the coelenterates—involving many more cell and tissue types—that a considerable regulatory increase must have occurred. Unfortunately, there is no fossil record of this step.

The last two steps in the evolution of the regulatory portion of the genome are associated with fossil records and take us to a level that embraces at least the primitive members of all the living phyla. The first step is the evolution of a fluid-filled body cavity—the coelom—within the middle tissue layer. The primitive function of the coelom was evidently as a hydrostatic skeleton, and its development permitted regular burrowing by moderately large worms employing peristaltic motion (Clark 1964). Thus the appearance of this step probably occurred about 700 million years ago, when burrows appear. The final step involved the evolution of coelomates as surface-dwellers and swimmers. These surface forms must have developed chiefly or entirely from the burrowing coelomate worm lineages, which were soft-bodied. Mineralized skeletons were developed to enhance adaptation to life upon rather than within the sea floor, beginning approximately 570 million years ago (Valentine 1973a). Judging from the anatomical plans of the living representatives of these new sea-floor dwellers, four or five major ancestral worm groups are implied (Fig. 3). The new groups are in general more complexly organized than their ancestral worm stocks and each must have required an enlargement of the regulatory genome. Some deuterostome lineage gave rise to the cephalochordates and eventually to primitive fishes, which were internally skeletonized as a locomotory adaptation.

There are other major groups of marine phyla besides the ones we have discussed, and each forms a fascinating problem in adaptation. However, they do not have significant fossil records and do not seem to be on the main line of descent of higher metazoans, so we shall not consider them here. One problem with fossil evidence, even for the groups that we do consider, is that the first appearances of phyla do not always correspond with their probable times of origin. Burrows presumed to represent coelomates antedate the remains of coelenterates and flatworms, which must have evolved earlier. This is easily understandable, as burrows are readily preserved, while soft-bodied organisms are among the rarest types of fossils. Once mineralized skeletons have become common, however, it is likely that the fossil record forms a reasonable approximation of the times of origin of well-skeletonized major groups.

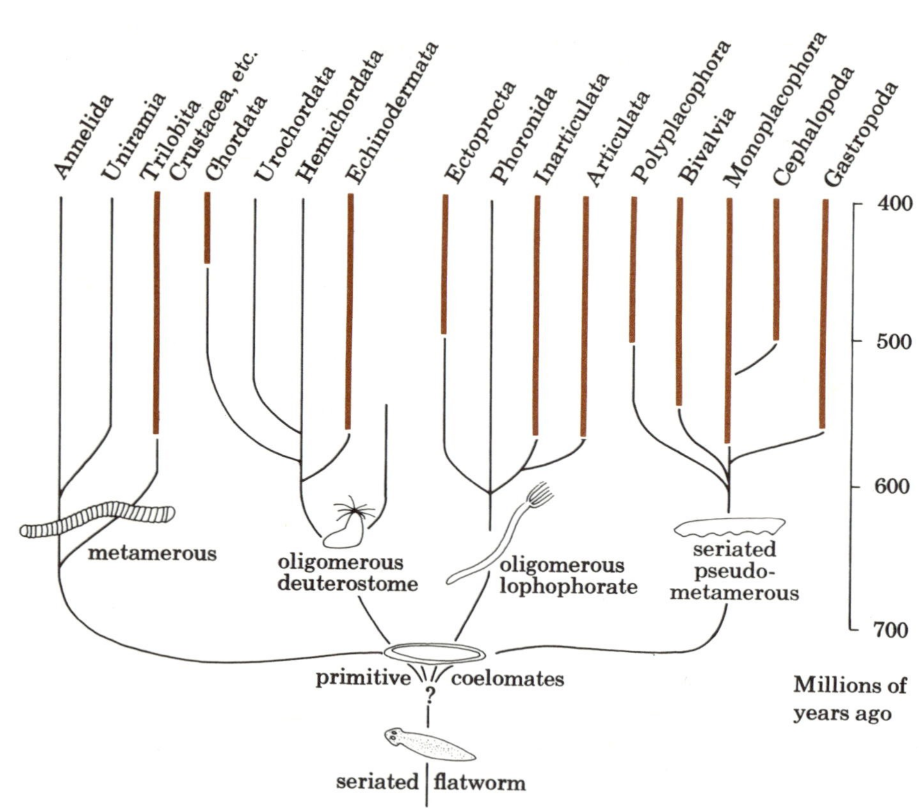

Figure 3. A model of coelomate evolution. The wide bars indicate that the lineage of the phylum had acquired a mineralized skeleton; most skeletons appeared around 570 million years ago.

This is particularly true with the surface-dwellers that first appear during the Cambrian, for the distinctive anatomical plans that cause us to regard them as new phyla usually involve skeletons as integral parts of the coadapted complexes of features that permitted new ways of life on the sea floor (Cloud 1949; Valentine 1973b). From groups that have the better fossil records, then, it appears that major anatomical plans were evolved relatively rapidly and that such developments sometimes occurred nearly contemporaneously in independent lineages.

Genome evolution and the origin of phyla

The rapid evolution of higher taxa is rendered genetically plausible by considering the evolution of the regulatory portion of the genome during the development of new anatomical plans. The development of coelomates, for example, must have begun with a flatwormlike stock which contained the genetic information required for the development of muscles, nerves, digestive glands, reproductive and excretory organs, and so forth. These tissues and organs were so integrated as to constitute a successful biological architecture, structurally and functionally. The development of a coelom and the elaboration of organ systems that followed involved chiefly the addition of new information to the regulatory apparatus and the creation of batteries of structural genes in novel combinations to associate organs in a novel architecture. The structural genes themselves need have changed only relatively little. While the regulatory additions were being assembled by natural selection, the old systems still functioned to produce perfectly viable organisms. This feature alone must have permitted relatively rapid development of useful new characters.

Once a coelomate worm was evolved, a variety of new modes of life could be achieved. Many restrictions on size—such as those imposed on flatworms by a lack of circulatory and respiratory systems—were now removed. It is likely that the coelomates radiated into several major types, each with a special range of functions (see Fig. 3). One type continued to burrow for detritus and developed multisegmented coelomic compartments. Another became a sessile burrow-dweller and filtered the bottom water, removing plankton and detritus before it became worked into the bottom sediment; this worm was chiefly divided into two or three coelomic compartments. Still others gathered detritus from the surface of the sea floor or crept and burrowed at the surface; they had few or no longitudinal coelomic compartments. For each of these cases, a considerable repatterning of the regulatory genome is indicated, in order to create modifications in the basic coelomate anatomy through a repatterning of the timing and extent of development of the tissue and organ systems. No doubt there were changes in the structural genes also, as well as an enlargement of the regulatory genome. Surface-dwelling phyla also must have radiated from these worm stocks. Additional information must have been required in the genome to encode for new anatomical additions, commonly including mineralized skeletal systems. Then each new phylum radiated in turn into a variety of classes and orders, and each of these new branches must have involved a regulatory repatterning.

Thus we can infer a whole series of events during genome evolution—those primarily of regulatory gene growth alternating with those primarily of regulatory gene repatterning, accompanied by the evolution of structural loci as well. New grades of complexity were attained and then exploited. While these major novelties evolved, microevolutionary processes produced arrays of species appropriate to the environments and to the potential of each lineage, through processes of genetic isolation and change. These accommodated the physiological variations required by spatial variations in the conditions of life within the adaptive zones of the new lineages. If the morphological distance between phyla were evolved simply by the accumulation of the slight differences that distinguish genera and families, at rates comparable to those that obtained from the fossil record it can be estimated that the common metazoan ancestor of modern phyla would probably have lived over one billion years before the first appearance of metazoa in the fossil record (Durham 1971). But new grades of organization must have evolved chiefly through development of new regulatory gene systems, and while the steps would still have been gradual, they would have proceeded in large measure through recombinations of features that were already present. The invention of new features through the evolution of structural genes would be minimal, and much of it might follow after the establishment of new anatomical plans. Progress could be rapid.

This model raises the question of timing. Why did the major groups evolve just when they did, and why at times did numbers of lineages independently give rise to major novelties nearly contemporaneously? Perhaps an answer is furnished by the 570 million years or so for which there is a good fossil record. During this interval, major changes in the structure of the biosphere and major waves of extinction and diversification correlate with major changes in the environmental regime that are in turn due to continental drift (Valentine and Moores 1972; Flessa and Imbrie 1973; Valentine 1973b). The processes of global tectonics, now well established, have caused continents and ocean basins to enlarge, to shrink or fragment, and to radically change their geographic patterns, slowly but more or less steadily for the last billion years, and probably for twice that time. The environmental effects of such major geographic changes are clearly profound and often involve changes in environmental variability—exactly the sort of change that must elicit changes in adaptive strategy among the biota, to judge by today's world (Valentine 1973b).

The major environmental patterns of late Precambrian time are still uncertain. Evidence that may well solve this problem is now accumulating, and we may soon have a model that enables us to evaluate the relation between the origins of phyla and changing environmental regimes. An additional factor may have been that rising amounts of free oxygen in the atmosphere, released by plants during photosynthesis, reached a succession of levels sufficient to sustain an increasing complexity of metazoan life (Berkner and Marshall 1964; Rhoads and Morse 1971). Whether or not this was the case, we still must explain the particular adaptive pathways pursued by the primitive lineages that seem related to changing environmental regimes.

Additional evolutionary factors

In addition to helping to explain problems of the origin of new grades and anatomical plans, certain changes in the systems of gene regulation may clarify other evolutionary phenomena. One trend that must involve regulatory evolution is the development of complex life cycles, as when organisms have two or more distinctive modes of life at different life stages. This would usually require more regulatory genes. A possible variation involves parasites, which commonly have become morphologically simplified as adults while developing highly complex and specialized life cycles with stages adapted to their various vectors and hosts. They may represent cases of regulatory repatterning in which morphological complexity is traded for developmental complexity.

Neoteny is another example. This is a process that involves the shifting of the characters that are present in the juvenile stage of ancestors into the adult stages of descendants. A famous hypothesis of chordate origins, due chiefly to Garstang, is that the swimming larvae of some primitive deuterostome (probably urochordates or sea squirts) achieved the ability to reproduce and thus became neotenous swimming adults, leading eventually to the evolution of the vertebrate skeleton to exploit this locomotory opportunity. Such a change in developmental timing is precisely the sort of phenomenon that might be expected from regulatory gene evolution in the presence of appropriate selective pressures.

Regulatory changes must frequently accompany the origin of new species as well as of higher taxa. In the fossil record, most species appear suddenly, because of the scarcity of intermediate forms. Their direct ancestry may be unknown, or even if an ancestor is suspected, descendants appear without known intermediates. Furthermore, many such species do not exhibit any significant skeletal changes during their known duration; eventually they simply disappear, perhaps to be replaced in the record by another species that also seems to appear suddenly. A few species do exhibit gradual morphological change through time, of course. It has been assumed that the former types have evolved in geographic isolation from their parental stocks—certainly a common mode of speciation (Mayr 1963). The descendant species may spread quickly—in just a few thousand years—when favorable environmental changes occur; their origins are simply missing from the spotty fossil record (Eldredge and Gould 1972). Regulatory mutations may in fact create the genetic isolating mechanisms between allopatric populations (Wilson et al. 1974). It may be that the episodic morphological "jumps"—gradual but localized in time—are commonly achieved by regulatory evolution. Any subsequent evolutionary changes among structural genes may involve physiological tolerances but need not alter the skeletal morphology, which is chiefly adaptive to a mode of life within a certain habitat range.

Indeed, the extent to which the mode of life of a lineage undergoes evolutionary change may well be a key to the extent to which the genetic regulatory system is involved, regardless of the taxonomic level achieved by the descendants. Primitive flatworms, burrowing coelomates, and surface-dwelling coelomates each developed an increased genetic and morphological complexity as they evolved novel modes of life. In many cases (as arthropods), whole new locomotory mechanisms had to be developed, and this involved a general change in the body plan. In other cases (as brachiopods), anatomical modifications appear to have been related to increased feeding efficiency in new life modes. Important evolutionary novelties have arisen when organisms with wide evolutionary potentials were provided with a broad ecological opportunity owing to environmental change. At such times, major alterations in the biological architecture are achieved through evolution of the regulatory genomes.

There thus appear to be three major modes of evolutionary change as reflected by changes in the genome. One involves change in the quality of structural genes (and in the alleles of nonstructural regulatory genes) and change in their frequencies within populations. A second involves the repatterning of genes in new combinations. And a third involves the expansion of the regulatory apparatus. All three modes must ordinarily operate to produce those series of anatomical changes that lead to the development of a wholly new animal type, and all three modes could operate to produce new species. Nevertheless, growth of the regulatory genome may usually predominate during evolution of new grades of complexity; gene repatterning may predominate during the radiation of major new variations within grades; and structural gene frequency changes may be especially important during phyletic evolution within species lineages.

The evolution of the regulatory portion of the genome can evidently achieve morphological changes rapidly, owing in large part to three aspects of regulatory change. First, as new regulatory patterns are evolving, the old ones may continue to function, thus maintaining adaptation of the lineage during important evolutionary trends. Second, repatterning of the regulatory apparatus may achieve new architectural relations by employing components that are already well developed, altering the relationships among structural genes, batteries of genes, and sets of gene batteries through a hierarchical regulatory apparatus. Third, the morphological changes that arise from spatial or temporal additions or repatterning of gene batteries commonly involve the groundplans of the organisms, and the resulting morphological distances between ancestor and descendant are therefore likely to be large. This is especially true when such changes are compared to the morphological results of structural gene evolution, which tends to develop only variations on a general morphological theme. Even though the changes proceed in small steps in both cases, the results of regulatory evolution appear greater and more rapid because they can achieve novel biological architectures. Our understanding of many evolutionary problems seems likely to increase dramatically as the mechanisms of gene regulation are deciphered.

References

Berkner, C. V., and L. C. Marshall. 1964. The history of growth of oxygen in the earth's atmosphere. In C. J. Brancuzio and A. G. W. Cameron, eds., *The Origin and Evolution of Atmospheres and Oceans.* New York: John Wiley & Sons.

Britten, R. J., and E. H. Davidson. 1969. Gene regulation for higher cells: A theory. *Science* 165:349–57.

Britten, R. J., and E. H. Davidson. 1971. Repetitive and non-repetitive DNA sequences and a speculation on the origins of evolutionary novelty. *Quart. Rev. Biol.* 46:111–33.

Britten, R. J., and D. E. Kohne. 1968. Repeated sequences in DNA. *Science* 161: 529–40.

Clark, R. B. 1964. *Dynamics in Metazoan Evolution: The Origin of the Coelom and Segments.* Oxford: Clarendon Press.

Cloud, P. E. 1949. Some problems and patterns of evolution exemplified by fossil invertebrates. *Evolution* 2:322–50.

Cloud, P. E. 1968. Pre-metazoan evolution and the origin of the Metazoa. In E. T. Drake, ed., *Evolution and Environment.* New Haven: Yale University Press.

Cloud, P. E. 1973. Pseudofossils: A plea for caution. *Geology* 1:123–27.

Cloud, P. E., G. R. Licari, L. A. Wright, and B. W. Troxel. 1969. Proterozoic eucaryotes from eastern California. *Proc. Nat. Acad. Sci., USA* 62:623–30.

Davidson, E. H., and R. J. Britten. 1973. Organization, transcription, and regulation in the animal genome. *Quart. Rev. Biol.* 48:565–613.

Dobzhansky, Th. 1970. *Genetics of the Evolutionary Process.* New York: Columbia University Press.

Durham, J. W. 1971. The fossil record and the origin of the Deuterostomia. *Proc. N. Am. Paleont. Conv.*, Chicago, 1969, Proc. H:1104–32.

Eldredge, N., and S. J. Gould. 1972. Punctuated equilibria: An alternative to phyletic gradualism. In T. J. M. Schopf, ed., *Models in Paleobiology.* San Francisco: Freeman, Cooper.

Flessa, K. W., and J. Imbrie. 1973. Evolutionary pulsations: Evidence from Phanerozoic diversity patterns. In D. H. Tarling and S. K. Runcorn, eds., *Continental Drift, Sea Floor Spreading and Plate Tectonics.* London: Academic Press.

Glaessner, M. F. 1969. Trace fossils from the Precambrian and basal Cambrian. *Lethaia* 2:369–93.

Glaessner, M. F. 1971. Geographic distribution and time range of the Ediacara Precambrian fauna. *Bull. Geol. Soc. Amer.* 82:509–14.

Holliday, R., and J. E. Pugh. 1975. DNA modification mechanisms and gene activity during development. *Science* 187:226–32.

Jacob, F., and J. Monod. 1961. Genetic regulatory mechanisms in the synthesis of proteins. *Jour. Molec. Biol.* 2:318–56.

Kauffman, S. 1971. Gene regulation networks: A theory for their global structures and behaviors. In A. A. Moscona and A. Monroy, eds., *Current Topics in Developmental Biology.* New York: Academic Press.

Mayr, E. 1963. *Animal Species and Evolution.* Cambridge, Mass.: Belknap Press.

Ohno, S. 1970. *Evolution by Gene Duplication.* Heidelberg: Springer-Verlag.

Rhoads, D. C., and J. W. Morse. 1971. Evolutionary and ecologic significance of oxygen-deficient marine basins. *Lethaia* 4: 413–28.

Schopf, J. W., and J. M. Blacic. 1971. New microorganisms from the Bitter Springs Formation (late Precambrian) of the north-central Amadeus Basin, Australia. *Jour. Paleontology* 45:925–60.

Simpson, G. G. 1944. *Tempo and Mode in Evolution.* New York: Columbia University Press.

Simpson, G. G. 1953. *The Major Features of Evolution.* New York: Columbia University Press.

Stein, A. S., J. S. Stein, and L. J. Kleinsmith. 1975. Chromosomal proteins and gene regulation. *Scientific Amer.* 232(2):46–57.

Valentine, J. W. 1973a. Coelomate superphyla. *Syst. Zool.* 22:97–102.

Valentine, J. W. 1973b. *Evolutionary Paleoecology of the Marine Biosphere.* Englewood Cliffs, N.J.: Prentice-Hall.

Valentine, J. W., and E. M. Moores. 1972. Global tectonics and the fossil record. *Jour. Geology* 80:167–84.

Wilson, A. C., C. R. Maxson, and V. M. Sarich. 1974. Two types of molecular evolution: Evidence from studies of interspecific hybridization. *Proc. Nat. Acad. Sci., USA* 71:2843–47.

David M. Raup

Probabilistic Models in Evolutionary Paleobiology

A random walk through the fossil record produces some surprising results

Most interpretation of the fossil record has centered around finding specific causes for specific evolutionary events. Why was there a sudden diversification of life in the late Precambrian? Why did one group of corals replace another? Why did the dinosaurs go extinct? Why did the human species evolve when it did? And so on. The approach has been highly deterministic; each event has been treated as unique, and although generalizations have been made there have been few attempts to look at groups of events in a probabilistic way.

Out of all the paleontological work, there has developed a strong consensus on a couple of major generalizations: first, that there has been a marked and broad increase in number of species over the last 600 million years, interrupted only by a few periods of mass extinction; and second, it is generally agreed that the fossil record shows steady improvement or progressive optimization of fitness of biologic structures. Optimal structures are common, but well-documented examples of the steps leading to the optima are hard to find.

David M. Raup is Professor of Geology at the University of Rochester. He received a B.S. from Chicago and M.A. and Ph.D. from Harvard and taught at Caltech and Johns Hopkins before joining the Rochester faculty in 1965. His varied background in research has included extensive studies of biocrystallography of echinoderms and mathematical analysis of the molluscan shell. In 1973, he was the first recipient of the Paleontological Society's Schuchert Award for contributions to paleontology by a person under the age of 40. He is the author, with Steven M. Stanley, of the widely used textbook Principles of Paleontology. *Professor Raup is currently President of the Paleontological Society. This article is based on a talk presented at a symposium on "Stochastic Processes in Evolution" at the annual meeting of the Society for the Study of Evolution in New Orleans on 1 June 1976.* *The research is supported in part by the Earth Sciences Section, National Science Foundation, NSF Grant DES75-03870.* *Address: Department of Geological Sciences, University of Rochester, Rochester, NY 14627.*

Let me say a bit more about the first generalization—that of increased species diversity. Approximately 200,000 fossil species have been described since Linnaeus (Raup 1976a), and the vast majority are invertebrate animals, and most are marine. Plants and terrestrial vertebrates make up a small fraction. Figure 1 shows the distribution of invertebrate fossils through geologic time—based on a tabulation of about half of the 140,000 species described since 1900. It shows a Paleozoic peak in the Devonian and a post-Paleozoic rise to a maximum in the Cenozoic. The rise toward the present is not nearly as great as has been previously estimated, but more important, most of the change may be an artifact of sampling. Younger rocks are much better exposed than older ones and inevitably yield more species. Figure 2 shows variation in the exposed area and volume of sedimentary rocks. These plots—particularly the one for area—are very similar to the species plot. The raw data for diversity are statistically correlated with both area and volume, and the correlations are significant at the 1% level. Ironically, area and volume are correlated with each other only at the 5% level! Thus it is quite possible that the number of marine invertebrate species has been reasonably constant throughout most of the last 600 million years. Apparent diversity may just be tracking sample size.

My reason for bringing this up is that it opens the way for thinking of marine species diversity as having been in dynamic equilibrium for a long time. This, in turn, makes it possible to explore species/area relationships, for example, on a world-wide scale. Simberloff (1974) has already done this for fossil groups of the Permian and Early Triassic.

Figure 3 shows the changes over time in the percentage composition of the invertebrate fossil record—again based on a sample of 70,000 species. Some of the changes are striking: the domination of the early Paleozoic by trilobites was followed by the virtual takeover of the invertebrate world by molluscs and protozoans after the Permian. Keep in mind that insects do not play a large role here because of lack of preservation. Must we find specific causes for all these changes? Because so much time is involved, it may be that small random fluctuations in numbers of taxa can accumulate as a Markov chain—a sequence in which future states of a system are influenced by preceding states—to produce changes of this magnitude.

Probabilistic approaches

The fossil record as a whole provides many possibilities for probabilistic models—with ample opportunity for making predictions and testing them. Although we have only one historical record, it is a large one and it contains many reasonably independent histories: biologic groups that have occupied widely separated continents and have had little interaction. Above all, we have very long time series—so long that events with vanishingly small probabilities become likely, and this includes rare physical events that have biological impact.

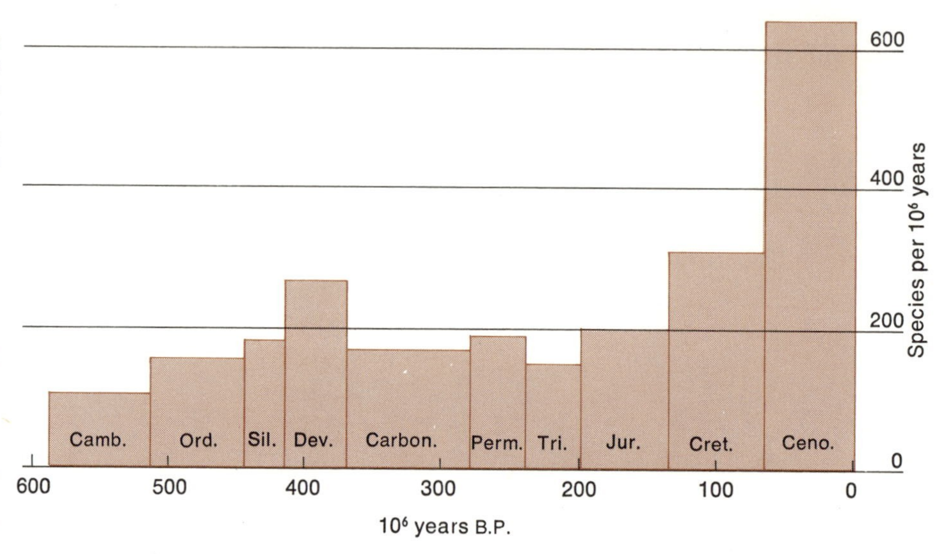

Figure 1. The variation in apparent numbers of fossil invertebrate species during the Phanerozoic (the past 600 million years) can be seen from this chart, which is based on published citations to about 70,000 of the species described between 1900 and 1970. (Raup 1976a.)

It may be that the length of time involved is such that changes in the adaptive relationships of species and in the relationships between species and their environment are so complex and multifactorial that natural selection may behave (mathematically) as a random variable. To be sure, there have been monotonic trends of environmental change, and these have had evolutionary consequences—but such trends reverse and go forward and back in an apparently unpredictable fashion. To describe such changes as "random" does not deny cause and effect; it just means that the sequence of events can be predicted only in a statistical sense. Each event has a cause, but the distribution of these causes in geologic time may be essentially random.

Recently, several frankly probabilistic models have been proposed for application to the fossil record of evolution. One is Steven Stanley's (1975) idea of *species selection*: the proposal that speciation may be a largely random process—with characteristics of the founder individuals in small populations being very important—and that a lot of selection actually takes place between species—in other words, selection within genera or families rather than within populations.

Another model of current interest is Leigh Van Valen's (1973) Red Queen Hypothesis, which he uses to explain fossil data that he contends show the probability of extinction to be stochastically constant within a taxonomic group. Van Valen thinks it possible to talk about species as having a "half-life"—descriptively analogous to the half-life of a radioactive isotope.

Let me return to Markov chains. Evolution certainly should be viewed in a Markovian framework, at least as far as time series are concerned. Figure 4 presents three simple random walks—time series characterized by small steps, showing values of, say, a morphologic character or a genetic frequency as it changes over time. There is uncertainty about whether the line will go up or down at a given point, but its position after the move is heavily constrained by where it was before the move: it can only be a small distance from the previous position. The analogy to evolution is that each species owes more to its phylogenetic legacy than to the genetic changes that took place during its speciation event. It need not concern us now *why* the line goes up or down; suffice it to say that such patterns of change are very common in the fossil record.

Imagine for a moment that we were to look at only the 500 points on each of

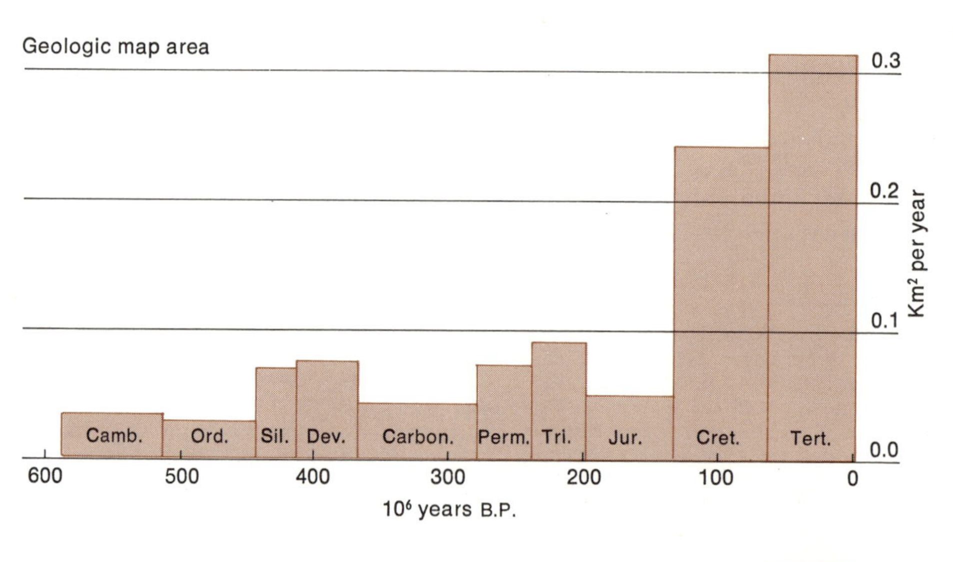

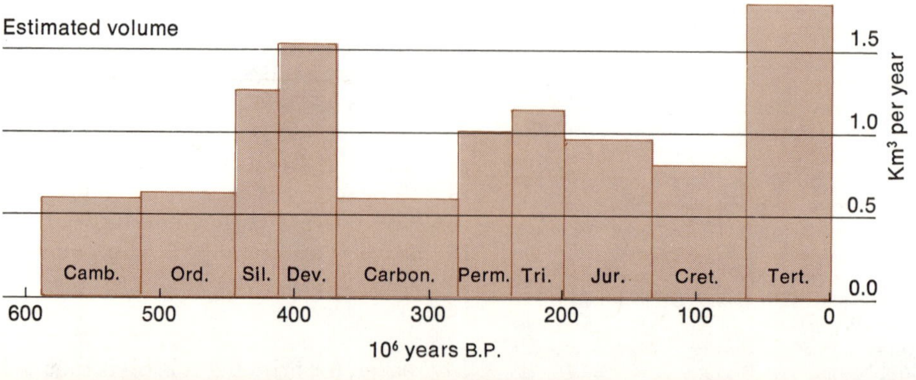

Figure 2. These charts show the surviving areas and volumes of sedimentary rocks (for Cambrian–Tertiary). *Top*, geologic map area. (Blatt and Jones 1975 and Blatt, pers. comm., 1975.) *Bottom*, sedimentary volume. (Gregor 1970; deep-sea deposits have been excluded from Gregor's data.)

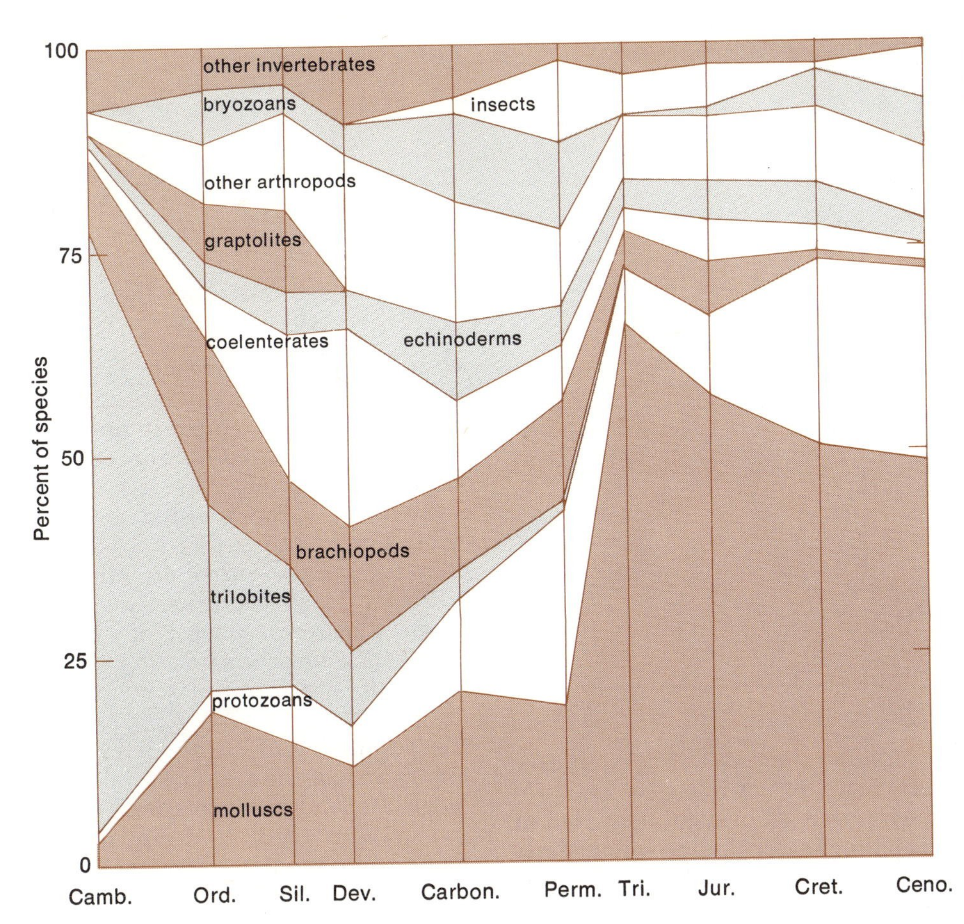

Figure 3. Considerable variation is evident in the taxonomic composition of the invertebrate fossil record through the Phanerozoic. (Raup 1976a.)

these plots—ignoring the lines connecting them—and think of a scatter diagram of time versus some attribute (the vertical coordinate). If conventional correlation coefficients are calculated, it turns out that the majority are statistically significant at the 1% level. In a sense, therefore, we can say that the attribute is significantly correlated with time in most cases even though we *know* that we are dealing with a totally random process. This is an important warning for those who are quick to describe evolutionary trends and interpret them in terms of single causes. Thanks to mathematicians such as William Feller, it is possible to test these trends using the random walk as a null hypothesis. None of these three shows significant departures from statistical expectations (see Feller 1968 for methodology).

Figure 5 shows two fossil time series from a recent paper by Hayami and Ozawa (1975), who used them as examples of orthoselection—that is, long-term response to single selective factors. Although the top one *does* depart significantly from expectations, and thus is a candidate for biological interpretation, tests using the Markov chain model on the bottom example demonstrate that the change could easily have occurred by chance—and the orthoselection explanation model is not justified (Raup, in press). The obvious trend here could have resulted either from

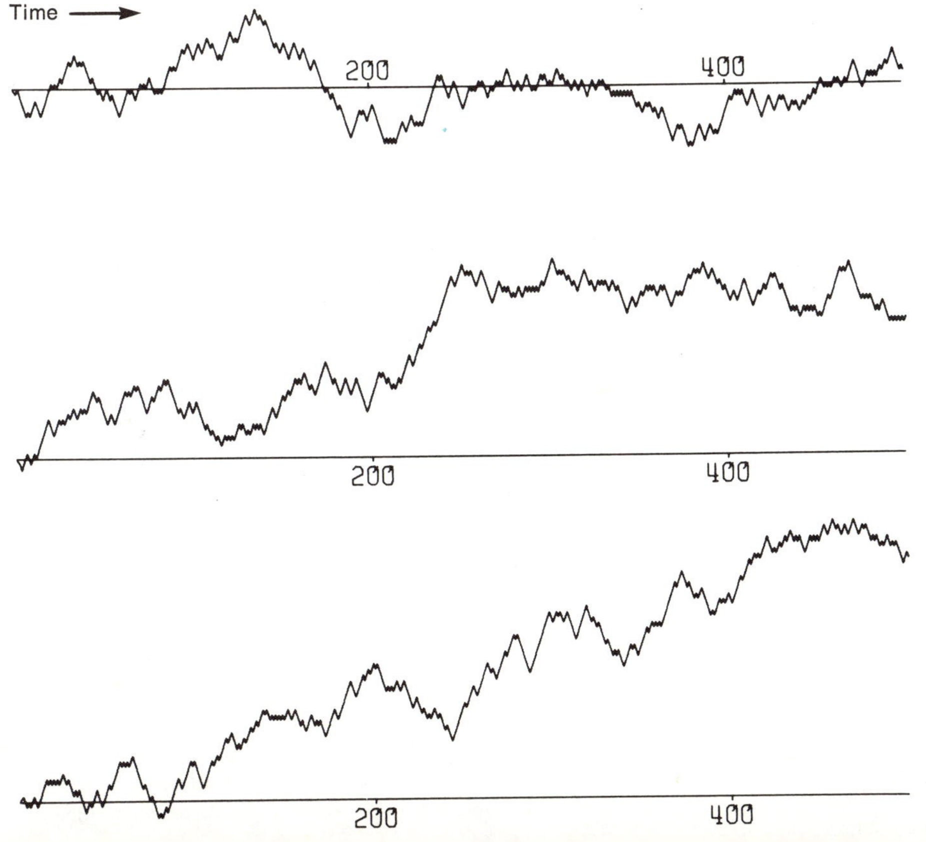

Figure 4. Three computer-generated random walks show the evolution of a morphological character or a genetic frequency. The three random walks, each 500 steps long, were selected from 20 that were constructed. The upper one comes closest to the normal idea of a random walk yet is a relatively uncommon type; the two lower ones are actually more typical because of the Markov property of random walks. In random walks that are reasonably long, the line of steps will cross the starting point (horizontal axis) in the second half of the walk in only 50% of the cases; in 20% of random walks, the line will stay on one side 97.6% of the time; in 10% of random walks, there will be no crossing after the first 3% of the walk. (Raup in press.)

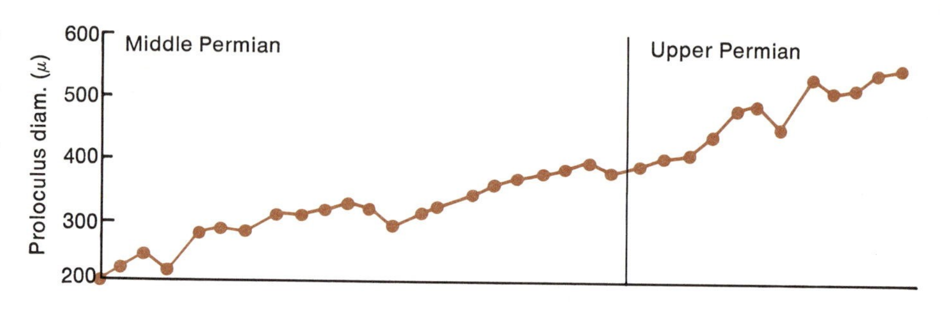

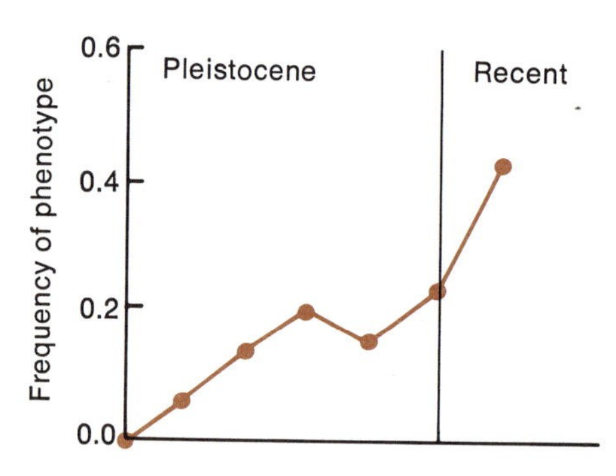

Figure 5. Fossil time series expressed in the format of a random walk. *Top:* This size variation through time for a Permian protozoan departs significantly from chance expectations, and is thus a candidate for biological interpretation. *Bottom:* Variation in frequency of one phenotype in a Pleistocene and Recent scallop could easily have resulted from chance. (After Hayami and Ozawa 1975.)

genetic drift or from a chance combination of independent selection events.

To summarize, the first question for the paleobiologist faced with an evolutionary trend should be: Does the trend represent a statistically significant departure from chance expectations? Only if the answer is positive is there justification for proceeding to look for a specific biologic or geologic cause for the trend. In evolutionary time series, the random walk model (as a kind of Markov chain) is the appropriate null hypothesis.

A Monte Carlo model

Let me now turn to a much more ambitious attempt at probabilistic modeling. Based on my recent collaboration with Stephen Gould, Thomas Schopf, and Daniel Simberloff (Raup et al. 1973; Raup and Gould 1974), it involves an application of branching processes to phylogenetic patterns and, as such, represents a somewhat more complex version of the Markov model. The work uses computer simulation and Monte Carlo methods to generate imaginary evolutionary trees. As shown in Figure 6, a branching pattern of lineages is built up using random numbers. In effect, we are asking What would evolution have looked like if it were a completely random process—without natural selection and adaptation and without predictable environmental change? The simulations are thus a giant null hypothesis for comparison with the real world.

Figure 6 illustrates the basic framework of the program. It operates in an arbitrary time scale, and it deals with a series of computer-generated evolutionary lines or lineages. The program is written in such a way that at each time interval, the evolving lineage may go extinct or persist to the next time interval; if it persists, it may branch to form a second lineage. These choices are based entirely on pre-set probabilities of one outcome or the other. Once a second lineage has been produced by the branching process, it too is subject to the same set of choices for survival or extinction or branching. In the case shown in Figure 6, the starting lineage branched after three time intervals but became extinct after six. The newly formed branch lasted considerably longer and before it became extinct gave rise either directly or indirectly to all the other lineages shown. The average length of a lineage will, of course, be a function of the probability assigned to extinction.

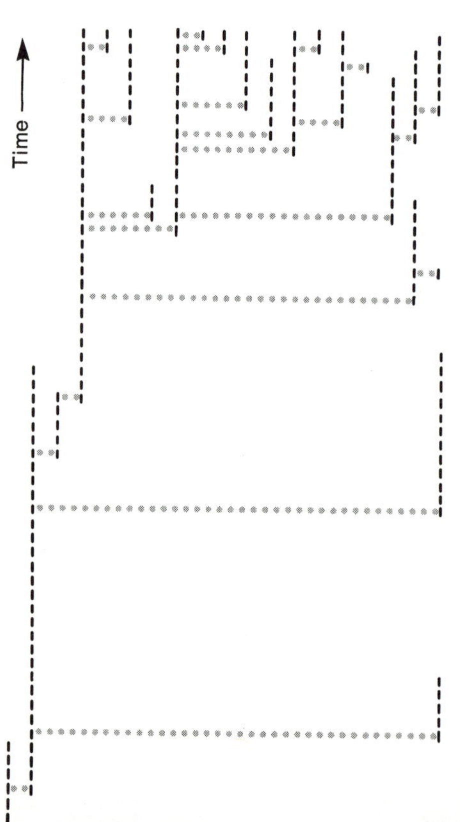

Figure 6. This simulated evolutionary pattern starts on the lower left with a lineage that lasted a half dozen time units. Each vertical dash is a time unit in the computer system, and each series of vertical dashes is a lineage; branching (*dots*) and extinction are signaled by machine-generated random numbers. (Raup and Gould 1974.)

The resulting output is an evolutionary tree of the sort most paleontologists spend much of their time developing or trying to decipher. As the program is written, several hundreds of these evolutionary lineages are generated in the course of a single run, so the output of a single run is several times larger and more complex than the one illustrated in Figure 6.

Now, if the probabilities of branching and extinction are the same and constant throughout the run, the number of co-existing lineages will wander up and down as a random walk. If, on the other hand, the probability of branching is higher than the probability of extinction, the number of co-existing lineages will increase geometrically. In most runs, we elected to compromise between these two strategies and established, in advance, an equilibrium number of lineages and had probabilities of extinction and branching continuously monitored on the basis of the number of lineages present, so that the number of lineages goes up to the equilibrium value and then fluctuates around it as a dynamic equilibrium.

What we have established therefore is a kind of evolution machine that is an extremely stylized expression of one evolutionary model. It is deterministic in that the branching continues until the equilibrium is reached, then it fluctuates about that equilibrium. It is stochastic in the sense that the fate of any lineage is determined only in a probabilistic sense.

We were not really attempting to simulate the real world; in fact, we did not expect very close correspondence with the real world. Rather, we wished to use this means to separate those kinds of events that are readily amenable to a stochastic explanation from those that require a more deterministic explanation. In short, we were looking at the differences between our simulations and the real world rather than the similarities. As it turned out, we found much greater correspondence between our model and the real world than we had anticipated.

Figure 7 shows the basic format we used for the computer simulations. In order to make the results more applicable to paleontology as it is practiced, we wrote into the program routines for grouping the lineages into higher taxa, such as families, orders, or phyla. This was done by purely mechanical means based on the size and persistence of monophyletic groups of lineages. Figure 8, a simplified form of the output of four computer runs, shows one level of computer-generated classification grouping. The patterns bear remarkable resemblance to many in the paleontological literature. It is a very easy matter indeed to pick out well-known groups in this array; for example you can find patterns that resemble that of the trilobites, like #16 in row C, where there was a very rapid radiation and diversification early in the history of the group followed by a gradual decline toward extinction. At the opposite extreme is one like #6 in run B which might simulate the insects: diversification over quite a long period of time so that the living diversity is extremely high.

In the computer output it is very common for several groups of lineages to go extinct at the same time—just as a matter of chance. In row B of Figure 8, for example, groups 4, 15, and 17 "died out" at precisely the same time (the 73rd time interval). These three represented 20% of the groups existing at that time, and thus the extinction event was a significant one. Also, there were several other groups in row B that became extinct either shortly before or after the 73rd time interval. In the real world, it is usually assumed that extinctions that coincide in time (actually or approximately) have a common cause. In the simulations, we know there was no common cause! Are the so-called mass extinctions of the geologic record real? Or is it statistically likely that completely independent extinction events coincided occasionally to produce the appearance of mass extinction? At present, a definitive answer to these questions is not available. Some of the mass extinctions appear to be real while others could easily be coincidences.

Figure 9 shows a real-world phylogeny plotted in the fashion of Figure 8. The main differences between them are in the range of variation in the shapes of the individual parts of the diagram, and are mostly quantitative rather than qualitative, however. Minor tinkering with the constants in the computer model will produce patterns like this.

We have made considerably more rigorous comparisons between the world of simulation and the real world (Gould et al., in press), but space does not permit detailed treatment here. Some of the matches are good, and some are not. In a few cases, events such as the major extinctions near the Cretaceous-Tertiary boundary stand out as being clearly "off-scale" with respect to the random model and apparently require a deterministic explanation.

Morphological change

So far I have not said anything about the morphology of the "organisms" produced by the computer simulations. Partly for the fun of it, we introduced morphology into the system in a completely random manner. That is, as an independent part of the program, we assigned to the starting lineage a morphology based on 10 to 20 completely arbitrary characters. Character states were expressed on integer scales starting at 0. At each branch point, each trait or character could change by one unit in either direction—that is, from 0 to +1 or from 0 to −1, or if already at +1 it could go back to 0 or to +2—so that as the program developed the complex

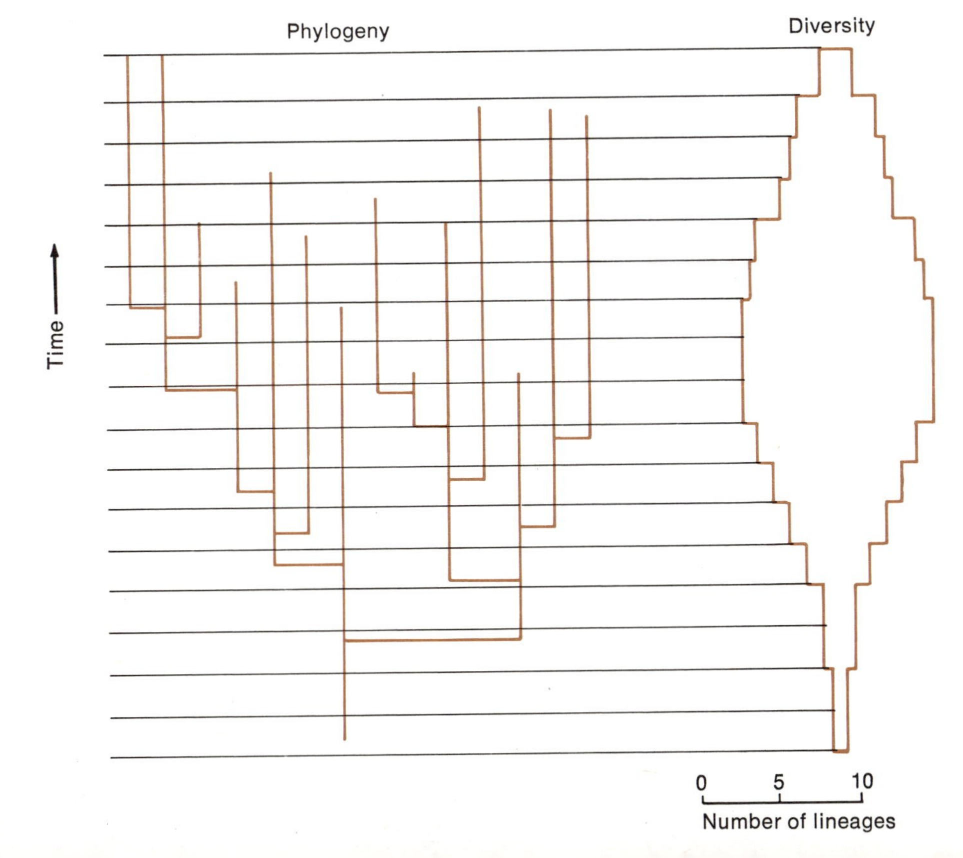

Figure 7. The hypothetical evolutionary pattern (*left*) is similar to what the computer produces; the lineages are all interrelated in the sense that they are derived from a single ancestor. Its summary (*right*) is a paleontologically common means of expressing the change in abundance of forms within an evolving group; the width of the pattern reflects the number of co-existing lineages through time. (Raup et al. 1973.)

branching network, it also allowed the several traits to wander independently or drift as a random walk.

We expected to find complete chaos because we were modeling the evolution of morphology without benefit of natural selection, and it is conventional wisdom to credit adaptation through natural selection with most of the order we observe in the biological world. But to our surprise we found a very high degree of order in the morphological results and found examples of a great many of the trends and patterns that we are used to seeing in the evolutionary record.

Figure 10, which shows some of the results, reveals that although there is quite a bit of variation in shape among these imaginary animals we call triloboids, departing in several directions from the starting configuration, there is also considerable order. Those in group A, for example, are characterized by having a relatively wide head and a long, narrow tail, in contrast to those in group D, which tend to have long, narrow heads and wide, short tails. The groups in Figure 10 are quite consistent internally: the groups of lineages appear to be "natural" collections of like shapes. This has been confirmed by applying conventional numerical taxonomy to the morphological data and comparing the resulting classification with the known branching pattern of the simulation (Schopf et al. 1975).

As mentioned above, we can see in the computer output many trends and patterns that are common in the fossil record and that carry with them a host of purely deterministic explanations. It is often claimed, for ex-

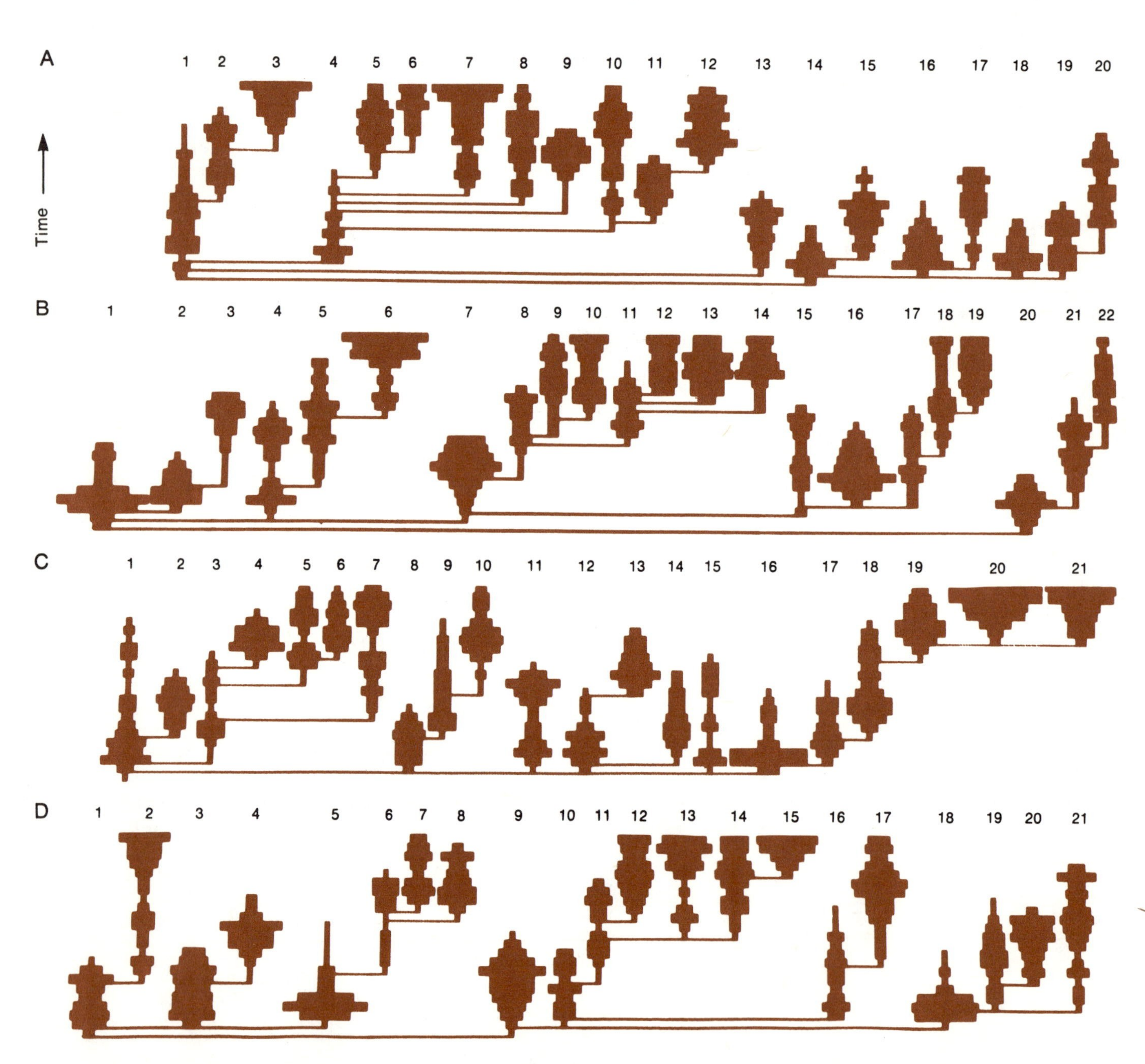

Figure 8. All runs of the simulation program—from which these samples were taken—used the same input constants (i.e. probabilities of branching and extinction); the differences in pattern are therefore due to chance. Each color area shows the diversity history of a monophyletic group of lineages. The ancestry of the groups relative to one another is shown by horizontal lines. In each run, there were up to 500 lineages segregated into about 20 larger groups. The groups at the top tend to be very flat-topped—this doesn't indicate extinction; it is simply where the computer run ended and is analogous to the present-day end of the fossil record. (Raup et al. 1973.)

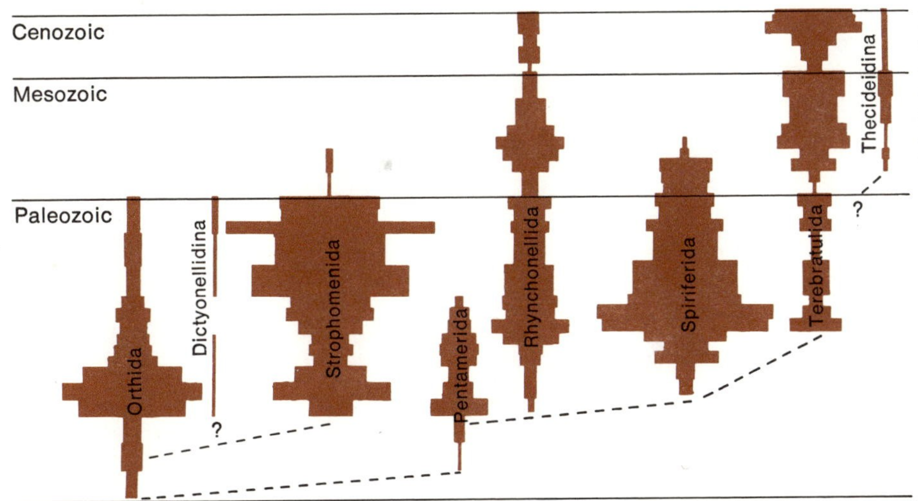

Figure 9. The real-world phylogeny of the articulate brachiopods is plotted in the fashion of Figure 8. The main difference between this display and the computer-generated ones is in the range of variation in the shapes of the individual parts of the diagram. (Williams 1965.)

ample, that there is an inevitable increase in specialization with time, and certainly the more specialized or bizarre forms occur late in the computer sequence—that is, near the top of the diagram. They are either very large or very small, and have other peculiarities, with respect to the average. In the simulations, we know that this is not caused by biological processes. Rather, random drift away from the starting point practically insures that the bizarre forms will be concentrated at the greatest distance from the starting point.

Figure 11 illustrates slightly more rigorously one of the kinds of results we got from the morphological modeling—a simple plot of the values of two characters for the 200 lineages in one of the runs. The two are highly correlated—overwhelmingly significant at the 1% level. It clearly suggests that the two characters here are correlated with each other so that, as one changes, a predictable change occurs in the other. But we know that this is not causal from the way in which the computer program was designed: the two characters here were generated in a completely independent fashion. This sort of correlation in the real world is automatically interpreted as indicating either genetic linkage or a functional or adaptive coordination.

The high degree of correlation observed in Figure 11 is not an exception. In fact, in approximately 75% of the possible combinations between characters a statistically significant correlation was found. The reasons for this have to do with the branching pattern of evolution, and although in hindsight it is easy to see that such correlations are in fact inevitable, we did not anticipate them (see Raup and Gould 1974 for further discussion). It is in this sense that the computer simulation has been extremely valuable as an exploratory tool. We were able to see, for example, that a selection-free model will inevitably produce certain patterns that we have always assumed were possible only as a result of natural selection.

This does not of course mean that we can argue persuasively against natural selection. It does mean that some of the patterns that are interpreted as resulting from selection cannot *in themselves* be used as evidence for selection: natural selection is only one

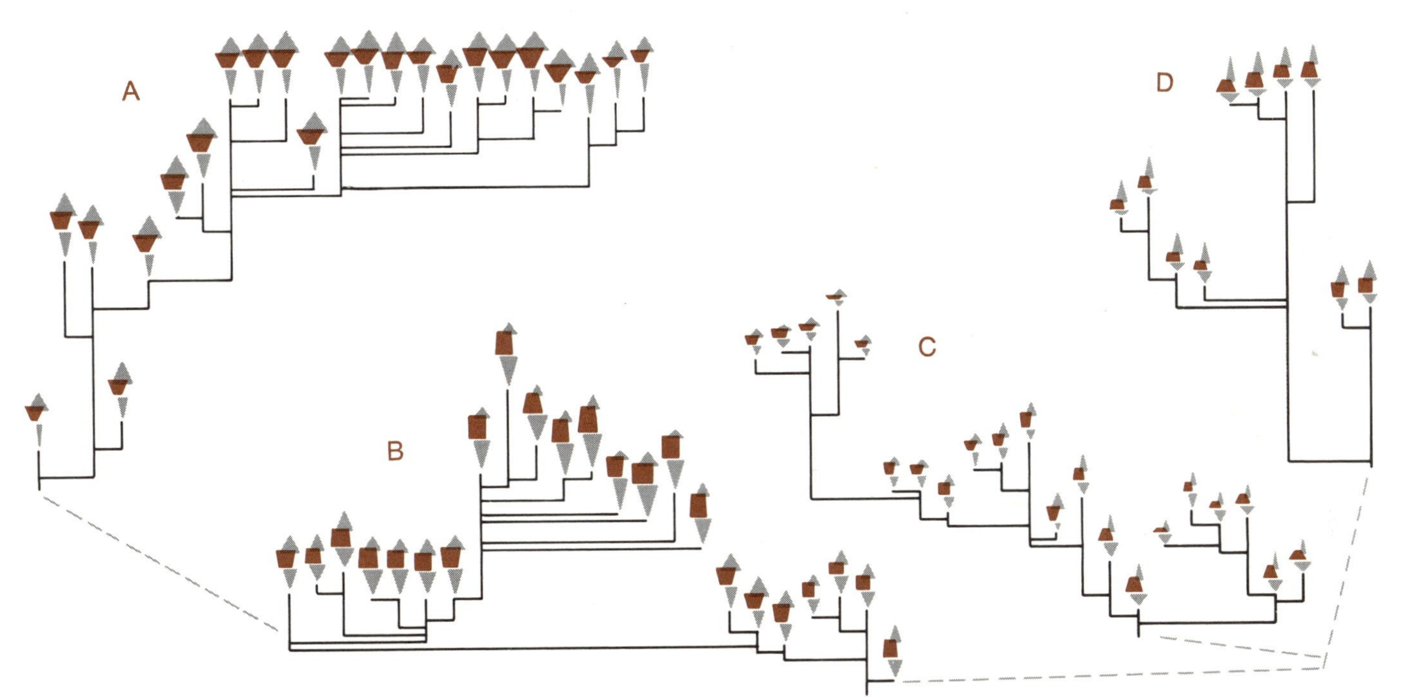

Figure 10. In this part of the morphologic output of the simulation program, the shapes of the glyphs reflect the states of five of the independently evolving morphologic characters. (Two define the shape of the head, one the length of the midsection, and two the shape of the tail.) The five traits are expressed as dimensions of these fictitious animals, which we call triloboids. The four regions are higher taxonomic groupings generated by the program on the basis of lineage branching patterns. The morphologic data did not contribute to the classification, yet the groupings are clearly reasonable ones. (Raup in press.)

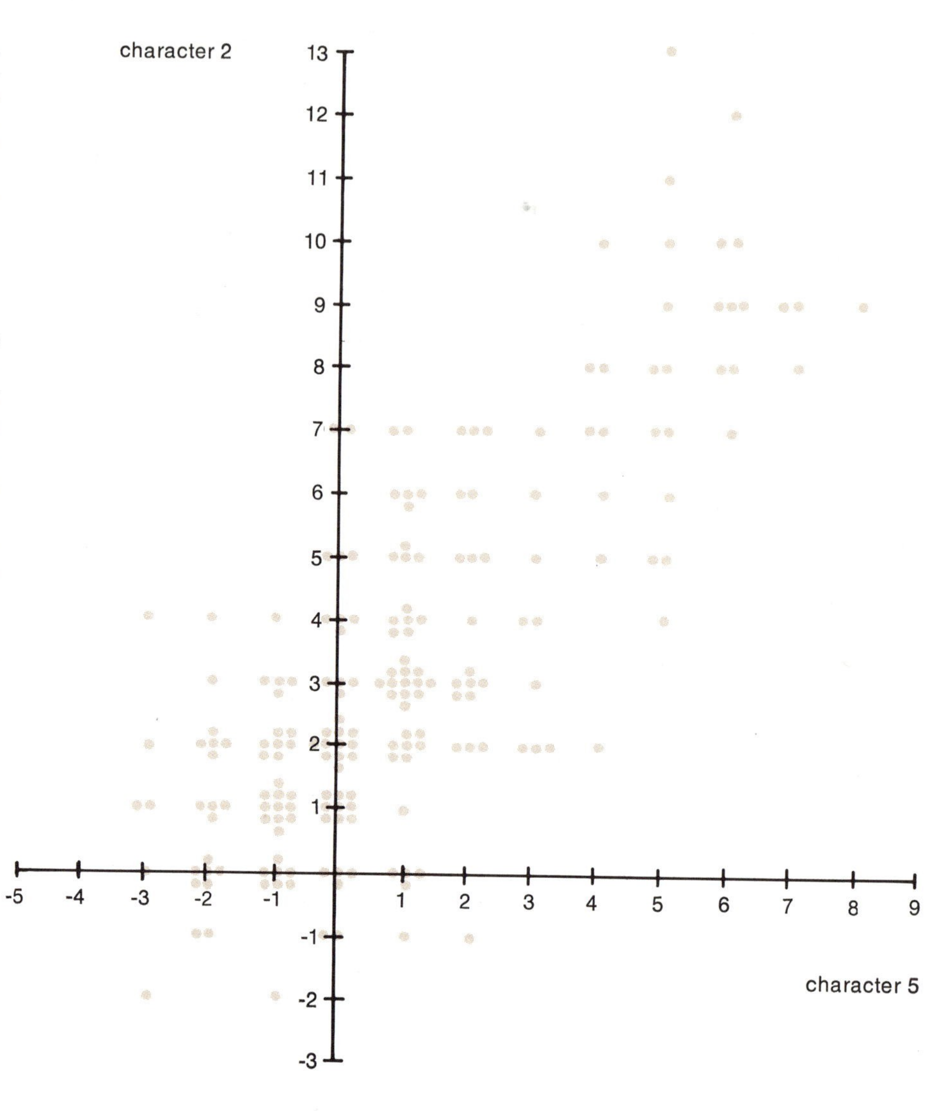

Figure 11. There is a considerable correlation between two morphologic characters from the 200 lineages of one computer run. Each point represents one lineage. The correlation is high, and though this might be interpreted in the real world as having a genetic or adaptive cause, the two characters developed independently, both starting at 0,0. (Raup and Gould 1974.)

way in which orderly patterns can be developed. The main difference between the integrated morphology seen in the real world and that found in the simulated world is that, in the real world, the collections of traits "make sense" in terms of the functioning of the organism. It is thus the functional analysis of morphology rather than the simple fact of correlation between characters that provides the basic evidence for adaptive evolution.

Perspective

The fossil record of evolution is amenable to a wide variety of models ranging from completely deterministic to totally stochastic. The computer program described here is crude and stylized and is most useful as an exploratory tool. It provides a source of hunches and works well as a test of one's own logic by showing what is possible and what is impossible under a given set of circumstances. The next step is to devise more rigorous tests for random variables in the fossil record. Some of these may employ Monte Carlo techniques using large samples. Others may make use of direct, analytical solutions based on probability arguments.

The approaches discussed in this paper are bound to be controversial. The real world of the fossil record probably contains a mixture of two types of events (or sequences of events): those caused by specific, nonrecurring phenomena and those so complex and unpredictable that they are best treated probabilistically in groups. We do not yet know the proportions of the two types, let alone the behavior of the second. It may turn out that the second type shows more order and predictability than is now apparent.

The concepts of "chance" and "randomness" are anathema to many scholars. To paleobiologists in particular, these concepts often seem unscientific and unmanageable and imply a defiance of cause and effect. Nothing could be farther from the truth, as has been demonstrated so often in other fields. It is perhaps ironical that some of the most effective applications of probabilistic models have been in fields traditionally close to paleobiology: geology (especially geomorphology) and evolutionary biology.

References

Blatt, H., and R. L. Jones. 1975. Proportions of exposed igneous, metamorphic, and sedimentary rocks. *Geol. Soc. Amer. Bull.* 86: 1085–88.

Feller, W. 1968. *An Introduction to Probability Theory and Its Applications.* NY: Wiley.

Gould, S. J., D. M. Raup, J. J. Sepkoski, Jr., T. J. M. Schopf, and D. S. Simberloff. In press. The shape of evolution: A comparison of real and random clades. *Paleobiol.*

Gregor, C. B. 1970. Denudation of the continents. *Nature* 228:273–75.

Hayami, I., and T. Ozawa. 1975. Evolutionary models of lineage zones. *Lethaia* 8:1–14.

Raup, D. M. 1976a. Species diversity in the Phanerozoic: A tabulation. *Paleobiol.* 2(4): 279–88.

———. 1976b. Species diversity in the Phanerozoic: An interpretation. *Paleobiol.* 2(4):289–97.

———. In press. Stochastic models in evolutionary paleontology. In *Patterns of Evolution,* ed. A. Hallam. Chapter 3. Amsterdam: Elsevier.

———, S. J. Gould, T. J. M. Schopf, and D. S. Simberloff. 1973. Stochastic models of phylogeny and the evolution of diversity. *J. Geol.* 81:525–42.

———, and S. J. Gould. 1974. Stochastic simulation and evolution of morphology—towards a nomothetic paleontology. *Syst. Zool.* 23:305–22.

Schopf, T. J. M., D. M. Raup, S. J. Gould, and D. S. Simberloff. 1975. Genomic versus morphologic rates of evolution: Influence of morphologic complexity. *Paleobiol.* 1:63–70.

Simberloff, D. S. 1974. Permo-Triassic extinctions: Effects of area on biotic equilibrium. *J. Geol.* 82:267–74.

Stanley, S. M. 1975. A theory of evolution above the species level. *Proc. Nat. Acad. Sci.* 72:646–50.

Van Valen, L. 1973. A new evolutionary law. *Evol. Theory.* 1:1–30.

Williams, A. 1965. Stratigraphic distribution. In *Treatise on Invertebrate Paleontology (Brachiopoda),* ed. R. C. Moore, pp. H237–H250. Lawrence, KS: Univ. of Kansas Press.

PART 2 *Invertebrate Paleontology*

Preston Cloud
James Wright
Lynn Glover III

Traces of Animal Life from 620-Million-Year-Old Rocks in North Carolina

The oldest well-dated metazoan fossils from North America are preserved beneath the deposits of a waning turbidity current that flowed downslope from an ancient volcanic borderland along the southeastern coast

The differentiation of simple, nonphotosynthetic but nucleated clusters of cells into specialized tissues and (later) organs was an event of great import in the history of the earth. It marked the transition from a protistan to a metazoan form of life, with all its potential for subsequent variations in style leading up to man himself. It also marked a dramatic change in the nature of the marine sedimentary record. Before the existence of differentiated multicellular animal life, the only means by which aquatic sedimentary regimes and their depositional products could be significantly altered were either purely physical, the result of chemical changes brought about by biological processes, or the sediment-trapping, sediment-binding, and microscopic boring activity of algae and bacteria. After the origin of the Metazoa, sediments were extensively affected by the direct physical and biochemical activities of these organisms that burrowed, bored, fed, and moved about beneath and on the depositional interface. The deposit feeders passed sediment through their intestinal tracts; the skeletal forms added their hard parts to sedimentary deposits; and some forms produced aggregate structures within which sediments were trapped.

Preston Cloud, a geologist with the U.S. Geological Survey, is also Professor Emeritus of Biogeology in the Department of Geological Sciences at the University of California at Santa Barbara. Since 1961 his research has focused on the primitive earth, especially the early biosphere and its interactions with other aspects of crustal evolution. James Wright is a Ph.D. student in the Department of Geological Sciences at UCSB, having completed his M.S. and thesis work on the area of this report with Glover at Virginia Polytechnic Institute and State University, Blacksburg. Lynn Glover III, a geologist with the USGS both before and after taking his Ph.D. at Princeton in 1967, has been Professor of Geology at VPI&SU since 1970. His research focuses on the regional geology and geochronology of the Carolina slate belt. *The authors are grateful to Dr. and Mrs. Warren Vosburgh and to Dr. Elizabeth C. Umstead, local landowners, for the courtesy of access to the site, to Drs. R. B. Neuman and Duncan Heron for organizing a quarrying operation there, and to James Vigil for supervising the removal and reassembling the large slab at Reston, VA. Fred Collier assisted by pointing out some interesting features of the structures. Research for this paper was supported partly by the USGS, partly by NSF Grant No. BMS 74-20046 and NASA Grant No. NGR 05-010-035 to Cloud, and partly by a VPI&SU Research Division grant to Glover. This is Biogeology Clean Lab Contribution No. 58.* *Address: Department of Geological Sciences, University of California, Santa Barbara, CA 93106.*

In this paper we describe impressions on 620-million-year-old rock surfaces from North Carolina that may represent the most ancient yet known Metazoa in the United States, if not in North America—*Vermiforma antiqua* Cloud n. gen., n. sp. Their visible traces indicate that they were large, soft-bodied, wormlike forms that were most likely detritus feeders. Associated sedimentary textures and structures indicate that their place of burial was beneath rather deep water on a gentle marine slope adjacent to a volcanic borderland that then fringed the eastern coast of North America, or perhaps the western coast of an adjacent continent (Glover and Sinha 1973).

Lithology, field relations, and age call to mind different and much more varied metazoan faunas that have been described from similar rocks of roughly similar age from southeastern Newfoundland (Anderson and Misra 1968; Misra 1969; Anderson 1972) and Midland England (Ford 1958, 1972). The English example clearly represents the globally distributed, soft-bodied metazoan fauna that ushered in the Paleozoic Era and Phanerozoic Eon, and the ones in Newfoundland and North Carolina may do so as well. This fauna is now widely referred to as the Ediacarian, after its most spectacular and best-known occurrence in ancient tidal-flat deposits of South Australia (Glaessner and Daily 1959; Termier and Termier 1960). No distinctively Ediacarian taxa are known to be associated with the form here described, however, and it appears to be some 60 million years more recent than the English and some of the Soviet occurrences with good Ediacarian affinities and ages of ~680 m.y. The North Carolina fossils, nevertheless, may be as old as or older than the Newfoundland ones, estimated by Anderson (1972) to be between 610 and 630 m.y. old but not well dated.

This discovery underscores the potential importance of the Carolina "slate belt" for the early evolutionary history of the Metazoa and the transition from pre-Phanerozoic to Phanerozoic. Because its marine volcaniclastic sediments are both fossiliferous and capable of being radiometrically dated at many points, this area is a promising region for delineating the first appearance of the Metazoa, the base of the Paleozoic and Phanerozoic, and the relations between Ediacarian and Cambrian.

Age and outcrop characteristics

A dozen or more large, curving, wormlike impressions were found by Glover and Wright on a roughly 2 × 3 m bedding-surface exposure in laminated, greenish, volcaniclastic, metasedimentary rocks on the south

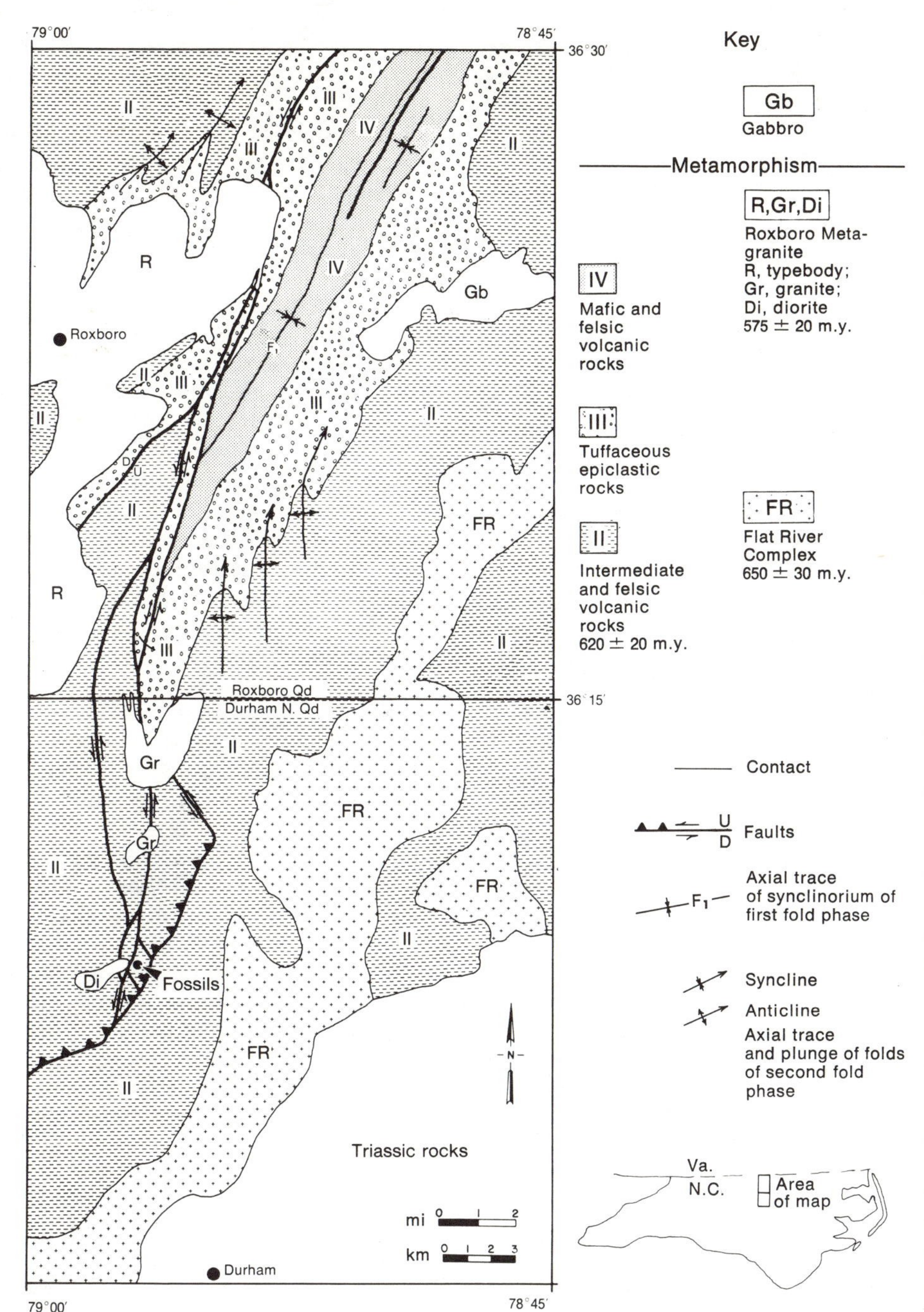

Figure 1. Imprints of the oldest metazoans yet discovered in the U.S. were found in the Carolina slate belt, about 16 km northwest of Durham, North Carolina, at the point indicated by the arrow. The map shows the geologic context of the fossils.

bank of the South Fork of Little River, about 16 km north and slightly west from Durham, North Carolina, in the Carolina slate belt (Fig. 1). The beds strike N15°E and dip 22°NW. Although these rocks occur within a small fault block whose exact stratigraphic position in the slate-belt sequence is not unequivocally established, they are assigned with a high level of confidence (Wright, 1974 M.S. thesis) to the upper part of map unit II of Glover and Sinha (1973) (Fig. 2). Zircon Pb-U isotope dating of pyroclastic rocks from the upper part of unit II gives an age of 620 ±20 m.y. (Glover and Sinha 1973). Folding and faulting of these rocks was followed by intrusion of a granodiorite pluton at Roxboro, with a Pb-U zircon concordia age of 575 ±20 m.y. (Glover and Sinha 1973; Briggs, Gilbert, and Glover, in press; confirmed by a second zircon sample recently dated by Sinha and Glover, unpublished). Another pluton, the Moriah, that intruded and erupted to supply pyroclastic sediments in the lower part of unit II, gives a zircon concordia age of 650 ±30 m.y. (McConnell, 1974 Ph.D. diss.). Thus the age of the metasedimentary rocks in which these structures occur is between 650 ±30 and 575 ±20 m.y. and most likely close to 620 m.y.

The general configuration of the structures discussed is sketched in Figure 3. They occur as interconnected J- and U-shaped impressions on the tan-weathering top surface of a 2- to 3-cm-thick, flat-laminated interval of very fine-grained, vitric, lithified and metamorphosed tuff (metatuff) that adheres tightly to the top of a 26- to 29-cm-thick, convolute-bedded stratum of similar rock (Fig. 4). The imprint-bearing surface is overlain by a 6-cm-thick bed of compositely graded metatuff, of which the bottom 2 cm is a single, graded unit (Fig. 5). The imprinted surface displays a complex of faint, anastomosing, subparallel microridges that resemble patterns produced by creep of soft sediment when a load is placed on it, as by dumping of an overlying graded bed. The U- and J-shaped loops show a striking preferred orientation N55°E to S55°W, and there is a hint of a faint current lineation parallel to the long axes of the impressions.

Fresh outcrop is sparse in the region and the occurrence described is one of the few good bedding-surface exposures known. The lithology, nevertheless, is typical for the stratigraphic unit: fine-grained marine tuffs, normally graded bedding, fine-scale composite grading, and soft-sediment deformation contribute to a general picture of pelagic sedimentation, with episodic introduction of coarser graywackelike sediments by turbidity currents that flowed west-southwest down the slopes of volcanic lands to the east and northeast. Indeed, the prominently and compositely graded nature and mineral composition of the immediately overlying 6-cm bed imply that the structures described were buried by sediments of volcanic origin from a waning turbidity current in which the dumping of the bottom 2 cm was followed by a succession of weaker pulses of the same event.

Biogeological investigations

Photographs and one specimen of a J-shaped loop sent to Cloud in early 1974 led to an inconclusive search for similar published markings and for living organisms that might produce such markings. On a joint visit to the outcrop in June 1974 we observed a striking preferred orientation of what appeared on the lichen-covered and unevenly lighted outcrop as discrete, large, J- and U-shaped surface forms, noted evidence of pelagic sedimentation and sediment-bearing density currents, and hypothesized that the structures might be the imprints of the bodies or tubes of soft-bodied organisms, transported to their sites of burial by a weak or waning density current that, on deposition, wafted them into the shapes observed Potential analogs considered at that time included, besides a large uniserial alga of some kind, a synaptid holothurian, an elongate ctenophore, the Pogonophora, the Echiurida, the Sipunculida, the Phoronida, and the Annelida. An algal origin was judged improbable, but studies made at that time were not successful in ruling it out or in narrowing the possibilities among the Metazoa.

Microscopic examination of surface peels and residues revealed no dermal ossicles, jaw parts, or remains of tubes or bristles that might have helped in identifying the systematic position of the structures. Nothing definitive was revealed by study and photography under a variety of lighting and coating conditions, including rubber molds and plaster casts, of specimens remaining in the outcrop. Light and electron microscopy of chips, thin sections, and maceration residues has, as yet, produced no information of interest.

The possibility that these might be the oldest metazoan fossils yet known in the United States, if not in North America, called for steps to ensure their preservation. Arrangements were made to quarry out the entire 2 × 3 m slab bearing the markings and move it to a permanent repository for safe-keeping, display, and further study. Most of it was removed in May 1975, and the better half (Fig. 3) was placed on display in the U.S. Geological Survey's National Center at Reston, Virginia; the remainder was housed in the U.S. National Museum of Natural History in Washington, D.C.

All units were restudied and photographed, both dry and wet, by Cloud in the following October. Good low-angle light on the wetted display slab, now nearly free of lichens, revealed previously unrecognized details. Individual impressions are shown in Figures 6–11.

Characterization and interpretation

Close examination revealed a number of interesting features. The J- and opposite-facing U-shaped loops are not separate units but continuous, asymmetrical structures that extend beyond the U-bend into a complex coil of overlapping smaller loops. In some instances there is a small, subcircular depression (1 to 1.5 cm in diameter) near the center of the coil of smaller loops. In specimens 1 and 2 of Figure 3 (see also Figs. 6–7), this small depression coincides with the beginning of the larger trace in a way that suggests a connection and invokes the image of something long and wormlike or whiplike emerging from or returning to a burrow or attached to a holdfast. In the same two specimens, the coil of small loops has an essentially identical pattern: they seem to overlap one another at similar sites and to be folded on themselves like a kinked hose at two places (short arrows on specimens 1 and 2 of Fig. 3).

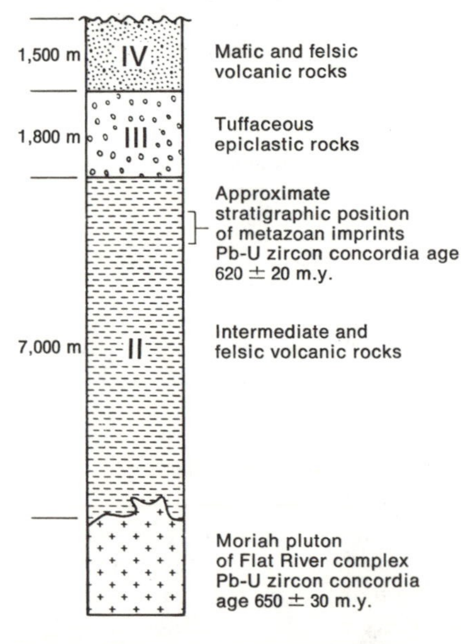

Figure 2. The approximate stratigraphic position of the metazoan fossils discussed here is in the upper portion of unit II of the generalized stratigraphic column for the Roxboro-Durham area, North Carolina. Pyroclastic rocks from this unit have been dated at ~620 m.y. old.

The markings are large—0.5 to locally as much as 2 cm wide and up to 1.1 m long from the beginning of the coil of small loops to the free end of the J-shaped loop of specimen 1 (Figs. 3, 7). The scaly-seeming texture is produced by unevenly crescentic to scalloped irregularities, convex away from the initial coil and toward the free end of the J-shaped loop. On the J-shaped limb shown in Figure 7, in fact, the "scales" seem to overlap. The margins of the impressions are sharply defined, and they bulge and contract irregularly along their lengths. Here and there the scalloped pattern fades, especially in the nearly straight limb that connects the U- and J-shaped loops, where the imprints tend to become dim and narrow as if being stretched. Points of apparent breakage (short arrow, specimen 4 of Fig. 3) and dual impression are discernible, and on specimens 3 and 4 (see also Fig. 11), it looks as if a linear body had rested briefly at one place and then shifted locally outward to a second position. Occasional bits of similar-textured, looping impressions (lower right of Fig. 3; Fig. 10) and faint incomplete imprints (left of specimen 3, Fig. 3) may represent fragments and temporary resting places of structures similar to those that made the main imprints. At the end of one small J-shaped impression is a faint possible extension that could be the imprint of a prostomium or folded crown of tentacles (center of Fig. 8).

In addition to these more prominent markings there are also some much narrower, long, curvilinear, but otherwise characterless impressions that may well be of biologic origin but about whose affinities it would be fruitless to speculate.

What can we infer from the above? The absence of branching would rule out most but not all algae of this size. In our experiments with unbranched stipes of kelp, the simulated action of the current failed to produce similar morphologies; apparently kelp stipes are both too stiff and too resilient to acquire through natural processes, or to retain, a coil of loops like those seen at the ends of several of our structures (top center Fig. 6, top right Fig. 7). Indeed it is hard to imagine how such a coil of loops could form except as a result of muscular contraction or the compression of a coiled burrow, for although coiled plants do exist, coiled algae of such dimensions are unknown to us, and these are most unplantlike coils. An algal origin thus seems highly improbable.

The orientation of the seemingly overlapping subcrescentic markings is reminiscent of polynoid surface scales, although the latter are limited to the dorsal surface. (Indeed, the observed markings seem very scalelike if we suppose the free end of the J-loop to be the tail of a wormlike animal.) The imprints also resemble a pattern of systematically backsweeping, chevron-shaped or subcrescentic tracks, such as are made by advancing sipunculids and other burrowing and crawling forms. A casual observer might simply call them "worm tracks" or horizontal burrows and let it go at that.

There is a superficial similarity between our structures and the aligned horizontal trace-fossil *Rhizocorallium irregulare,* illustrated by Fürsich (1975, his fig. 12B) from the upper Jurassic of Dorset, England. The alignment of that form, however, is presumably related to nutrient concentrations in ripple troughs and not to orientation by currents. The Dorset specimens, like all *Rhizocorallium,* show well-defined shift marks or "spreite" (Seilacher 1967), produced *within* the sediment between the limbs of the U-shaped burrow. Spreite are lacking in the North Carolina structures, which appear to be surface features. *Rhizocorallium* also generally displays arthropodlike scratch-markings, packing of ellipsoidal fecal pellets along the burrow walls, a mostly simple, straight-sided U-shape, and an association with relatively shallow-water sands, all of which have led to its being considered a crustacean burrow (Abel 1935; Lessertisseur 1955; Seilacher 1967; Sellwood 1970; Häntzschel 1975). Our fossils lack these features and are actually nothing like *Rhizocorallium.*

The striking preferred orientation of our structures might be analogized with the regular browsing patterns or feeding orientations of certain infauna, bathyal browers, and current-oriented suspension feeders whose feeding patterns are efficiently adapted to the exploitation of a dispersed or transient source of food. One can also imagine that the coils of loops at the ends of specimens 1 and 2 of Figure 3 might have been the result of compression of a nearly flat, upwardly spiraling burrow, like a flat *Daimonhelix.* And the poor marginal definition and scratchlike texture at the left side of the U-loop in Figure 9 are very suggestive of a crawl track.

All such interpretations, however, run into difficulty with the systematic asymmetry of these impressions, the substantial widening and narrowing along their length, the tapering at the free end of the J-shaped arm where it is preserved, the local evidence of stretching and rupture in the long limbs that join J- and U-bends, the superposition of specimen 4 on specimen 3 of Figure 3 (also Fig. 11), indicating bodily movement of the

Fig. 3. The configuration of the impressions discussed is as shown in this sketch, which was prepared from a mosaic of vertical photographs of the wet rock surface. The 0.9 × 1.3 m slab bearing the imprints illustrated was removed to the National Center of the U.S. Geological Survey, Reston, Virginia, where it is now on display. Numbered specimens, discussed in text, correspond to photographs in Figures 6–11.

Figure 4. The fossils were found on the surface of a 28- to 31-cm-thick bed of lithified and metamorphosed tuff, shown here in vertical profile. Note the convolute bedding of the lower 26–29 cm and the planar lamination of the upper 2–3 cm.

whole of specimen 4 over specimen 3, and the impressions that look like fragments (Fig. 10) and temporary or transient contacts (just left of specimen 3 of Fig. 3, under the numeral). This combination of characteristics would be most unusual for burrow, crawl track, feeding track, or resting imprint, for such structures do not undergo natural breakage, and evidence of nontectonic deformation or shifting could occur only in actual bodies or their imprints. Burrows do not pinch and swell, and burrows of the same kind rarely overlap.

Searching for alternative explanations, one is struck by the "current-swept" appearance of the structures and the fact that the thin overlying bed is conspicuously and compositely graded. The sorting characteristics of this bed imply that it was deposited by a waning turbidity current, the first pulse of which buried and perhaps imposed a preferred orientation on the organisms represented. Could these curious curvilinear impressions then be the current-swept remains of the soft bodies, proboscides, or slime tubes of organisms that remained attached within burrows at the center of the coil of small loops while ingesting detritus from the surrounding surface?

If they were material things and not peculiar tracks, they would have had to be fixed in some way or drag at both ends for the action of the current to generate the orientations observed. Flume experiments performed for us by David Pierce indicate that joined U- to J-shaped forms are seen temporarily when a long, limp structure like overcooked spaghetti, pinned at one end, drags toward the other (although when stipes of the kelp *Macrocystis* of similar diameters were the experimental material similar

Figure 5. Overlying the fossil-bearing surface was a 6-cm bed of compositely graded metatuff. The vertical-profile photograph clearly reveals that the bottom 2 cm is a single graded unit.

structures were not formed at the scale and current strengths observed). All experimental materials streamed directly down-current under continuing exposure to a strong flow. A waning turbidity current, however, might sweep soft and flexible tubular objects into the configurations observed, perhaps after dropping them from suspension, introduce fragments of others from upslope, and bury all beneath a lethal blanket of sediment that entombed a permanent record of the event in the form of the imprints described—provided there were a way to pin one end.

Such an origin would be consistent with the observed characteristics of both fossils and sediments, and it seems somewhat more probable than that trails or burrows would assume these strangely asymmetrical but consistently oriented patterns. The flume experiments indicate that the current would have been from N55°E to S55°W, which accords with paleogeographic data suggesting a volcanic borderland to the northeast. The pattern of coils and kinks, moreover, suggests that the organism may have coiled convulsively at one end, perhaps in resistance to further dislodgment, thus providing the terminal pinning mechanism.

The consistent shape and orientation of the J-shaped ends is another puzzling feature. Unless this is a primary feature of the imprints, it seems to call for a local pinning at the apex of the J, leading to deformation by the same current that arranged the other end into opposite-facing U's. Down-current movement of an entire unit is,

Figure 6. The photograph corresponding to specimen 2 of Figure 3 shows the typical organization of the fossil into an asymmetrical, omega-shaped loop, connecting at one end to a J-shaped loop and at the other to a coil of overlapping smaller loops. The arrow points to the small, subcircular depression near the center of the coil of loops. Coin is 2.5 cm in diameter.

of course, indicated by the superposition of specimen 4 on specimen 3 of Figure 3 (also Fig. 11). The pinning called for at the apexes of the J's could be a function of surface irregularities.

To investigate the nature of fixation at the coiled end, a 5-cm vertical core was centered on the depression at the middle of the overlapping coils of specimen 2 and sliced through it vertically. Contrary to our expectation, no evidence of a burrow was found. The top few millimeters show signs of disturbance, but, beginning at a level 7 mm below the surface, fine laminations in the rock directly beneath the depression show no sign of penetration down to the level of the convolute-bedding illustrated in Figure 4. It seems highly probable, therefore, that fixation at the coiled end was achieved not by actual attachment in a burrow, but by the muscular activity of a wormlike organism throwing itself into a coil of loops to resist dislodgment by external forces (and conceivably as an initial stage of burrowing back into sediments from which it had been dislodged upslope).

What kind of organism might produce such impressions? Of the possible analogs originally considered and not previously ruled out, we can now eliminate synaptid holothurians and ctenophores, because neither of them is capable of throwing itself into such a coil. Possible analogy with the Phoronida is seriously weakened by the characteristically small size (<20 cm long) of phoronids, the fact that they are not capable of coiling in the manner observed, and their restriction to shallow water. That leaves the Sipunculida, Annelida, Echiurida, and Pogonophora, all of which have deepwater representatives and attain sizes comparable to or exceeding those of our imprints (Hyman 1959; Grassé 1959; McIntosh 1922; MacGinitie and MacGinitie 1968; and Barnes 1974). The Sipunculida reach a maximum length of 55 cm with introvert extended (Grassé 1959, p. 786), extend only their short, muscular, reversible proboscis or introvert from their burrows while feeding, and retract it rapidly and directly into a stubby muscular body at the slightest disturbance. It is doubtful that they could leave such long, looping imprints.

Both the Echiurida and the Pogonophora attain lengths comparable to those of our fossils, and both also show variations in surface morphology that could produce a dragging effect such as might account for the J-bends. Both have simple and probably primitive anatomies, while certain slender "chitinous" tubes of early Paleozoic age may actually be the tubes of Pogonophora (Sokolov 1967). The prostomiumlike extension at the center of Figure 8 might even be interpreted as the impression of a tentacular cluster similar to that of the Pogonophora—although this is far too remote a possibility to be based on a single uncertain occurrence. A nearly straight, narrow groove extending some 13 cm across the rock to the left from this feature may be interpreted either as an extension of the prostomiumlike feature or as a groove in the sediment produced by the dragging of this feature across it. Alternatively, the entire structure to the left of the crescentically marked area may be a drag mark.

Even though the Pogonophora live in deep water, attain great lengths, and

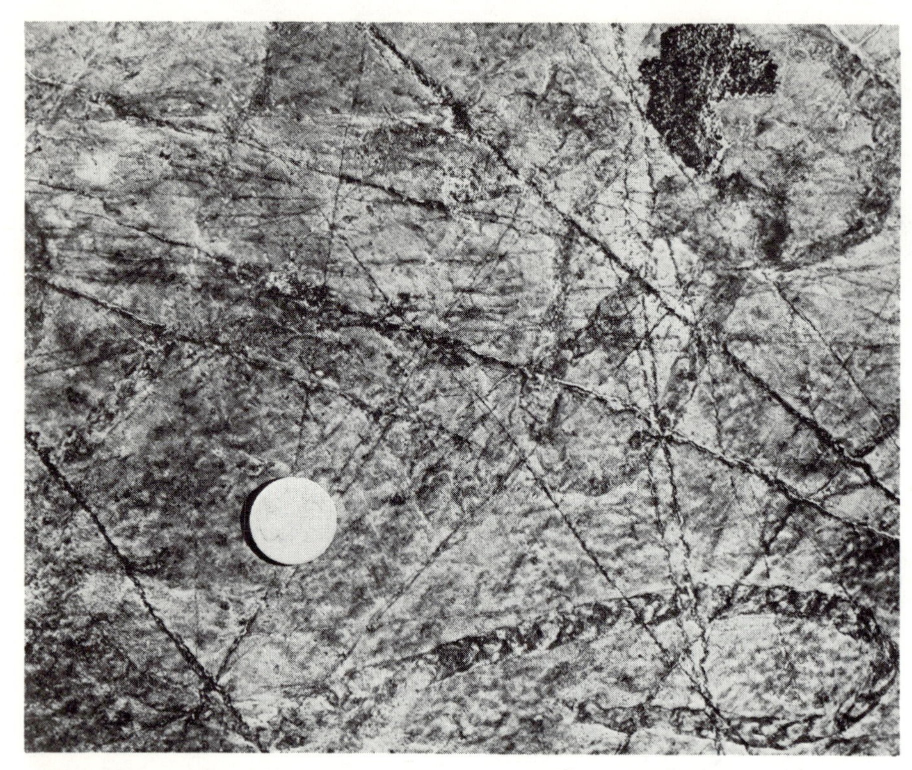

Figure 7. Corresponding to specimen 1 of Figure 3, the photograph illustrates the type of the species *Vermiforma antiqua* Cloud. The J-shaped limb clearly shows the scaly-seeming texture of the fossil imprints.

display a grossly suitable morphology, the fact that they are very slender, delicate animals seems to rule them out. The largest attain diameters of only about 2.5 mm (Hyman 1959), whereas our compressions reach widths of 1.5 to locally as much as 2 cm and presumably had maximal diameters before compression of slightly greater than 1.2 cm. Moreover, given the abundance of these creatures suggested by this locality, if they had lived in preservable tubes, one might expect to find some remains of the tubes on the surface or in the residues of hydrofluoric-acid macerations of the rock.

Analogy with the Echiurida remains plausible. The echiurids have a very extensible but noneversible proboscis, and some of them secrete remarkable mucus tubes of substantial length while feeding. Individuals of the Japanese species *Okeda taenioides*, with a trunk only 40 cm long, have a proboscis more than 1.5 m long, and some species of *Bonellia* with a trunk only 7 to 8 cm long have proboscides which, on expansion, are as much as 1 m long (Grassé 1959). The feeding position and pattern suggested for *Tatjanellia grandis* by Zenkevitch (in Grassé 1959) is not impossibly unlike one that might be envisaged for our fossils, and the proboscis morphology illustrated by Grassé and other authors for several echiurids allows much variation. Nevertheless, a burrow limited to the top 7 mm of rock hardly seems capable of accommodating the trunk of such an organism, and we find the analogy remote.

We also considered the possibility of affinity with some combination of worms and their mucus tubes. The feeding behavior of *Urechis* is described as follows by MacGinitie and MacGinitie (1968): "The worm begins secreting mucus, and as it continues to secrete more mucus it slowly backs down its burrow. This leaves a tube . . . with the proboscis and anterior end of the worm projecting into the mucus tube." It then strains water and particulate matter through the tube, from which it eventually withdraws, swallowing the tube for its contained nutrients. If echiurids, feeding in this way, were washed or shaken out of their burrows, carried in suspension by a turbidity current to their sites of burial, and there dropped, might they and their mucus tubes be swept by waning current flow into morphologies similar to

Figure 8. At the end of one of the J-shaped loops were found traces of what might be a prostomium or folded crown of tentacles, or perhaps simply a drag-mark descending to lower left from center of photograph.

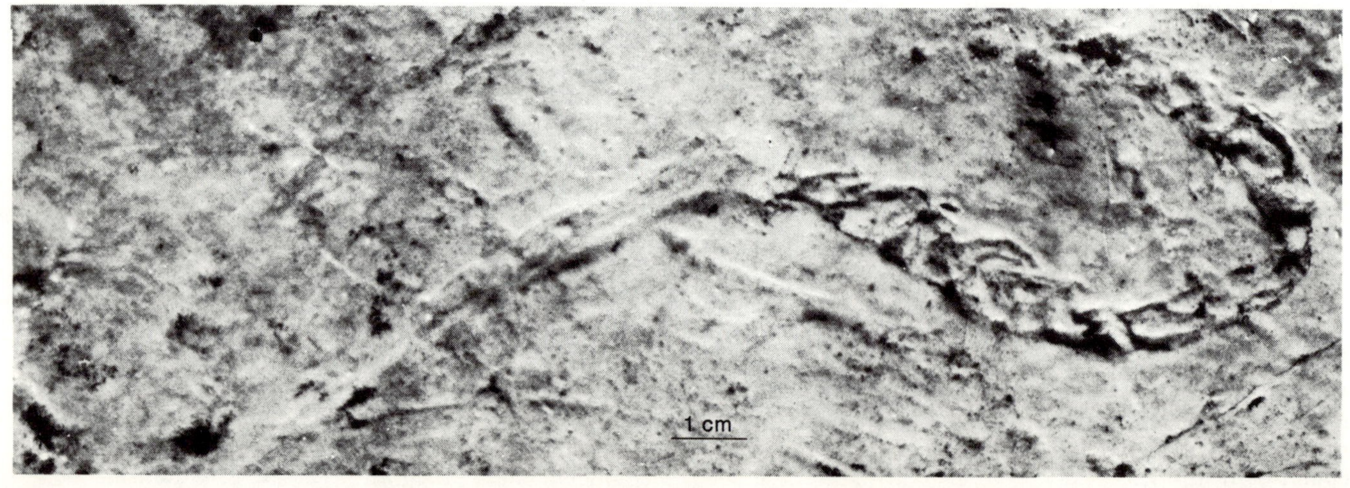

those observed? Most unlikely, it seems to us.

There remain the possibilities that these structures were annelid worms or that they represent a now-extinct phylum. In either instance, the features described would be consistent with their having been carried downslope when the sediments around them were thrown into suspension by the turbidity current whose energizing sediments now constitute the overlying bed, then dropped and buried as they set themselves to resist further displacement while the waning current first swept them into the alignments seen and then dumped the remainder of its load on them. Indeed it is possible that the convolute bedding shown in Figure 4 and the covering 6-cm graded unit shown in Figure 5 were both the result of a single seismic event, related to the demonstrable volcanism of the time.

In the end we find ourselves confronting a problem that is simply not solvable in detail with the information available. Although the probability seems very high that the structures do, in fact, represent the traces of ancient soft-bodied marine invertebrates, and that they are the imprints of the current-swept bodies of such organisms rather than similarly oriented burrows or tracks, we can only guess at systematic affinities. An annelidan affinity might best explain the segmentation suggested by the pattern of crescentic markings, although the echiurids are segmented in earlier ontogenetic stages and may have had segmented ancestors. An extinct phylum, of course, can have any morphology called for by the characteristics observed.

Whatever these organisms were, they evidently lacked preservable jaw parts, but they may have possessed a scaly surface ornamentation and may have been burrowers (although probably not tube-builders). Such an array of characteristics, although consistent with placement in the Annelida, does not dictate assignment to any specific living class, order, or family. Nor is it obvious why such large "worms," unless dead or greatly weakened, could not have escaped upward through perhaps 10 or 12 cm of uncompressed wet sediment to resume life at or beneath a new depositional interface.

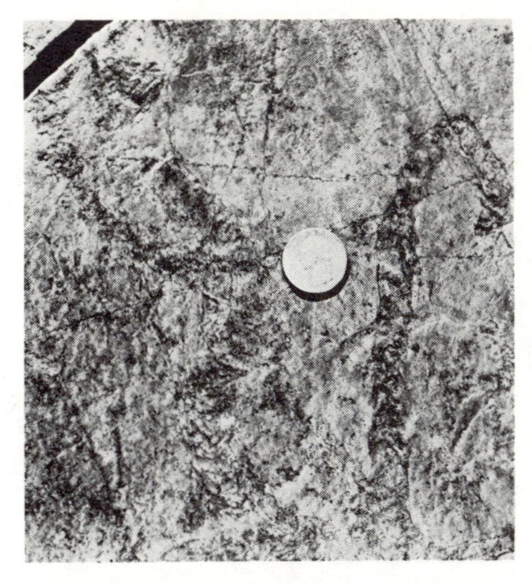

Figure 9. The scratch-like texture of the left side of the U-shaped loop in the photograph seems more like a track than an imprint, but the structure has the general configuration of the other imprints. This is specimen 6 of Fig. 3.

Figure 10. The impressions shown, from specimen 7 of Fig. 3, seem to be fragments of larger impressions.

Other Metazoa of comparable age

In 1948, when Cloud challenged the then-current dogma that Metazoa *must* have had a very long sub-Cambrian history, there were still many records of alleged Metazoa of great antiquity. Most of these have since been eliminated as either not Metazoa or not sub-Cambrian (e.g. Cloud 1968; Glaessner 1972; Stanley 1976). Nevertheless, the subsequent discovery of the Ediacarian fauna suggests that there was indeed an interval in earliest Phanerozoic and Paleozoic time during which the soft-bodied Metazoa evolved, and which preceded the general appearance of a typical Cambrian fauna of calcareous shelly invertebrates.

We earlier noted the near-global distribution of Ediacarian types of fossils and the age of about 680 m.y. of dated occurrences in England and the USSR. In North America the record is more uncertain. Although it contains no recognizable Ediacarian elements, the remarkable soft-bodied "Mistaken Point" fauna reported by Anderson and Misra (1968; see also Misra 1969), from the as yet unsatisfactorily dated Conception Group of southeastern Newfoundland, may be of about the same age. This fauna seems to occur stratigraphically above a diamictitic ("mixtitic") and probably ice-related sequence similar to those that elsewhere around the Atlantic margin appear to be mainly >600 and <800 m.y. old. This relationship is consistent with a very early Paleozoic age, comparable to the earlier-noted 680-m.y.-old fossiliferous volcaniclastic sediments of Leicestershire and the 620-m.y.-old fossils here described. Anderson (1972) estimates the age of the Mistaken Point fauna to be between 610 and 630 m.y.

Another report of Metazoa of possibly comparable age is the recent announcement by Allison (1975) of a flatworm from the Tindir Group of east central Alaska—a mainly or wholly pre-Phanerozoic unit. Questions arise, however, concerning both the age and the affinities of this fossil.

The geology of the discovery locality, on Washington Creek about 10 km airline upstream (south) from the Yukon River, is shown in general terms on the U.S. Geological Survey's 1/250,000 "Geologic Map of the Charley River Quadrangle, East-Central Alaska," by Earl E. Brabb and Michael Churkin, Jr. (1969). This map shows the outcrop area to be in the so-called "Basalt and red beds" unit of the Tindir Group, referred to as "Upper Precambrian." No outcrop of the "Basalt and red beds" unit has been observed, however, between the Washington Creek exposures and the supposedly equivalent Yukon River outcrops of that unit—a distance of over 16 km in a structurally complex region.

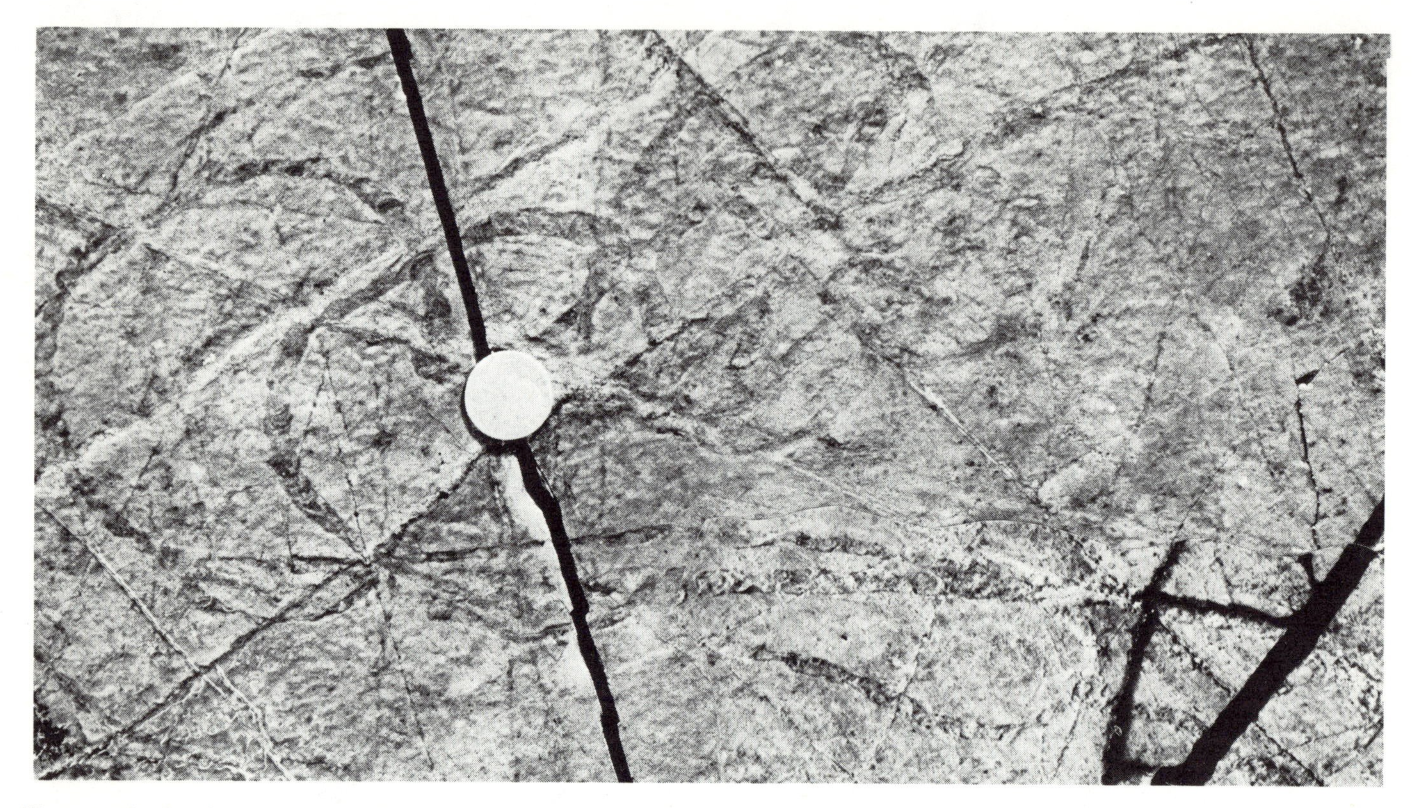

Figure 11. In the photograph of the two superimposed specimens labeled 3 and 4 in Figure 3, the straight limb connecting the loops becomes dim, narrow, and offset, as if being stretched.

There is, of course, a degree of similarity between the outcrops on Washington Creek and the type Tindir, otherwise two geologists as experienced as Brabb and Churkin would not have correlated them. Both the Washington Creek outcrops and the type Tindir contain basalts, dark shales, and cherts; but pillow lavas are typical of the "Basalt and red beds" unit of the Tindir, whereas neither pillows nor red beds are seen in the supposed Washington Creek equivalents. Thus, on visiting the Washington Creek outcrops after a 12-day field review of the type Tindir with Gary Kline, who is restudying that sequence, Cloud concluded, as did Kline on a later visit, that, *if* they are Tindir, they are either a basinal facies of it or perhaps younger than any part of the type Tindir.

The tiny, tubellarian-shaped fossil described by Allison (1975) as *Brabbinthes* and interpreted by her as a flatworm poses another kind of problem. Although Cloud saw this 0.5-mm-long fossil briefly in Fairbanks, reviewed Allison's paper, found no reason at the time to question her identification, and indeed urged her to publish, he now doubts that this fossil is a flatworm. Sections from his own collections in the same area show specimens of similar tone and texture that suggest joined, multiple, tapering, subcylindrical spiculelike structures with circular central canals.

Reviewing Allison's paper and excellent photographs of *Brabbinthes* raised some puzzling questions—about the shape and length of the "gut," for instance. Although it is not impossible that this might be a very primitive feature, the configuration is atypical for any living flatworm (D. E. Costello, pers. comm.). A thin section that just happened to cut close to the axis of one ray of certain hexactinellid spicules while cutting obliquely through three other, perhaps curving, rays of the same plane and centering on the intersection of vertical rays would yield a morphology similar to *Brabbinthes*. Although the peculiar medial bulging and distal tapering of *Brabbinthes* is not a spicular configuration familiar to Cloud, on examination of the photograph at his request, Robert Finks (pers. comm.) supported a hexactinellid affinity and noted that such spicules do, in fact, commonly show similar swelling and tapering. A hexactinellid assignment is also supported by Willard Hartman (pers. comm.). Finks adds that, although the axial canals of recent hexactinellid spicules are square in cross section, they become enlarged and rounded on fossilization.

At Allison's invitation, Cloud reexamined her thin section, in which she drew attention to the presence of other yet undescribed structures including delicate needlelike stauractines of a type that is fairly common in Cambrian rocks elsewhere (e.g. Finks 1970; Rigby and Gutschick 1976). Thus, although the possibility cannot be completely ruled out that *Brabbinthes* really is some strange primitive flatworm whose only known example just happened to be cut by a thin section that centered almost exactly on the gut and medial axis, that seems unlikely. It is statistically much less probable than that the alignment of the thin section was along the axial canal of one ray of a sponge spicule which is one of several from the same locality.

On the other hand, the interpretation here preferred—that the organism is a hexactinellid spicule—is of only marginal assistance with an age assignment. Although the earliest *known* hexactinellids are early Cambrian (Finks 1970), neither Finks nor Cloud would be greatly surprised to find them in sub-Cambrian (i.e. Ediacarian) rocks. Many thin sections and other preparations were studied in search of conclusive evidence, but no Radiolaria, dinoflagellates, Chitinozoa, or other organisms clearly indicative of a Phanerozoic age were found. We did find, in one thin section, a thin sliver of laminated, chitinous-looking material with a reticulate or chambered structure that was examined for us by J. M. Schopf, who expressed the opinion that it was most probably metazoan and would suggest a Cambrian or later age.

The matter of age then, remains unsettled. Assuming that it is possible to obtain a good radiometric age on the chloritized, fragmental, and weathered volcanic rock on Washington Creek, and that the rock turns out to be >680 or even >600 m.y. old, the biologic affinities of all organisms present will become of very great interest. The presence of some sponge spicules is the one thing that seems well established, whatever *Brabbinthes* may be. Although we find a post-Ediacarian Paleozoic age most probable, therefore, it is not possible at this writing to say even that with a high degree of confidence.

It remains to note briefly two other reports of very ancient North American Metazoa. Young (1972) records trace fossils from the upper Miette Group of east-central British Columbia that are probably both metazoan and very old—probably as old as or a bit older than those here described (Cloud 1973; Javor and Mountjoy 1976). Unfortunately, however, we do not yet have a firm radiometric age for these Miette fossils. The last record we wish to note here is the report by Chapman (1975) of possible burrows from the Longarm Quartzite, low in the Ocoee Supergroup of North Carolina. A sub-Cambrian or even a pre-Phanerozoic age seems likely for this occurrence. It is hard to imagine what these gigantic (to 2.5 cm wide and 2 m long), reportedly branching structures might be without seeing them in outcrop. However, the fact that they occur in parallel on a joint face (Chapman, pers. comm. and photographs) along which there could have been slippage suggests the possibility that they might be secondary, perhaps chatter-mark-like, features.

Phylum unknown, *cf.* Annelida
Class unknown
Order and Family unknown
Genus *Vermiforma* Cloud, n. gen.
Figures 3, 6–11

Description: Wormlike surface impressions up to 1.1 m long from end to end and as much as 1.5 to 2 cm wide display scalloped, perhaps scaly texture and local attenuation and bulging of cross section. Impressions with 1 to 2 mm negative relief imply some body thickness. Coiled at one end, the impressions pass beyond a sharp flexure into an omega-shaped loop and continue through a contiguous J-shaped loop to a narrowing free end. They are interpreted as compressed, soft-bodied, originally tubular, probably detritus-feeding, toothless marine invertebrates, perhaps related to annelids. One flatworm-shaped specimen only a few centimeters long may be a juvenile or possibly a different taxon.

Type: Vermiforma antiqua Cloud.

Age and occurrence: Pb-U zircon concordia age 620 ± 20 m.y. Basal Phanerozoic and Paleozoic (Ediacarian) in the sense of Cloud (1973 and in press). Found on bedding surface of thinly bedded metatuff of map unit II of Glover and Sinha (1973), on the south (right) bank of the South Fork of Little River, about 16 km N. and slightly W. from Durham, North Carolina (Durham N. 15-minute quadrangle).

Species *Vermiforma antiqua* Cloud, n. sp.
Figures 3, 6–11

Description: As for the genus, except for one flatworm-shaped specimen only a few centimeters long, comprising a single J-shaped loop, tentatively included as a juvenile of *V. antiqua* but possibly representing a wholly different taxon.

Type: Specimen illustrated in Figure 7, U.S. National Museum of Natural History number USNM 216221. This is associated with other samples (as indicated in Fig. 3 and detailed in Figs. 6–11) on a single large block currently on display at the National Center of the U.S. Geological Survey in Reston, Virginia.

Age and occurrence: As for the genus.

Figure 12. Taxonomy of *Vermiforma antiqua* Cloud n. gen. n. sp.

Thus the Carolina slate belt fossils, described here as *Vermiforma antiqua* Cloud n.gen. n.sp. (Fig. 12), constitute the oldest record yet known from North American rocks that is both well dated and convincing as to metazoan (eumetazoan) biology.

References

Abel, O. 1935. *Vorzeitliche Lebensspuren.* Gustav Fischer.

Allison, C. W. 1975. Primitive fossil flatworm from Alaska: New evidence bearing on ancestry of the Metazoa. *Geology* 3(11):649–52.

Anderson, M. M. 1972. A possible time span for the late Precambrian of the Avalon Peninsula, southeastern Newfoundland, in the light of worldwide correlation of fossils, tillites, and rock units within the succession. *Canadian J. Earth Sci.* 9:1710–26.

———, and S. B. Misra. 1968. Fossils found in the Precambrian Conception Group of southeastern Newfoundland. *Nature* 220: 680–81.

Barnes, R. D. 1974. *Invertebrate Zoology,* 3rd ed. W. B. Saunders.

Briggs, D. F., M. C. Gilbert, and L. Glover III. In press. Petrology of the Roxboro Metagranite, North Carolina. *Geol. Soc. America Bull.*

Chapman, J. J. 1975. Possible fossil burrows from the Precambrian of North Carolina. *Geol Soc. America, Abstracts* 7(4):477.

Cloud, P. 1948. Some problems and patterns of evolution exemplified by fossil invertebrates. *Evolution* 2(4):322–50.

———. 1968. Pre-metazoan evolution and the origins of the Metazoa. In *Evolution and Environment,* ed. E. T. Drake, pp. 1–72. Yale Univ. Press.

———. 1973. Possible stratotype sequences for the basal Paleozoic in North America. *Am. J. Sci.* (273):193–206.

———. In press. Major features of crustal evolution. *Trans. Geol. Soc. S. Africa.*

Finks, R. M. 1970. The evolution and ecologic history of sponges during Paleozoic times. *Zool. Soc. London, Symp. no. 25,* pp. 3–22.

Ford, T. D. 1958. Pre-Cambrian fossils from Charnwood Forest. *Proc. Yorkshire Geol. Soc.* 31:211–17.

———. 1972. The oldest fossils. *New Scientist* 15(297):191–94.

Fürsich, F. T. 1975. Trace fossils as environmental indicators in the Corallian of England and Normandy. *Lethaia* 8:151–72.

Glaessner, M. F. 1972. Precambrian palaeozoology. Univ. Adelaide, Centre for Precambrian Research, Spec. Paper no. 1, pp. 43–52.

———, and Brian Daily. 1959. The geology and late Precambrian fauna of the Ediacara fossil reserve. *South Australian Mus. Rec.* 13(3): 369–401

Glover, L., III, and A. K. Sinha. 1973. The Virgilina deformation, a late Precambrian to Early Cambrian (?) orogenic event in the central Piedmont of Virginia and North Carolina. *Am. J. Sci.* 273-A:234–51.

Grassé, P.-P. 1959. *Traité de Zoologie,* vol. 5. Masson.

Hantzschel, W. 1975. Trace fossils and problematica: Treatise on invertebrate paleontology, Part W, Miscellanea, Supplement 1.

Hyman, L. H. 1959. *The Invertebrates,* vol. 5. McGraw-Hill.

Javor, B. J., and E. W. Mountjoy. 1976. Late Proterozoic microbiota of the Miette Group, southern British Columbia. *Geology* 4(2): 111–19.

Lessertisseur, J. 1955. Traces fossiles d'activité animale et leur signification paléobiologique. Soc. Geol. de France, Mém. 74.

MacGinitie, G. E., and N. MacGinitie. 1968. *Natural History of Marine Animals,* 2nd ed. McGraw-Hill.

McConnell, K. J. Geology of the late Precambrian Flat River Complex and associated volcanic rocks near Durham, North Carolina, Ph.D. diss., Virginia Polytechnic Inst. and State Univ., 1974.

McIntosh, W. C. 1922. *A Monograph of the British Annelids, vol. 4, pt. 1, Polychaeta—Hermellidae to Sabellidae.* Roy. Soc. London.

Misra, S. B. 1969. Late Precambrian (?) fossils from southeastern Newfoundland. *Geol. Soc. America Bull.* 80:2133–40.

Rigby, J. K., and R. C. Gutschick. 1976. Two new lower Paleozoic hexactinellid sponges from Utah and Oklahoma. *J. Paleont.* 50(1):79–85.

Seilacher, A. 1967. Bathymetry of trace fossils. *Mar. Geol.* 5:413–28.

Sellwood, B. W. 1970. The relation of trace fossils to small-scale sedimentary cycles in the British Lias. In *Trace Fossils,* ed. T. P. Crimes and J. C. Harper, pp. 489–504. Seel House Press.

Sokolov, B. S. 1967. Ancient pogonophorans. (In Russian.) *Doklady Akad. Nauk USSR* 177(1):201–04.

Stanley, S. M. 1976. Fossil data and the Precambrian-Cambrian evolutionary transition. *Am. J. Sci.* 276:56–76.

Termier, H., and G. Termier. 1960. L'Ediacarien, premier étage paléontologique. *Rev. Gen. Sci. et Bull. Assoc. Français Avan. Sci.* 67(3–4):79–87.

Wright, J. E. Geology of the Carolina slate belt in the vicinity of Durham, North Carolina. M.S. thesis, Virginia Polytechnic Inst. and State Univ., 1974.

Young, F. G. 1972. Early Cambrian and older trace fossils from the southern Cordillera of Canada. *Canadian J. Earth Sci.* 9:1–17.

John Pojeta, Jr.,
Bruce Runnegar

Fordilla troyensis and the Early History of Pelecypod Mollusks

Early Cambrian fossils from New York State provide important clues to the evolution of the class

Fordilla troyensis is the oldest known pelecypod, or bivalve mollusk. A diverse and ancient branch of the animal kingdom, pelecypods now appear to have arisen from rostroconch mollusks about 540–570 million years before the present. They began their first major adaptive radiation and diversification about 500 million years ago.

Pelecypods are characterized by a shell made up of right and left valves which are hinged along the dorsal side of the animal; typically the dorsal margin has several crenulations known as teeth and sockets. The shell is opened by a ligament and the foot and is closed by adductor muscles. The foot is controlled by pedal and intrinsic muscles. The shell is composed of calcium carbonate secreted by flaps of tissue, known as the mantle, which surround the other soft parts and are attached to each valve along insertion areas called the pallial line.

John Pojeta, Jr., a geologist for the United States Geological Survey, joined the Survey's Branch of Paleontology and Stratigraphy after receiving his Ph.D. from the University of Cincinnati in 1963. His primary research interests are the systematics and phylogenetic relationships of Paleozoic pelecypods and Ordovician biostratigraphy. He is a Research Associate in paleobiology, Smithsonian Institution, and an Associate Professorial Lecturer at George Washington University.
Bruce Runnegar, Associate Professor of geology at the University of New England, Armidale, Australia, joined the faculty of the university after receiving his Ph.D. from the University of Queensland in 1967. His major research interests are in Permian pelecypods and stratigraphy and Cambrian mollusks. He was a Post-Doctoral Fellow in paleobiology, Smithsonian Institution, 1969–70.
Address for Dr. Pojeta: U.S. Geological Survey, E-501 U.S. National Museum, Washington, DC 20244.

Using the scars left on the shell by the various kinds of muscles and the pallial line, paleontologists can reconstruct at least some of the soft parts of fossil pelecypods.

Estimates of the number of living species of pelecypods range from 6,300 (Boss 1971) to 15,000 (Nicol 1969); there are about 7,000 known post-Cambrian Paleozoic species, over 15,000 known Mesozoic species, and at least as many Cenozoic species as Mesozoic forms. Thus, throughout most of Phanerozoic time, when fossils are abundant in the rock record, pelecypods were an important element of the world's biota. They are especially abundant in marine Mesozoic and Cenozoic rocks. In the Paleozoic they are less common in shelly deposits, although in certain beds and at certain places they are as numerous as in younger rocks.

Exciting new discoveries have recently led to breakthroughs in our understanding of early pelecypod evolution and diversification. It has been established (Pojeta 1971) that all the major lineages (subclasses) of pelecypods had evolved by Middle Ordovician time (about 450 my; see Fig. 1). By the end of Ordovician time (about 430 my), the class Pelecypoda had undergone a major adaptive radiation through which most of the modes of life seen in younger forms had been developed. The only two life habits that are not yet known in the Ordovician are cementing of the shell to the substrate and swimming (Fig. 3). Because the Ordovician was the time of primary radiation and diversification of the Pelecypoda, the pelecypod fossils of this period are especially important in understanding the phylogeny, systematics, and general development of the class.

Earliest Ordovician (Tremadocian—about 500 my) pelecypod faunas are known only from Afghanistan, Argentina, and France. They are all probably members of the subclass Palaeotaxodonta, having numerous small vertical or nearly vertical teeth along the hinge line, and are related to such living forms as *Nucula* Lamarck and *Yoldia* Möller (see Fig. 2). By the end of Early Ordovician time (Arenigian—about 475 my) seven families of pelecypods belonging to three subclasses are known from Argentina, France, Malaysia, Sweden, and Wales. By the end of Middle Ordovician time all known subclasses of pelecypods—Palaeotaxodonta, Heteroconchia, Isofilibranchia, Pteriomorphia, and Anomalodesmata—are present in the fossil record, and the class had a worldwide distribution. Common names of living examples of these subclasses include, respectively: nut clams; cockles, Venus clams, razor clams, river mussels, soft shell clams, surf clams, and jewel boxes; marine mussels; ark shells, spiny oysters, pearl oysters, scallops, pen shells, and oysters; and Pandora clams.

The importance of *Fordilla*

The diversity of the pelecypods in the Ordovician suggests two possible alternatives for their early phylogeny: (1) the group may have undergone a more or less long period of Cambrian evolution or (2) a newly evolved group whose morphology represented an adaptive breakthrough may have diversified

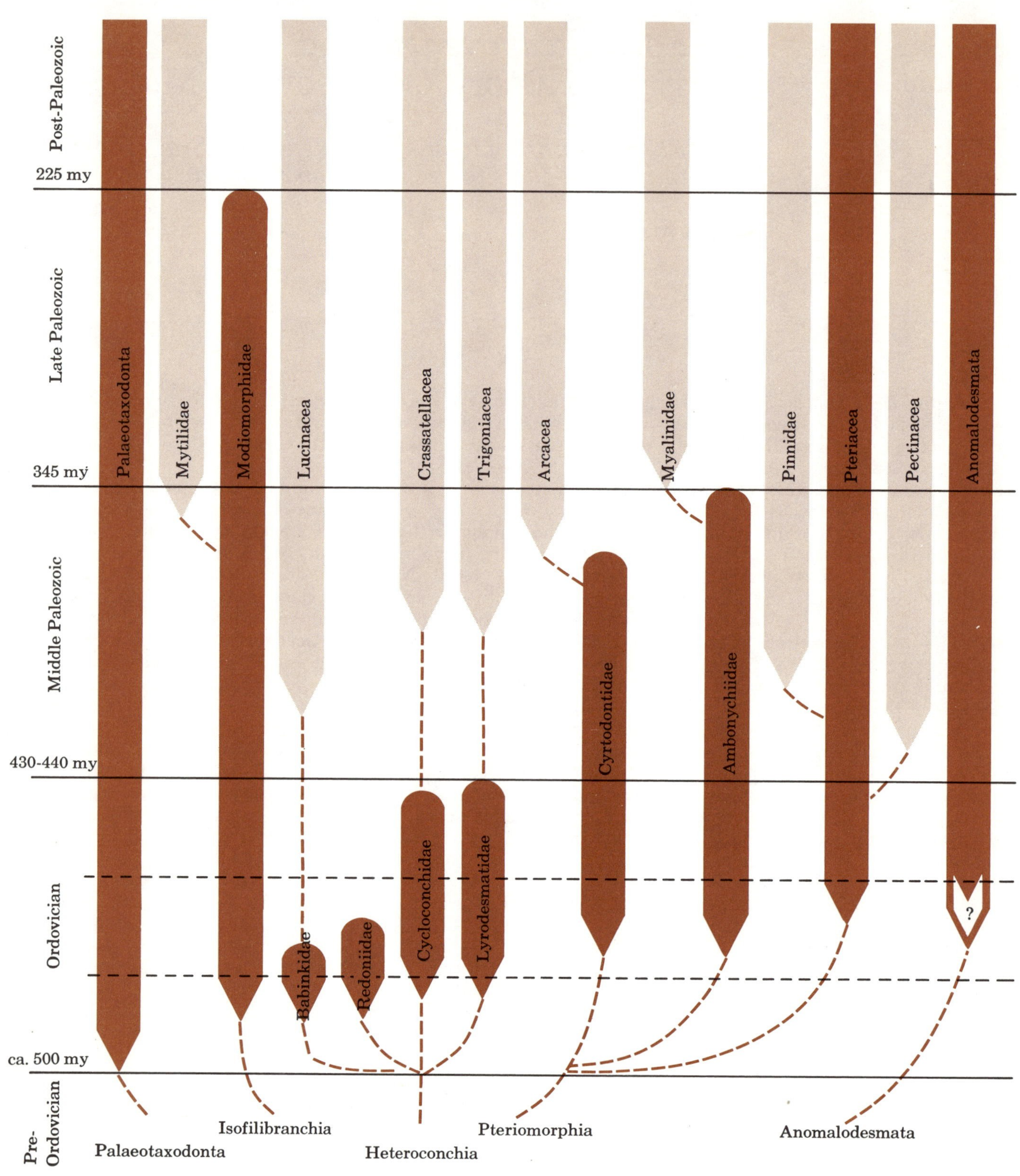

Figure 1. This sketch of the phylogenetic relationships of Paleozoic pelecypods reveals that all major lineages (subclasses) had evolved by Middle Ordovician time. Lineages in dark color originated in the Ordovician; light-colored lineages have post-Ordovician origins. (Adapted from Pojeta 1971.)

without a long period of Cambrian evolution. Until recently the second possibility seemed the more likely, as there was no known unequivocal record of Cambrian (pre-Tremadocian) pelecypods.

Pelecypods other than *Fordilla* have been reported from the Cambrian by various authors, beginning as early as 1855. However, all such reports have been discounted (Pojeta, in press) for a variety of reasons. Some of the reported occurrences were later found to be Ordovician in age, some were misidentifications of other bivalved animals, and most are too poorly preserved to show conclusive pelecypod characteristics. Thus the recent discovery of specimens of *Fordilla troyensis* Barrande, with muscle scars clear and intact, in the Lower Cambrian rocks of the Taconic Mountains of

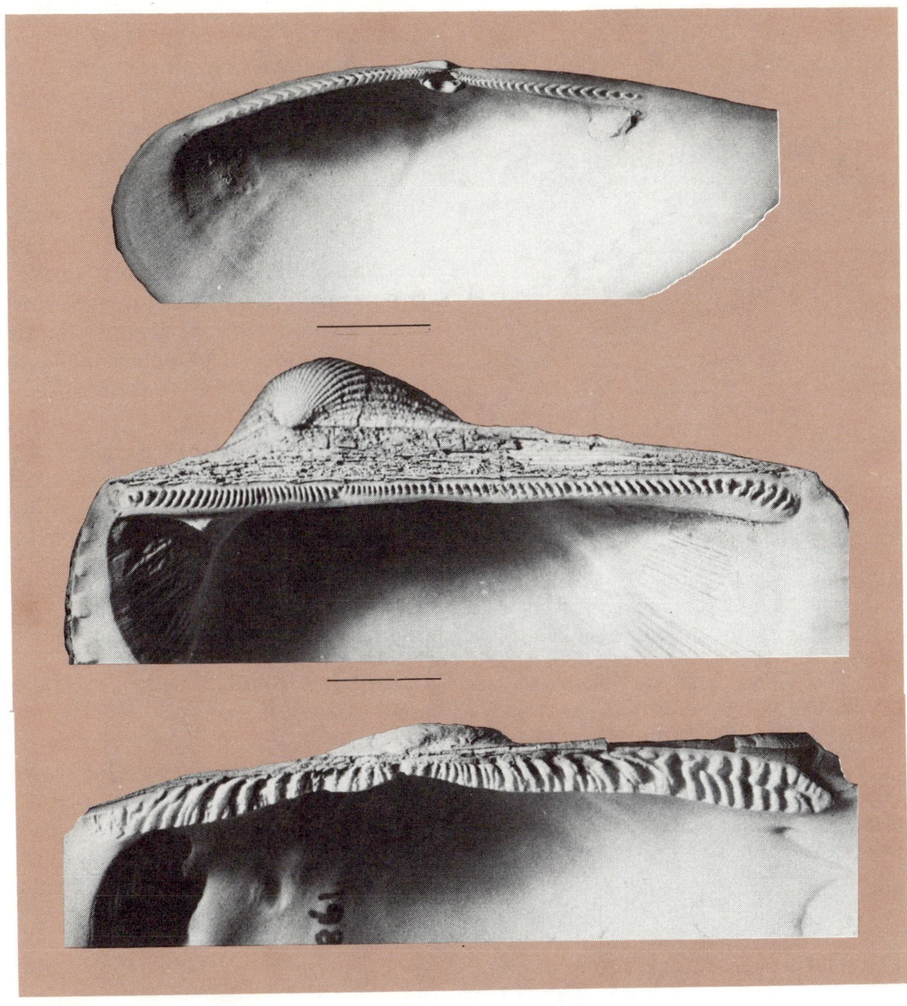

Figure 2. A recurrent feature of pelecypod evolution, taxodont dentition developed independently in the three major lineages shown here. *Top to bottom:* Recent palaeotaxodont *Yoldia limatula* Say; Recent arcacean *Anadara notobilis* Röding; Recent unionacean *Iridina ovata* Swainson. All valves shown are right valves; bars = 1 cm.

eastern New York State, provides the first unambiguous documentation of the presence of the class Pelecypoda in the Cambrian system (Pojeta, Runnegar, and Kříž 1973).

Fordilla is a small bivalved animal that possesses a musculature consisting of anterior and posterior adductor muscles, pedal retractor muscles, and a pallial line. The shell is equivalved, is ornamented only with comarginal growth lines, has erect ligamental areas, and lacks anterior and posterior lateral teeth (Figs. 4 and 5). Most authors (Ulrich and Bassler 1931; Raymond 1946; Kobayashi 1972) had previously regarded *Fordilla* as a bivalved crustacean arthropod (conchostracan). In the hierarchy of classification, the phylum Arthropoda is equal in rank to the Mollusca and includes all those animals having a jointed body and appendages encased in a scleritized exoskeleton, such as the trilobites, crustaceans, insects, spiders, centipedes, and millipedes. Some of the crustaceans are bivalved and externally look a great deal like pelecypods. Thus, until the musculature of *Fordilla* was known, it was not possible to place it unequivocally in the Pelecypoda. Now that the musculature of *Fordilla* is known, it is clear that it is a pelecypod; further, it is the oldest known pelecypod found to date, and all known occurrences are from Lower Cambrian rocks (540–570 my).

In the United States *Fordilla* is presently known from the *Elliptocephala asaphoides* assemblage of the Taconics of eastern New York State, which is regarded as late Early Cambrian in age by Lochman (1956) and Rasetti (1967). Elsewhere in North America it occurs in the Lower Cambrian of Newfoundland and the Lower Cambrian Bastion formation of Greenland (Poulsen 1932). Outside of North America the only well-documented occurrence of *Fordilla* is from the Lower Cambrian rocks of Bornholm, Denmark (Poulsen 1967). *Fordilla* has been tentatively identified from the Lower Cambrian rocks of England (Cobbold 1919) and Portugal (Delgado 1904), but these occurrences are uncertain. Rozanov et al. (1969, p. 35) indicated the occurrence of *Fordilla* in the Lower Cambrian rocks of Siberia in a diagram. Of all

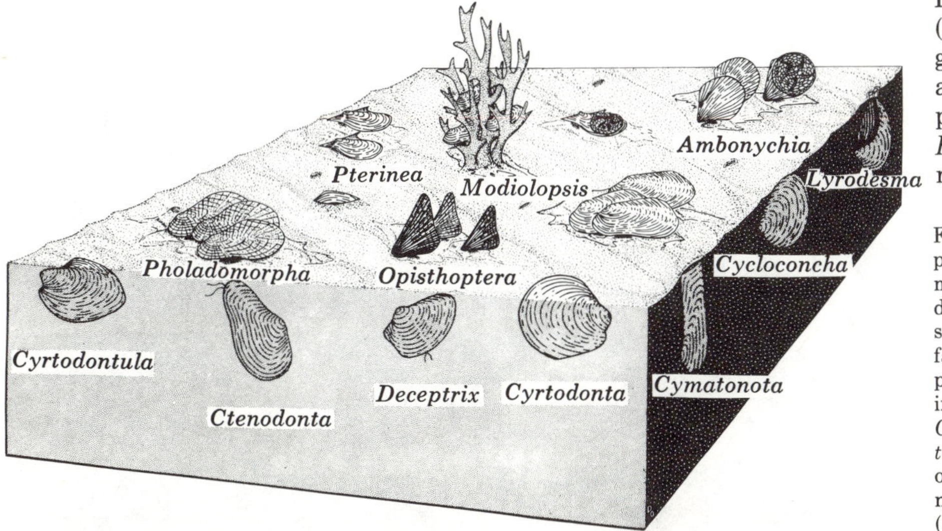

Figure 3. By the end of the Ordovician, pelecypods had developed most of the modes of life that characterize the class down to the present. This life-habit reconstruction of the Late Ordovician pelecypod fauna shows representative genera in their probable life positions. Infaunal or burrowing forms are represented by *Cyrtodontula, Ctenodonta, Deceptrix, Cyrtodonta, Cymatonota, Cycloconcha,* and *Lyrodesma.* The others are epifaunal forms. Specimens are not drawn to relative or actual scale. (Adapted from Pojeta 1971.)

the reported occurrences of *Fordilla*, however, only the specimens from New York State preserve the muscle scars.

Fordilla's descendants

Morphologically *Fordilla* resembles younger pelecypods that are adapted for infaunal (burrowing) life. *Fordilla* is laterally compressed, has a well-developed anterior end with attendant musculature, and shows no sign of a byssus—a structure composed of a series of threadlike filaments used by pelecypods for attaching themselves to the substrate; pelecypods which are epifaunal (living on the substrate) often possess a byssus. The posterior portion of the pallial line of *Fordilla* is expanded and moniliform and is similar to that of some younger forms, all of which live infaunally. The general morphology of *Fordilla* also suggests a suspension-feeding mode of life rather than a deposit-feeding one. Suspension-feeders remove food particles from the water in which they are immersed, whereas deposit-feeders obtain their food from the sediment of the substrate.

There is a stratigraphic gap of at least 40 million years in the fossil record of pelecypods between the occurrence of *Fordilla* in the Early Cambrian and the Early Ordovician, when pelecypods underwent a major adaptive radiation and taxonomic diversification. There are no known Middle and Late Cambrian pelecypods. The existence of gaps in the fossil record of various groups of organisms is not uncommon; among the pelecypods there is a significant gap in the history of lucinaceans of over 100 million years between their occurrence in the Devonian and their reappearance in the Mesozoic. The limpet-shaped monoplacophoran mollusks show a gap of over 300 million years in their record, between the youngest known Devonian fossils and living forms. The coelacanth crossopterygian fishes are last known as fossils in the Cretaceous although they still exist in modern seas, a gap in their record of over 60 million years. Perhaps the gap in the Cambrian record of pelecypods can best be explained by thinking of the Cambrian forms as living in deeper, cooler waters; because the rock record is dominated by shallow-water sediments, there would be less chance of finding Cambrian pelecypods. Following this line of reasoning, the pelecypods become abundant in the Ordovician record because this represents the time when they invaded warm shelf and epicontinental seas.

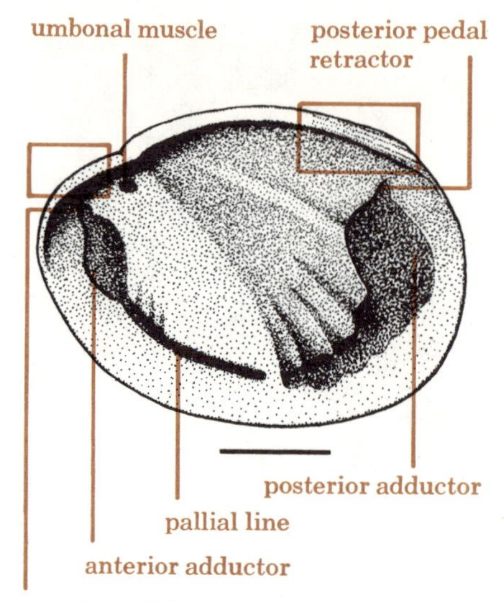

Figure 4. This composite reconstruction of the interior of a right valve of *Fordilla troyensis* shows the muscle insertion areas. Portions of the hinge within colored rectangles have been seen and are known to lack teeth; the uncolored portion was missing from the specimens we have seen. Bar = 1 mm. (Adapted from Pojeta, Runnegar, and Kříž 1973; © 1973 by the American Association for the Advancement of Science and published with their permission.)

The general shell shape of *Fordilla* is similar to such Arenigian cycloconchid heteroconchs as *Actino-*

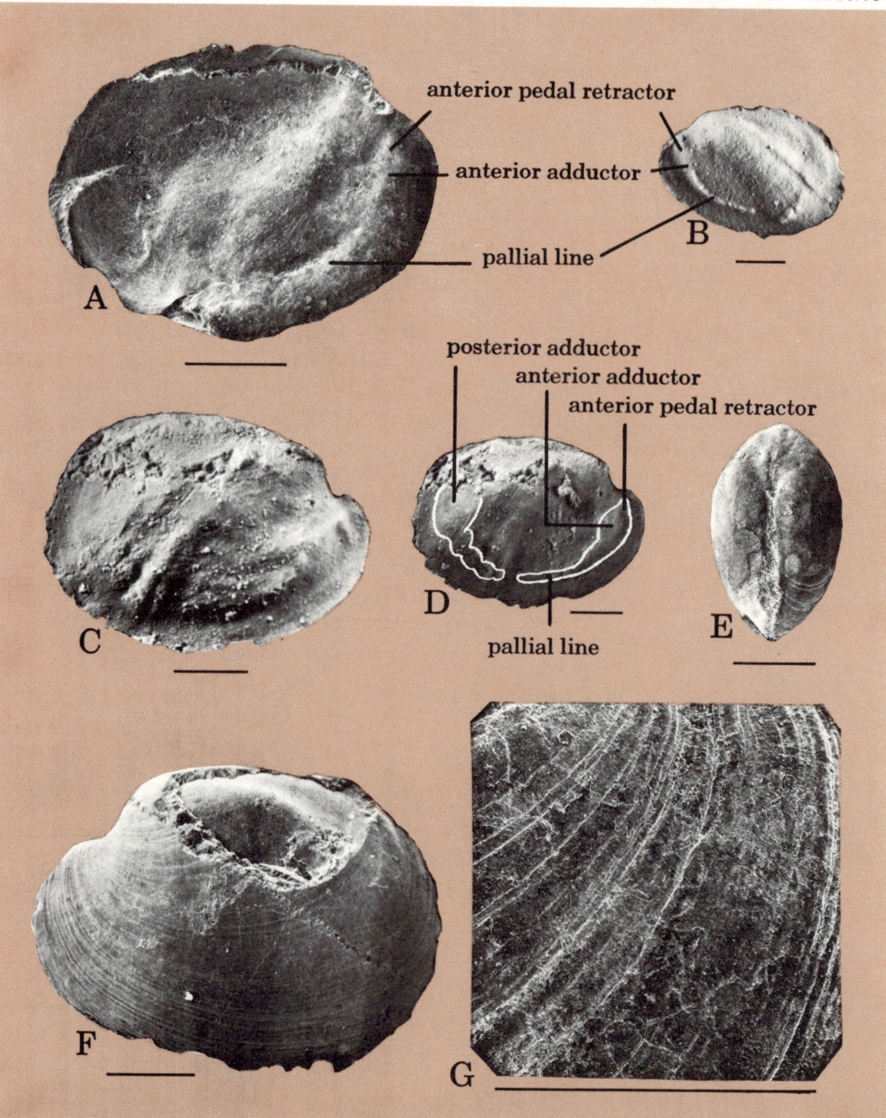

Figure 5. Specimens of *Fordilla troyensis* Barrande from Lower Cambrian rocks, Troy, N.Y., illustrate the morphology of the species. *A.* Interior of a left valve showing muscle scars, anterior portion of the pallial line, and lack of anterior and posterior lateral teeth. *B.* Internal mold of a left valve, showing muscle insertion areas. *C.* Internal mold of a right valve with muscle scars. *D.* Same specimen as in *C*, with muscle scars outlined. *E.* Dorsal view of the only known articulated specimen, showing equal convexity of the two valves. *F.* Exterior of a left valve showing growth lines. *G.* Enlargement of the anterior portion of a right valve exterior, showing growth lines. Bars = 1 mm.

donta secunda (Salter) and such Middle and Upper Ordovician forms as *A. naranjoana* (de Verneuil and Barrande) and various species of the genus *Cycloconcha* Miller (Fig. 6). In addition to shell shape, some cycloconchids have a broadly inserted pallial line reminiscent of the broadly inserted pallial line of *Fordilla*. Because of these similarities *Fordilla* is regarded as being ancestral to the family Cycloconchidae. Pojeta (1971) related the families Lyrodesmatidae, Redoniidae, and Babinkidae to the Cycloconchidae, all of which were united in the order Actinodontoida; the 4 families constitute the known record of Ordovician heteroconchs.

Relating the other known Ordovician pelecypods to *Fordilla* is more difficult. It has been postulated by Yonge (1962) that the retention of the byssus in adult pelecypods is a neotenous or paedomorphic feature (involving the maintenance of a juvenile characteristic in the adult stage). Members of the subclass Pteriomorphia have a byssus as adults and could be derived neotenously from the actinodontoids, as most Ordovician pteriomorphs have a well-developed dentition comparable to the teeth of some actinodontoids. Isofilibranch pelecypods also have a byssus, but Ordovician forms do not have well-developed teeth and it seems unlikely that they were derived from the actinodontoids. Isofilibranchs may have arisen directly from *Fordilla* by the neotenous retention of the byssus in the adult, or they may be descended from as yet unknown Cambrian forms.

The development of taxodont dentition is a recurrent feature of pelecypod evolution, occurring independently in the palaeotaxodonts, arcaceans, and unionaceans (Fig. 2). In the arcaceans (Cox 1959) taxodont teeth developed from a dentition consisting of cardinal and lateral teeth. Many unionaceans also have dentition consisting of cardinal and lateral teeth. By inference, the taxodont dentition of palaeotaxodonts could have arisen in the same way from the cardinal and lateral teeth of actinodontoid heteroconchs. The morphology of the Ordovician representatives of the fifth subclass, the Anomalodesmata, is poorly known and their relationship to other Ordovician pelecypods and *Fordilla* is obscure; some have a modioliform shape and may be descended from the isofilibranchs.

Figure 6. During the Cambrian and Ordovician the Pelecypoda evolved through *Fordilla troyensis,* as shown in this postulated phylogeny. Names at the top of the figure indicate the modern pelecypod groups that arose from various Ordovician ancestors.

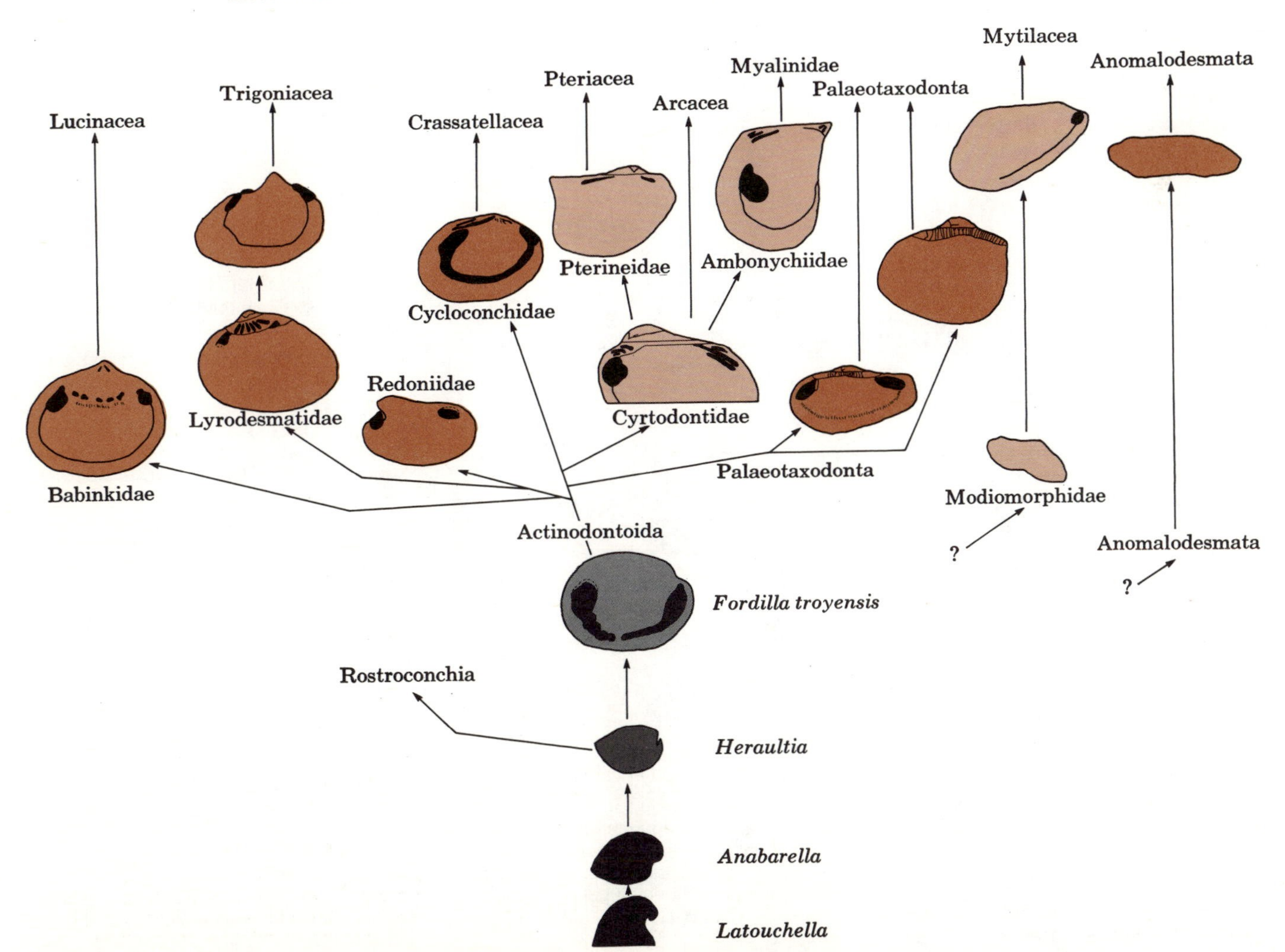

In summary, it is postulated that *Fordilla* gave rise to the cycloconchids and through them to the rest of the Actinodontoida. The byssate ambonychiid, cyrtodontid, and pterineid pteriomorphs of the Ordovician have well-developed cardinal and lateral teeth and could have been derived from the actinodontoids paedomorphically by the neotenous retention of the larval byssus throughout life. Ordovician byssate isofilibranchs may have been derived from the actinodontoid stem stock before dentition became well developed, by the neotenous retention of the byssus, directly from *Fordilla* by the same process, or from as yet unknown Cambrian forms. The palaeotaxodonts may have been derived from the actinodontoids by a rearrangement of dental elements. The origin of the Anomalodesmata remains obscure; it may be related to the isofilibranchs.

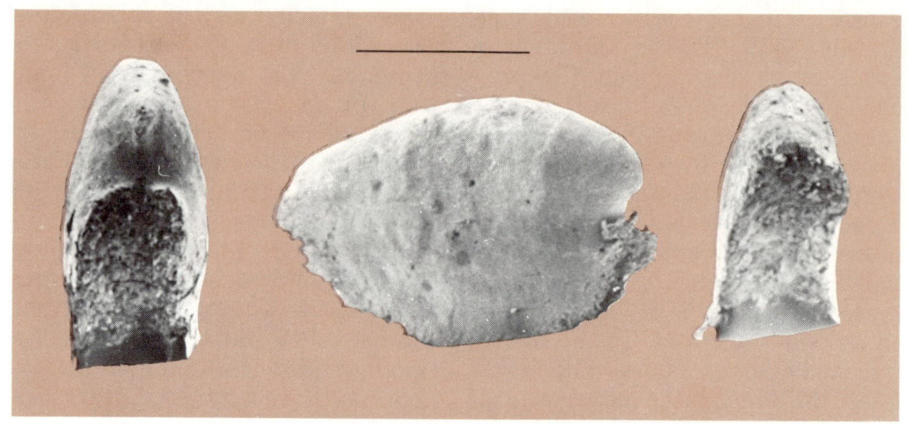

Figure 7. Three views of *Heraultia varensalensis* Cobbold show, *left to right*, anterior face with gape; right lateral face; and posterior face with gape. Both *Heraultia* and *Fordilla* have laterally compressed shells of about the same size and with similar lateral profiles, and both are Early Cambrian in age. Bar = 1 mm.

Forerunners

Runnegar and Pojeta (in press) note the similarities of *Fordilla* to the oldest known rostroconch mollusk *Heraultia* Cobbold (Fig. 7). Rostroconchs are a separate class of bivalved Paleozoic mollusks which differ from pelecypods in their univalved larval shell, valve articulation, and musculature. Both *Heraultia* and *Fordilla* have laterally compressed shells of about the same size, with similar lateral profiles, and both are Early Cambrian in age. They differ in that *Heraultia* has a univalved larval shell, lacks a dorsal commissure as an adult, and is thus pseudobivalved.

Heraultia is in effect a laterally compressed univalve with anterior, posterior, and ventral shell gapes. It can be related to such laterally compressed monoplacophorans as *Anabarella* Vostokova (Fig. 8), which have a curved ventral margin resulting in the development of enlarged anterior and posterior shell gapes. We envisage such forms as *Anabarella* giving rise to *Heraultia,* and *Heraultia* in turn, by mutation, producing a flexible hinge in the larval shell, thereby giving rise to *Fordilla*. The main difference between pelecypods and rostroconchs is that the pelecypod shell is bivalved from the beginning of its formation. Once a flexible ligament was established in the larval shell, all adult features of a pelecypod would follow in short order. This change probably occurred in the Early Cambrian when *Fordilla* evolved from *Heraultia* or from some closely allied form.

Figure 8. In this view of the left side of *Anabarella* sp., lateral compression and the extent of the shell gape along anterior and posterior faces are indicated by the curved ventral margin. Bar = 1 mm. (Photograph courtesy of Robin Godwin and S. C. Matthews.)

References

Boss, K. J. 1971. Critical estimate of the number of Recent Mollusca. Harvard Univ. Mus. Comparative Zoology, *Occ. Papers on Mollusks* 3:81–136.

Cobbold, E. S. 1919. Cambrian Hyolithidae, etc., from Hartshill in the Nuneaton District, Warwickshire. *Geol. Mag.* 56:149–57, pl. 4.

Cox, L. R. 1959. The geological history of the Protobranchia and the dual origin of taxodont Lamellibranchia. *Mal. Soc. London Proc.* 33:200–09.

Delgado, J. F. N. 1904. Faune Cambrienne du Haut-Alemtejo (Portugal). *Communicacões da Commissão do Servico Geologico de Portugal* 5:307–74, pls. 1–6.

Kobayashi, T. 1972. On the discontinuities in the history of the order Conchostraca. *Japan Acad. Proc.* 9, pt. 1:725–29.

Lochman, C. 1956. Stratigraphy, paleontology, and paleogeography of the *Elliptocephala asaphoides* strata in Cambridge and Hoosick Quadrangles, New York. *Geol. Soc. America Bull.* 67:1331–96, 10 pls.

Nicol, D. 1969. The number of living species of Molluscs. *Systematic Zool.* 18:251–54.

Pojeta, J., Jr. 1971. Review of Ordovician pelecypods. *U.S. Geol. Survey Prof. Paper* 695:46, 20 pls.

———. In press. *Fordilla troyensis* Barrande and early pelecypod phylogeny. *Bulls. American Paleont.*

Pojeta, J., Jr., B. Runnegar, and J. Kříž. 1973. *Fordilla troyensis* Barrande: The oldest known pelecypod. *Science* 180:866–68.

Poulsen, C. 1932. The Lower Cambrian faunas of east Greenland. *Medd. om Grønland* 87, no. 6:66, 14 pls.

———. 1967. Fossils from the Lower Cambrian of Bornholm. *Mat. Fys. Medd. Danske Vid. Selsk* 36, no. 2:48, 9 pls.

Rasetti, F. 1967. Lower and Middle Cambrian trilobite faunas from the Taconic sequence of New York. *Smithsonian Misc. Colls.* 152:111, 14 pls.

Raymond, P. E. 1946. The genera of fossil Conchostraca—An order of bivalved Crustacea. Harvard Univ. *Mus. Comparative Zoology Bull.* 96:217–307, 6 pls.

Rozanov, A. Yu., et al. 1969. Tommotian Stage and the Cambrian lower boundary problem. *Acad. Sci. USSR Trans.* 206:319, 55 pls.

Runnegar, B., and J. Pojeta, Jr. 1974. Molluscan phylogeny: The paleontological viewpoint. *Science* 186:311–17.

Ulrich, E. O., and R. S. Bassler. 1931. Cambrian bivalved Crustacea of the order Conchostraca. *U.S. National Mus. Proc.* 78:1–130, 10 pls.

Yonge, C. M. 1962. On the primitive significance of the byssus in the Bivalvia and its effect in evolution. *Marine Biol. Assoc. United Kingdom Jour.* 42:113–25.

Robert M. Linsley

Shell Form and the Evolution of Gastropods

Snail shells coil in many ways, reflecting varied patterns of adaptation to the problems of internal and external water flow

Several phyla, including echinoderms, arthropods, and mollusks, reveal during the first period of their recorded history a large assortment of bizarre forms. These fossils from the Cambrian period (500–600 million years ago) have long been regarded as "experiments" on the basic plans of animals, too peculiar to be understood by direct comparison with modern taxa and too ephemeral to shed much light on the "successes" in the history of life. Recently, however, newly discovered or reinterpreted mollusks from this period have been arranged into a phylogenetic scheme, with several of them put close to the ancestry of living forms. The fossils have been considered largely as geometrical forms that, by elongation, flattening, coiling, or notching of the edges, produced the molluscan array of the Cambrian, including the extant classes.

But the varied shapes of these shells cannot be understood apart from the functional adaptations of the animals that secreted them. Understanding of the functional significance of shell form is necessary to evaluate the plausibility of the transitions suggested by the proposed phylogenies. And although our conception of that significance in living gastropods is very crude, generalizations based on the study of living snails allow us to see the earliest known snails as more than mere geometries.

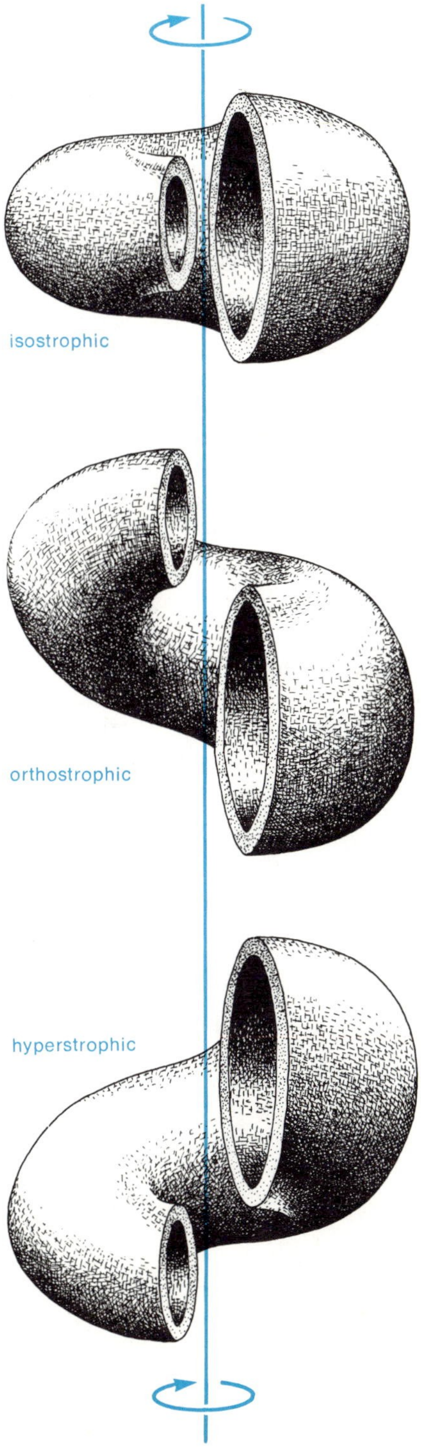

Gastropods are set apart from other mollusks by torsion, which is an embryological rotation of the viscera and the coiled shell with respect to the axis of the foot of the animal. Torsion in primitive mollusks is presumed to have been a 180° rotation of the shell and visceral hump relative to the axis of the foot. In more advanced gastropods, torsion may be as little as a 90° rotation; in fact, in some gastropods, such as sea slugs, it may be only a vestigial, larval process, completely disappearing in adults. Torsion is accomplished by the abortion (or differential development) of one of a pair of embryonic retractor muscles, frequently with the aid of subsequent differential growth (Crofts 1937). Because one muscle begins pulling before the other, the shell and its included flesh are rotated relative to the fleshy foot. The mantle cavity, which initially points posteriorly, is redirected anteriorly, and opens over the head. The result is that the anus, which is initially posterior, is swung to an anterior position, and organs that were on the right side of the animal end up on the left side.

isostrophic

orthostrophic

hyperstrophic

Figure 1. There are three principal kinds of coiling in gastropod shells: isostrophic, where movement of the generating curve along the axis of coiling is zero; orthostrophic, where movement is *down* the axis of coiling; and hyperstrophic, where movement is *up* the axis of coiling. Isostrophic coiling forms bilaterally symmetrical shells. The orthostrophic and hyperstrophic forms illustrated here are right-handed variants; left-handed variants—mirror images of the right-handed ones—also occur.

Robert M. Linsley holds the Whitnall Chair of Geology at Colgate University, having served on the staff of the Geology Department there since 1955. He received his training at the University of Michigan. Most of his research has focused on Paleozoic gastropods, but he has spent the last four years photographing and observing living gastropods in hopes of gaining insight into the fossils. The author is particularly indebted to William Kier for the colored photographs in this article. Address: Department of Geology, Colgate University, Hamilton, NY 13346.

The disposition of soft parts after torsion lays down the architectural facts of life for snails. The ways in which snails accommodate (and take advantage of) the twisted condition produced by torsion, in how they respire, move, protect themselves, and rid themselves of waste, can be read in the ways they build their shells. And the aperture of the shell is particularly expressive of the snail's story. All of a snail's business with the outside world and all of its size increase during growth are confined to that often differentiated margin. It is the hole through which we get to know the donut.

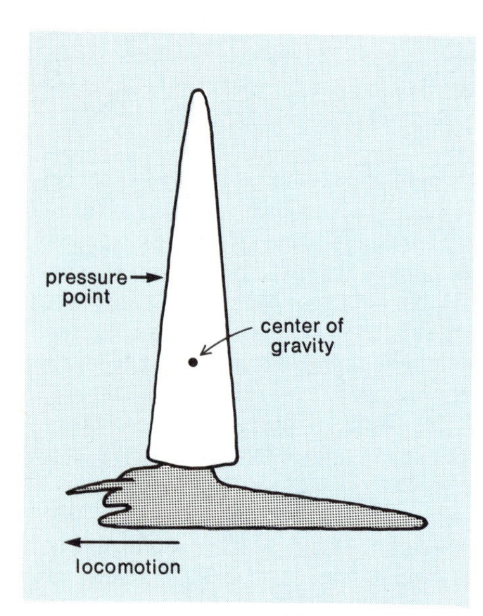

Figure 2. This imaginary mollusk with an uncoiled shell would have great problems moving through water, since both the pressure point and the center of gravity are very high and the frontal cross section is very large.

Shell form

Within regularly coiled gastropods we can recognize two principal coiling types: isostrophic forms, which are bilaterally symmetrical; and anisostrophic forms, which are not bilaterally symmetrical (Fig. 1). All living gastropods have anisostrophic shells, although a few converge toward isostrophism. Anisostrophic shells may in turn be divided into two major groups. The vast majority of living gastropods have orthostrophic shells, in which translation—the direction of the coiling—is *down* a central axis. Hyperstrophic shells, in which translation is *up* the axis of coiling, are relatively rare in modern gastropods, but were far more common in the early history of mollusks. The trivial geometric point of whether translation is up or down the axis of coiling results in rather significant biological differences.

Each of the anisostrophic forms exists in dextral (right-handed) and sinistral (left-handed) variants. The concepts of dextral and sinistral originally had the same meaning in reference to coiling in gastropods that they had in the physical sciences, but have evolved to mean something quite different. Dextral and sinistral gastropods are mirror images of each other both internally and externally, whereas hyperstrophic and orthostrophic forms are not. The "handedness" of gastropods is ultimately determined by the way torsion occurs. In a dextral gastropod it is the right retractor muscle that remains active and causes torsion, swinging the spire of the shell around in a counterclockwise direction when viewed from above the animal. In a sinistral gastropod it is the left retractor muscle that activates torsion, pulling the spire in a clockwise direction. Relatively few modern gastropods show sinistral coiling, and for the sake of simplicity, all forms discussed in this paper are dextral unless otherwise noted.

The primary function of the gastropod shell is to provide protection, be it from predation, from desiccation, from excessive water motion, or from noxious features of the environment. Probably the primitive molluscan ancestor could not withdraw deeply into its shell, but adhered to the substrate with its foot and pulled the shell down over exposed parts of its body against a solid substrate for protection. All modern gastropods that move actively have this ability to "clamp" their shell for protection. To be effective in clamping the shell, the plane of the aperture must be tangential to the ventral portion of the body whorl (see Fig. 3); this is indeed the geometric shape seen in the majority of living snails. In addition, most gastropods are capable of a second line of defense, withdrawing the soft parts deep into the shell and sealing the aperture with a plate (the operculum). A third defensive technique is locomotion, and escape locomotion is frequently a modified form of the normal locomotor pattern.

We are accustomed to regarding snails as the exemplar of nature's sluggish creatures. In fact, this large class of mollusks includes a wide variety of activity levels and diverse mechanisms of locomotion (Miller 1974). I have clocked some snails at the "mind-boggling" speed of more than 1 cm/sec, while others cement themselves to the substrate and are immobile throughout their adult life.

Gastropods use either cilia or a wide

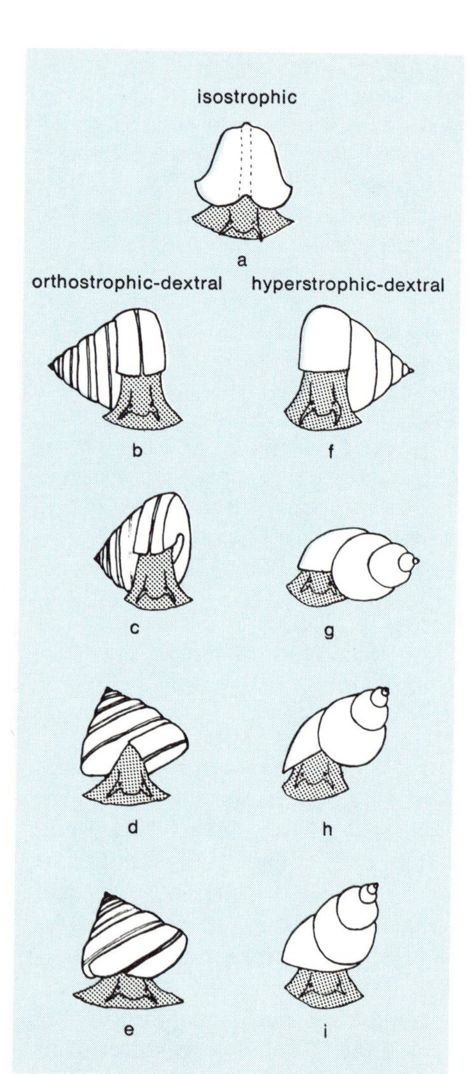

Figure 3. Isostrophic shells (a) pose no problem of balance to the snail. Anisostrophic shells do, however, and these problems are solved differently by orthostrophic and hyperstrophic forms. In a right-handed orthostrophic shell (b) the spire is displaced to the snail's right side. To bring it into balance, the spire is first swung backward by regulatory detorsion (c), then tipped upward by inclination of the axis of coiling (d). The aperture is then modified so that it is tangential to body whorl (e). In a right-handed hyperstrophic shell, in which the spire is initially displaced to the snail's left side (f), detorsion swings the spire forward (g); again, inclination and formation of a tangential aperture complete the process (h and i). (From Linsley 1977.)

Figure 4. *Haliotis assimilis* Dall. is a fairly fast-moving snail. Its shell is positioned almost entirely by inclination, but the slow translation rate, the rapid whorl-expansion rate, and the large displacement of the whorl from the coiling axis give it a hydrodynamically efficient shell. The shell was photographed over a mirror so that both sides are visible.

Figure 5. *Entemnotrochus rumphii* Schepmann, a living pleurotomariacian with an exceptionally deep anal slit, has a medium-spired shell and is therefore a relatively slow snail. During the Paleozoic era Pleurotomarians were abundantly distributed throughout the shallow waters of the world's seas, but today they are restricted to depths in excess of 300 m.

range of muscular contractions of the foot to propel themselves. Oddly enough, the varied mechanisms of locomotion do not correlate well with rate of motion in gastropods. A better correlation can be made between shell form and speed than between locomotion mechanism and speed. Intuitively it seems unlikely that shape-induced drag on a snail shell should be significant; yet an empirical correlation has been obtained between streamlined shape and the speed of the animal (Linsley, in press). I have made casual analyses of the main geometric properties of shells and kinds of ornaments that relate to "streamlining." In general, the more ornate the shell, the slower the snail; the more streamlined the shape, the faster.

Streamlining is partly dependent on the way the shell is carried during locomotion. In order to appreciate some of the problems of carrying a shell through a fluid medium, even at low speeds, imagine a preposterous mollusk carrying a long, conical shell straight up over its back (Fig. 2). This shell obviously offers a very large cross-sectional area, relative to its volume, for the fluid medium to act against. As a result, the pressure point of the water acting against the shell (essentially the geometric center of the cross-sectional area of the shell) will be located far above the substrate, and even well above the center of gravity of the organism. A great deal of energy will therefore be required to prevent that shell from tipping over behind the animal during motion. The tendency to topple will be enhanced by the shearing effect between the locomotor organ (typically the sole of the foot) and the inertial mass of the shell, located far above the foot.

Coiling of the shell will obviously help resolve these problems (Fig. 3). It will lower the center of gravity and the pressure point, reduce the total surface area of the shell subject to fluid drag, reduce the shearing effect, and thus reduce the energy necessary to move the shell forward. A bilaterally symmetrical isostrophic shell with the axis of the coiling parallel to the substrate would have its center of gravity positioned over the midline of the foot automatically. But in right-handed orthostrophic coiling the shell spire projects to the right side of the foot and introduces imbalance. Shell balance may be restored by one of two processes: the spire can be tipped up (inclination) or the spire can be twisted toward the rear so that torsion is reduced (regulatory detorsion). In fact, all orthostrophically coiled gastropods use a combination of these two processes to move the center of gravity over the midline of the foot and to the lowest possible position (Linsley 1977).

Figure 6. The shell of *Terebra maculata* Linne is so high spired that it can't be held over the body, and the animal drags the shell behind itself instead.

The various modes of shell coiling are not all equal in their effects. To analyze some problems attendant upon shell form, consider three examples of shells with a circular generating curve: very low-spired shells (haliotiform) (Fig. 4); medium-spired shells (turbiniform) (Fig. 5); and high-spired shells (turritelliform) (Fig. 6). David Raup and coworkers have provided the conceptual framework for quantitative analysis of the four elements that constitute shell form: (1) shape of the generating curve, (2) rate of expansion of the generating curve, (3) distance of the generating curve from the axis of coiling, and (4) translation of the generating curve up or down the coiling axis. In Raup's parameters the major difference in the three examples would be the amount of translation down the axis of coiling, but there are attendant differences in rate of expansion of the generating curve and distance of the generating curve from the axis of coiling. The greater the translation rate, the less the whorl expansion rate, and the closer the whorls are to the axis of coiling.

In the lowest-spired forms, such as *Haliotis* (Fig. 4), the shell is positioned almost entirely by inclination, with slight amounts of regulatory detorsion. The center of gravity and the cross-sectional area presented in forward passage of the shell are both very low, and the pressure point is at about the same level as the center of gravity. *Haliotis* is capable of relatively high speeds. The major adaptive restriction of this form of shell is that the rate of expansion is so large and the shell is therefore so wide open that *Haliotis* cannot withdraw into it and seal the aperture with an operculum. Clamping is thus its only means of defense, which restricts it to a hard substrate.

A medium-spired snail, such as *Entemnotrochus* (Fig. 5), balances its shell with about a 60° inclination and 20° detorsion. As a result, the center of gravity of the shell is quite high, the pressure point is even higher, and its front profile is asymmetrical, causing torque because the spire side of the shell offers greater resistance (and high-placed resistance) than the basal side. The speed achieved by these gastropods can only be modest at best. Viewed in this light, shells of intermediate shape would seem to have the worst of all possible worlds, but speed of movement is only one factor of survival in the world of gastropods.

As the spire height of the shell increases it is possible for the plane of the aperture to converge on the axis of coiling, thus reducing the amount of inclination necessary to maintain a tangential aperture and increasing the amount of regulatory detorsion. All of these factors serve to lower the center of gravity and the pressure point and to reduce the amount of forward cross-sectional area and the amount of torque caused by unequal pressure acting on the asymmetrical shell. Thus snails with moderately high-spired shells are speedier than those with medium-spired shells.

As spire height continues to increase, however, the center of gravity of the shell moves progressively farther and farther from the aperture (or columellar muscle) until it is no longer practical for the animal to hold the shell up off of the substrate, as in *Terebra* (Fig. 6). The apical part of the shell then rests on the substrate and the snail becomes a shell dragger rather than a shell carrier. These are among the slowest of gastropods.

Terebra actually represents only one category of shell draggers. Others include forms such *Conus,* in which the aperture is positioned off to one side so that the shell rests on the substrate beside the foot, and *Architectonica* (Fig. 7), which has a radial aperture. In all of these locomotion is accomplished by the animal pushing its foot ahead on the substrate and then dragging the shell up to the foot. These are the real slowpokes among the snails. For long periods of time they do not move, and when they do, locomotion is very labored.

Therefore, from these geometrical considerations, helically coiled snails with round apertures cannot be counted among the fastest of the gastropods. An exception to this generalization is the naticid gastropods, which, when moving, essentially internalize their shell by enveloping it with their foot and mantle. As a result they become perfectly bilaterally symmetrical animals when in motion, with a low center of gravity, small frontal cross section, and low pressure point.

Two other groups of gastropods fall outside this geometrical sequence: the motionless snails and the leapers. Motionless snails usually are recognized easily because either there is an attachment scar where the shell was cemented to the substrate (*Petaloconchus,* Fig. 8) or the shell has disjunct or open coiling (*Siliquaria,* Fig. 9) which makes it too awkward to carry around (Peel 1975).

The remarkable leapers, such as the *Xenophora* on the cover or *Strombus,* are diverse in shape, but in all of them the shell rests on the substrate and supports the apertural opening above the substrate. The animal normally lives hanging from the aperture without touching the seafloor, the stilts forming a protective cage around it (Linsley and Yochelson 1973). When the animal moves, it thrusts its foot against the substrate, lifts the shell up, and then lets it fall forward (Berg 1972)—a most effective means of locomotion that allows higher levels of activity (Berg 1975) than are found in other atypical gastropods.

For most snails, however, higher speed lies in lower positioning of the axis of coiling. In most of the living fast snails the geometrical solution to this problem is the elongation of the aperture along the dimension parallel to the axis of coiling, causing the elevation of the spire to decrease and allowing the shell to be balanced more by regulatory detorsion than by inclination. Most frequently this is accomplished by the construction of a siphonal canal (as in *Fasciolaria,* Fig. 10), or by a decrease in translation, which allows the aperture to extend almost the entire length of the shell (as in *Oliva,* Fig. 11).

For some snails, of course, speed is not a priority, and the adaptations that allow greater speeds would be disadvantages. Snails in surf zones,

Figure 7. *Architectonica nobilis* Röding has a shell with a radial aperture, which does not provide any protection to the animal when the shell is clamped against the substrate, so the animal must withdraw into its shell for protection. As a result the animal is most careless about how it holds its shell when moving. Sometimes it is carried over the back, sometimes off to the side, sometimes even upside down.

Figure 8. *Petaloconchus* sp. is a motionless snail, as is evident from the scar over the bottom of the shell where it was cemented to the rock substrate.

Figure 9. *Siliquaria* sp. is also a motionless snail. Its irregularly coiled shell would be awkward to carry.

for example, may spend as much time clinging to the sides of rocks or beneath overhangs as they spend crawling along the tops. For them, the long narrow foot that frequently results from a long narrow aperture is a disadvantage, because only a broad foot provides the surface area necessary to adhere effectively to rocks in an active surf. Moreover, a large amount of inclination allows the columellar muscle to contract at almost right angles to the substrate, effectively clamping the tangential aperture to the surface. In forms with little inclination and at least a moderate amount of translation, on the other hand, the columellar muscle must bend sharply between its insertions on the shell and its foot (Fig. 12). In this case, the muscle cannot resist the pull of gravity on the high spire of the shell. Thus forms with large amounts of inclination, though poorly suited to rapid motion, have advantages in certain environments that make them viable forms.

Although the subject has not yet been studied, I presume correlation will be found between speed and other parameters of behavior, such as food preference. Unfortunately, the nice old standard classification "carnivore" is not useful here, since many snails are carnivorous on stationary animals such as anemones, tunicates, and sponges. You don't have to be particularly fast to catch a sponge. Yet in preying actively on other gastropods, as do *Oliva, Natica,* and *Fasciolaria,* speed may be significant. It would not seem accidental that these three are the fastest of the seventy species for which I have been able to obtain data (Linsley, in press).

Internal water flow

While the flow of water outside the shell has considerable selective significance, the flow of water inside the shell through the mantle cavity is at least equally important. All modern gastropods that exhibit motion take in water from the anterior portion of the shell, whereupon it enters the mantle cavity, passes the osphradium (a water-testing organ), gills, hypobranchial gland (which removes sediment), and finally the kidney aperture and anus just before leaving the shell (Fig. 13). Taking in water at the front of the shell aperture allows the animal to sample water from the direction in which it is moving. A smooth, straight flow of water over the organs and out of the aperture near the anus is advantageous in that the possibility of contamination from eddies is eliminated. The problem of direction of water currents through the mantle cavity was thus one of the important factors that guided much of the evolution of the Gastropoda (Fig. 14).

Ancient gastropods and their ancestors

Although we have a fairly simple biological definition for *gastropod* and can obviously describe the varieties of living gastropods in some detail, classification of early fossil mollusks can be quite difficult. There has been particularly active debate in recent literature concerning the isostrophically coiled shells found in Paleozoic rocks. Two kinds of shells are found: those with multiple paired muscle scars, the Monoplacophora, and those with a single pair of scars, the Bellerophontacea. The debate is primarily centered around whether the Bellerophontacea should also be considered untorted Monoplacophora, or whether they were torted and should be considered gastropods.

Monoplacophora are primitive mollusks that were believed to be extinct until the discovery of a few live specimens, called *Neopilina,* in 1952. Originally, Monoplacophora were probably adapted to a habitat similar to that of modern limpets—that is, to a high-energy wave environment. The cap-shaped shell is well adapted to this environment because, first, it is low enough to offer a minimal resistance to the water, and second, the apex of the shell is usually directed upward into the waves and the wave energy in effect presses the shell more firmly to the substrate. In forms such as *Scenella,* water probably flowed into the mantle cavity on both sides of the head, and supplementary currents may have entered the shell laterally, as is the case in many Polyplacophora. This in turn leads me to believe that, as in the living *Neopilina,* the ancestral forms had multiple pairs of gills hanging in the laterally situated mantle cavities (see Fig. 15).

The evolutionary line that concerns

us probably involved adaptation of the organism to an environment of quiet water and a hard substrate. Minimal currents removed the need for a low-profile shell and permitted the development of a more slowly expanding cone-shaped shell, a deeper shell that would allow elaboration of the soft parts. The deepening of the shell had three immediate results. First, it crowded the multiple paired organs, thus giving selective advantage to the elimination of some pairs and eventually the development of more efficient single pairs of organs. The most important change would be the reduction of the number of paired gills to one or possibly two pairs (Starabogatov 1970). Second, it caused flexure (the bending of the digestive tract into a **U**-shaped tube). And third, it necessitated a modification of the pathways of the water currents (Fig. 15).

As the Monoplacophora developed a higher conical shell, there would be room for increased elaboration of the digestive and reproductive organs, but the problem of balancing the shell would intensify. Presumably water currents would still enter from the anterior, though now they would have to move up into the deepened mantle cavity and then make a **U**-turn to pass the anus before leaving the mantle cavity at the posterior. One solution, of course, would be for the shell to be coiled in many volutions over the head, as in *Cyrtolites.* In mollusks of this sort the anterior portion of the aperture was now obstructed both by the earlier volutions and by the fleshy stalk leading from the foot to the visceral mass. As a result, the mantle cavity would now be restricted to the lateral and posterior portion of the shell, and the inhalent water would be displaced to the lateral shell margins.

The shells of coiled Monoplacophora that did not develop an anal sinus or slit (such as *Cyrtolites*) tend to have a diamond-shaped aperture. I suggest that the angulations at the sides of these shells were to accommodate incurrent water flow (Starabogatov 1970). Thus the diamond-shaped aperture may well be indicative of an untorted Monoplacophora, and might be useful for class assignment of genera for which the musculature is unknown.

The main argument for believing the bellerophontaceans to be torted, on the other hand, hinges on the distinctive features of their own apertures. Knight's interpretation of *Knightites* (1947) and Peel's interpretation of trilobed forms (1974) argue convincingly that the indentations in the aperture of these bellerophonts were to accommodate inhalent water flow, and it is unlikely that inhalent water entered from the posterior. Thus, these indentations would not have developed where they did, had not torsion occurred.

As torsion is a result of differential muscle development, finding a shell with a single, asymmetrically placed adductor muscle scar is surely an indication of a gastropod. Contrarily, a shell with multiple paired adductor muscle scars is surely indicative of an untorted animal and not a gastropod (Horný 1965a, 1965b; Rollins and Batten 1968). Unfortunately for solving the Bellerophon problem, the true bellerophontacean shells have muscle scars that fit neither of these conditions (Fig. 16): their small single pair of scars is implanted on the columella deep within the shell, almost a full volution from the aperture (Knight 1947; Peel 1972).

True bellerophontaceans were thus probably capable of retracting the body deep into the shell. In contrast, the isostrophically coiled monoplacophoran shells have their retractor muscles located directly above the aperture; presumably these animals could not withdraw into the shell, but could only clamp the shell down on top of their body. But in fact, it seems clear from an analysis of muscle mechanics that deep withdrawal into the

Figure 10. *Murex ternispina* Lamarck has a long siphonal canal, the extension on the right side of the animal shown here. This canal helps direct water currents into the shell; it also allows the shell to be balanced more by detorsion than by inclination, thus permitting a lower spire and a lower center of gravity.

Figure 11. The shell of *Oliva miniacea* Röding has an aperture that extends the length of the animal, with the inhalent siphon anterior and the anus posterior. Again, the shell is balanced primarily by regulatory detorsion. Its shape actually converges on the bilaterally symmetrical, and *Oliva* is thus one of the fastest of modern gastropods.

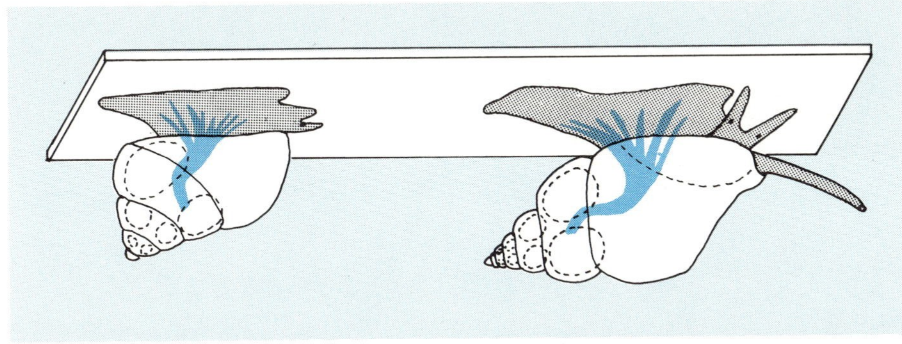

Figure 12. The retractor muscle in a medium-spired snail is well positioned for holding the shell against the substrate even while the animal is hanging from the side or bottom of a rock, since the muscle can pull in basically a straight line. In a form with the axis of coiling closer to parallel with the substrate, however, the muscle must bend rather sharply between its insertions on the shell and the bottom of the foot. When the animal is hanging below a rock, gravity pulls the spire of the shell down; the muscle is poorly positioned to prevent this.

shell was a necessary condition for torsion. Since the muscles of an isostrophically coiled monoplacophoran like *Cyrtolites* are placed close to the aperture, abortion of all but a single muscle would only have caused a lopsided clamping of the shell. Only when the muscles are located at least 180° in from the aperture will the contraction of a single muscle cause torsion.

Thus the position of the muscle scars and the shape of the aperture both lend themselves to interpretation of the bellerophonts as torted. I therefore conclude that the Bellerophontacea were the primitive gastropods, derived from the coiled Monoplacophora by torsion, and ancestral to all other gastropods.

Torsion and the gastropod condition

Torsion had selective significance for the gastropod on many planes. First of all, it provided increased space (the mantle cavity) for the head when the shell was clamped down for protection. Furthermore, once torsion had developed, it was possible for the inhalent currents to swing from a lateral position to the advantageous anterior position. That this occurred is obvious from Knight's study (1952) of *Knightites* and Peel's study (1975) of the trilobed bellerophonts. I am assuming that an osphradium or comparable water-testing organ, which would make anterior incurrents advantageous, was present even in these primitive groups.

With its attendant advantages, however, torsion brings the concomitant disadvantage of locating the anus over the head of the animal. This is not the devastating problem that our anthropocentric reactions might imply, or torsion would never have occurred in the first place. All living mollusks have both gills and anus located in the mantle cavity, and all fecal material is nicely packaged in a mucuslike material and discharged in pellets that drop out of the mantle cavity without fouling the gills. The advantage of moving the anus from over the head is thus not so much a problem of sanitation as one of recycling. Dropping the fecal material off to one side would reduce the chances of the snail reconsuming this material with its now low energy potential. But it is this problem more than any other that may have guided the early evolution and radiation within each group and convergence between the groups. When the problem was finally solved by the more advanced gastropods, the group underwent an explosive radiation that allowed it to become one of the dominant invertebrate macrofaunal elements of the world's seas.

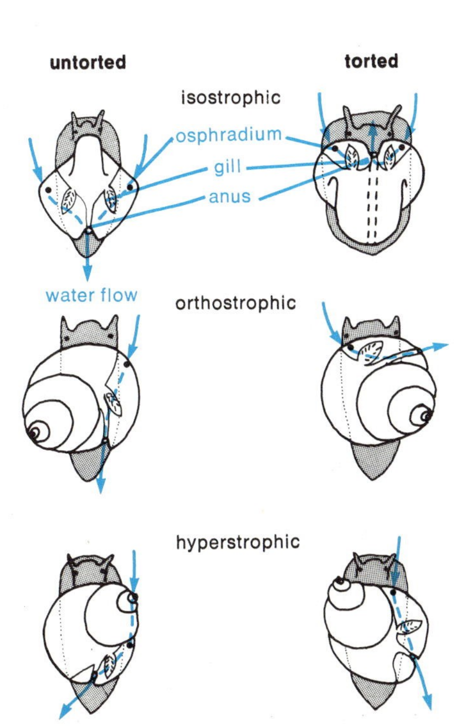

Figure 13. Each of the six possible configurations of shell and soft parts shown here represents a different compromise between simplification of internal water flow and effective balancing of the shell when the animal moves through water.

Three early solutions

The bellerophontacean solution was, very simply, the development of an anal slit and the perfection of internal water flow. (I presume that the monoplacophoran ancestor did not have a slit, but if Rollins and Batten, 1968, are correct in their thesis that the bellerophontaceans are polyphyletic, then perhaps later bellerophontaceans evolved from monoplacophorans preadapted by having a slit.) Once torsion occurred, there would have been a strong selective pressure for the two inhalent areas to migrate toward the anterior. However, as the two inhalent currents converged, strong reflexed currents would have developed, causing increased hazard of eddies. Development of an indentation in the shell at the anus would allow the currents to be straightened out, thus eliminating the danger of eddies. Therefore, while an anal sinus, or slit, was functionally important for an isostrophically coiled monoplacophoran, it was almost essential for an isostrophically coiled gastropod because of the convergence of inhalent and exhalent water pathways.

Within the gastropods in general, but the bellerophontaceans in particular, two alternate strategies seem to have determined the form of the anal slit. The slit may be either broad and shallow or narrow and deep (Fig. 17). A broad, shallow slit, such as that seen in *Bellerophon*, is generally associated with a broad shell with either a narrow umbilicus or none at all. The breadth of the anterior portion of the shell permitted the inhalent currents to be placed at the lateral extremities of the anterior margin of the aperture, and the mantle cavity could be relatively shallow, with the gills situated quite close to the apertural margin

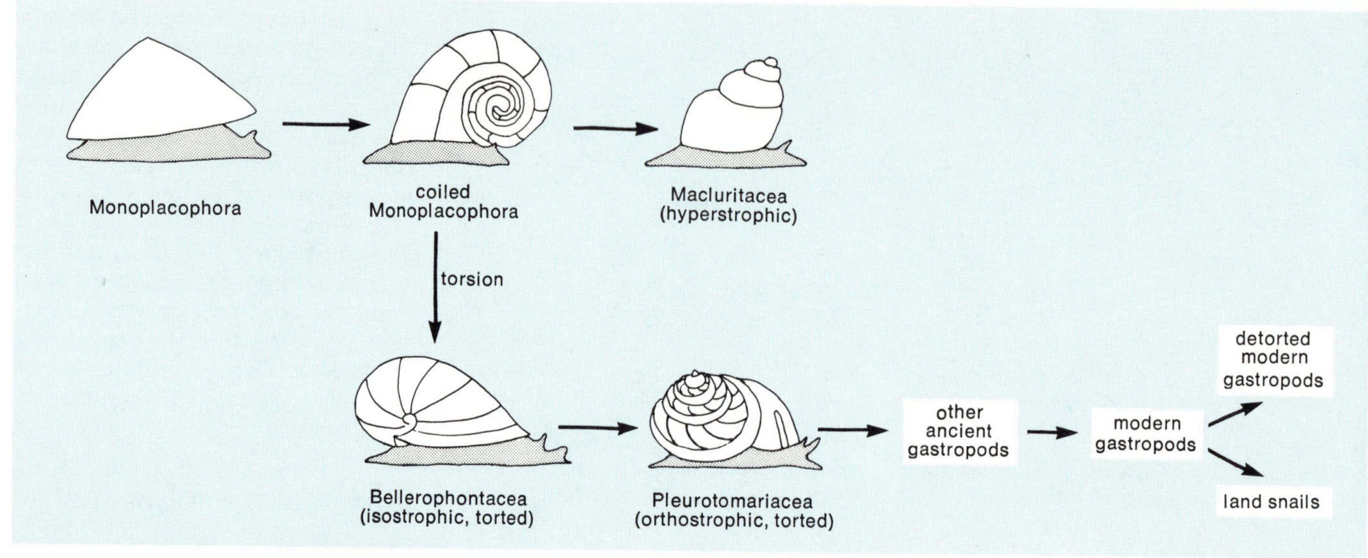

Figure 14. Only recently have phylogenetic trees for the early gastropods been based on interpretations of the functional significance of shell forms as well as on geological and geometrical data.

(see Peel 1974). As a result, the water currents through the mantle cavity would follow skewed but not sharply deflected pathways and the exhalent area would be at least minimally offset from the inhalent area.

In contrast, bellerophontaceans with a deep, narrow slit, such as *Tropidodiscus,* tend to have a very slender shell with wide, open umbilici. The slenderness of the shell perforce moves the inhalent area directly adjacent to the slit, and thus the separation of the inhalent and exhalent areas must be accomplished by deepening the slit and correspondingly deepening the mantle cavity. In *Tropidodiscus* the slit is exceedingly narrow and may be as deep as one-half volution, effectively creating a posteriorly placed anus.

Within the Bellerophontacea evolved a diverse number of morphotypes that were adaptations to a number of different life modes: there were burrowers, slow grazers, perhaps even active carnivores. The only good solution to the problem of the anus being located over the head was the narrow, deep slit developed by *Tropidodiscus* and its allies. The major disadvantage was the weakness of the shell caused by the very deep slit, but this was partly compensated for by corrugation of the flanks of the shell. This solution was not appropriate for the other members of the superfamily, however, because the high center of gravity of this form and the narrow shell and foot restrict it to very quiet water.

The pleurotomariacean solution was, simply stated, orthostrophic coiling, with regulatory detorsion moving the anus to the right side of the snail's head. This process placed the left gill in a favorable anterior position, while the right gill was placed in a position of lesser importance, off to the right side of the animal. The movement of the anus from over the head was immediately advantageous, but the positioning of one gill and one osphradium in a directly anterior position provided the greatest long-term benefit. No longer did the gastropod receive water—and the chemical information about the environment contained therein—from off to the sides; now it came from directly ahead. This shift gave the left gill and its associated osphradium-hypobranchial gland an ascendant position

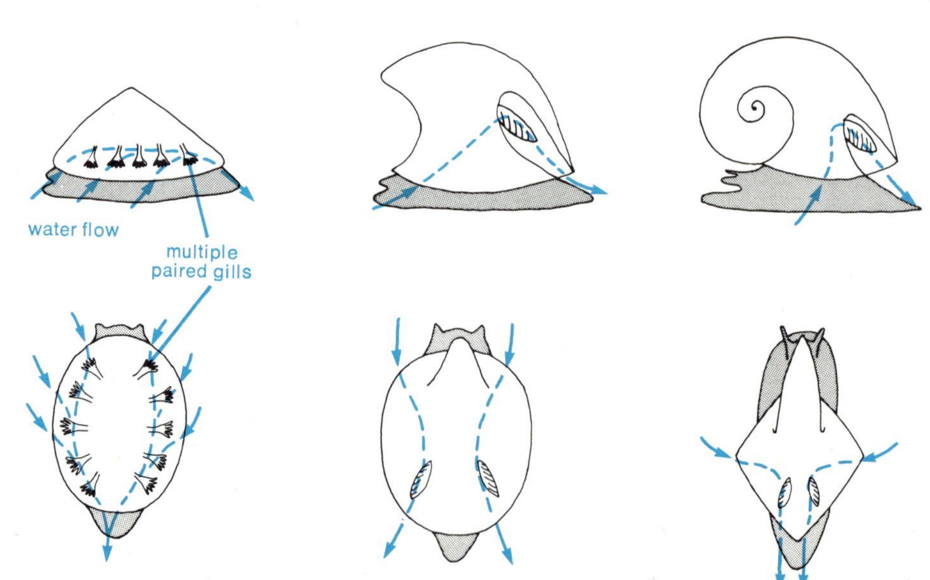

Figure 15. Internal water flow was one of the guiding factors in the evolution of Monoplacophora. In low cap-shaped forms water flow probably resembled that in living Polyplacophora, with a dominant anterior-to-posterior flow supplemented by weaker lateral inhalent flow. As the shell became deeper, the gills were reduced to a single pair and the mantle cavity deepened, forcing the water into a more U-shaped path. When the shell coiled over the head, as in *Cyrtolites,* the inhalent currents were displaced to a more lateral position.

relative to the right gill, probably in three or four different lineages, thus ultimately giving rise to the modern gastropods.

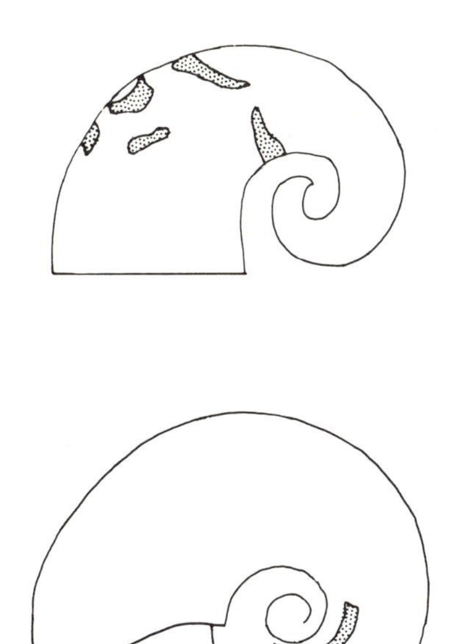

Figure 16. Fossil shells of the Monoplacophoran *Cyrtolites ornatus* Conrad (*above*) show five pairs of muscle scars situated close to the aperture on the inside of the shell. Fossil shells of *Bellerophon* (*below*), on the other hand, show a single pair of muscle scars, about 180° from the aperture. (*Cyrtolites* after Horný 1965a; *Bellerophon* after Peel 1972.)

Although the benefits that were gained by orthostrophic coiling outweighed the costs, the costs were not insignificant. The primary cost was a reduction of mobility because of the loss of symmetry. Concomitant with the detorsion that accompanied orthostrophic coiling was the elevation of the center of gravity and the pressure point, an increase in frontal area, and an increase in torque. I believe that this increase in general resistance, relative to its bellerophontacean ancestor, imposed a less active life style on the pleurotomariaceans, and that the great majority of them were algae grazers on a relatively firm substrate, without the variety of activity levels available to the bellerophontaceans.

What little we know about living pleurotomariaceans supports this interpretation. Our knowledge is scant because they are restricted to water over 300 m in depth. Dr. Conrad Neumann has reported seeing them from a submarine, however, and says they were clinging to rocks.

Along with the Bellerophontacea and the Pleurotomariacea, a third group, the Macluritacea, has long been considered among the dominant Paleozoic gastropods. But the aperture of these shells is shaped rather strangely when compared with the apertures of other gastropods. A recent unpublished study by myself, W. Kier, and Cathy McNair suggests that this difference in aperture shape makes sense if these forms did not undergo torsion, and that Macluritacea therefore should not be considered gastropods. They exhibit anisostrophic coiling, however, and almost undoubtedly competed with the gastropods, at least during the early portion of their evolution. Macluritacean shells can thus be interpreted by the same principles as those of gastropods, and can help us understand the selective pressures on gastropods.

The apertures of the earliest Macluritids, such as *Matherella*, are distinctly elongate, suggesting that these two forms had only one inhalent and one exhalent stream and probably only a single gill. The process of shell balancing in these hyperstrophic forms resulted in the spire being swung forward, over the head of the animal (as it is swung back in gastropods). This process served to put one gill in the preferred anterior position and allowed the elimination of the other gill and its associated inhalent stream.

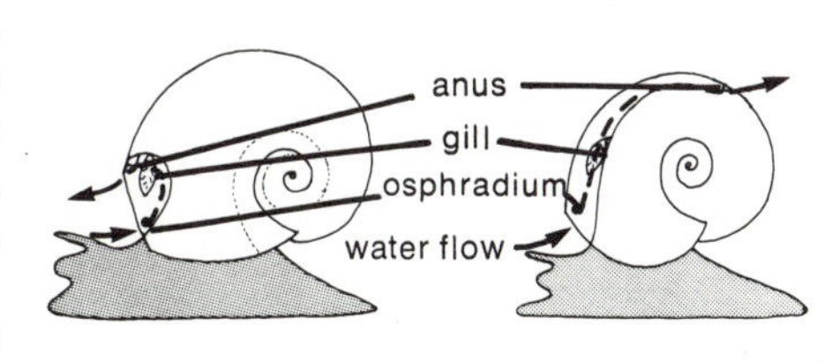

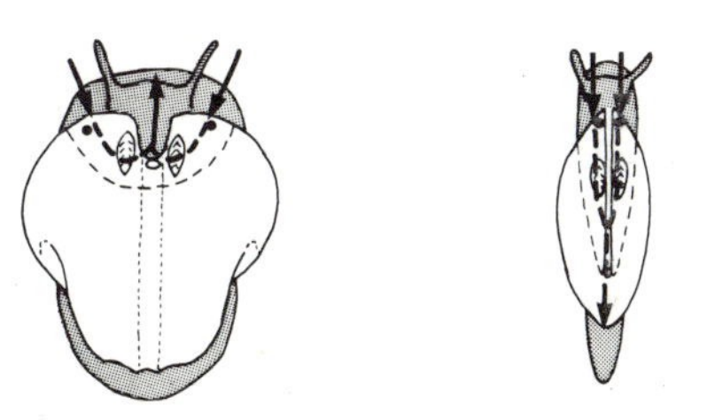

Figure 17. In bellerophontaceans with a short, broad slit the paired inhalent currents are displaced laterally to straighten them out, though they still retain an essentially anterior placement. In bellerophontaceans with a deep, narrow slit, the water currents are essentially straight, with an anterior inhalent stream and a posterior exhalent stream.

The shells of the early Macluritids have tangential apertures and thus probably carried their shells on their backs, as did most of the gastropods of that time. But with the spire projecting out over the head of this animal, there is little doubt that mobility was greatly reduced in this group. The descendants of these early forms, such as *Maclurites* and *Euomphalus*, show evidence of an even further reduction in mobility. These later forms all have radial apertures, which strongly suggests that the advanced Macluritids did not normally live with the aperture parallel to the substrate, and did not creep on the seafloor with the shell positioned dorsally above the foot, for if this were the normal mode of locomotion, the aperture would not have offered protection. Rather, *Maclurites* probably spent most of its time lying on the left side of the shell with the aperture perpendicular to the substrate. The curve of the operculum fits the curve of the outside of the body whorl, the flattened left side of the shell makes a solid base on which to sit, and the curving sides of the shell would allow currents to sweep over the shell, even to press it down more firmly against the substrate, which was probably firm mud, swept with moderate currents.

Only rarely did *Maclurites* move across the seafloor, and then it probably dragged its shell. For protection it could pull its massive calcareous operculum in to seal up the aperture. The only modern gastropods with this degree of immobility are all filter feeders, and thus I would suggest this as a mode of life for *Maclurites*. It might have been a deposit feeder, but the majority of modern deposit feeders show more activity than I envisage for *Maclurites*.

Subsequent evolution

Throughout the Paleozoic the two major groups of Gastropoda—the Bellerophontacea and the Pleurotomariacea—and the Macluritacea formed the dominant forms in the geologic records, each with its own adaptations. The Bellerophontacea were the dominant active grazers and possibly carnivores; the Pleurotomariacea, the dominant less active grazers; and the Macluritacea (and their euomphalid descendants), the dominant filter feeders. These groups were little affected by the mass extinctions of the Permo-Triassic. Shortly afterward, however, an event (or multiple events) occurred that rendered these ancient gastropods obsolete and relegated the survivors to a position of relatively minor importance: the rise of more advanced gastropods with one very important improvement, the pectinibranch gill (Yonge 1947).

This gill had gill filaments only on one side, and was therefore much more efficient, especially in resisting fouling and entangling of the filaments. All snails with a pectinibranch gill have eliminated the posterior gill of the primitive pair, enabling the anus to migrate to the right side of the aperture. This in turn made possible tremendous simplification of internal water flow, with the inhalent current entering one end of the aperture and leaving the other.

In addition, the shape of the aperture was modified (generally by its elongation, and most frequently by means of an extended siphonal canal), which effectively swung the axis of coiling so that it was subparallel to the axis of the foot and subparallel to the substrate. If one considers *Oliva* (see Fig. 11), it is clear how effectively this epitome of the modern gastropod condition has solved the problem of being a gastropod. Inclination has been reduced to less than ten degrees, and regulatory detorsion is essentially ninety degrees. The right gill has been lost and the anus has moved to the "right" side of the aperture, which through detorsion is now posterior. Hence the anus has achieved a completely posterior position. The inhalent water enters from the anteriormost portion of the shell in a unidirectional current. This allows water sampling by the osphradium to be localized so that it can be used directionally to sample water, to detect prey, and to locate enemies. The form of the shell has reduced torque effectively to zero, so *Oliva* is essentially a bilaterally symmetrical animal. As a result it is a highly mobile animal, well adapted to an active carnivorous mode of life.

References

Batten, R. L., H. B. Rollins, and S. J. Gould. 1966. Comments on "the adaptive significance of gastropod torsion." *Evolution* 21: 405–6.

Berg, C. J., Jr. 1972. Ontogeny of the behavior of *Strombus maculatus* (Gastropods: Strombidae). *Am. Zool.* 12:427–33.

———. 1975. Behavior and ecology of conch (superfamily Strombacea) on a deep subtidal algal plain. *Bull. Mar. Sci.* 25:307–17.

Crofts, D. R. 1937. The development of *Haliotis tuberculata,* with special reference to organogenesis during torsion. *Phil. Trans. R. Soc. Lond.* (*B*) 288:219–68.

Ghiselin, M. T. 1966. The adaptive significance of gastropod torsion. *Evolution* 20:337–48.

Horný, R. 1962. New genera of Bohemian lower paleozoic Bellerophontina. *Vest. Ústřed. úst. geol.* 37:473–76.

———. 1963. Lower Paleozoic Bellerophontina (Gastropoda) of Bohemia. *Sb. geol. Ved. Paleontologie* 2:57–164.

———. 1965a. On the systematical position of *Cyrtolites* Conrad, 1838 (Mollusca). *Casopsis Narodniho Muzeum,* odd. prir, 134: 8–10.

———. 1965b. *Cyrtolites* Conrad, 1838 and its position among the Monoplacophora (Mollusca). *Sbornik Narodniho Muzea V. Praze* 21B, no. 2: 57–70.

Knight, J. B. 1947. Bellerophont muscle scars. *Jour. Paleont.* 21:264–67.

———. 1952. Primitive fossil gastropods and their bearing on gastropod evolution. *Smithsonian Misc. Coll.* 117, no. 13, pp. 1–56.

Knight, J. B., R. L. Batten, and E. L. Yochelson. 1960. Mollusca 1. In *Treatise on Invertebrate Paleontology,* ed. R. C. Moore, part I. Geol. Soc. Am. and Univ. of Kansas Press.

Lemche, H. 1957. A new living deep-sea mollusc of the Cambro-Devonian class Monoplacophora. *Nature* 179:413–16.

Linsley, R. M. 1977. Some "laws" of gastropod shell form. *Paleobiology* 3:196–206.

———. In press. Locomotion rates and shell form in the gastropoda. *Malacologia* 17(2): 193–206.

Linsley, R. M., and E. L. Yochelson. 1973. Devonian carrier shells (Euomphalidae) from North America and Germany. U.S.G.S. Professional Paper 824.

Miller, S. L. 1974. The classification, taxonomic distribution and evolution of locomotor types among prosobranch gastropods. *Proc. Malacol. Soc. Lond.* 41:233–72.

Moore, R. C., C. G. Lalicker, and A. G. Fischer. 1952. *Invertebrate Fossils.* McGraw-Hill.

Peel, J. S. 1972. Observations on some Lower Paleozoic tremanotiform Bellerophontacea (Gastropoda) from North America. *Paleontology* 15 (pt. 3):412–22.

———. 1974. Systematics, ontogeny and functional morphology of Silurian trilobed bellerophontacean gastropods. *Bull. Geol. Soc. Denmark* 23:231–64.

———. 1975. A new Silurian gastropod from Wisconsin: The ecology of uncoiling in Palaeozoic gastropods. *Bull. Geol. Soc. Denmark* 24:211–21.

Pojeta, J., Jr., and B. Runnegar. 1976. The Paleontology of the Rostroconch mollusks and the early history of the phylum Mollusca. U.S.G.S. Professional Paper 968.

Raup, D. M. 1962. Computer as aid in describing form in gastropod shells. *Science* 138:150–52.

———. 1966. Geometric analysis of shell coiling: General problems. *Jour. Paleont.* 40: 1178–90.

Raup, D. M., and J. A. Chamberlain, Jr. 1967. Equations for volume and center of gravity in ammonoid shells. *Jour. Paleont.* 41: 566–74.

Raup, D. M., and A. Michelson. 1965. Theoretical morphology of the coiled shell. *Science* 147:1294–95.

Rollins, H. B., and R. L. Batten. 1968. A sinus-bearing Monoplacophoran and its role in the classification of primitive molluscs. *Paleontology* 11 (pt. 1):132–40.

Runnegar, B., and P. A. Jell. 1976. Australian middle-Cambrian molluscs and their bearing on early molluscan evolution. *Alcheringia* 1:1–30.

Runnegar, B., and J. Pojeta, Jr. 1974. Molluscan phylogeny: The paleontological viewpoint. *Science* 186:311–17.

Simroth, H. 1906. Mollusca. In *Klassen und Ordnungen des Tier-reichs,* ed. H. G. Bronn, 2nd ed., vol. 3, pp. 85–89. Leipzig: C. F. Winter.

Starabogatov, Y. I. 1970. Systematics of early Paleozoic monoplacophora. *Paleontological Journal* 3:6–17 (trans. from Russian by the American Geological Institute).

Termier, H., and G. Termier. 1950. Sur l'organisation palléale des Bellerophontides. *C. R. Ac. Sc.* 230.

Thiele, J. 1935. *Handbuch der systematischen Weichtierkunde,* vol. 2. Jena: G. Fischer.

Vagvolgyi, J. 1967. On the origin of molluscs, the coelom, and coelomic segmentation. *Systematic Zool.* 16:153–68.

Wenz, W. 1938. Allgemeiner Teil und Prosobranchia. In *Handbuch der Palaeozoologie,* ed. O. H. Schindewolf, vol. 6, pt. 1, pp. 1–240.

Yochelson, E. L. 1967. Quo vadis, Bellerophon? In *Essays in Paleontology and Stratigraphy,* ed. C. Teichert and E. L. Yochelson. Dept. of Geol., Univ. of Kansas, Special Publ. 2, pp. 141–61.

Yochelson, E. L., H. Flower, and G. F. Webers. 1973. The bearing of the new late-Cambrian monoplacophoran genus *Knightoconus* upon the origin of the *Cephalopoda. Lethaia* 6:275–309.

Yonge, C. M. 1947. The pallial organs in the aspidobranch Gastropoda and their evolution throughout the Mollusca. *Phil. Trans. R. Soc. Lond.* (*B*) 232:443–518.

F.M. Carpenter

The Geological History and Evolution of Insects

The lengthy title of this paper is an indication of the current view of insect evolution. Were I a vertebrate zoologist discussing the evolution of the vertebrates, I would have considered it sufficient to use just that phrase for a title and would have expected everyone to know that the discussion would center around the fossil record. The same would also be true of some groups of invertebrates. However, I have referred to the geological record in my title because I suspect that few biologists, including some who work with insects, realize that the insects have had a geological history and even an extensive fossil record.

Since this subject is one in which relatively few biologists have been interested, I shall begin the discussion with a comment about the literature and material bearing on the subject. My index to the publications on fossil insects includes some 3000 papers, contributed by 700 authors. Only two general, compilative treatises have appeared; one by Samuel Scudder in 1886 [1], and the other by Anton Handlirsch in 1906-1908 [2]. Both of these authors had unique ideas on insect evolution, especially Handlirsch, whose views unfortunately are the ones usually found in textbooks of zoology, palaeontology, and evolution. The material which forms the basis for the extensive literature in this field is the countless thousands of specimens, perhaps 500,000,[1] contained in the museums and university collections in Europe and North America. Up to the present time about 13,000 species of fossil insects have been formally described. The geological formations which have produced these specimens range from the Upper Carboniferous through to the present.

The first aspect of the evolution of insects which I shall consider is a general one. I believe we can recognize four important stages in their history, our present insect fauna consisting of some representatives of all stages. The first of these stages was a wingless insect, exemplified in our existing fauna by two orders, the *Thysanura* (silverfish) and the *Entotrophi*. The existence of such a phylogenetic group of wingless insects, termed the Apterygota, is based on the premise that wings evolved *after* the origin of insects and not *with* their origin—a conception that has been almost universally accepted by zoologists for fully sixty years. The opposite view, based on the belief that the first true insects were winged and that all wingless species are secondarily so, was advocated by Handlirsch [2,3]; it was a corollary to his conviction that insects arose directly from trilobites, the lateral lobes of which became functional wings. So far as I am aware no one who has given serious thought to the subject, with the exception of Handlirsch, has accepted this idea. It is true that the wingless Apterygota are known only as far back as the Triassic period and that the winged insects, or Pterygota, extend to the Upper Carboniferous. However, apart from a few Baltic amber inclusions, only two specimens of Apterygota have been found in all geological strata. Their fragility and the very absence of wings, of which most fossil insects consist, make their chances of preservation as fossils very slight indeed. This is an instance in which the structure of living material furnishes more evidence than the geological record.

Development of wings

The second stage in the evolution of the insects began with the development of wings. The time when these appendages started to appear is not established, but three specimens of insects with fully developed wings have been found in the lowest of the Upper Carboniferous period. However, even if the Upper Carboniferous record is accepted as the time of wing development, it is clear that the insects attained flight fully 50 million years before the reptiles and birds did—a period of time during which the insects, so far as is known, were the sole inhabitants of the air as aerial creatures. By the time flying reptiles and birds had evolved, the insects were well established in their new environment. It is intriguing, though futile, to reflect on the possibility that if the insects had not taken to the air before the vertebrates, they might never have successfully attained flight. The significance of flight for insects was undoubtedly great during the late Palaeozoic. This was the age of amphibians and small reptiles. Scorpions, spiders, and spider-like arachnids, belonging to extinct orders, were abundant. All these predators unquestionably subsisted to some extent, and probably to a great

[1]This figure includes the one hundred thousand or more amber insects originally at the Albertus University of Konigsberg, but apparently entirely destroyed by bombing during World War II.

Frank M. Carpenter was Agassiz Professor of Zoology and Curator of fossil insects at the Harvard Museum of Comparative Zoology. In this article, based on the Sigma Xi address given at the meeting of the American Institute of Biological Sciences at Cornell University in 1952, his views on insect evolution derived from his study of the fossil record are expounded.

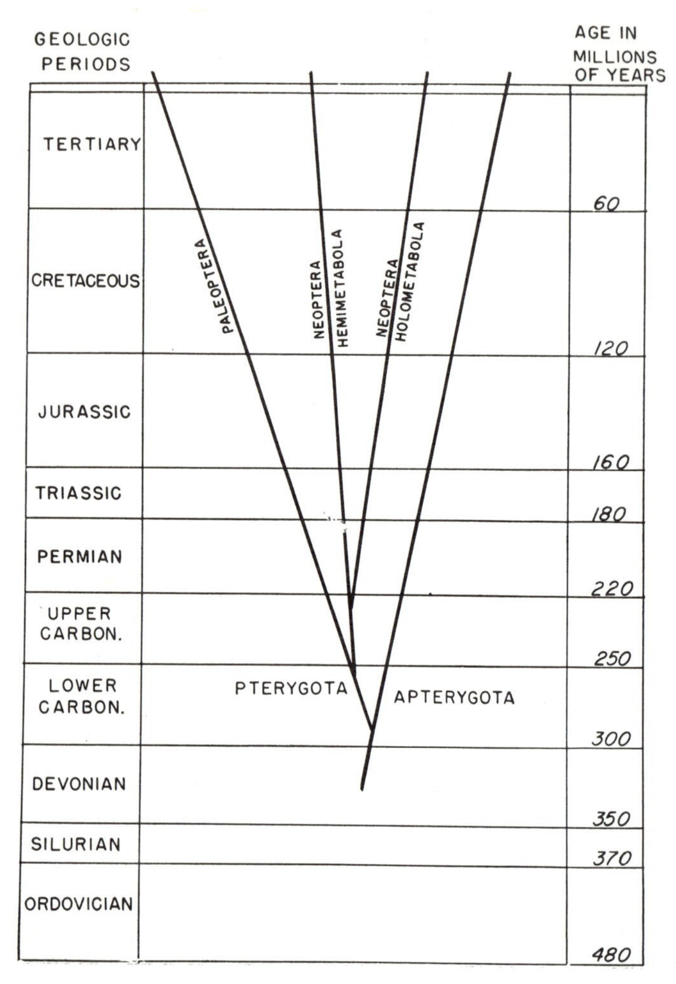

Figure 1. Main lines of insect evolution, as described in the text.

extent, on the wingless insects, which had no means of escape. It is not surprising, therefore, that the ability to fly changed the direction of insect evolution and that in our present insect world only one-tenth of one per cent of the species are Apterygota.

The process by which wings were acquired by insects[2] has been a question of much speculation, for they are not modifications of previously existing appendages. However, significant evidence has been provided by the study of fossils. All of the more generalized Pterygota of the Carboniferous period, and even some species of the Permian, possessed a pair of membranous flaps, arising from the dorsum of the first thoracic segment. These flaps contained veins and were covered with minute hairs like those of the true functional wings borne on the second and third thoracic segments. There is every indication that true wings began, like membranous prothoracic flaps, as lateral tergal expansions. However, so far as we know, the prothoracic flaps never developed into functional wings. In most insects the flaps have completely disappeared and in others they have been absorbed into a pronotal disc.

The first winged insects, or *Palaeoptera*, which we have been considering, had a simple wing articulation and were incapable of flexing their wings back over the body at rest; hence, they were preserved as fossils with their wings outstretched. Dragonflies and mayflies—the sole living representatives of the *Palaeoptera*—exhibit the same limitations in wing structure. The third stage in insect evolution began with the modification of certain plates of the wing articulation so as to permit wing flexing; these insects are known as the *Neoptera.*[3] The survival value of wing flexing was great, for it enabled the insects, between flights, to hide among foliage or under objects on the ground. The fossil record shows that this stage was reached by early Upper Carboniferous time, when many of the palaeopterous insects were predaceous and of great size, though no flying vertebrates had yet appeared. Later, when the flying reptiles, or pterosaurs, and birds appeared in the Mesozoic, the neopterous insects had all the advantage. The palaeopterous insects, which had been dominant during the Carboniferous and Permian, began to wane and *Neoptera* to flourish. This trend in insect evolution has continued up to the present time to such an extent that 90 per cent of the existing orders, including 97 per cent of the species, are now neopterous.

Metamorphosis

The first of the neopterous insects were closely related to stone flies and locusts and possessed incomplete metamorphosis, that is, they passed through a series of nymphal stages which gradually approached the adult form. These are designated as the hemimetabolous *Neoptera*. The fourth step in insect evolution was the development of a more complex type of metamorphosis, in which the insects pass through a series of larval stages bearing little resemblance to the adult. Eventually, they enter into one or two quiescent stages, during which extensive morphological and physiological changes take place. These, the holometabolous *Neoptera*, presumably had several advantages over the hemimetabolous types. The immature forms, being very different from the adults, could occupy different environments and feed on different types of foods. The tissues of other organisms, both animal and plant, were thus invaded by larval forms as internal parasites, the adult insects remaining free-living and capable of flight. The holometabolous insects make their first appearance in the Lower Permian strata. The existence of two orders, the scorpion flies or *Mecoptera*, and the *Neuroptera* in the Lower Permian shows that complete metamorphosis must have begun before the end of the Upper Carboniferous period. Starting from the beginning of the Permian, when only about 5 per cent of the known species of insects had complete metamorphosis, the percentage of species has progressively increased to the present maximum of 88 per cent.

A simple phylogenetic diagram, shown in Figure 1, superimposed on the geological time scale, serves to summarize this general aspect of insect evolution. The three modifications—origin of wings, wing-flexing, and com-

[2]That the Pterygota are of monophyletic origin seems almost certain. Lemche, however, has advocated [4,5] a polyphyletic origin, even claiming that such insects as the *Grylloblattidae* and the females of *Zoraptera* and of certain Lampyrid beetles are primitively wingless (non-alate). The evidence for his conclusion seems insufficient (see, for example, Carpenter, 1948 [6]).

[3]The phylogenetic groups which are here termed the *Palaeoptera* and *Neoptera* were recognized independently by Martynov [7,8] and Crampton [9].

plete metamorphosis—mark the points of separation of the phylogenetic lines. Since holometabolous insects are known to have existed from the Lower Permian strata, the upper phylogenetic division must have taken place before the end of the Upper Carboniferous; and since neopterous insects are known from the lowest of the Upper Carboniferous strata, the middle division, or wing-flexing, must have taken place in the Lower Carboniferous, which is beyond the present record of the insects. The first phylogenetic division must have occurred even earlier.

Fossil record

Turning from this phylogenetic treatment of the insects, I shall next consider their history as it is now actually known from the fossil record. This discussion will involve some mention of extinct orders and an explanation of my point of view on this controversial subject. The artificial and arbitrary nature of higher taxonomic categories is well known to systematists. Such categories are established for dealing with organisms in a very limited period of geological time, not with the whole geological record of a group, with annectent forms appearing at intervals. This is an elementary concept for vertebrate paleontologists, most invertebrate paleontologists, and paleobotanists. In other words, most paleontologists have come to identify these higher categories by trends or tendencies in a group, recognizing that some of its members might even lack the specific structures indicated in most of them. Unfortunately, many students of fossil insects have not followed such a concept and have erected taxonomic categories, such as families and orders, on single fragmentary specimens. Accordingly, some extinct orders of insects have been established on either very vague features or peculiar structures that might not occur in another species. Altogether, as a result of such practices, 44 extinct orders of insects have been established—almost twice as many orders as are usually recognized as now existing. From an extended study of most of the material on which these extinct orders have been based, I am convinced that only ten of them deserve ordinal status into which the other orders can be combined or merged in one way or another. In the following discussion, I shall refer only to these ten orders.

The insect fauna of the Upper Carboniferous period was basically primitive, for although some neopterous orders were present, they were in the minority. This was the only period in the history of the insects, so far as is known, when this was the case. The palaeopterous orders, of which there were five, included three main types. One of these types, comprising mayfly-like insects, was a complex of three extinct orders—the *Palaeodictyoptera*, *Protephemerida*, and *Megasecoptera*. Of these the *Palaeodictyoptera* were the most generalized; they had prothoracic wingflaps and in general the Carboniferous species showed a lack of specializations. Unfortunately, nothing at all is known of the immature stages of this order.

The little known *Protephemerida* require no comment here, but the *Megasecoptera* show several unusual features. They had very long abdominal cerci, lacked prothoracic flaps, and had more highly modified wings and body structures than the *Palaeodictyoptera*. In some

Table 1. Geological ranges of existing orders.

	Name of order	Earliest geological record
1.	*Collembola* (Springtails)	Devonian [?]
2.	*Entotrophi* (Bristletails)	Late Tertiary
3.	*Thysanura* (Silverfish)	Triassic
4.	*Odonata* (Dragonflies)	Early Permian
5.	*Ephemerida* (Mayflies)	Early Permian
6.	*Perlaria* (Stone flies)	Late Permian
7.	*Orthoptera* (Grasshoppers, crickets)	Triassic
8.	*Blattaria* (Roaches)	Late Carboniferous
9.	*Isoptera* (Termites)	Early Tertiary
10.	*Dermaptera* (Earwigs)	Jurassic
11.	*Embiaria* (Embiids)	Early Tertiary
12.	*Corrodentia* (Book lice)	Early Permian
13.	*Mallophaga* (Bird lice)	[No fossils known]
14.	*Hemiptera* (Bugs)	Early Permian
15.	*Anoplura* (Sucking lice)	Pleistocene
16.	*Thysanoptera* (Thrips)	Late Permian
17.	*Mecoptera* (Scorpion flies)	Early Permian
18.	*Neuroptera* (Ant lions, Dobson flies)	Early Permian
19.	*Trichoptera* (Caddis flies)	Jurassic
20.	*Diptera* (Flies, mosquitoes)	Jurassic
21.	*Siphonaptera* (Fleas)	Early Tertiary
22.	*Lepidoptera* (Butterflies, moths)	Early Tertiary
23.	*Coleoptera* (Beetles)	Late Permian
24.	*Strepsiptera* (Stylops)	Early Tertiary
25.	*Hymenoptera* (Bees, ants, wasps)	Jurassic

species the wings were falcate (Fig. 2), in others petiolate; in still others the prothorax was armed with spines. Noteworthy, also, was the presence of wing markings, which are evident even in specimens preserved in black shale. What colors were originally in the wings is not known, but a definite color pattern is indicated in the fossils.

As palaeopterous insects, the *Megasecoptera* presumably developed by incomplete metamorphosis; the presence of true nymphal forms definitely associated with adults, in the British coal measures, substantiates this conclusion. It should be noted, on the contrary, that Forbes [10] has expressed the belief that the *Megasecoptera* were actually holometabolous. It is true that, although most species of *Megasecoptera* are found preserved in palaeopterous fashion with wings outspread, a few families included species that unquestionably held their resting wings over the abdomen. That these latter species represent the beginnings of the neopterous line of evolution seems very doubtful to me in view of their several specializations; also, that they represent a distinct order, quite removed from the rest of the *Megasecoptera*, seems equally unlikely. I am led to believe, therefore, that the species of this order which were able to hold the wings over the abdomen developed this ability independently of the true neopterous types.

Another order of palaeopterous insects, very different from the three just mentioned, was the *Protodonata*, which closely resembled dragonflies. Like the latter, they were predaceous and had spiny legs and large mandibles. All of the *Protodonata* were large and some members of one family, with a wing expanse of two and one-half feet, were the largest insects known.[4] Nymphs of the *Protodonata* are entirely unknown, but in view of the

[4]These particular insects are the only extinct insects, so far as is known, that were larger than existing species. The inference has been drawn from the *Protodonata* that all Palaeozoic insects were very large, but this is not the case.

Figure 2. Photograph of *Mischoptera nigra* Brongniart, a megasecopteron from the Carboniferous of France. ×1.

similarity of the adults to true *Odonata*, we infer that the immature stages could not have been very different.

The third type of palaeopterous insect in the Carboniferous fauna has no counterpart in an existing order. Although named the *Protohemiptera*, they were closely allied to the *Palaeodictyoptera*, since they possessed prothoracic wing-flaps and other characteristics of the latter. But the mouth-parts of the *Protohemiptera* were modified to form a long suctorial beak, resembling that of certain *Diptera*, or flies, though differently formed. The Carboniferous members of this order were so much like the *Palaeodictyoptera* that some of the insects whose head structure is unknown and which have been considered *Palaeodictyoptera* were, I believe, *Protohemiptera*. The members of this order presumably fed either on plant juices from the large club mosses and tree ferns, or on the blood of amphibia and reptiles.

The neopterous insects of the Carboniferous include a vast and confusing assemblage related to the locusts and stone flies. Most of the species belong in the extinct order *Protorthoptera*, with a few aberrant ones in the *Caloneurodea*, and still others, obviously true roaches, in the *Blattaria*. The *Protorthoptera* show great diversity of structure. The more generalized species had membranous forewings and cursorial legs; others had leathery wings and either saltatorial or prehensile legs. Essentially, the *Protorthoptera* possessed the same amount and the kind of diversity that exists among the true *Orthoptera*, yet it is highly doubtful that any of these Carboniferous forms gave rise directly to the living groups they resemble.

The roaches were another interesting order in the Carboniferous. Although in numbers of individuals and described species they exceeded all other carboniferous insect orders, I am convinced that their abundance is very misleading. The swampy areas inhabited by the roaches supplied the best of conditions for their preservation as fossils, whereas other insects might encounter such optimum conditions only rarely. This condition would account for a disproportionately large number of roaches preserved as fossils. The extensive series of described species of roaches is due to the fact that Handlirsch and others have ignored the extreme instability of wing venation in both living and extinct types. Apart from their numbers, the most notable feature of the Carboniferous roaches was their close resemblance to species now living. Recently, however, an unexpected structure has been discovered in some Carboniferous roaches from Belgium: a long, projecting ovipositor, fully as long as the abdomen [11]. In all living roaches the ovipositor is vestigial or rudimentary, and the eggs are either laid in large capsules or else they hatch and form nymphs in the body of the female parent. The ovipositor in some of the Carboniferous species indicates a very different method of egg-laying.

From this survey of the Carboniferous fauna it is apparent that the insects had acquired surprising diversity and specializations by the Upper Carboniferous period, though some really generalized species were also included. Nevertheless, I am convinced that we have not yet begun to appreciate the extent of the Upper Carboniferous insect fauna. This conviction is based in part on the nature of the fauna in the lowest Permian strata and in part on the known diversity of the Carboniferous insects, even though represented by relatively few species.

Figure 3. Photograph of *Clatrotilan andersoni* McKeown, an orthopteron with a large stridulatory organ, from the Triassic of New South Wales. ×0.7.

If the same number of living species were collected at a few isolated localities over the world, we could not expect to obtain from them a good idea of the complexity of the world fauna as it exists today. It is not beyond the limits of possibility, therefore, that the extinct orders of Carboniferous insects were in their time comparable in extent to the major orders now living.

The insect fauna of the early Permian period was distinctive, for it was a combination of nine extinct orders and seven living ones. None of the extinct orders, except the *Protodonata*, are known to have lived beyond the Permian. The *Palaeodictyoptera* and *Protohemiptera* had apparently reached their maximum development in the Upper Carboniferous, only a very few having been found in Permian strata. The *Megasecoptera*, on the contrary, flourished all through the period. The *Protodonata*, also, were more numerous than in the preceding period, and very large species, like those previously noted, have been found in Permian beds in Kansas, Oklahoma, and several parts of Europe. Since no flying vertebrates were yet in existence, these large predatory insects must have ruled the air for many millions of years, for they persisted well into the Mesozoic. They may have been an important factor in the extermination of soft-bodied and weak-flying insects, such as the *Palaeodictyoptera* and *Megasecoptera*. The Permian *Protorthoptera* continued to show diversity of form. Among them, for example, are some whose cerci or posterior appendages in the male were modified to form pincers, or claws, resembling those in some of the living *Orthoptera*.

Three additional extinct insect orders make their first appearance in the early Permian. One of these, the *Protelytroptera*, which were related to the earwigs, were the first insects known to develop true protective forewings, or elytra; also the hind wings were greatly expanded and contained hinges which enabled the wings to fold up beneath the overlying elytra. Another extinct order of the early Permian, the *Protoperlaria*, was related to the stone flies; the adults were generalized with prothoracic wing-flaps, but the nymphs were adapted for an aquatic life. The final extinct order, the *Glosselytrodea*, appeared in the late Permian; they were characterized by highly modified elytra with unique venation.

The living insect orders of the early Permian, in addition to the roaches, comprised the may flies, dragonflies, bark-lice, true bugs, flies, and *Neuroptera*. With the exception of the *Neuroptera*, these Permian representatives were more generalized in most respects than any existing members of their orders. The Permian may flies, for example, had homonomous wings, whereas in all living species the posterior pair of wings are much reduced, both in size and venation. The Lower Permian was apparently close to the time of origin of most of these orders, for basic characteristics of related orders are combined in some species. A surprising feature is that these first insect representatives of existing orders are smaller in size than most present species of their orders, and some of the fossil species are as small as the smallest now living.

Before the end of the Permian, three more living orders of insects appeared. One of these, the stone flies, included a species which can be assigned with confidence to a living family. The other two orders comprise the thrips and the beetles. The dominant insects of the late Permian were true bugs, or *Homoptera*, which were clearly adapted for feeding on plant juices.

As is evident from this survey, the Permian insects were a remarkable assemblage. During no other geological period has such a diverse insect fauna existed. A striking contrast is found in the Triassic, at the beginning of the Mesozoic, in which the disappearance of all extinct orders, except the *Protodonata*, transformed the facies of the Triassic fauna to a semblance of that at present. True orthopteran insects first appear in Triassic beds. Among them were several species having well-developed stridulatory structures on the forewings of the males. The insects had a wing expanse of about 9 inches, and the stridulatory area of the wing was fully as large as that in any living insect, as shown in Figure 3.

By the beginning of the Jurassic the *Protodonata* became extinct, possibly because of the flying reptiles, or pterosaurs, which appeared early in the period. Earwigs, caddis flies, true flies, and the *Hymenoptera* are found in middle Jurassic strata. The flies, or *Diptera*, were almost exclusively midges or cranefly-like, there being none of the higher *Diptera*, many of which are now conspicuously associated with flowering plants. Similarly, the *Hymenoptera* were either relatives of sawflies or parasitic types; the aculeates, such as bees and wasps, were absent. Many of the Jurassic insects belonged to families now living (see Fig. 4). Looking at such specimens one finds it difficult to realize that they were contemporary with the pterosaurs, dinosaurs, and *Archaeopteryx*.

The Cretaceous insect fauna is virtually unknown, since very few specimens have been found. The gap is an un-

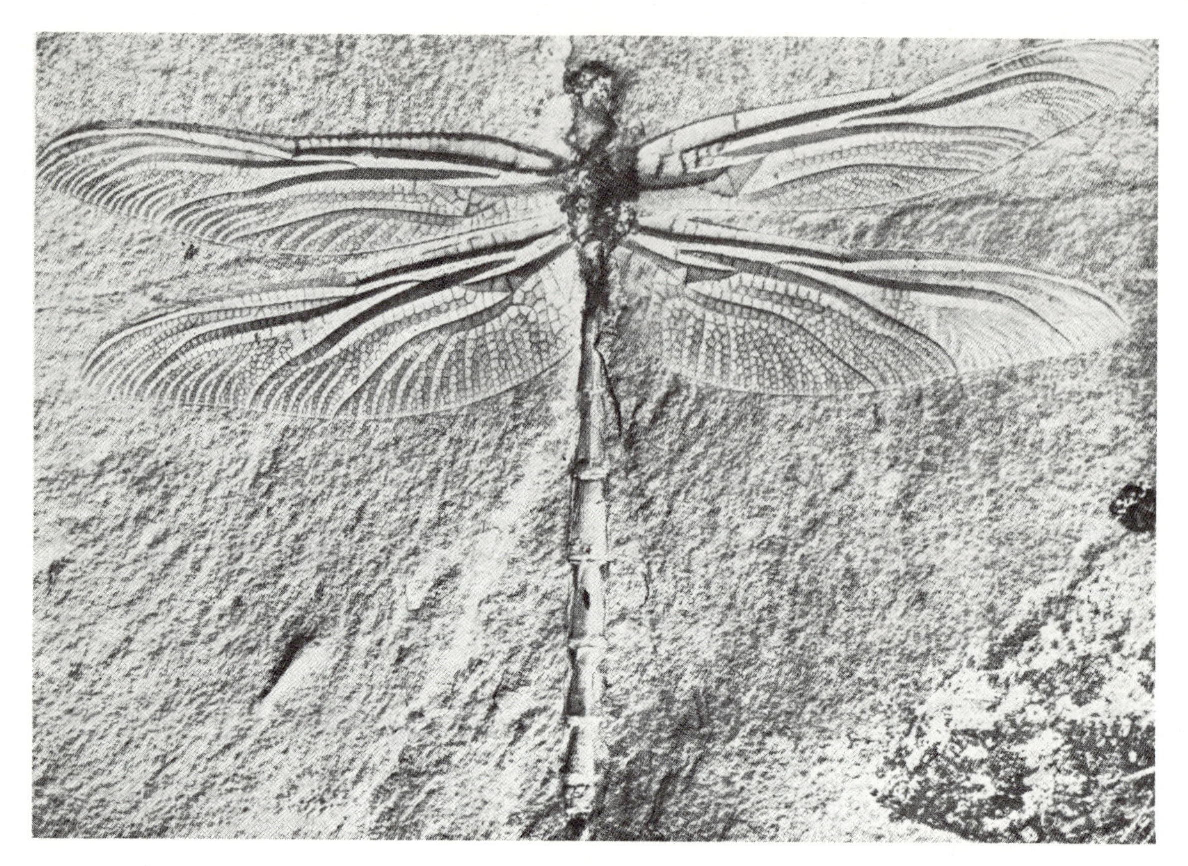

Figure 4. Photograph of *Protolindenia wittei* (Giebel), an odonatan from the Jurassic of Bavaria. ×1.7.

fortunate one, for a rapid development of the flowering plants and of the vertebrates took place during this long period. It is not surprising that insects of the early Tertiary period consist almost exclusively of families now living and to a large extent of living genera. The *Lepidoptera* and *Isoptera* first appear in early Tertiary rocks, but the nature of the earliest representatives shows that these groups arose in the Mesozoic. The insects of the Tertiary are better known than those of any equivalent interval of geologic time, largely because of the Baltic amber, which was formed from the resin of pine trees about 50 million years ago and which has preserved types of insects that would almost certainly not occur in rock formations. For example, two specimens of fleas, presumably from a rodent inhabiting the amber forest, have been found in the amber. The amber inclusions have also enabled more exact comparisons with living insects than ordinary preservation would permit. There are several instances of genera being recognized and established for amber species and subsequently being found in existence. More remarkable still is the occurrence in the amber of certain species of insects, mostly ants, which are apparently identical with some species now living. The Baltic amber has also furnished proof of the existence of social habits among the insects at that time, for the ants that occur there include, in addition to males and females, major and minor workers. The extent to which the complex habits of living ants had already been acquired in the early Tertiary is shown by the presence of plant lice attended by ants in search for honey dew, and by the presence of mites attached to the ants in the same manner as is characteristic today. It is worth noting, however, that by no means all of the families of insects had acquired such evolutionary stability by the

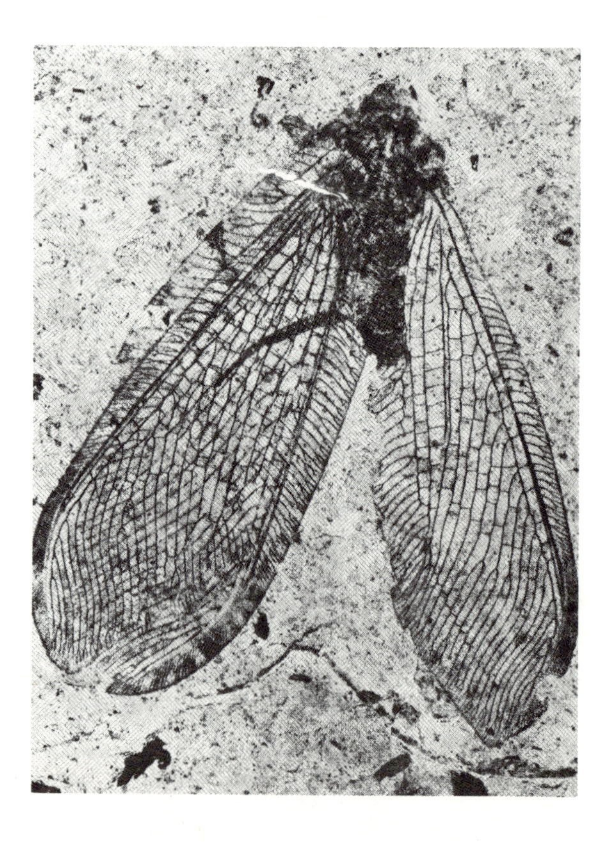

Figure 5. Photograph of *Lithosmylus columbianus* (Cockerell), an osmylid fly *(Neuroptera)* from the Miocene shales of Colorado. The family Osmylidae does not now occur in North America. ×3.

Figure 6. Photograph of *Raphidia moriua* (Rohwer), a snakefly from the Miocene shales of Colorado. The genus *Raphidia* does not now occur in North America. ×5.

Figure 7. Photograph of *Holcorpa maculosa* Scudder, a scorpion fly from the Miocene shales of Colorado. ×2.

early Tertiary period. The bees preserved in the amber, for example, belong to extinct genera.

A study of Tertiary insects also contributes to our understanding of the geographical distribution of living families and genera, many of which occupied very different regions from those now inhabited (see Figs. 5, 6). An example of this is shown in Figure 7, which depicts a peculiar scorpion fly from mid-Tertiary shales in Colorado; it belongs to a group now restricted to parts of Asia. Hundreds of examples of such changes could be given [12]. The best known of these is the occurrence in the Colorado Tertiary of tsetse flies (*Glossinidae*), now confined to Africa. Incidentally, the suggestion has been made by several mammalogists that trypanosomiasis, a protozoan disease now transmitted by the tsetse flies in Africa, might have been a factor in the extermination of some of the Tertiary mammals in North America.

A number of inferences might be drawn from the geological history of the insects as we now know it, only a few of which have been indicated above. Certainly there is one justifiable conclusion, namely, that our existing insect fauna is but a small fragment of the total insect aggregation that has occupied the earth during the past 250 million years. Understanding of insect evolution depends to a large extent on a knowledge of the extinct insect population. The investigation of the fossil record has only begun, and progress is slow, but the significance of the record increases with each discovery.

References

1. Scudder, S.H. Systematic review of our present knowledge of fossil insects. *Bull. U.S. Geol. Surv. 31,* 1–128, 1886.
2. Handlirsch, A. Die fossilen Insekten und die Phylogenie der rezenten Formen. W. Engelmann, Leipzig, 1906–1908.
3. Handlirsch, A. Neue Untersuchungen über die fossilen Insekten. *Ann. naturh. Mus. Wien,* Teil I. *1, 48,* 1–140, 1937.
4. Lemche, H. The origin of winged insects. *Vidensk. Medd. fra Dansk naturh. Foren, 104,* 127–168, 1950.
5. Lemche, H. The wings of cockroaches and the phylogeny of insects. *Vidensk. Medd. fra Dansk naturh. Foren, 106,* 288–318, 1942.
6. Carpenter, F.M. The supposed nymphs of the *Palaeodictyoptera. Psyche, 55,* 41–50, 1948.
7. Martynov, A.B. The interpretation of the wing venation and tracheation of the Odonata and Agnatha. *Rev. Russie Ent., 18,* 145–174, 1923 (in Russian). Also English translation in *Psyche, 37,* 245–280, 1930.
8. Martynov, A.B. Über zwei Grundtypen der Flügel bei den Insekten und ihre Evolution. *Zeit. Morph. Ökol. Tiere, 4, 3,* 465–501, 1925.

9. Crampton, G.C. The phylogeny and classification of insects. *Pomona Journ. Ent. Zool., 16,* 33–47, 1924.
10. Forbes, W.T.M. The origin of wings and venational types in insects. *Am. Midl. Nat. 29,* 381-405, 1943.
11. Laurentiaux, D. Le problème des blattes paléozoiques à l'ovipositeur externe. *Ann. Paleont., 37,* 187-195, 1951.
12. Ander, K. Die Insektenfauna des Baltischen Bernsteins nebst damit verknüpft zoogeographischen Problemen. *Kung. Fysiografiska Sällskapets Handlingar, N.F., 53,* 1-82, 1942.

Léo F. Laporte

Paleoenvironments and Paleoecology

The study of the ecologic relationships of fossil animals and plants helps to determine the role of environment in their evolution

Several years ago, during the "ecological crisis," bumper stickers demanding "Ecology now" were commonly seen. This prompted a colleague to suggest that paleontologists put on their cars bumper stickers proclaiming "Paleoecology then." Whimsical though this suggestion was, I believe there was serious motivation behind it: any genuine and profound understanding of the complex interrelationships of living organisms with their environments must include knowledge of past ecological patterns. This notion has recently been clearly articulated by Andrew Hill (*1*):

Past ecosystems mark stages in the evolution of the present-day situation, and paleoecology has the potential of providing a time perspective for modern ecological studies. A knowledge of the succession of fossil ecosystems offers insight into those factors that control their structure and cause change. Profound ecological . . . modifications . . . accompanied the evolution of man, and at present man's impact upon environments and animal communities is considerable. Thus, information regarding the dynamics of ecological change is particularly important where man has been involved over the time period of his recent evolution.

Dr. Laporte is Professor of Earth Sciences and Fellow of Cowell College, University of California, Santa Cruz. After receiving his undergraduate and graduate education at Columbia University, he taught at Brown University for 12 years, until 1971 when he joined the faculty at Santa Cruz. Trained as a paleontologist, Dr. Laporte's particular research interests are concerned with the paleoecology of Paleozoic marine organisms and environments. He is author of an introductory text on paleoecology, Ancient Environments; *recently he published* Encounter with the Earth, *an introduction to environmental geology, which has been a side interest of his for several years. Address: Cowell College, University of California, Santa Cruz, CA 95064.*

In this essay, I will attempt to summarize the progress of developments in paleoecology in the context of the present state of the art. Let me point out immediately that, while major advances in a scientific field are often accomplished by new paradigms (to use Thomas Kuhn's language), the impetus behind paleoecology's success has been, instead, the application of a method: the examination and analysis of modern interactions of organisms and environments to infer principles and explain processes that are relevant to the interpretation of ancient biotas and their habitats as now recorded by fossils in their surrounding rock matrix. This approach, of course, follows the famous dictum that "the present is the key to the past," distilled from the writings of the founders of modern geology in the late eighteenth and nineteenth centuries—James Hutton, John Playfair, and Charles Lyell. As just noted, however, the inverse of this is also true—the past may be a key to the present. We might expect, therefore, continuous interplay between what we learn about the ancient and what we observe in the recent, to the end that human occupation of the planet will foster, not preclude, life in all its richness and abundance.

The record of life on earth is a complex sequence of fossil animal and plant assemblages. Paleontologists, in their efforts to describe and interpret the history and evolution of life, must explain the sources of variation among these different assemblages. The major factors that determine the composition of a particular fossil assemblage are *age*—that is, the stage of evolution that the various life forms in the assemblage had reached at a particular time; *environment*—the physical, chemical, and biological influences exerted by the environment where the assemblage lived; and *geography*—the broad, regional distribution of life, of which the assemblage is part, that existed at that place at that time.

Thus, a Silurian molluscan assemblage is easily distinguished from a Miocene one because of the differences in evolutionary stages reached by Silurian and Miocene molluscs during their temporal separation, some 400 million years. Fossils from a Devonian reef (primitive corals, calcareous algae, and brachiopods) are distinct from Devonian tidal flat fossils (ostracodes and algal structures) because of different environmental requirements found in reefs as opposed to tidal flats. Even fossil assemblages that are from the same environment and are of the same age can be discriminated on the basis of differences in broad geographic distribution of the organisms. For example, Permian marine invertebrates that occupied the Tethyan Sea, which stretched from Gibraltar through the Mediterranean to east Asia 250 million years ago, can be separated from Permian marine faunas that occupied environmentally similar but geographically distinct seas of the North American midcontinent.

Paleoecology and biology

Biology has clarified the major processes of evolution through its contributions on the genetics, embryology, comparative anatomy, functional morphology, and ecology of living organisms. Paleontology has provided the historical evidence of those processes operating over enormous spans of time by establishing the temporal sequences of fossil animals and plants, interpreting ancestor-descendant relationships, and documenting particular trends in morphological characters or character complexes. Thus biology, on the one hand, tells us how evolution proceeds while descriptive paleontology, on the other hand, tells us what the results of those processes have been at some specific time in the past.

But there is an important gap here. Paleontology may provide evidence that a particular group of fossils has evolved in a certain way, and biology may provide the principles according to which such evolution must occur. But we want to know more than this: we want to know how come this evolution occurred at that time and place in the past. In short, we wish to understand not only the "what," "when," and "how" of evolution, but also the "why." (We are not speaking of a metaphysical "why" but an evolutionary "why"—that is, what biological purpose did a particular evolutionary event or novelty serve?) It is here that paleoecology can contribute significantly to the study of organic evolution—by bridging the gap between processes of evolution and their products as seen in the geologic record.

The modern theory of organic evolution stresses the integration of a population of organisms with its environment. This integration is ultimately achieved by natural selection, as expressed through differential reproduction, so that the adaptive characteristics of the population meet the various physical, chemical, and biological demands of the local environment. George G. Simpson has described this adaptive interaction of organisms and environment as the extent of overlap between the "prospective functions of organisms and the prospective functions of the environment." The biologist and paleontologist can describe and evaluate the prospective functions of recent and ancient organisms, whereas the ecologist and paleoecologist attempt to describe and evaluate the prospective functions of recent and ancient environments. Together these specialists contribute to the full understanding of life history by interpreting the adaptive interaction of organisms and environments through time.

An illustration from the marine fossil record will clarify this. An interesting group of relatively large, bottom-dwelling, calcareous protists—the fusulinids—flourished in the shallow seas of the late Paleozoic throughout the world, but they are now extinct. During their 100-million-year history, the fusulinids, starting from a late Mississippian ancestral stock, evolved into four separate families including some 60 genera and more than 200 species. At their peak of development, fusulinids often outnumbered all other contemporary shelly invertebrates in some fossiliferous strata. Major morphological trends during this diversification have been traced in great detail by a number of paleontologists and include overall increase in size; change from a discoidal to an elongate, subcylindrical shape; increased complexity of the shell wall; and fluting of the wall, which subdivided the shell interior into a number of chambers. These trends, along with others confined to individual families, occurred relatively rapidly, and nine distinct temporal subdivisions within the Pennsylvanian and Permian periods are currently recognized. Despite the abundance of fusulinids and the detailed knowledge of their descriptive evolutionary history, virtually nothing is known about the adaptive significance of these trends.

Thus, even though we have adequate theory relating to evolutionary processes and elaborate information about the evolutionary products, we do not understand just what evolutionary opportunities fusulinids were exploiting. There are two reasons for this lack of understanding. First, we do not now have sufficient knowledge about the various environments that fusulinids inhabited beyond the fact that "they seem to have been restricted to off-shore, open water environments" (*2*). Second, we do not know the biological functions performed by most fusulinid morphological characters.

The answers, of course, might be had by detailed paleoecological studies of the sedimentary rocks in which fusulinids are found in order to establish the prospective functions of the environment. The prospective functions of the organisms might be elucidated by examining the biology and ecology of modern living alveolinids, a group of benthic calcareous protists that, although phylogenetically separate and distinct, share many striking morphological similarities with fusulinids. Integration of these approaches would certainly go a long way in explaining fusulinid adaptations and, by extension, answer the "how come" of their evolutionary history.

Early views

Although paleoecology has rather recently become a major field of study within the broader area of paleontology, paleoecologic inquiry is really as old as paleontology itself. For example, the evidence brought to bear on the "fossil controversy" during the Renaissance and Enlightenment was often paleoecological in nature. To prove one way or another whether fossils were merely sports of nature, works of the devil, the result of spontaneous generation, or whether they were indeed the actual remains of once-living organisms involved the demonstration that the "figured stones" had virtually identical modern counterparts. In the case of marine invertebrate fossils, the further conclusion that they were deposited in an ancient sea—although now found on dry land—merely added fuel to the fire. For not only did this imply the acceptance of fossils as once-living organisms, but it also indicated multiple migrations of ancient seas across the lands. The last conclusion made matters worse for the rigidly orthodox biblical interpreters, because it was clearly stated in Genesis (1:9-10) that only *after* the lands were divided from the waters on the third day did the Creator make animals and plants.

Jean-Étienne Guettard (1715–86), a French naturalist, wrote one of the earliest paleoecological treatises, in 1759, in which he stated that through careful observation of the accidental peculiarities of fossil shells their marine origin can be demonstrated.

> One finds [among fossil shells] all the classes, all the genera, and even many of the species with which we are familiar and which are fished up today: who cannot recognize . . . the Purpuras, the Murexes, the Oysters, the Cowries, and so many other fossil shells? . . .
>
> I believe, however, I ought to support yet another proof by bringing more attention to the accidents that one notes among fossil bodies, [something] which hasn't

Diagram of Regions of Depth in the Ægean Sea.

Sea-Bottom = deposits forming.	Region.	Depth in fathoms.	Characteristic Animals and Plants.
Extent—12 feet. Ground various. Usually rocky or sandy (conglomerates forming).	I.	2	Littorina cœrulescens. Fasciolaria tarentina. Cardium edule. Plant:—Padina pavonia.
Extent—48 feet. Muddy. Sandy. Rocky.	II.	10	Cerithium vulgatum. Lucina lactea. Holothuriæ. Plants:—Caulerpa and Zostera.
Extent—60 feet. Ground mostly muddy or sandy. Mud bluish.	III.	20	Aplysiæ. Cardium papillosum.
Extent—90 feet. Ground mostly gravelly and weedy. Muddy in estuaries.	IV.	35	Ascidiæ. Nucula emarginata. Cellaria ceramioides. Plants:—Dictyomenia volubilis. Codium bursa.
Extent—120 feet. Ground nulliporous and shelly.	V.	55	Cardita aculeata. Nucula striata. Pecten opercularis. Myriapora truncata. Plant:—Rityphlœa tinctoria.
Extent—144 feet. Ground mostly nulliporous. Rarely gravelly.	VI.	79	Venus ovata. Turbo sanguineus. Pleurotoma maravignæ. Cidaris histrix. Plant:—Nullipora.
Extent—156 feet. Ground mostly nulliporous. Rarely yellow mud.	VII.	105	Brachiopoda. Rissoa reticulata. Pecten similis. Echinus monilis. Plant:—Nullipora.
Extent—750 feet. Uniform bottom of yellow mud, abounding for the most part in remains of Pteropoda and Foraminifera.	VIII.	230	Dentalium 5-angulare. Kellia abyssicola. Ligula profundissima. Pecten hoskynsi. Ophiura abyssicola. Idmonea. Alecto. Plants:—0.
Zero of Animal Life probably about 300 fathoms.			
Mud without organic remains.			

True Scale of the above Diagram.

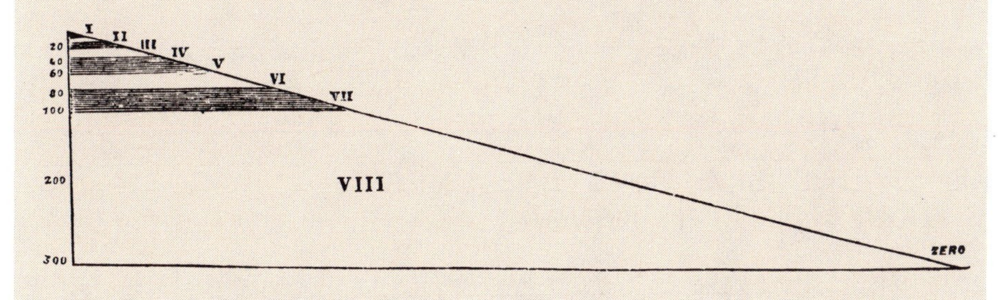

Figure 1. Forbes's depth zonation of invertebrates and algae in the Aegean Sea was a landmark effort of relating marine species to particular environmental factors—in this case water depth and substrate. As a geologist, Forbes wanted to use such data for interpreting depositional environments of ancient rocks. Note that Forbes reported no bottom life beyond 300 fathoms, no doubt owing to inadequate sampling (From ref. 5)

been done so far.... These accidents [which] seem to me to prove that the fossils have belonged to bodies which have had life... are reduced to four forces; they concern [the fossils'] attachments, their preservation, their destruction, or their deformation....

[Modern] shells which are attached to other shells or especially to other bodies are bivalves... or multivalves. The first ordinarily are oysters of different species, the second are barnacles or acorn shells, to which one can perhaps add worm tubes. ... The adherence of oysters occurs on [other] oysters, ... on corals, on tree branches, or on pebbles.

Guettard then proceeds to show similar sorts of attachments among fossil specimens including encrusting oysters, barnacles, and serpulid worms. He summarizes his "attachments" argument by saying:

One can no longer refuse this truth that fossil shells have really enclosed animals that had life because their accidents, their attachment or adherence, indicate a succession of time in their formation, and that this can only be done in the way that shells taken from the sea [today] do it, which in this aspect entirely resembles fossil shells [3].

By analogous reasoning Guettard goes on to prove that the modes of preservation, disarticulation, and fragmentation seen among fossil shells are exactly like those found among sea shells seen along the coasts today. Thus Guettard, by using observations from living marine organisms to interpret fossil occurrences, anticipated one of the fundamental modes of analysis of the modern paleoecologist.

Still earlier, Leonardo da Vinci (1452–1519) used what might be termed paleoecological reasoning to counter the argument that the flood of Noah could explain the origin of all fossil shells.

The cockle is a creature incapable when out of water of more rapid movement than the snail, or even is somewhat slower, since it does not swim, but makes a furrow in the sand, and supporting itself by means of the sides of the furrow will travel between three and four braccia a day [equal to about 6 to 8 feet]; and therefore, with such a rate of motion it would not have travelled from the Adriatic Sea as far as Monferrato in Lombardy, a distance of 250 miles, in forty days.... If you should say that the shells were empty and dead when carried by the waves, I reply that where the dead ones went the living ones were not far distant, and in the mountains

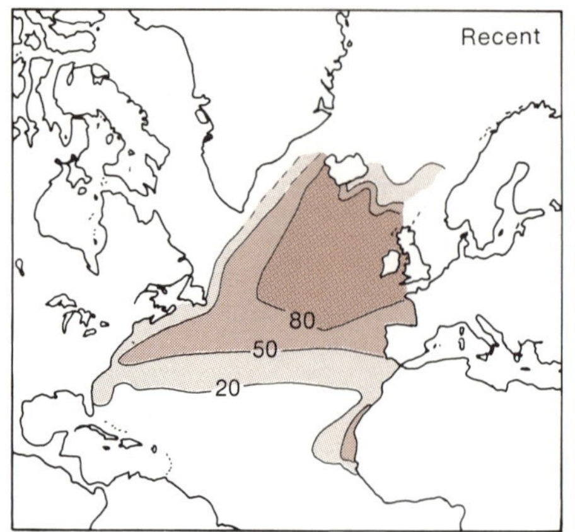

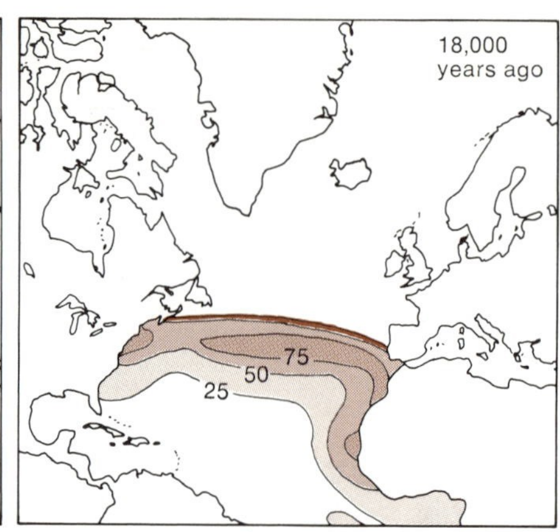

Figure 2. (*left*) The relative abundance of planktonic microfossils (the darker the color, the more abundant the microfossils) belonging to present-day subpolar assemblages is plotted from tops of cores raised from the Atlantic sea floor. (*right*) Samples of sediment from these same cores that was deposited during the last glaciation, some 18,000 years ago, show these species extending farther south, indicating that subpolar water masses reached much lower latitudes at that time. (After ref. *8*)

are found all living ones, for they are known by the shells being in pairs, and by their being in a row without any dead, and a little higher up is the place where all the dead with their shells separated have been cast up by the waves. . . . And if the shells had been in the turbid water of a deluge they would be found mixed up and separated one from another, amid the mud and not in regular rows in layers as we see them in our own times [*4*].

Thus, da Vinci, like Guettard, realized not only that fossils were the geologic remains of marine life but also that inferences could be made about how these fossils came to be incorporated in sedimentary rocks. Admittedly, looking back from our present state of knowledge, these early paleoecological writings seem naive. Nevertheless, it is clear that, virtually from the beginning, inquiry into the true nature of fossils had, for some naturalists at least, paleoecological leanings.

A major milestone in the development of paleoecology was the recognition that studies of living marine biotas could contribute significantly to the interpretation of the depositional environments of fossiliferous marine strata. Particularly important in this regard was the British naturalist Edward Forbes (1815–54), who is seen as a principal founder of modern paleoecology.

Two papers published by Forbes are

Figure 3. In 1937, Elias used the depth zonation of algae and invertebrates in modern seas (*the top section*) to infer the depth zonation of marine fossils for part of the early Permian period in Kansas, the so-called Big Blue time (*The bottom section*). (From ref. *9*)

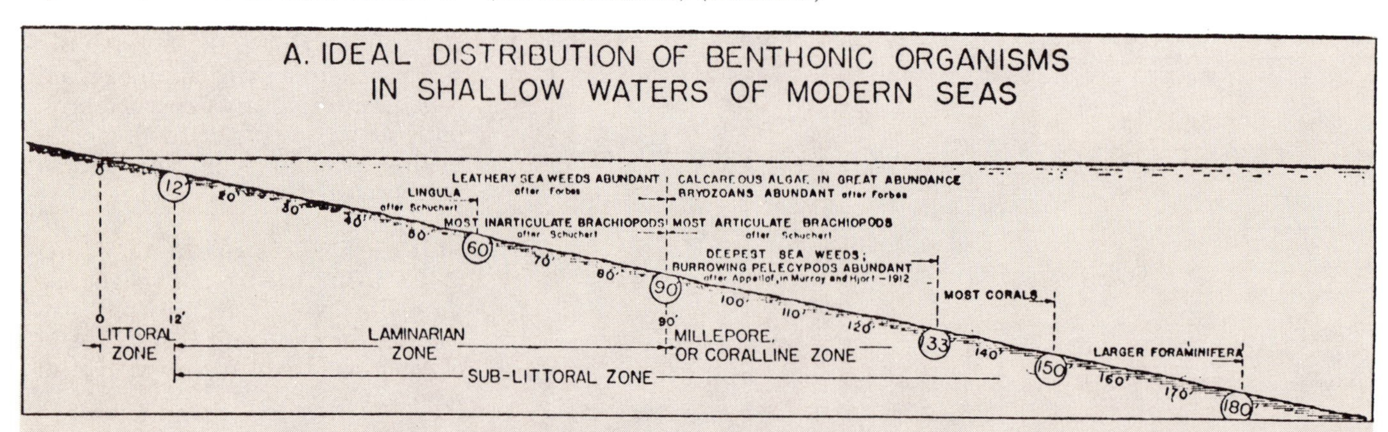

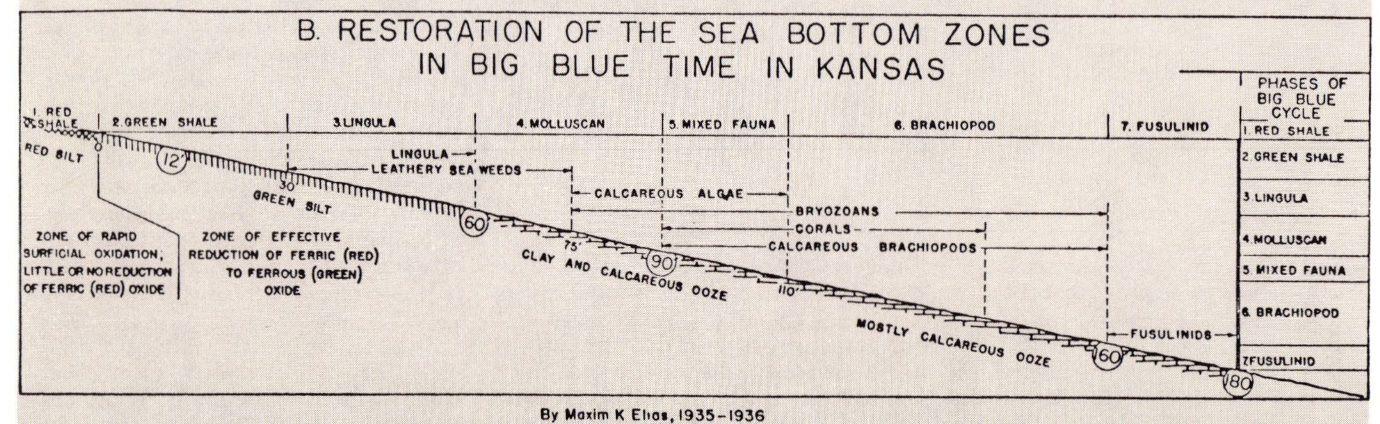

especially seminal. In 1843, Forbes published a report on the distribution of marine invertebrates and algae in the Aegean Sea in which he attributes the presence and abundance of individual species to the primary influences of climate, salinity, and water depth, with secondary influences being substrate, tides and currents, and influx of fresh water (Fig. 1). After presenting his data and their interpretation, Forbes says:

The importance of these results must be obvious to the geologist. The inductions as to climate or distribution, which he may draw from his examination of the Testacea [or shelly invertebrates] of a given stratum, will vary according to the depth in which those Testacea lived and the ground [or substrate] on which they lived. ... By carefully observing the mineral character of the stratum in order to ascertain the nature of the former sea-bottom, by noticing the associations of species and the relative abundance of the individuals of each in order to ascertain the depth, and by calculating the percentage of northern or southern forms separately for each tribe, our conclusions [as to the original environment] will doubtless approximate very nearly to the truth [5].

Forbes goes on to predict how the sediments and biotas would appear as fossiliferous marine rocks "were the bottom of the Aegean Sea, with its present inhabitants, to be elevated and converted into dry land."

Forbes was apparently so taken with the importance of this new approach for interpreting fossils and rocks that one year later, in 1844, he summarized and refined his earlier conclusions in a paper entitled "On the Light Thrown on Geology by Submarine Researches" (6). He begins this paper with gentle scolding: "the geologist has been fully occupied above water, and the naturalist has pursued his studies with too little reference to their bearing on geological questions, and on the history of animals and plants *in time*."

At least one naturalist-geologist shared Forbes's convictions, for in the same volume immediately following his paper, there is one by William Rhind, entitled "The Geological Arrangement of Ancient Strata Deduced from the Condition of the Present Ocean Beds." In general, however, it seems that, like so many original thinkers, Forbes was well ahead of his time, for his advice was

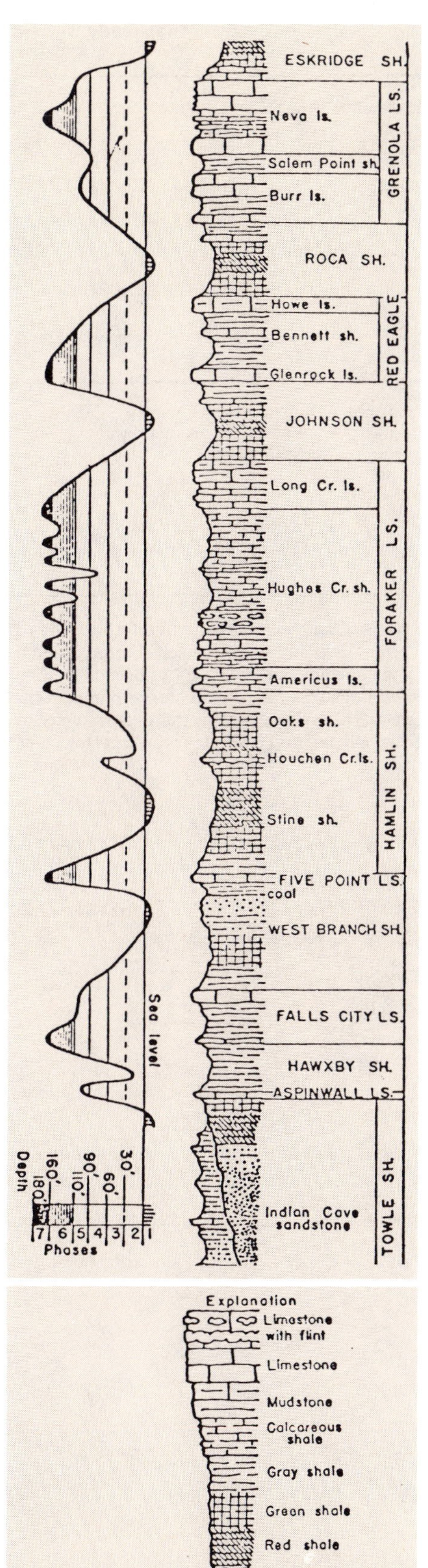

Figure 4. Using the inferences about depth distribution of fossil organisms as shown in Fig. 3, Elias interpreted the temporal fluctuations of sea level recorded by early Permian sedimentary rocks, some 275 million years ago. (From ref. 9)

not followed by geologists or naturalists. It wasn't until almost a century later that American paleontology began to move in the direction of paleoecological inference and interpretation. Two papers published in the 1930s seem now, in retrospect, classic examples of Forbes's original point of view.

More recent approaches

In 1933, M. L. Natland described five foraminiferal assemblages from the waters off the southern California coast that were depth- and temperature-dependent. Using these same species Natland then interpreted the depositional environments of some upper Tertiary foraminiferal-bearing sedimentary rocks from the nearby Ventura Basin (7). Natland emphasized that his results demonstrated that fossils were not only guides to past time—a conclusion overly stressed by his contemporaries—but also guides to past environment.

A direct descendant of Natland's methodology is that followed by CLIMAP (Climate: Long-Range Investigation, Mapping, and Prediction), part of the National Science Foundation's International Decade of Ocean Exploration program. As shown in Figure 2, the distribution of present-day subpolar planktonic microfossils is used to determine the position of subpolar waters during the last glaciation, some 18,000 years ago, based on their presence in sediment cores raised from the Atlantic sea floor (8).

In 1937, M. K. Elias published a paper in which he not only used precisely the scientific tactics recommended by Forbes but even had to rely on some of Forbes's Aegean Sea data, owing to the lack of more modern studies of shallow marine organisms and environments (9). Like Natland, Elias made explicit the connection between the contemporary and the ancient (Fig. 3). Unlike Natland, however, Elias referred to the full spectrum of bottom-dwelling life, including algae, protists, and the higher invertebrates, as well as to the sediments to reconstruct early Permian, epicontinental sea environments. Elias then used this model to interpret temporal shifts in depositional environments as recorded by early Permian rocks in eastern Kansas (Fig. 4). Although Elias's interpretations seem a little too simple now, his approach was highly original and powerful.

One predictable outcome of looking at living marine invertebrates in their natural habitats was the recognition of the importance of feeding type and relation to the substrate of individual taxa (Fig. 5). The role of the substrate in limiting the distribution and abundance of benthic invertebrates is especially relevant for paleontologists, because this is one parameter of the original environment that is directly recorded in the geologic record. The sediments—which were the substrate for the bottom-dwelling fossils—are preserved, even though somewhat altered, as the enclosing rock matrix of the fossils. Moreover, the texture and composition of the substrate can be related to other important environmental variables—salinity, water energy and circulation, organic content, stability, and so on—known to be ecologically significant.

An extremely interesting application of feeding type and substrate relation to fossil marine invertebrates was made by Steven Stanley, who explained the major, post-Paleozoic radiation of the bivalves as due to the evolution of the infaunal (buried in the sea floor), siphon-feeding way of life (Fig. 6). Prior to Stanley's work, conventional wisdom held that the post-Paleozoic expansion of bivalves reflected their occupation of niches left vacant by widespread extinctions of another group of invertebrates—the brachiopods—at the end of the Paleozoic. (At the end of the Paleozoic, the brachiopods crashed from some 200 genera to fewer than 50.)

By determining the mode of feeding and substrate relation for 28 extant superfamilies of bivalves, Stanley found that more than half are infaunal, siphonate-suspension feeders and that 14 of the 18 superfamilies that have appeared since the end of the Paleozoic pursue this way of life. He concluded that mantle fusion and siphon formation permitted significantly deeper burrowing and invasion of a whole new adaptive zone, and consequently had little or nothing to do with the massive extinction of the brachiopods, all of which lived on the substrate rather than deeply within it.

With the exception of Elias's Permian fossils, the studies I've mentioned all deal with closely related living representatives of the fossil taxa in question. Reconstructing the paleoecology of ancient forms by reference to the ecology of recent representatives has certainly seemed to work, especially in recent and ancient taxa not widely separated in time. But the farther one goes back in geologic time, of course, fewer and fewer fossil taxa have extant species or genera. Even in cases where they exist, we are uneasy about assuming that their ecological requirements have remained the same over, say, tens of millions of years. Instead, paleoecologists are forced to read out from the enclosing rock matrix—independent of the fossils themselves—as much environmental information as possible, using such evidence as size and composition of the sedimentary grains, primary structures like mud cracks and ripple marks, and systematic lateral and vertical changes in these and other characteristics.

In the last two decades geologists have learned a great deal about what kinds of sediments are deposited in a wide variety of terrestrial and marine environments. Consequently, the different internal characteristics of

		Filter-feeders	Sediment-feeders	Herbivores	Carnivores and scavengers
Epifaunal living on the surface of the sea floor	—Mobile	Crustacea		Gastropoda Echinoidea	Gastropoda Annelida Crustacea Asteroidea Ophiuroidea
	—Attached	Porifera Bryozoa Brachiopoda Bivalvia Crustacea Crinoidea			
Infaunal living buried in the sea floor	—Burrowing in soft sediment	Bivalvia Annelida	Bivalvia Annelida Ophiuroidea Echinoidea		Gastropoda Crustacea
	—Boring in hard rock or wood	Porifera Bivalvia			

Figure 5. In this chart of the life habitats of bottom-dwelling marine invertebrates, colored type indicates the major groups of each feeding type and substrate relationship. An important first step in determining the paleoecology of fossil marine invertebrates is to ascertain how a given group exploited its food and space resources. Inferences about the life habits of fossils can come from comparison with closely related living forms, from the morphology of the fossils themselves, and from the nature of the enclosing rock matrix. (After ref. *10*)

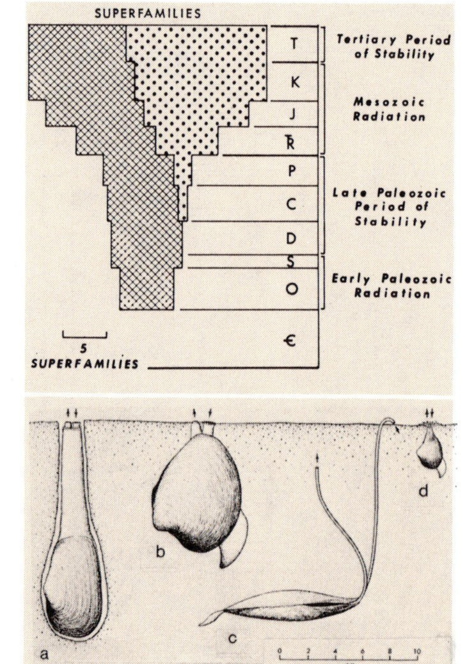

Figure 6. (*top*) The post-Paleozoic expansion of bivalues was thought to be connected to the extinctions of another group of invertebrates, whose niches were then left vacant. Stanley, however, explained their expansion in terms of an increase in superfamilies with infaunal, siphon-feeding forms (*stippling*), which exploited new niches, as compared with superfamilies without these forms (*cross-hatching*). (*bottom*) Siphon-feeders living in soft sediment include (*a*) *Mya*, (b) *Mercenaria*, (c) *Tellina*, and (d) *Cuspidaria*. (From ref. *11*)

sedimentary rocks provide telltale information about the specific environments in which they were deposited. Once that ancient environment is reconstructed, we can then begin to think about how the associated fossil organisms related ecologically to it. This, in turn, requires that paleoecologists understand the general principles of functional morphology and comparative physiology of the groups with which they are working in order to provide biologically reasonable hypotheses about how the fossils managed in these environments.

Because of the necessity of first defining and delineating the depositional environments of the rocks in which the fossils are found, most paleoecological research involves an activity commonly referred to as "environmental stratigraphy." That is, we must first understand the nature and three-dimensional distribution of the paleoenvironments of the rock strata; then in that context we can proceed to working out the paleoecology of the included fossils. This procedure obviously is meant to preclude circular reasoning whereby we might interpret an environment on the basis of the fossils found, and then explain the paleoecology of the fossils in terms of that environment.

Once we have established the paleoecology of one ancient environmental setting, we can compare it to a similar one from a different geologic epoch. The point of such a comparison is to examine how much evolution there might have been from the earlier situation to the later one. That is, while we certainly would expect the organisms in comparable habitats to evolve (a different cast of actors), we might wonder how much, if any, change there was in the organization of the community of organisms (completely different scenarios).

To test this notion from tidal flat and shallow subtidal environments of the

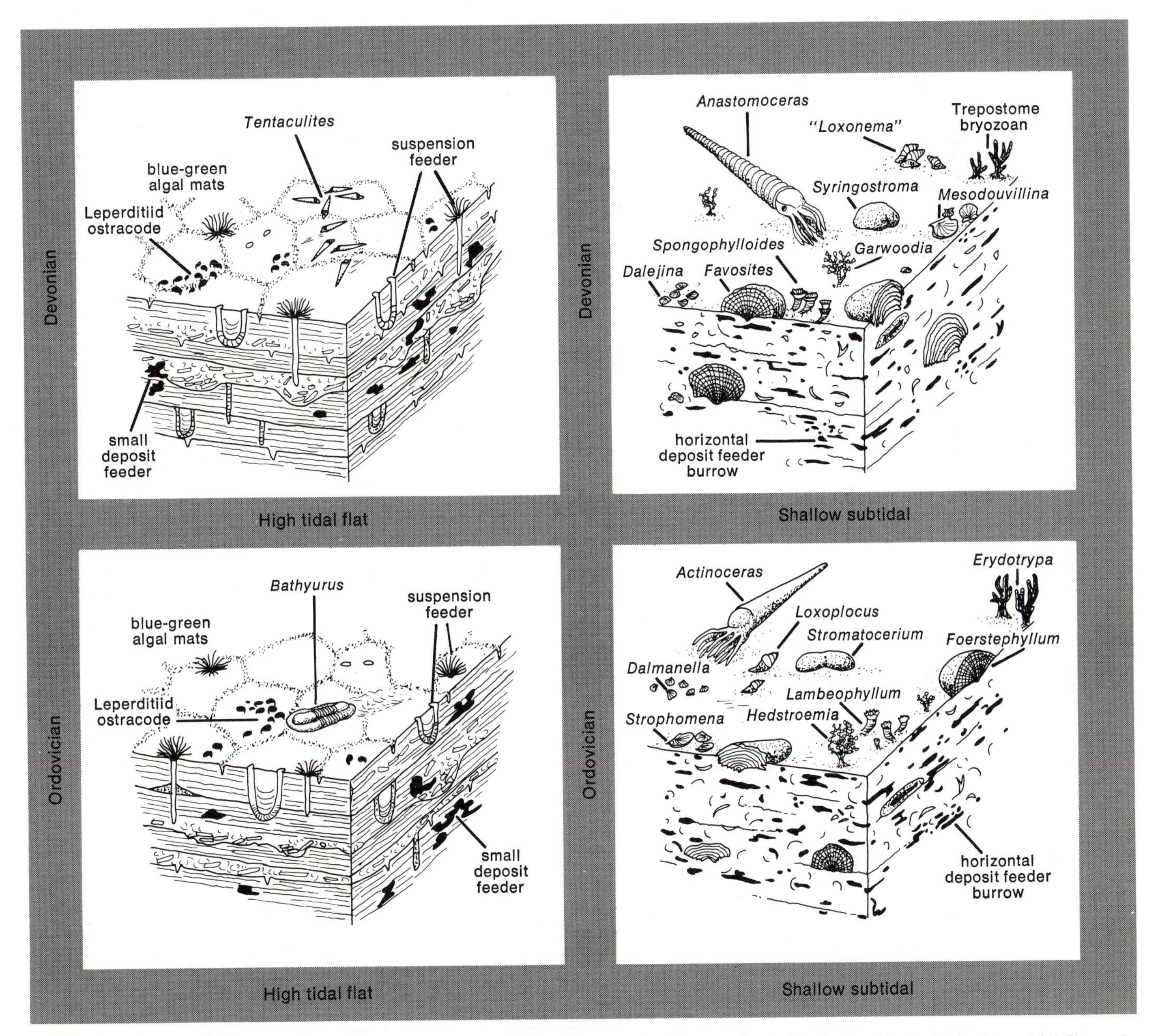

Figure 7. These schematic illustrations of nearshore marine communities from Paleozoic rocks of New York show that despite the 70-million-year separation in time from the Ordovician and Devonian, the high tidal flat and shallow subtidal marine communities are virtually identical with respect to basic types of organisms, even though particular genera and species vary. The only really significant difference appears to be the presence of trilobites (*Bathyurus*) in the Ordovician tidal flat environment and their absence in the Devonian, whereas the Devonian contains tubular benthic molluscs (*Tentaculites*), which are lacking in the Ordovician. (After ref. *12*)

Paleozoic, Kenneth Walker and I collaborated on a study of Ordovician and Devonian limestones from upstate New York that were separated by some 70 million years (*12*). Using independent criteria from the inorganic characteristics of the rocks (primary structures, composition, texture, and so on), we first established that the rocks had comparable environments, such as high and low tidal flat and shallow subtidal areas. We then determined which fossils occurred where and what their feeding types and relations to the substrate were. Surprisingly, we found that there was a striking similarity in kind and abundance of organisms in comparable Ordovician/Devonian environments (Fig. 7). It appears, in this example at least, that these tidal flat and shallow subtidal communities remained virtually identical for 70 million years, although there was considerable evolutionary turnover in particular species. Other paleoecologists have been pursuing such studies to see how other ancient communities have varied through time. This remains one of the especially exciting areas of paleoecological analysis.

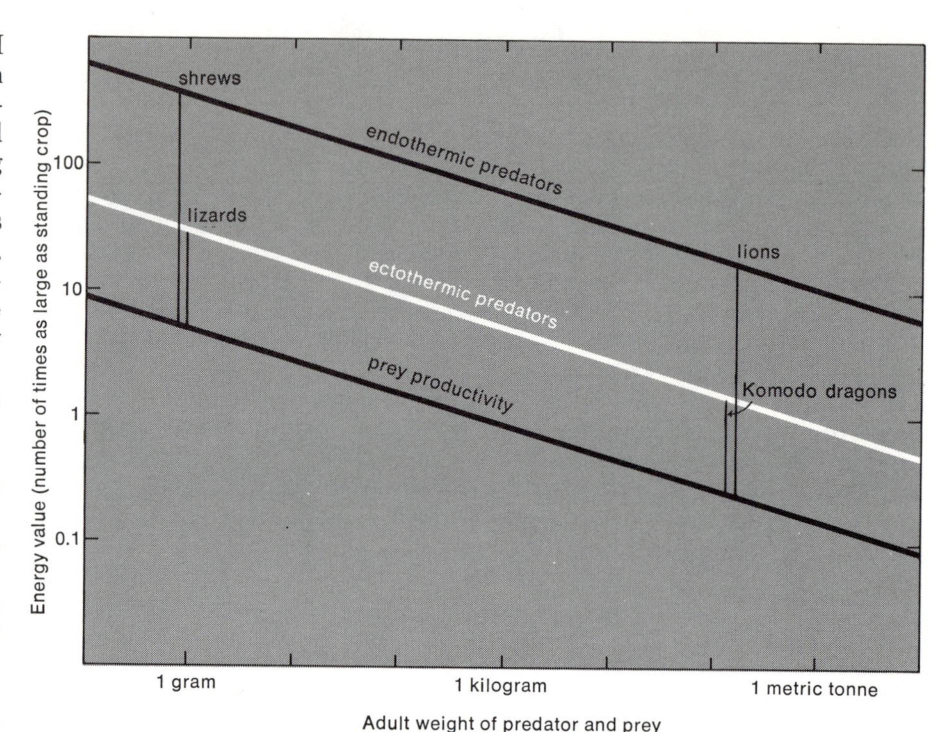

Figure 8. A long-settled issue—that dinosaurs were simply big cold-blooded lizards—has been opened again. Bakker uses predator-prey ratios to support his theory that dinosaurs were warm-blooded. Although the predator-prey ratio for endothermic and ectothermic predators remains the same irrespective of the sizes of the animals involved, there is an order of magnitude difference in the amount of energy from prey required to support endothermic predators as compared with ectothermic predators, whether considering lizards and shrews or Komodo dragons and lions. (Adapted from "Dinosaur Renaissance," by Robert T. Bakker. Copyright © Scientific American, Inc. All rights reserved.)

Hot-blooded dinosaurs

One of the methodologically most interesting and original paleoecological studies (and consequently still controversial) is that of Robert Bakker, who argues that the ruling reptiles of the Mesozoic era, commonly known as dinosaurs, or "terrible lizards," were endothermic (warm-blooded) rather than ectothermic (cold-blooded). That is, the dinosaurs were capable of internally generating and maintaining body temperatures necessary for sustaining high levels of biological activity (*13*). Bakker further concludes that endothermy was the key to success for these large land vertebrates of the Mesozoic—in fact, so much so that they were competitively superior to the contemporary early mammals. Rather than rehash all of Bakker's evidence, I wish only to indicate the nature of his paleoecological reasoning in determining that the dinosaurs had high basal metabolism, the key to their endothermy.

Three important lines of evidence are useful for interpreting dinosaurian paleoecology: their bone structure, their predator-prey ratios, and their paleogeographic distribution. The bone structure of dinosaurs is very similar to that of endothermic animals like birds and mammals: the density of spaces occupied by blood vessels is similar, as is the frequency of Haversian canals, which also contain blood vessels and nerves. Ectotherms have bones with fewer Haversian canals and a lower density of blood vessels, and the bones also show "growth rings" in species that live in climates with strong seasonal variations. By contrast, the bones of modern endotherms and dinosaurs lack such growth rings.

The second line of evidence is that of predator-prey ratios. Endothermic predator populations require ten times more energy from their prey than do ectothermic populations, all other things being equal (adult size, number of individuals, and so on), as shown in Figure 8. Because the prey population is about the same in energy value for either ectotherms or endotherms, then the number of endothermic predators that can be supported by the prey must be about one-tenth the number of ectotherms that can be supported. Bakker's calculations of predator-prey ratios for various communities of Mesozoic dinosaurs range from one to three percent, much like those of advanced (endothermic) mammalian communities of the succeeding Cenozoic era.

Finally, paleogeographic evidence, while more equivocal, supports Bakker's hypothesis of dinosaurian endothermy. Using data from environmental stratigraphy and paleomagnetism, Bakker (among others) argues that there was probably somewhat more latitudinal variation in climate in the latter half of the Mesozoic as compared to the early half, and so dinosaurs would have had to adapt to rather cold climates, given their wide paleogeographic distribution.

Whether Bakker's startling revision of dinosaur ecology in fact stands up remains to be seen. What I find especially interesting as a paleoecologist, however, is the way Bakker uses fairly routine approaches in biology and geology (comparative histology and physiology and environmental stratigraphy) and applies them most imaginatively to an issue that had long seemed settled—that dinosaurs were simply big cold-blooded lizards. This underscores again what I noted at the beginning: insights about the paleoecology of fossils have come about mostly through the application of a method, rather than from some new conceptual breakthrough or paradigm.

As in many other areas of science today, paleontologists find this an exciting time to be alive. Advances in biology—particularly in population genetics—and in geology—especially plate tectonics and paleoclimates—provide a useful context for interpretating the paleoecology of past life. Besides giving the "plain story" of what kind of fossil organism lived when, and what it looked like, paleontologists are now able to say something also about how the fossil managed as a once-living, complex, biological system. True, there will undoubtedly be many things hidden from us about life of the past, owing to the inevitable loss of information down through the geologic ages. But surely before the last chapter is written, paleontologists will have much more to say about paleoenvironments and paleoecology.

One area of fascinating paleoecological inquiry that lies in the future is that of our own human origins. Recent discoveries in East Africa strongly suggest that there may have been at least three contemporary species of hominids living side by side during the late Pliocene/early Pleistocene (*14*). The theory of competitive exclusion requires that each of these species must have had its own particular ecological niche. Moreover, two of these species seem to have become extinct without issue, while the third continued to evolve into our own. Here lies a truly marvelous paleoecological challenge: to define and trace the fine-grained paleoecologic preferences and distinctions of our hominid ancestors over the last several million years. What a story that should be!

References

1. A. Hill. In press. Taphonomical background to fossil man: Problems in paleoecology. *Geol. Soc. London Bull.*
2. M. L. Thompson. 1964. Fusulinaceae. In *Treatise on Invertebrate Paleontology,* ed. R. C. Moore, Part C: Protista 2, vol. 1, pp. 358–436.
3. J.-E. Guettard. 1759. Sur les accidens des Coquilles fossiles, comparés à ceux qui arrivent aux Coquilles qu'on trouve maintenant dans la Mer. In *Histoire de l'Académie Royale des Sciences,* 1765, pp. 189–226. Paris: Imprimerie Royale.
4. E. MacCurdy. 1958. *The Notebooks of Leonardo da Vinci,* vol. 1, pp. 331–32. George Braziller.
5. E. Forbes. 1843. Report on the Mollusca and Radiata of the Aegean Sea, and their distribution, considered as bearing on geology. *British Assoc. Adv. Sci.,* Rept. 13, pp. 130–93.
6. E. Forbes. 1844. On the light thrown on geology by submarine researches. *Edinburgh New Philosophical J.* 36:318–27.
7. M. L. Natland. 1933. The temperature and depth distribution of some recent and fossil Foraminifera in the southern California region. *Bull. Scripps Inst. Oceanography,* Tech. Series, 3 (10):225–30.
8. N. Kipp, A. MacIntyre, et al. 1976. In *Investigation of Late Quaternary Paleoceanography and Paleo-climatology,* ed. R. M. Cline and J. D. Hays. *Geol. Soc. America Mem.* 145:25, 51.
9. M. K. Elias. 1937. Depth of deposition of the Big Blue (Late Paleozoic) sediments in Kansas. *Geol. Soc. America Bull.* 48: 403–32.
10. A. L. McAlester. 1968. *The History of Life.* Prentice-Hall.
11. S. Stanley. 1968. Post-Paleozoic adaptive radiation of infaunal bivalve molluscs: A consequence of mantle fusion and siphon formation. *J. Paleontology* 42:214–29.
12. K. R. Walker and L. F. Laporte. 1970. Congruent fossil communities from Ordovician and Devonian carbonates of New York. *J. Paleontology* 44:928–44. The Society of Economic Paleontologists and Mineralogists.
13. R. T. Bakker. 1975. Dinosaur renaissance. *Sci. Am.* 232(4):58–78.
14. R. E. Leakey. 1976. Hominids in Africa. *Am. Sci.* 64:174–78.

"Let's face it. Evolution has passed us by."

PART 3 *Ancient Geographies*

Allison R. Palmer

Search for the Cambrian World

Sedimentological, geophysical, and paleontological data can be used to identify ancient continents and provide some limits to their placement on the globe

Information accumulated from many fields of geology, particularly in the last decade, creates a convincing case that our present world geography is both dynamic and transient. Historical geologists are thus faced with some major questions: Does the present geological record retain sufficient information to permit solution of the problem of the sequence of past geographies? If so, how far back in time can they be resolved?

In this paper I will examine these questions as they relate to the 70 million-year-long Cambrian period, which began about 570 million years ago. At that time, the oceans of the world began to teem with invertebrates, who left their shells in the sediments of the shallow seas that were beginning to flood the continents, and we now recognize rocks of the Cambrian age and its subdivisions by the preserved remains of these organisms. Most

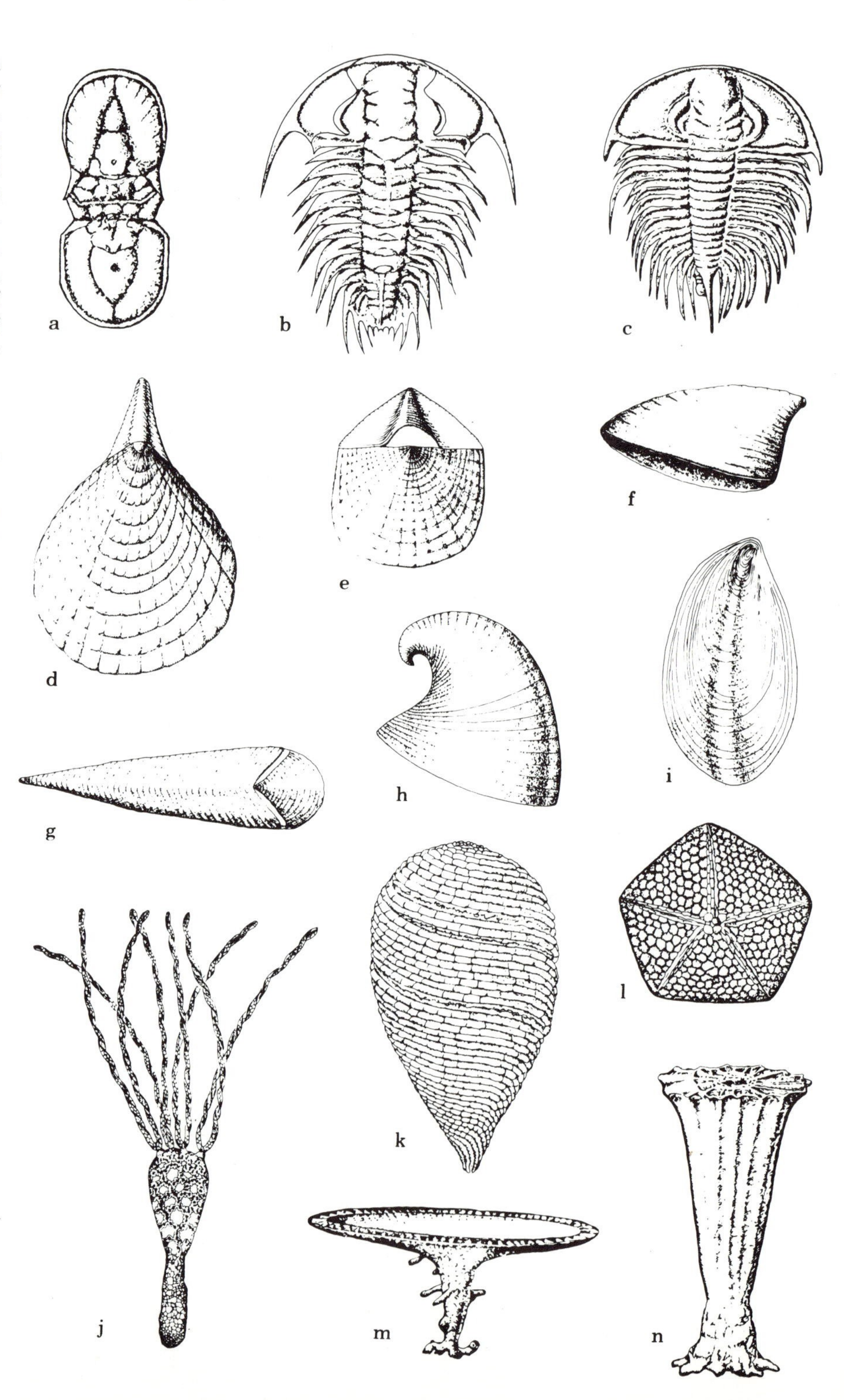

Figure 1. Typical Cambrian fossils: trilobites *(a, b, c)*; brachiopods *(d, e, f)*; molluscs *(g? h, i)*; echinoderms *(j, k, l)*; and archaeocyathids *(m, n)*.

Allison R. Palmer is professor of geology at the State University of New York at Stony Brook. For sixteen years, before coming to Stony Brook, he was the Cambrian paleontologist for the U.S. Geological Survey. During the past decade he has visited many of the major Cambrian areas of the world to gather information about Cambrian stratigraphy and trilobite faunas for comparison with North America. He is currently president of the Cambrian Subcommission of the International Stratigraphic Commission. In 1967 he received the Walcott Medal of the National Academy of Sciences for outstanding contributions to knowledge of Cambrian geology. Address: Department of Earth and Space Sciences, State University of New York, Stony Brook, NY 11790.

commonly they are trilobites—an extinct group of marine arthropods—and, to a lesser extent, brachiopods, molluscs, and echinoderms (Fig. 1). Rarer fossils indicate that most animal groups, with the principal exception of the vertebrates, have a record as far back as the Cambrian period. The plant record shows only a variety of marine algae. There is no record of plant or animal life of any kind on land at that time. Thus, the resolution of Cambrian geography rests on the extent of the recoverable record of the sediments and animal remains preserved in the Cambrian seas. Figure 2 shows that the general distribution of outcrops of Cambrian rocks is sufficiently widespread to make the problem of Cambrian geography worth pursuing.

Evidence for a dynamic geography

In order to search for the Cambrian world, however, it is first necessary to understand something about the present-day dynamic relations between major geological features of the earth—the oceans, continents, and mountain systems. The crust of the earth is made up of a mosaic of huge plates many tens of kilometers thick and often several thousands of kilometers in width. These plates are interacting with their neighbors, and the resulting crustal stresses are reflected in the global belts of earthquake activity that define their margins (Fig. 3). The significance of the interaction of these plates in past geologic time is incorporated in the theory of plate tectonics, which is less than a decade old.

According to the theory, some plate margins are zones of crustal spreading as two plates move apart and material from the underlying mantle rises between them to form new crust. On the confined surface of a sphere, enlargement of one plate margin can only be accommodated by a comparable reduction in plate size at some other plate margin. This can be accomplished either by buckling of the crust between plates or by having the edge of one plate pass beneath another in the zone of collision. Plate margins that are neither being created nor consumed are ideally zones of horizontal shear between passing plates.

The spreading zones, which are predominantly represented today by the mid-ocean ridges, are fairly straightforward regions reflecting tensional stresses. The other plate margins are often represented by complex combinations of shear and compressional stresses. Where compression predominates, there is often a zone, defined by the distribution of earthquake foci, which dips beneath one plate at an angle of about 45 degrees. This zone, called a subduction zone, marks the region where one plate overrides (or underrides) another. The exposed, generally lower-density parts of plates, which we term continents, are passive participants in the global tectonic drama. Only when the continental part of a plate reaches a subduction zone, or when a subduction zone passes beneath a continental margin, does the continent respond to plate-tectonic stresses.

The colliding margins of present-day plates (Fig. 3) are characterized either by geologically complex mountain systems, such as the Cordilleras of North and South America and the Alpine-Himalayan trend across southern Europe and Asia, or by belts of volcanic island arcs, such as those bordering the northern and western Pacific Ocean. Geological study of the mountain systems shows a history of stresses over most of the past 150 million years. During this time, they may have incorporated old marginal volcanic island belts and associated volcanogenic sediments, and even fragments of older ocean crust represented by rock complexes called ophiolites. These complexes are characterized by a sequence of

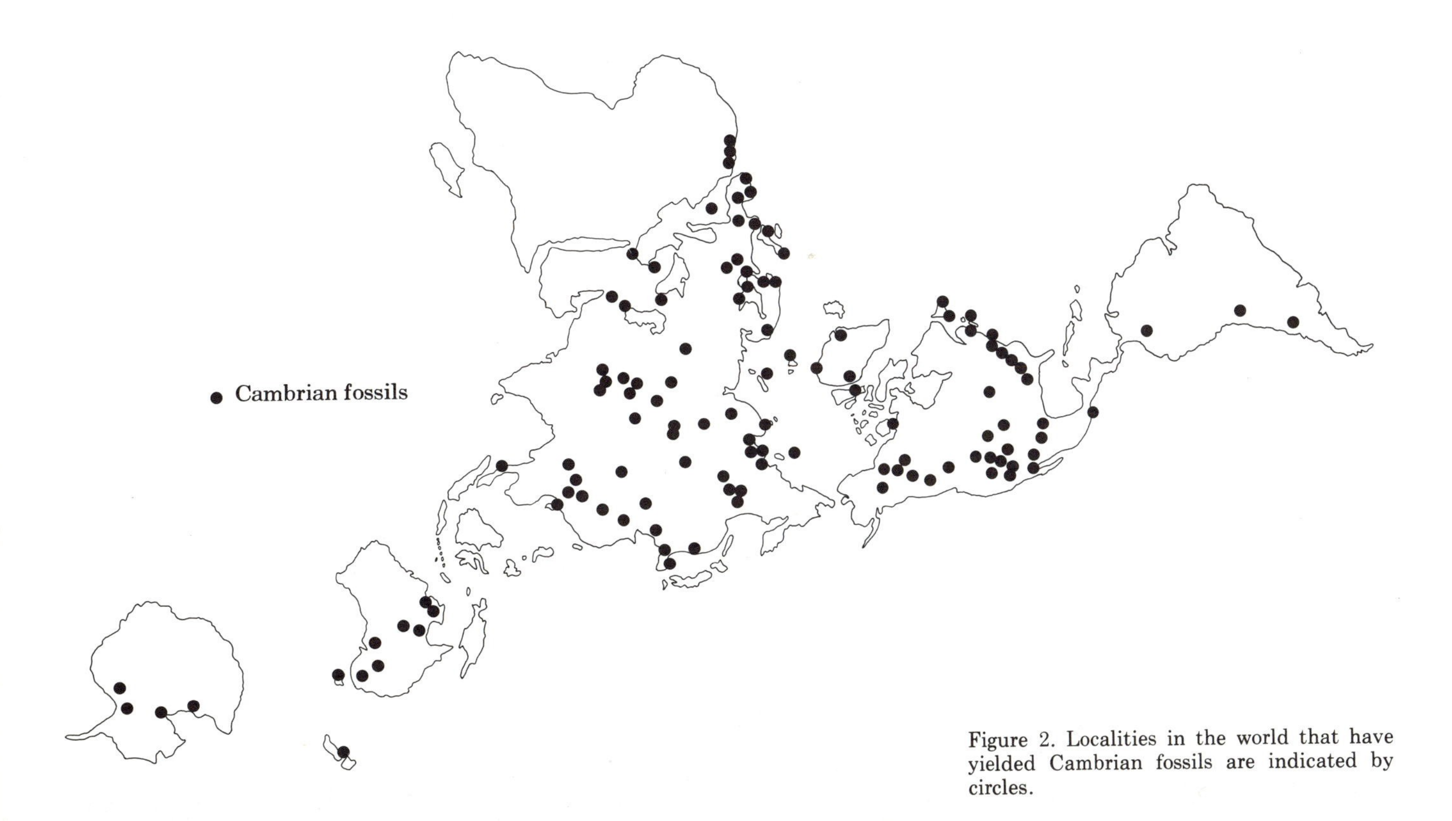

Figure 2. Localities in the world that have yielded Cambrian fossils are indicated by circles.

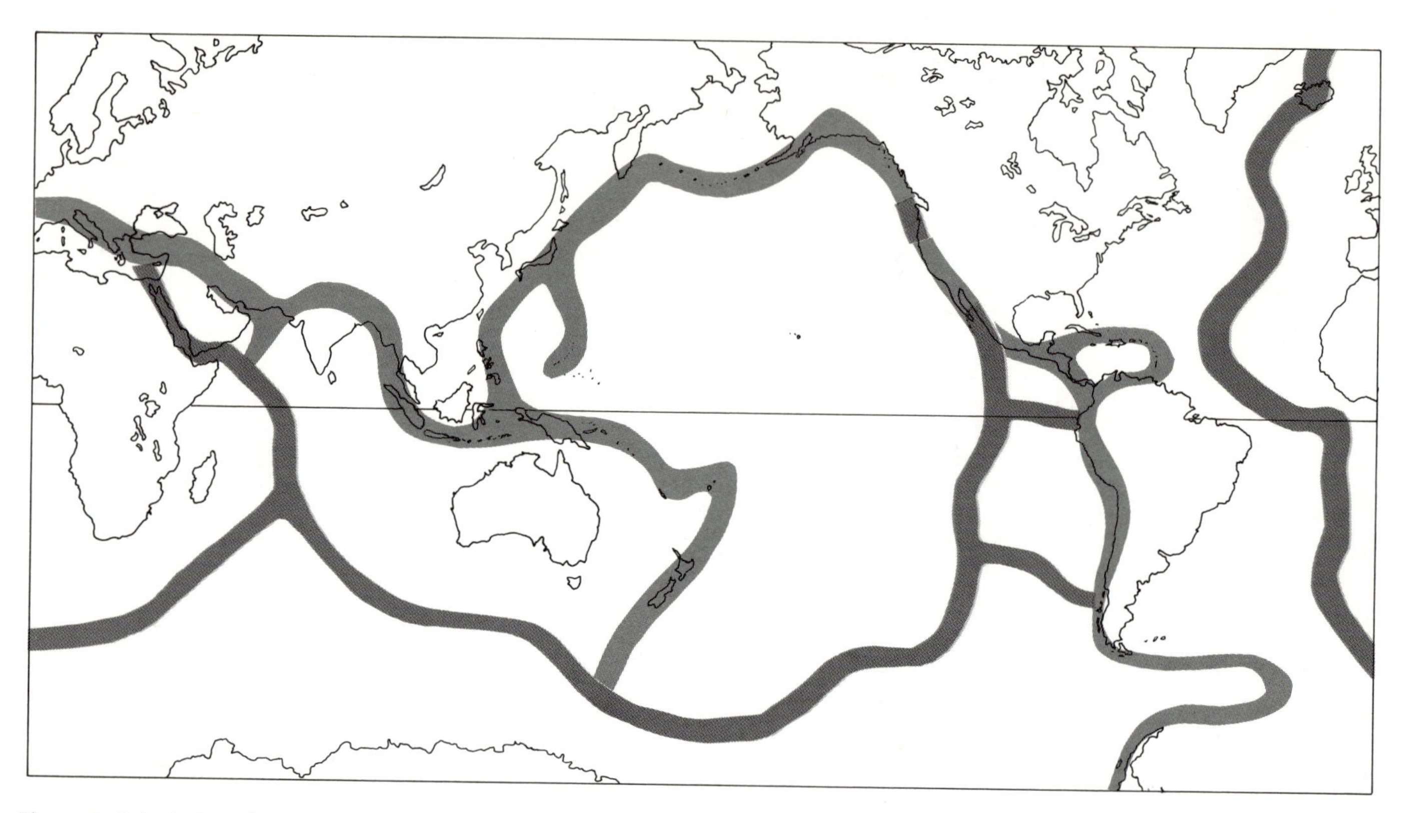

Figure 3. Principal earthquake belts of the world. *Dark grey,* regions of crustal extension between plates; *light grey,* regions of dominant compression or shear.

rocks beginning with high-density iron-magnesium-rich igneous rocks believed to be close in composition to the upper part of the mantle. They are overlain by lavas showing evidence of extrusion into sea water, and the lavas, in turn, are covered by siliceous sediments typical of those now found only in the deep sea.

Analysis of the depositional environments of the deformed sedimentary rocks in the mountain systems shows that many of the strata accumulated as marine sediments at or beyond the margins of a continent. The great vertical and horizontal stresses involved in a collision have raised or lowered these sediments many hundreds or thousands of meters, buckled them into complex folds, and in some places thrust them inland along great zones of subhorizontal shear sometimes for distances of many tens of kilometers.

In addition to the general increase of temperature with depth, heat produced by friction and compression along subduction zones has characteristically melted great masses of rock at depths of 150 to 500 kilometers below sea level and similar distances inland from the oceanic trenches that mark the surface expression of the zones. Along stressed continental margins, some of these masses have risen through the overlying rocks, causing additional buckling and, in some instances, great outpourings of lava. When the rocks cooled at depth, belts of granite such as those now exposed by erosion in the Sierra Nevada mountains and many other parts of the North and South American Cordilleras were formed. Also, heat from the rising magmas, added to that generated in the zones of greatest compression, has further contributed to the complex geology of mountain systems by metamorphosing some belts of former sediments and lavas into slates, schists, or gneisses. Thus, mountain systems preserve an important part of the history of plate-tectonic stresses in continental areas. Our search for the Cambrian world, however, begins with a consideration of the zones of spreading.

Back to Pangaea

As new crust is created by the cooling of magma along the spreading zones, it "freezes" within it a record of the earth's magnetic field strength. We now know that the magnetic field reverses its polarity at irregular intervals on scales of tens of thousands of years. As a result, the strength of the magnetic field, particularly in middle and high latitudes, is either enhanced or retarded from "normal" levels. Because the sea floor is moving symmetrically away from the zones of spreading, it records the varying field strengths as bands of low or high magnetic intensity (anomalies) parallel to the spreading zones. The duration for each of the periods of normal and reversed polarity is known for most of the past 200 million years, and the width of the anomaly bands on the sea floor, beginning at the spreading zone, is proportional to these time periods progressing backward from the present. Thus, the sea floor, like a tree, retains a permanent record of its "growth" history—each anomaly band has a known age. By replotting plate margins to bring the anomaly bands of particular ages back to the spreading zones, we can locate with moderate rigor the passive continental elements of a plate and the positions they occupied at times in the past—as far back as the "frozen" anomaly record permits.

Dietz and Holden (1970) presented such a restoration in a series of world maps from the Recent back to the end of the Permian period (about 225 million years ago). Their

Permian map (Fig. 4), slightly modified by newer work in the Mediterranean region, shows that the world then seemed to have only a single continent called Pangaea. The next questions in our historical quest concern Pangaea: Was it the primordial continent? and Is it therefore a representation of the earlier Cambrian world as well? To obtain answers, we must consider other characteristics of plates and some additional geophysical and geological evidence, because all sea floor older than about 200 million years has long since been consumed in the world's subduction zones, and thus the direct record of earlier plate motions through use of magnetic anomaly patterns has been lost.

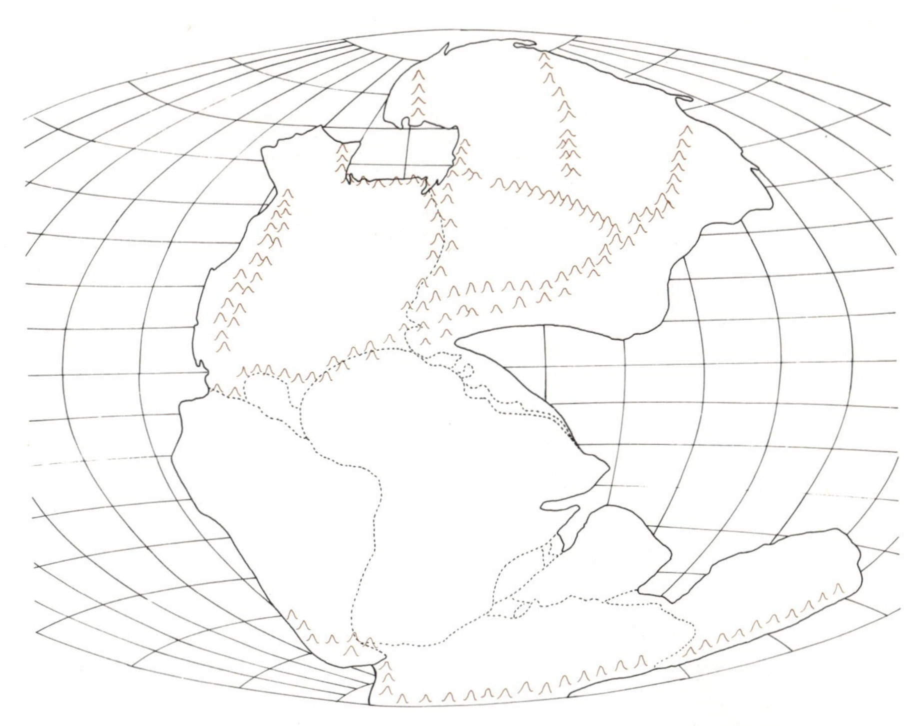

Figure 4. Outline of the world continent, Pangaea, of about 200 million years ago showing the general distribution of pre-Pangaea, post-Cambrian mountain systems (modified from Dietz and Holden 1970).

The obvious place to look for a record of older plate movements is the world's mountain systems, particularly those that traverse the interior of Pangaea. The eastern part of North America from Connecticut to northern Greenland preserves the characteristic record of a collision between continent and sea floor, and later between continent and continent, that spanned the time from Early Ordovician (about 500 million years ago) to Middle Devonian (about 370 million years ago) (Bird and Dewey 1970). This same history is recorded in the mountains of western Sweden, most of Norway, and northern England and Ireland. Elsewhere in Pangaea, the Ural Mountains, between the Europe and Asia of political geographers, record the stressing of the eastern margin of northern Europe by northern Asia from Devonian to Early Permian time (Hamilton 1970). A broad meridional system of pre-Alpine mountains of Carboniferous age extends across central and southwestern Europe and northwestern Africa, and it may have a continuation along the eastern and southern parts of North America. Finally, other meridional systems of this age, showing evidence of typical continent margin stress at least as far back as the early Cambrian, are recorded across the southern part of the USSR and central China. Thus, plate tectonic theory predicts that prior to Pangaea true oceans (i.e. topographically low, submerged parts of the earth's surface with higher-density, "oceanic," crust) isolated most of present North America from northwestern Europe, northern Europe from southern Europe, Europe from Asia, and southern Asia from northern Asia. Our question about Pangaea is at least partly answered. It probably was not the primordial continent.

The pre-Pangaea continents

We now have two new questions: Were the pre-Pangaea areas just defined Cambrian continents? and if so, where were they on the globe? To answer the first, we can begin with a look at the Cambrian rocks and faunas of pre-Pangaea North America. For many short intervals of time within the Cambrian period, the sedimentary environments at all localities with rocks of those ages can be plotted on a map. When this is done, three major sedimentary regions can be distinguished (Fig. 5) which are arranged about a central land area that diminishes in size throughout the Cambrian as the continent is inundated. These are (1) an inner region of marine sands and shales derived from the central land (inner detrital belt); (2) an intermediate region of relatively clean limestone and dolomite accumulation in generally shallow seas (carbonate belt); and (3) a region of dark-colored, thin-bedded siltstone, silty limestone, and occasional black chert that suggests somewhat deeper marine conditions seaward of the carbonate belt (outer detrital belt). Thus, pre-Pangaea North America seems to have been a distinct Cambrian continent, based on distribution of the major marine sedimentary environments.

Analysis of the distribution of the most abundant organisms, the trilobites, generally supports the sedimentary evidence from pre-Pangaea North America. The trilobites of the inner detrital belt and all except the seaward margin of the carbonate belt are known almost exclusively from North America (important exceptions are discussed below). The trilobites from the outer detrital belt include a majority of North American elements along with representatives of a distinctive, probably pelagic, group called agnostids (Fig. 1) and a few nonagnostid trilobites that have wide distribution in similar environments throughout the world.

In eastern Newfoundland, Nova Scotia, southern New Brunswick, and eastern Massachusetts, there are Cambrian rocks lithologically and faunally unlike those of the

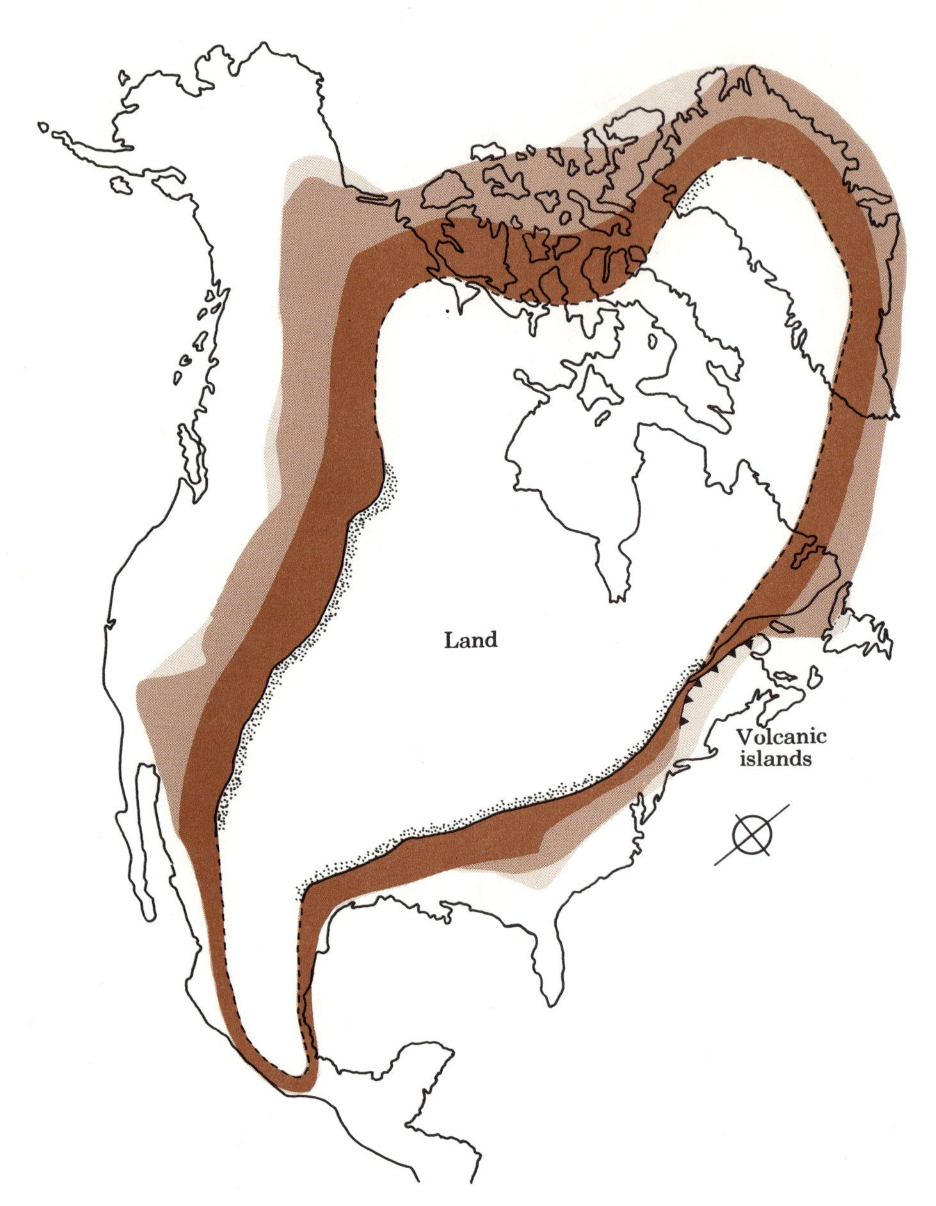

Figure 5. A generalized paleogeographic map of North America for Middle Cambrian time showing the distribution of the major marine sedimentary regions: *dark color,* an inner belt of detritus derived from the land; *medium color,* a middle belt of marine carbonates whose outer edge marks the limit of the shallow Cambrian seas; *light color,* an outer belt of deeper water shales and limestones.

rest of the continent. These rocks and faunas are almost identical to those found in central England, and either western France or Spain. Furthermore, they are found in comparatively undisturbed areas east of the old post-Cambrian, pre-Pangaea mountain system that extends from southern Connecticut through central Newfoundland and that has been interpreted as an old collision zone. Thus, Cambrian evidence supports plate tectonic theory that this part of modern North America should be excluded from the Cambrian continent.

Minor modifications of the North American faunal pattern are noted in the Cambrian of Alaska, where some Lower Cambrian levels have yielded trilobites typical of Siberia, and in southern Mexico, where a Late Cambrian fauna has very strong similarities to faunas from northwestern Europe and Bolivia. The possible significance of these will be discussed later.

Generally, the faunal and lithologic data from pre-Pangaea North America suggest a "model" for other Cambrian continents: a central land area flanked by inner detrital and carbonate belts characterized by shallow-water restricted marine environments containing predominantly endemic trilobites; and a contrasting area formed by the outer margin of the carbonate belt, facing the ocean, and by the slightly deeper environments of the outer detrital belt, which contain both American endemic trilobites and a large number of more widely distributed ones.

The other pre-Pangaea fragment that best fits this model is the present northeastern part of Asia, centered on the Siberian Platform (Fig. 6). Its pre-Cambrian core is less extensive than that of North America, but it contains a large region of limestones and dolomites contemporaneous with lithologically similar rocks of North America. In contrast, however, the carbonates bear almost totally different trilobite complexes. Flanking this region in the north, southeast, and south are rocks which are lithologically similar to those of the American outer detrital belt and which contain a mixture of Siberian and cosmopolitan elements. These rocks permit correlation between the totally distinct endemic faunas of the restricted seas of the North American and Siberian continental interiors.

Another fragment consistent with the North American model is represented by Mongolia, northern China, and Korea (Fig. 7). Here, a pre-Cambrian land was flooded from the south by Cambrian seas. Transects from the land area southeastward across Korea and southward to the Yangtze river show the typical tripartite sedimentary pattern. The faunas of the carbonate belt and inner detrital belt are almost totally distinct from those of both Siberia and North America, but the outer-detrital-belt rocks preserved in southern Korea and the Yangtze Basin contain both Chinese and cosmopolitan elements. North of the Mongolian continent is a strongly deformed, meridional, post-Cambrian, pre-Pangaea mountain system that represents a probable plate collision between Mongolia and Siberia. The stresses on the margin of Mongolia that faces Siberia have apparently destroyed most of the Cambrian record there, but recently discovered Cambrian rocks in north-central Mongolia preserve a record of marine sediments and interbedded volcanic elements.

In contrast to these areas, the Baltic continent in present northwestern Europe had a central land area that was gradually submerged by Cambrian seas, but significant areas of carbonate sedimentation never developed to separate the inner and outer detrital belts. Faunas containing predominantly pelagic elements are found in nearshore sediments in Scandinavia, and no clear-cut distinction between faunas from restricted marine environments and those from ocean-facing environments has yet been made. These Scandinavian faunas differ sharply from those of Siberia, Mongolia, and North America and subtly from those of southern Europe. These findings, which are consistent with the evidence of old collision zones, suggest the presence of oceanic regions to the west, south, and east of the Baltic continent in Cambrian time.

Details of the Cambrian geology of southern Europe have been obscured by post-Cambrian, pre-Pangaea deformation and by very limited present-day outcrops. In contrast to northern Europe, there are moderately to well-developed Lower Cambrian carbonates. These attain their greatest thickness in southern Spain, southern France, and Sardinia. In northwestern France and England, a distinctive red muddy limestone of early Cambrian age developed that is indistinguishable from contemporaneous rocks with nearly identical trilobites in southeastern Newfoundland and eastern Massachusetts.

The Middle Cambrian silty and shaly rocks of all these areas also share many common trilobites. Paleogeographic analyses of southeastern Newfoundland, England, northwestern France, and Spain all suggest that a Lower Cambrian land lay in the direction of the present Atlantic Ocean. When Pangaea is disassembled along the ancient collision zones running from the northeastern United States through Scandinavia and through central Europe, this Lower Cambrian land is seen as a possible linear "microcontinent" whose dimensions are very poorly known. Strong Early and Middle Cambrian faunal affinities between Spain and Morocco suggest that this "microcontinent" may have had a position relative to Africa analogous to the present-day position of New Guinea with respect to Australia.

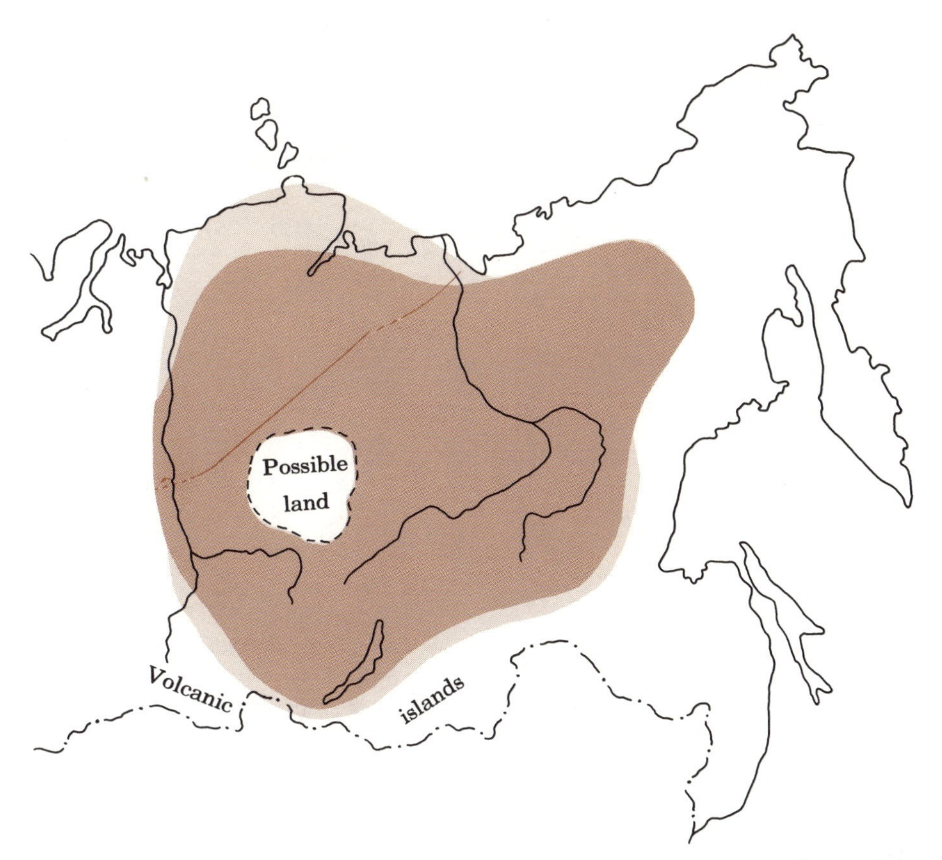

Figure 6. Generalized Middle Cambrian paleogeographic map for northeastern Asia showing distribution of sedimentary environments comparable to the carbonate belt and outer detrital belt of North America, as shown in Figure 5.

The largest pre-Pangaea continent, which includes Africa, South America, India, Australia, and Antarctica, has been called Gondwanaland (Fig. 8). Cambrian rocks are known from northwestern Africa, possibly southwestern Africa, a few scattered localities in the Andes, the Transantarctic mountains, the Middle East, and many areas in the eastern half of Australia, including Tasmania.

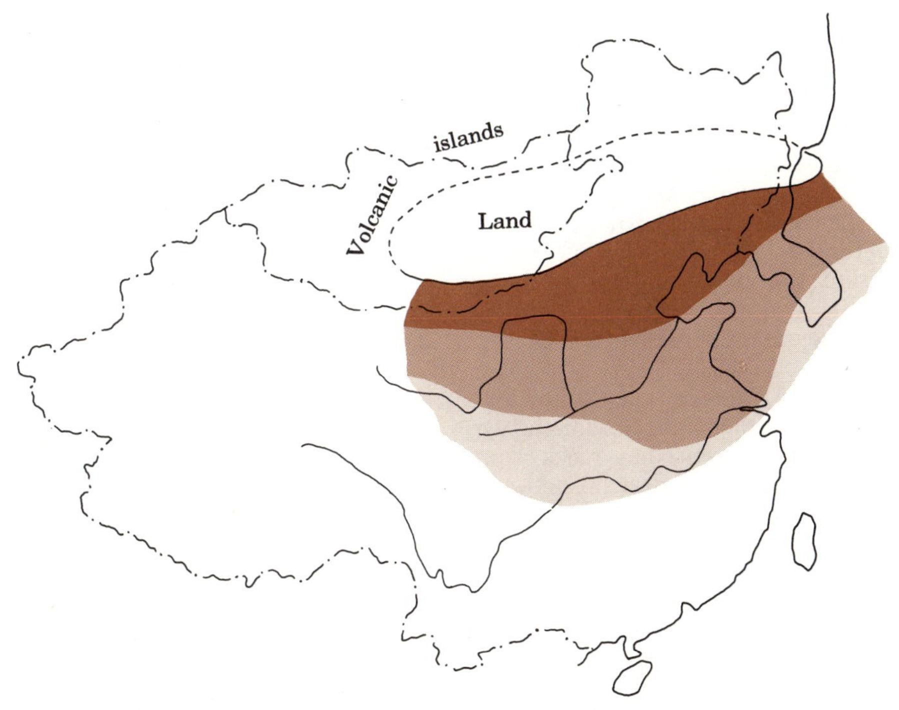

Figure 7. Generalized paleogeographic map for a part of northern China and Mongolia for Middle Cambrian time. Sedimentary environments are shown as in Figure 5.

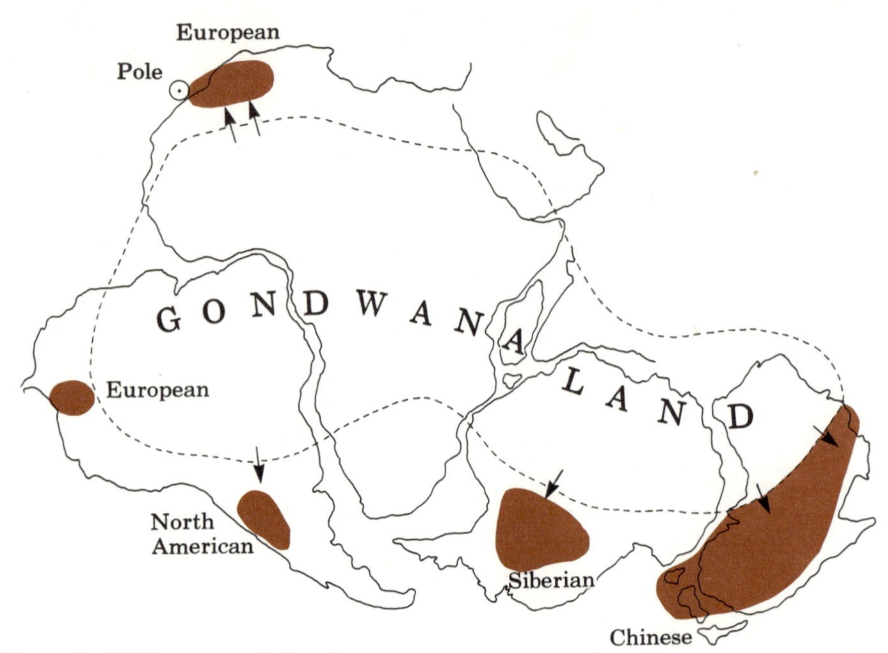

Figure 8. Outline map of the Gondwana continents showing *(color areas)* distribution and regional affinities of Middle and Upper Cambrian faunas and *(arrows)* general shoreward-to-seaward directions based on analysis of sedimentary patterns.

In northwestern Africa, the Lower Cambrian includes thick limestone sequences with large, well-developed reef-like structures built by algae and an extinct phylum of sponge-like organisms called Archaeocyatha (Fig. 1), partly contemporaneous with shales, siltstones, and sandstones to the southeast. The overlying Middle and Upper Cambrian, as in southern Europe, consists of siltstones or shales and completely lacks limestones. Thus, although an inner detrital belt and carbonate belt can be determined in the Lower Cambrian, there are no clear sedimentary belts in the younger Cambrian. While the trilobites throughout the Cambrian have strong similarities with those of southern Europe, the lateral relations to the Lower Cambrian sediments indicate that the main Cambrian land lay to the southeast in the African interior.

In South America, the Cambrian of western Argentina, although very limited in extent, has rocks representing the inner detrital belt, carbonate belt, and outer detrital belt and an indication of a land area in the South American interior to the northeast. The Cambrian faunas of Argentina in all sedimentary belts are nearly the same as those of North America. In Bolivia, however, where only late Upper Cambrian siltstones are known, the faunas are very similar to those of southern Mexico and western Europe and are less like those of North America. A single locality in Colombia has yielded a specimen of *Paradoxides*, a trilobite typical of western Europe and northwestern Africa and not found at all in the Cambrian of pre-Pangaea North America.

In Australia, extensive Cambrian outcrops show that eastern Australia was being stressed from the east, and extensive thick sequences of volcanic rocks and volcanogenic sediments with associated Cambrian trilobites are known in Victoria and Tasmania. Rocks representing the inner detrital belt and the carbonate belt are found in a broad north-south region traversing central Australia, clearly indicating that western Australia was Cambrian land. Many of the trilobites of the carbonate belt and inner detrital belt of Australia have affinities with those of China and southeast Asia. In the ocean-facing environments, in addition to the Australian-Chinese elements, there are many forms similar to those from the outer detrital belt environments of southern Korea, central China, western United States, and Argentina, as well as northwestern Europe.

In Antarctica, there are only a few localities from which paleogeography can be deduced. However, there are indications of igneous activity and crustal stress in the Cambrian in western Antarctica and parts of the Transantarctic mountains, and limestones with trilobites of Early and Middle Cambrian ages, representing a carbonate belt, are present in outcrops in the Transantarctic mountains and as boulders in glacial moraines derived from the interior of eastern Antarctica. The trilobite faunas have their greatest affinities with those of Siberia, less with Australia, and almost none with South America or northwest Africa.

Early Cambrian sandy rocks from the Salt Range in Pakistan represent the inner detrital belt and have faunas very much like those from comparable environments of Australia and China. Cambrian rocks from Turkey and the Middle East are still incompletely known, but they contain faunas of European, Chinese, and possibly American affinities.

The distribution of Cambrian faunas around Gondwanaland poses a distinct problem in terms of our North American continental model. From Australia to Argentina, although the three sedimentary belts seem to be recognizable through most of the Cambrian, the trilobite faunas from the carbonate or inner detrital belts of Australia–Antarctica and Argentina are almost totally distinct from one another rather than being nearly identical, as might have been expected. Furthermore, the faunas are almost completely unlike those on the other side of Gondwanaland from Bolivia to northwestern Africa.

One possible solution to the faunal contrasts around the periphery of Gondwanaland is that the continent was formed by collision of several Cambrian continents, each with its distinct faunas, before the Middle Paleozoic, when considerable stratigraphic evidence favors a single continent. Although clearly defined pre-Pangaea, post-Cambrian mountain systems traversing the interior of Gondwanaland are not apparent, radiometric dates from several belts of deformed unfossiliferous rocks in parts of Africa and South America show a range of igneous and metamorphic activity overlapping the time interval represented by the Cambrian. However, none of the trends of these zones seems to isolate the critical regions with different faunas (Fig. 9).

Thus, for Gondwanaland, the distributions of sedimentary environments and of vectors indicating the location of land areas are consistent with the interpretation of the interior as a single Cambrian continent. This is supported by the matching pre-Cambrian geology of the Atlantic margins of South America and Africa which strongly suggests that the present continents were together before the Cambrian. Likewise, the geology of the parts of Antarctica and Australia that are matched in the construction of Pangaea from sea-floor spreading data indicates that these present continents had a similar early Paleozoic history and probably were in contact in Cambrian time. Details of the fit of Africa–South America, Australia–Antarctica, and India are still being debated, but the differences between the various extant versions do not seriously affect the Cambrian story.

The paleontologic and stratigraphic data summarized above thus support the interpretation from plate tectonic theory that the mountain systems crossing Pangaea outline the margins of plates at least as old as the Cambrian. However, this has still not provided information useful for the former locations of these plates with respect to each other. Some help in solving the problem comes from geophysical studies of remnant magnetism.

Where were the plates?

Many igneous rocks, particularly lava flows, contain weakly magnetic minerals that align themselves in the earth's magnetic field as the magma cools, in the same way that iron filings align themselves in the field of a bar magnet. These minerals preserve not only a measure of the field strength, useful in studying the magnetic anomalies and sea-floor spreading, but also the magnetic declination and inclination (a measure of latitude) at the point where the magma cooled. Delicate geophysical instruments can detect this remnant field orien-

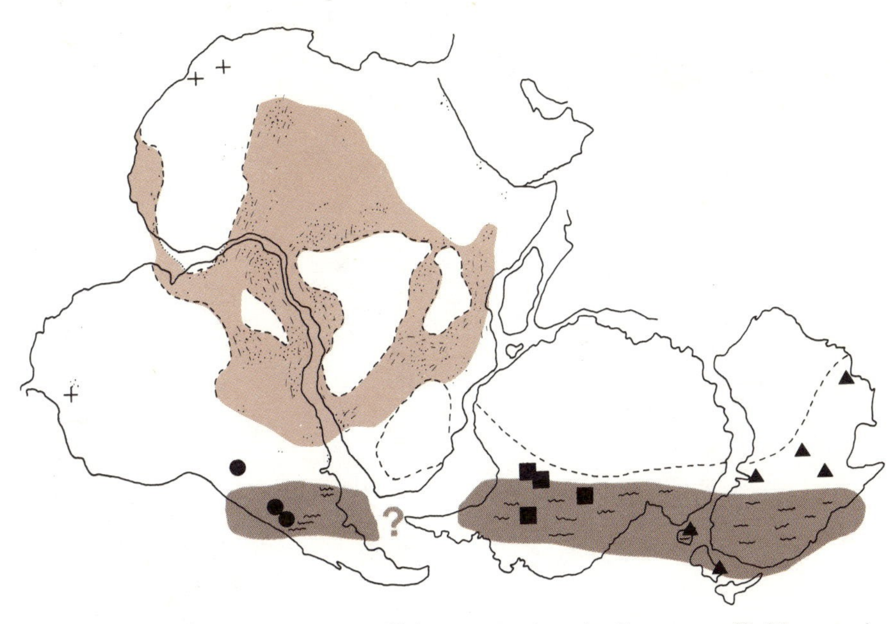

Figure 9. Gondwanaland, showing *(light color)* the distribution of regions of crustal deformation with radiometric ages of igneous and metamorphic rocks that overlap Cambrian time and *(dark color)* areas of definite continent-margin stress during the same period. +, localities of faunas with predominantly European affinities; ●, localities of faunas with predominantly North American affinities; ■, localities of faunas with predominantly Siberian affinities; ▲, localities of faunas with predominantly Chinese affinities.

tation in many igneous rocks, and also in some ancient sediments that have iron-oxide cement. Thus, it is theoretically possible to obtain the latitude of formation of many ancient rocks if they have not been reheated or deeply weathered subsequent to their formation.

Using this technique, paleomagnetists made some of the first geophysical contributions to the revival of the "continental drift" theory in the late 1950s. They discovered that the latitude of formation of igneous rocks older than about 100 million years on many continents was significantly different from that of their present location and that the older the rock, in general, the greater the latitudinal discrepancy.

At first, this had two plausible explanations: (1) the magnetic pole had "wandered" relative to a fixed continent, or (2) the continent "wandered" relative to a fixed pole. "Polar wandering curves"—the calculated position of a remnant magnetic pole, assuming a fixed continent—were determined for each continent. When they were found to be increasingly different as older rocks were examined, a third possibility arose: perhaps there were multiple magnetic poles in the past. However, when the Atlantic Ocean was "closed" by fitting the continents together like parts of a jigsaw puzzle, the poles of Permian time for the Atlantic continents (the age of Pangaea according to sea-floor spreading theory) essentially coincided! This was strong evidence that the continents rather than the poles had moved. We seemed to have a powerful tool to determine at least the ancient latitude of any part of the earth where suitable rocks are preserved.

Remnant magnetism has some drawbacks, however. It is very sensitive to many external events and it is very weak in older rocks. Thus, the 95% confidence limits for the calculated pole positions from many ancient rocks cover a circular or elliptical area of 10 or more degrees radius. Furthermore, because subtle geological events may have "reset" the magnetic orientation in an old rock to the latitude of a much later time, the actual age of some calculated poles has been disputed. On top of all of this, the rocks most suitable for remnant paleomagnetic study are found in continent-margin environments where chances for deformation are greatest and where the fossil record that helps to determine the age of a sample is often the poorest. Thus, data from remnant magnetism in

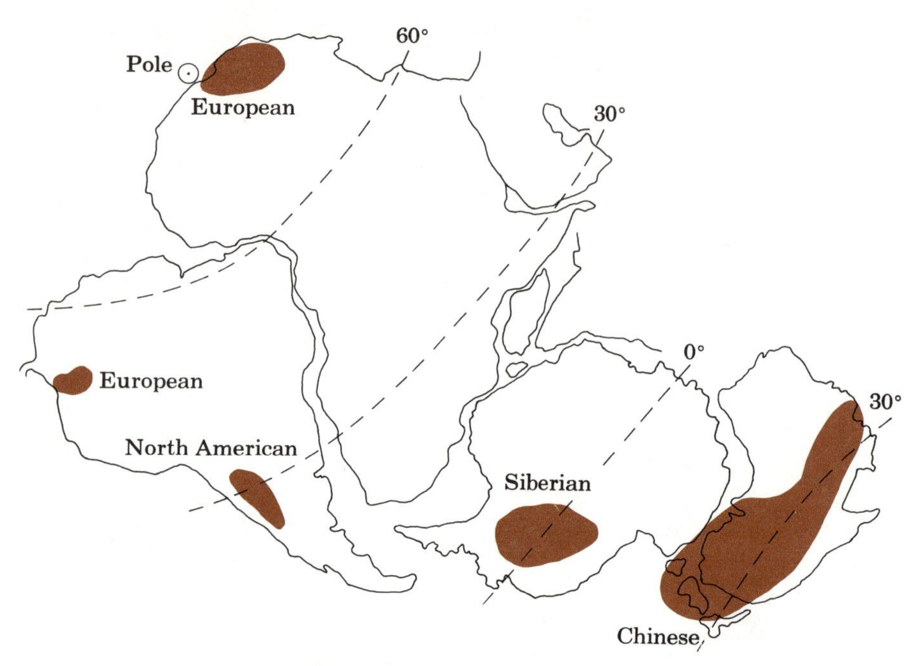

Figure 10. Map showing rough Middle Cambrian paleolatitudes for Gondwanaland superimposed on faunal distributional data.

Cambrian rocks give an important *sense* of latitudinal orientation but are often frustratingly imprecise, and they do not provide any clue about the second key coordinate for global location—longitude.

Help from paleontology

Limited Cambrian paleomagnetic data suggest that the South Pole was in the vicinity of northwestern Africa in Middle Cambrian time (Fig. 10). When Gondwanaland is placed on a globe with this orientation, the areas of contrasting Cambrian faunas discussed earlier seem to be distinctly correlated with latitude. Faunas with European affinities are in high latitudes of the southern hemisphere, those with North American affinities are in middle latitudes, those with strong Siberian affinities are in low latitudes, and those with Chinese affinities are in middle latitudes, but in the northern hemisphere!

The resulting hypothesis that latitude plays an important role in determining the global contrasts of Cambrian faunas, particularly in the restricted marine areas, is actually testable. Limited paleomagnetic data from nearly undisturbed rocks of the Russian and Siberian platforms permit latitudinal orientation of Eurasia in the Cambrian on geophysical evidence completely independent of evidence used to orient Gondwanaland. North America, for which reliable Cambrian paleomagnetic data are still disputed, can also be independently oriented with respect to Gondwanaland by aligning the trend of the mountain system in northeastern United States and maritime Canada with its continuation in Ireland, England, and Scandinavia. When this is done, we seem to have a good first-order confirmation of the hypothesis. Using equators derived from the data of either the Russian or Siberian platform, Siberia is in low latitudes, western Europe is in high latitudes, and China is in the hemisphere opposite that of Europe. In North America, the Early Cambrian faunas of Alaska with their trilobites of Siberian affinities are in the low latitudes, and the Late Cambrian faunas of southern Mexico with their strong European affinities are in the high latitudes.

A limited answer

Unfortunately for the reader who wishes a satisfactory conclusion to this story and an internally consistent picture of the Cambrian world, this is about as far as we can presently go. We have reasonably good evidence that the world consisted of at least five major continents: (1) most of present North America; (2) a Baltic continent comprising most of northern Europe; (3) a Siberian continent comprising most of northern Asia; (4) a Mongolian continent comprising most of northern China and Mongolia; and (5) Gondwanaland, comprising the present continents of South America, Africa, Australia and Antarctica, and the subcontinent of India. There is some evidence that similarities and differences among the Cambrian faunas of the world reflect latitudinal, and therefore climatic, controls on faunal distribution. This information puts some constraints on geographic reconstructions of the Cambrian world.

The Cambrian period was at least 70 million years in length, and we do not have enough information on a global scale about any one reasonably small subdivision of it to pin down with any degree of precision the location of even the larger Cambrian continents. We have even less information about the relations of the smaller fragments to the larger continents. Much better information is needed about the time-space relationships of the Cambrian rocks and faunas of South America, central and southeastern Asia, the Mediterranean region, the Middle East, and the Arctic and Antarctic regions before the search for the Cambrian world can be satisfactorily resolved. We have the potential tools and some sense seems to be appearing. However, it is quite clear that communication and cooperation between most major subdisciplines of geology will be increasingly required as the drama of the dynamic history of the earth unfolds.

References

Bird, J. M., and J. F. Dewey. 1970. Lithosphere plate-continental margin tectonics and the evolution of the Appalachian orogen. *Bull. Geol. Soc. America* 81:1031–60.

Dietz, R. S., and J. C. Holden. 1970. Reconstruction of Pangaea: Breakup and dispersion of continents, Permian to Present. *Jour. Geophys. Res.*75:4939–56.

Hamilton, W. 1970. The Uralides and the motion of the Siberian and Russian platforms. *Bull. Geol. Soc. America* 81:2553–76.

"I wouldn't worry. With continental drift, Africa or South America should come by eventually."

Richard K. Bambach
Christopher R. Scotese
Alfred M. Ziegler

Before Pangea: The Geographies of the Paleozoic World

Pre-Pangean configurations of continents and oceans can be reconstructed according to our knowledge of geologic processes still operating in the modern world

Although the earth is known to be 4,600 million years old, most of our geologic information is from rocks formed in the last 570 million years, the time during which organisms have left a good fossil record. This span is subdivided into the traditional geologic time scale (Fig. 1). During this time, life colonized the land, the vast majority of deposits of fossil fuels accumulated, and all of the present mountain ranges of the earth formed. While this was going on, continental drift—a result of plate-tectonic processes—caused the geography of the earth to change constantly.

The paleogeographic history of the last 240 million years, the Mesozoic and Cenozoic eras, is well understood, because this part of the geologic record is relatively well preserved and the pattern of dated magnetic reversal stripes on the ocean floor allows us to reposition the continents as they were during this interval. Reconstructions of continental positions by Dietz and Holden (1970a,b), Van der Voo and French (1974), and Smith and Briden (1977) show that about 240 million years ago the present continental blocks were grouped together into a "supercontinent" called Pangea. The most recent quarter billion years of geologic time has seen the breakup of Pangea, the formation of the "new" Atlantic and Indian Ocean basins, and the collision of some of the fragments of Pangea to form the Afro-Eurasian landmass, the nucleus of a new supercontinent. These events have been documented by data collected by projects such as the Deep Sea Drilling Project (Nierenberg 1978).

Richard K. Bambach is Associate Professor of Paleontology at Virginia Polytechnic Institute and State University. During 1978–79 he was Visiting Associate Professor at the University of Chicago. His research interests cover a wide range of topics in paleoecology. Christopher R. Scotese is a Research Assistant in the Department of Geophysical Sciences at the University of Chicago. He is a specialist in the use of computer graphics for depicting paleogeography and is currently studying the paleomagnetism of Paleozoic rocks. Alfred M. Ziegler is Professor of Geology at the University of Chicago. He has published in the areas of community paleoecology, stratigraphy, and paleogeography. All three authors have been working on an atlas of paleogeographic maps. Address: Richard K. Bambach, Department of Geological Sciences, Virginia Polytechnic Institute and State University, Blacksburg, Virginia 24061.

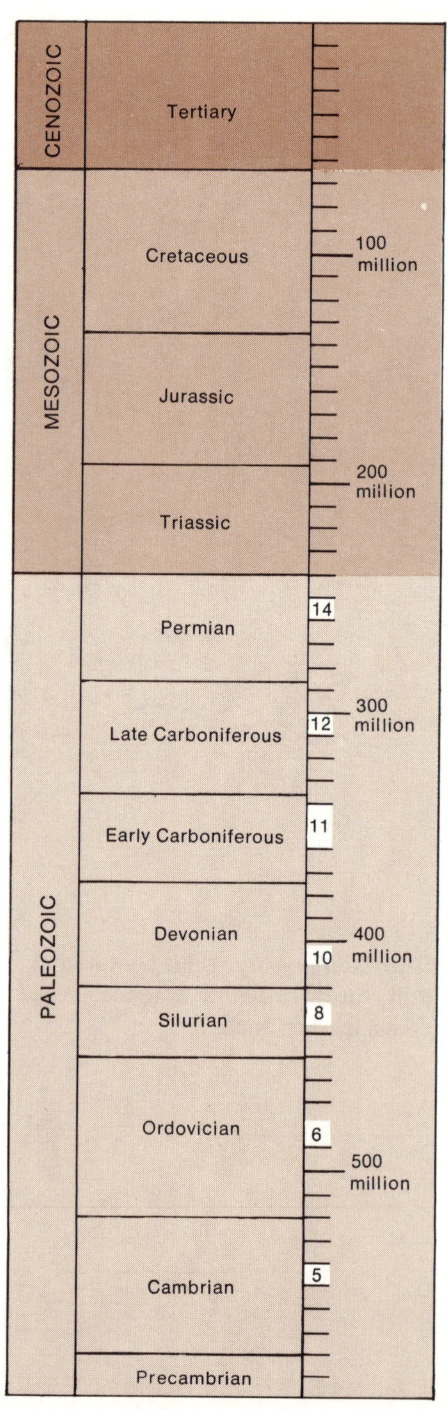

Figure 1. Geologic time. (White blocks represent time spanned by the indicated map.)

But because no pre-Pangean geographic relationships are preserved, the positions and paleogeography of the continents in the preceding 330 million years, the Paleozoic Era, is much more difficult to determine. The pioneering effort at mapping the positions of continental blocks in the Paleozoic was a set of four maps by Smith, Briden, and Drewry (1973), based solely on paleomagnetic information. The first reconstructions utilizing paleoclimatic and tectonic information combined with paleomagnetic data were presented in a study of the Silurian by Ziegler, Hansen, and others (1977) and another covering times from the start of the Paleozoic to the Cenozoic prepared in the Soviet Union (Zonenshayn and Gorodnitskiy 1977a,b). These maps identified most of the separate paleocontinents of the Paleozoic and proposed a logical, consistent sequence for Paleozoic plate motions.

The emerging pattern of geographic change during the Paleozoic allows us to return to a uniformitarian view of

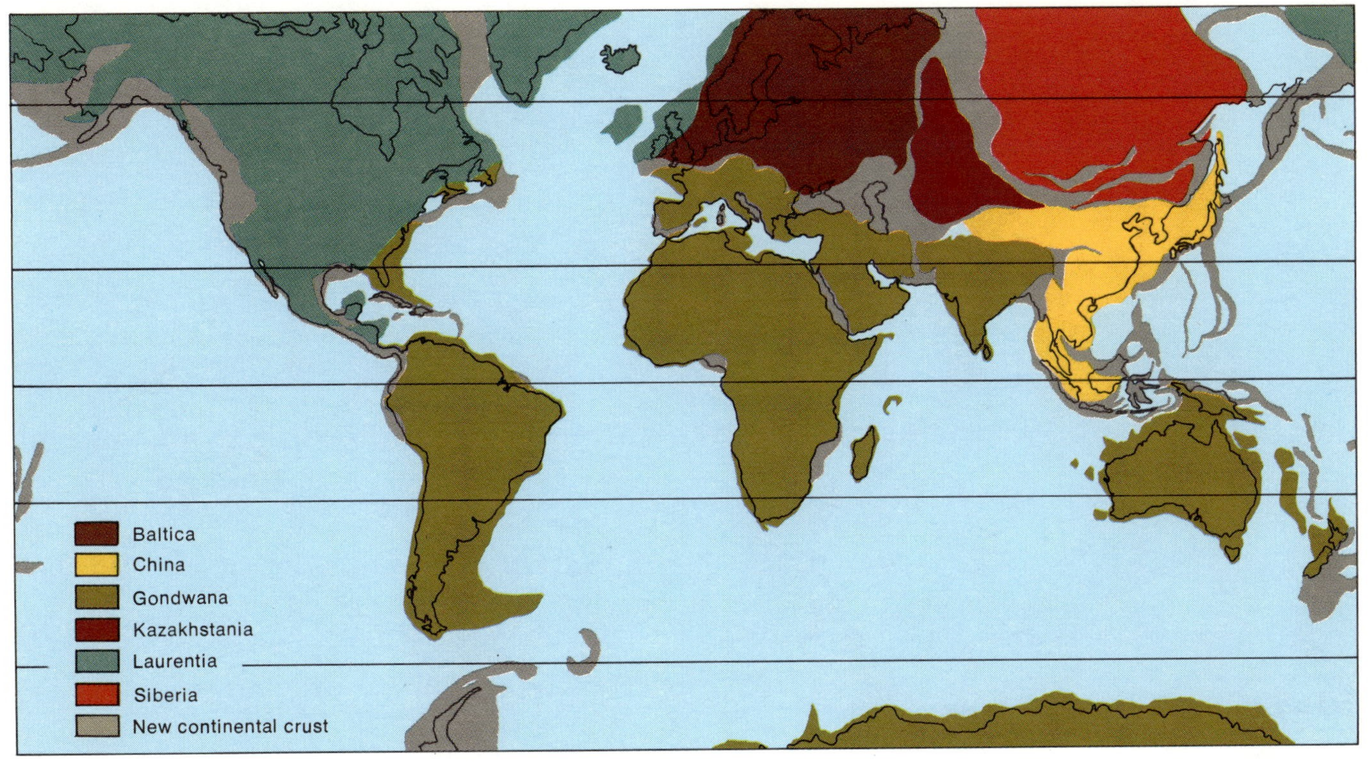

Figure 2. The modern location of Paleozoic continental pieces

earth history. No longer do we need to invoke improbable concepts of borderlands, land bridges, and worldwide shifts in climatic equilibrium to explain the seemingly odd locations in the modern world of ancient salt deposits, coral reefs, and evidence of glaciation or the widely scattered occurrence of similar fossil flora and fauna. Rather, we can understand how these features are the result of movements of the continents themselves, by geologic processes still operating today.

Making paleogeographic reconstructions

Reconstructions of the changing Paleozoic world are made by identifying areas that acted as separate continents, positioning these paleocontinents in their correct orientations, compiling data indicative of geographic and climatologic features, and interpreting the distribution of environmental conditions on each paleocontinent. The interpretation of Paleozoic geography requires synthesis of data derived from many fields of geology. Paleomagnetism has served as the cornerstone of this work (McElhinney 1973), but paleoclima-

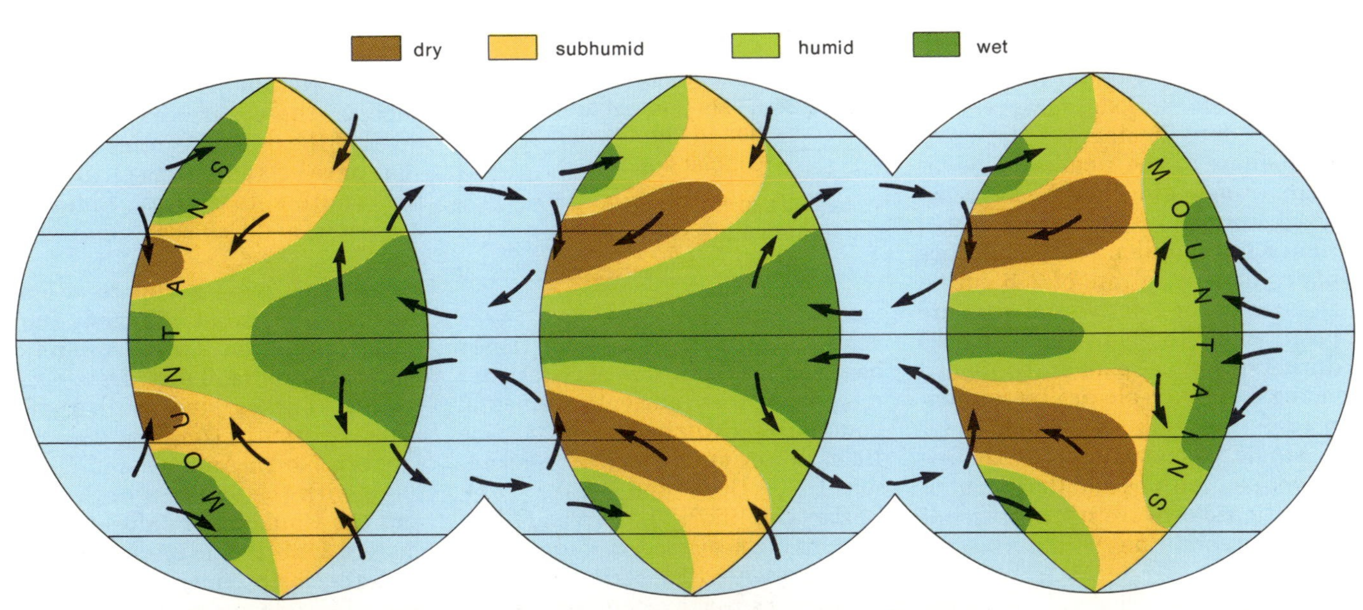

Figure 3. Idealized model of climatic conditions as a function of latitude and geographic configuration

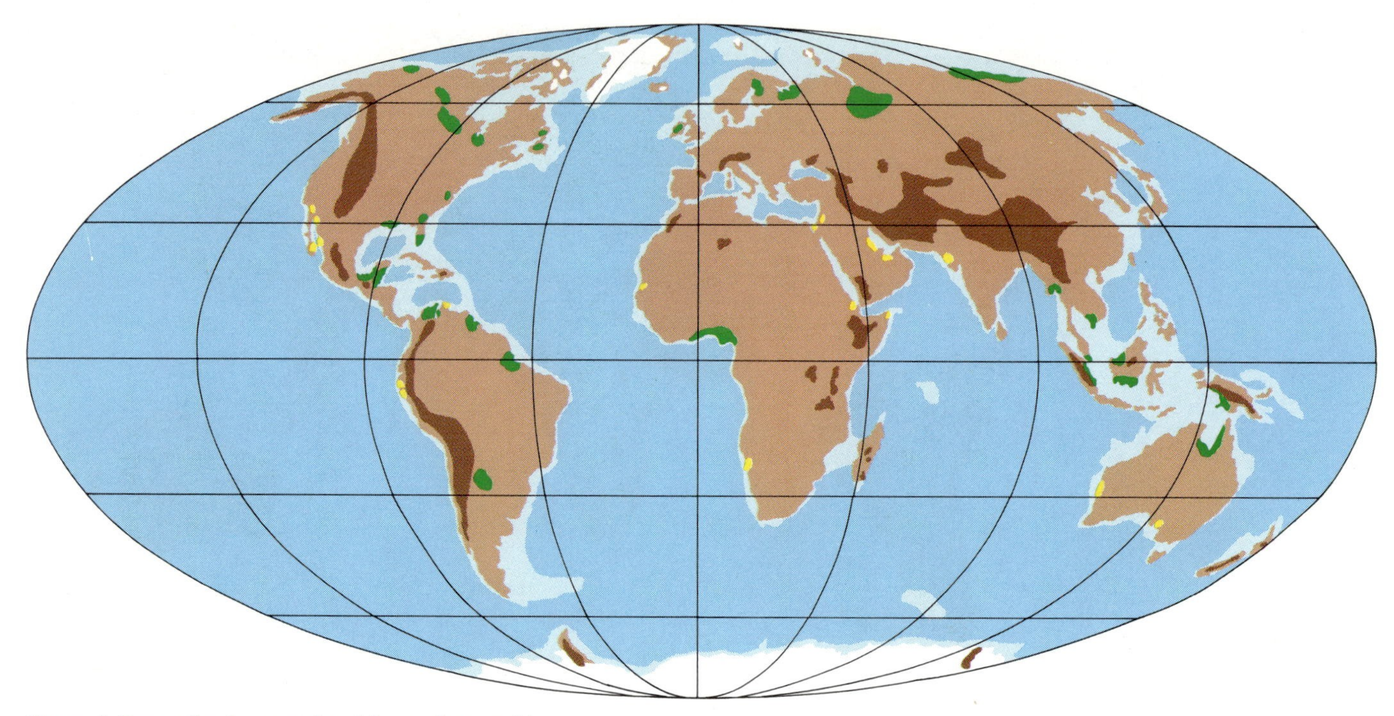

Figure 4. Generalized geography of the modern world

tic, paleobiogeographic, and tectonic data are also vital in compiling detailed reconstructions. Our methods and details of the results have been presented in a series of recent papers (Ziegler, Hansen, et al. 1977; Ziegler et al. 1977, 1979; Scotese et al. 1979), in which the basic data for the following reconstructions were presented.

Geography changes constantly. If too great a period is covered by a single reconstruction, the geographic conditions will be represented by too broad an average to give a clear picture of actual conditions at any particular time. The 12 geologic periods shown in Figure 1 are too long to be summarized by general reconstructions. These periods, however, are divisible into 75 stages, a stage corresponding to the smallest subdivision of time-equivalent rocks recognized worldwide. Each stage lasted from 5 to 15 million years, and since plates move at a rate of 2–8 cm/yr, changes in the relative and absolute positions of the continental blocks during such time spans are contained enough to be realistically summarized. A stage is thus a useful basis for a single reconstruction, though shoreline positions and the extent of glaciation shown are still averages of changing conditions within a stage.

The continental blocks of the Paleozoic that collided to form Pangea were not the same as the continents that formed as Pangea split apart during the Mesozoic or as those which exist today. Identifying ancient continents requires the recognition of features that mark the outlines or margins of these paleocontinents. A major achievement of plate-tectonic theory has been the definition of criteria for recognizing these ancient continental boundaries (Mitchell and Reading 1969; Dewey and Bird 1970; Dickenson 1970; Burk and Drake 1974; Burke et al. 1977). The torn and rifted margins of once-associated continents are represented by certain geologic features, especially belts of basaltic igneous rocks associated with elongate basins bordered by normal faults. Other features, especially belts of andesitic igneous rock and mountain belts with strongly folded rocks, represent the deformed margins of continental blocks under which oceanic crust was subducted as lithospheric plates moved together. Where such belts cross a continent, such as the Ural Mountains in Eurasia, two former continents appear to have collided and been sutured together.

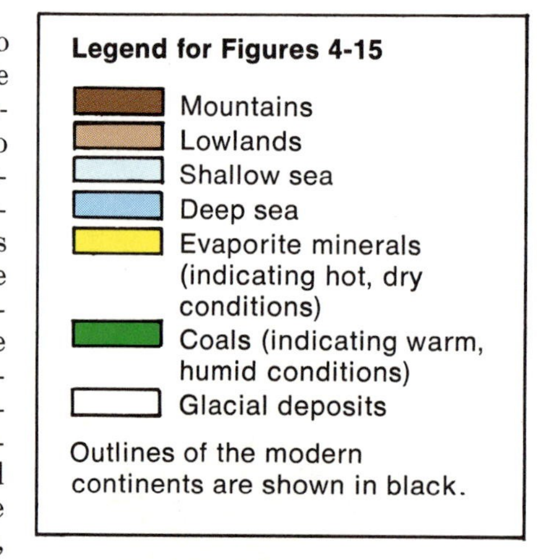

We recognize six major paleocontinents during various parts of the Paleozoic: Gondwana, Laurentia, Baltica, Siberia, Kazakhstania, and China. Figure 2 shows the parts of the modern continents that belonged to each of these paleocontinents and areas of the present continental crust that have accreted to the margins of the continents during Paleozoic, Mesozoic, and Cenozoic times.

Gondwana was a supercontinent composed of what are now South America, Florida, Africa, Antarctica, Australia, India, Tibet, Iran, Saudi Arabia, Turkey, and southern Europe. Laurentia comprised most of modern North America and Greenland, with the addition of Scotland and the Chukotski Peninsula of the eastern USSR. The missing parts of eastern North America were either part of Gondwana or associated with

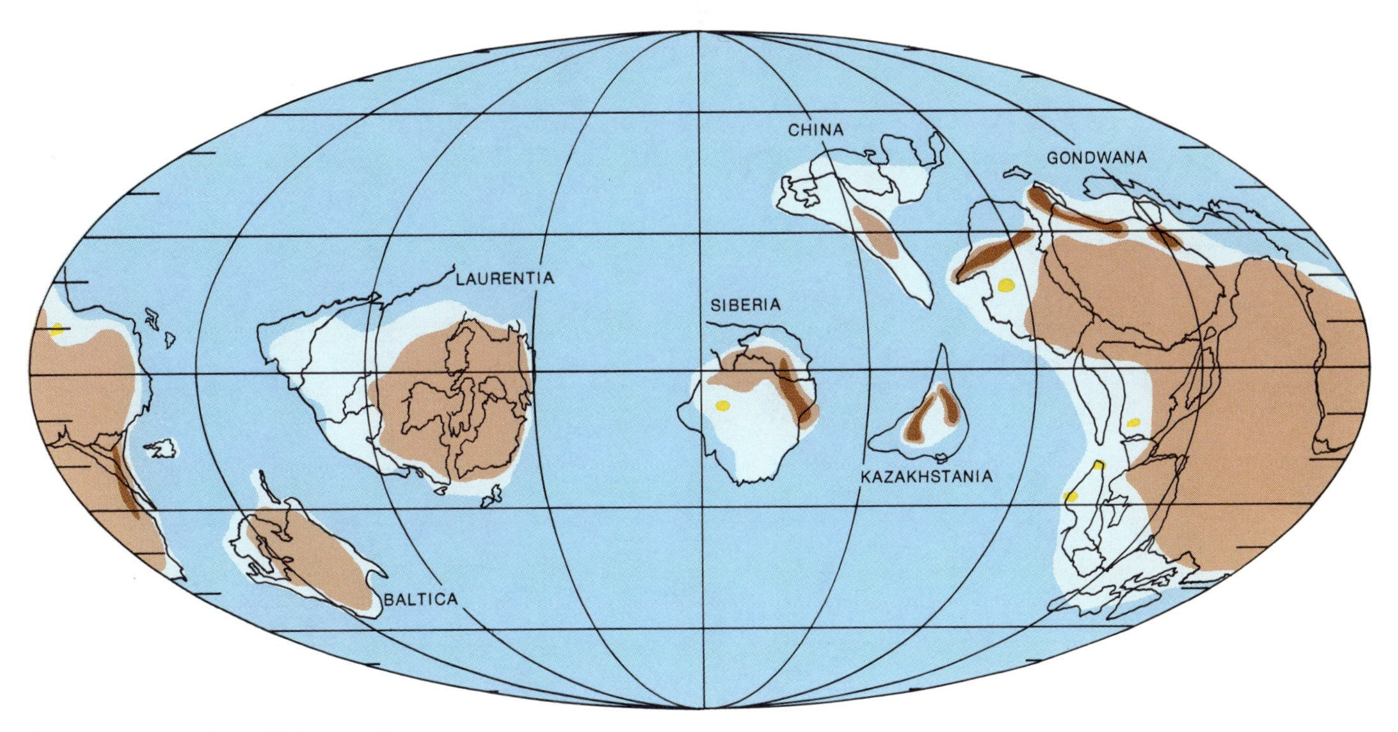

Figure 5. Late Cambrian (550–540 million years ago)

small microcontinents similar to modern New Zealand. Baltica was composed of Russia west of the Ural Mountains, Scandinavia, Poland, and northern Germany. Siberia from the Urals east to the Verkhoyanski Mountains was a separate continent in the Paleozoic. Its southern margin crossed Asia north of present Kazakhstan and south of Mongolia. Kazakhstania was a triangular continent centered on what is now Kazakhstan, with one part extending up between the Urals and southwestern Siberia and another part extending east between the Altai on the north and the Tien Shan Mountains on the south. China is a complex area that may have been subdivided into more than one block at times in the Paleozoic, but there are general similarities which imply that the pieces were not widely separated. For now, we treat all of southeast Asia, including China, Indochina, and part of Thailand and the Malay Peninsula, as a single continent.

Positioning the continents is at the heart of any reconstruction of world paleogeography. Assigning correct positions to paleocontinents involves both the latitude-longitude location and the orientation of the continents relative to the appropriate paleonorth direction. We believe, for instance, that in the Silurian period Siberia was centered at about lat. 30° north and was rotated 180° from its modern orientation, so that its present Arctic coast faced south.

Paleomagnetic information is the basis for all continental positioning. Paleomagnetism provides direct quantitative evidence for both latitude and the north-south orientation of the paleocontinents. Although the magnetic poles may stray from the rotational poles, the earth's magnetic field has a north-south polarity and maintains an alignment that over geologic time on the average parallels the rotational axis. The lines of magnetic force also vary in their inclination to the earth's surface as a function of latitude, being vertical at the magnetic poles and parallel to the surface at the equator. The remnant magnetism of adequately preserved rock samples indicates the azimuth direction to the north (or south) magnetic pole and the latitude of the rock at the time the magnetism was imposed. Thus, from oriented samples we can determine the latitude and orientation of continental blocks for times in the past. Paleoclimatic indicators serve as independent checks on latitude and as guides to latitude when paleomagnetic information is not available.

Longitudes are determined by integrating biogeographic relationships and plate-motion constraints. Although it is not possible to assign absolute longitude (relative to the prime meridian) in the Paleozoic, the whole interval is bracketed by times when relative longitude can be determined within narrow limits. At the time of our earliest reconstruction, in the Late Cambrian, all the continents except Baltica and China straddled the equator (see Fig. 5) and, with their surrounding oceans, occupied much of the space available in the equatorial belt. Since their relative order in sequence around the globe is clear from biogeographic evidence, space constraints alone fix their longitudinal positions within rather narrow limits.

At the end of the Paleozoic (Fig. 14), most of the continental blocks were grouped together in the supercontinent of Pangea. The spatial fit of the continents at this time tells us their relative longitudes quite precisely. Thus, with the endpoints of the Paleozoic well defined, we can follow a regular pattern of plate motion as the plates, bearing the separate continents which ringed the equator in the Cambrian, shifted and brought the continental blocks together to form the pole-to-pole mass of Pangea by the Permian.

The final task in preparing paleogeographic reconstructions is interpreting the geographic features and

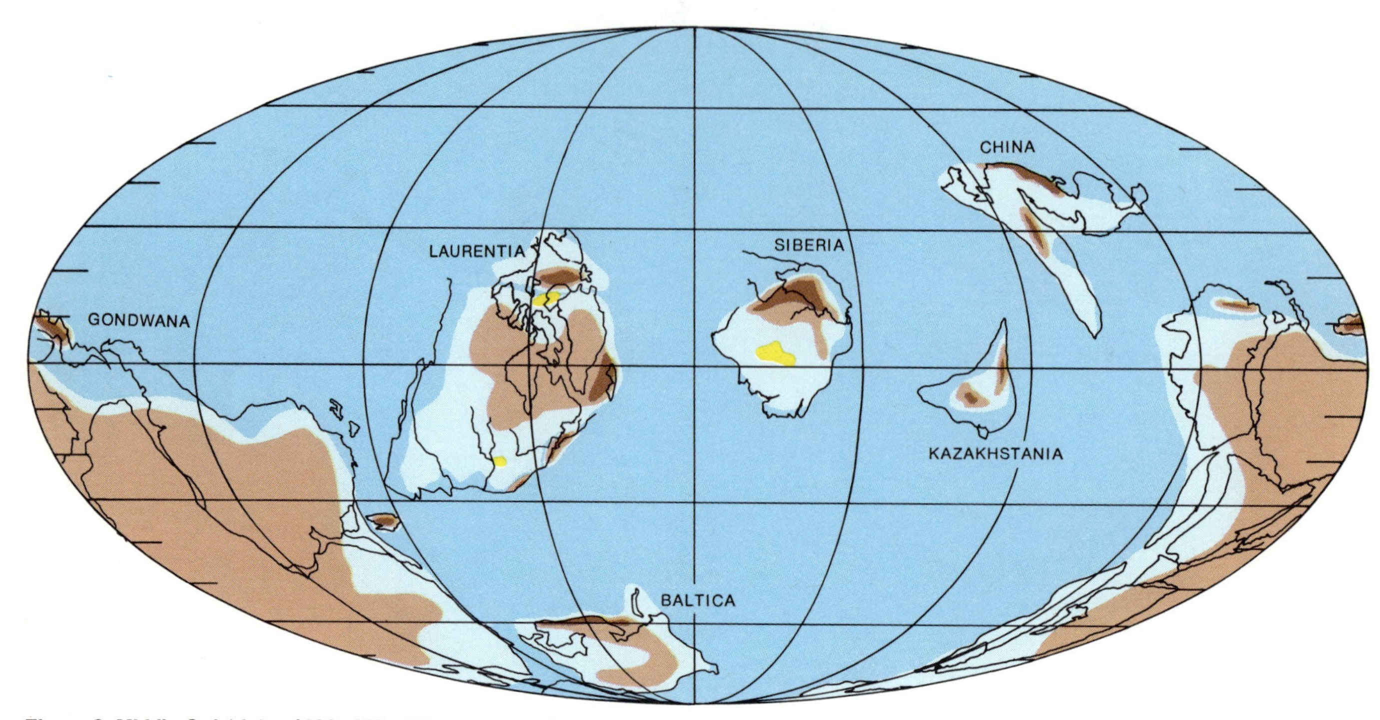

Figure 6. Middle Ordovician (490–475 million years ago)

distribution of environmental conditions on the continental blocks. In our reconstructions we identify regions that were highlands and mountains, lowlands and floodplains, coastal areas, shallow and deep marine platforms, and submarine-slope and deep-sea environments. Environmental conditions are interpreted from the processes known to have formed particular types of rocks. Sands and coarser-grained detrital sediments are deposited in high-energy environments such as stream channels and along beaches, while fine-grained sediments such as clays and muds accumulate in low-energy environments such as floodplains, lagoons, and deep offshore marine environments. Limestones typify warm, shallow marine conditions at some distance from a source of land-derived detrital sediments. Sands and muds, for example, are present along both the Atlantic and Gulf coasts of modern North America where rivers bring them to the shore, but in the Florida Keys and the Bahamas—areas far removed from sources of detrital sediments—limestones are accumulating from the skeletal secretions of abundant marine life. Fossils help differentiate terrestrial,

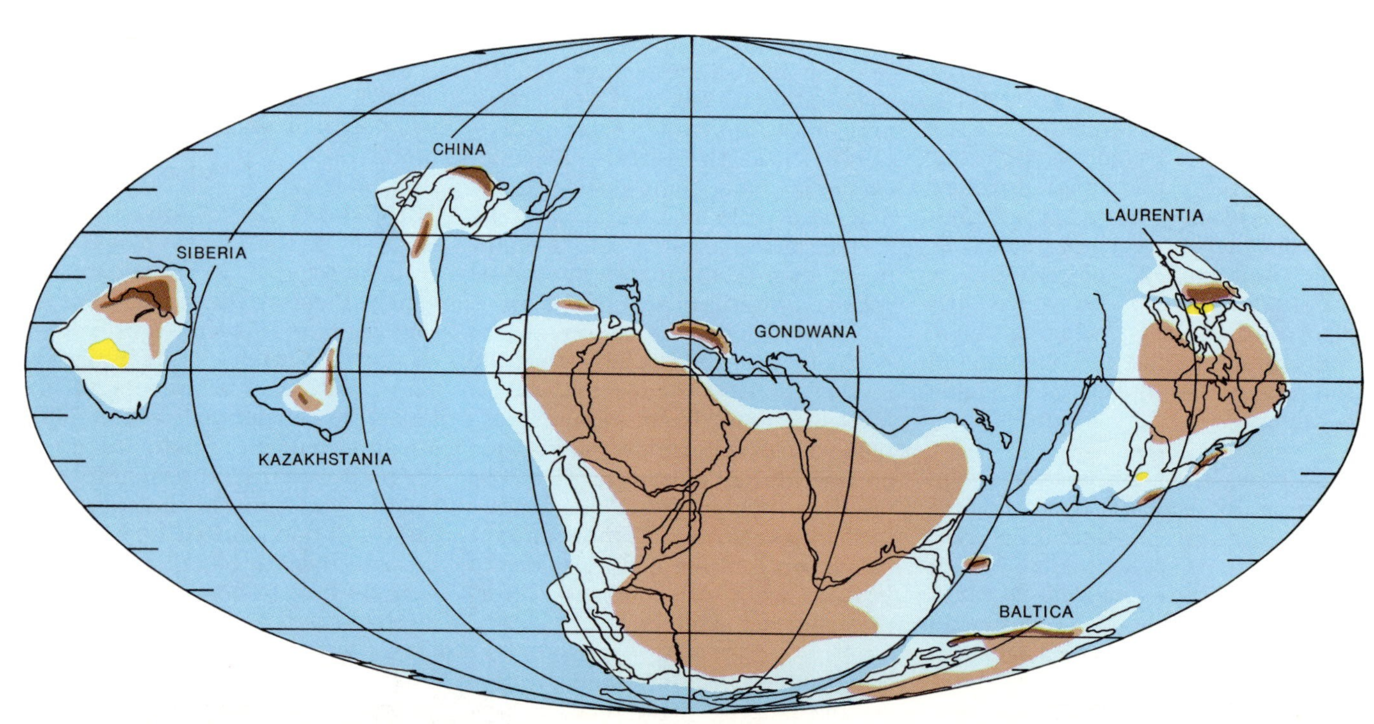

Figure 7. Middle Ordovician, view of earth rotated 180° from the view in Figure 6

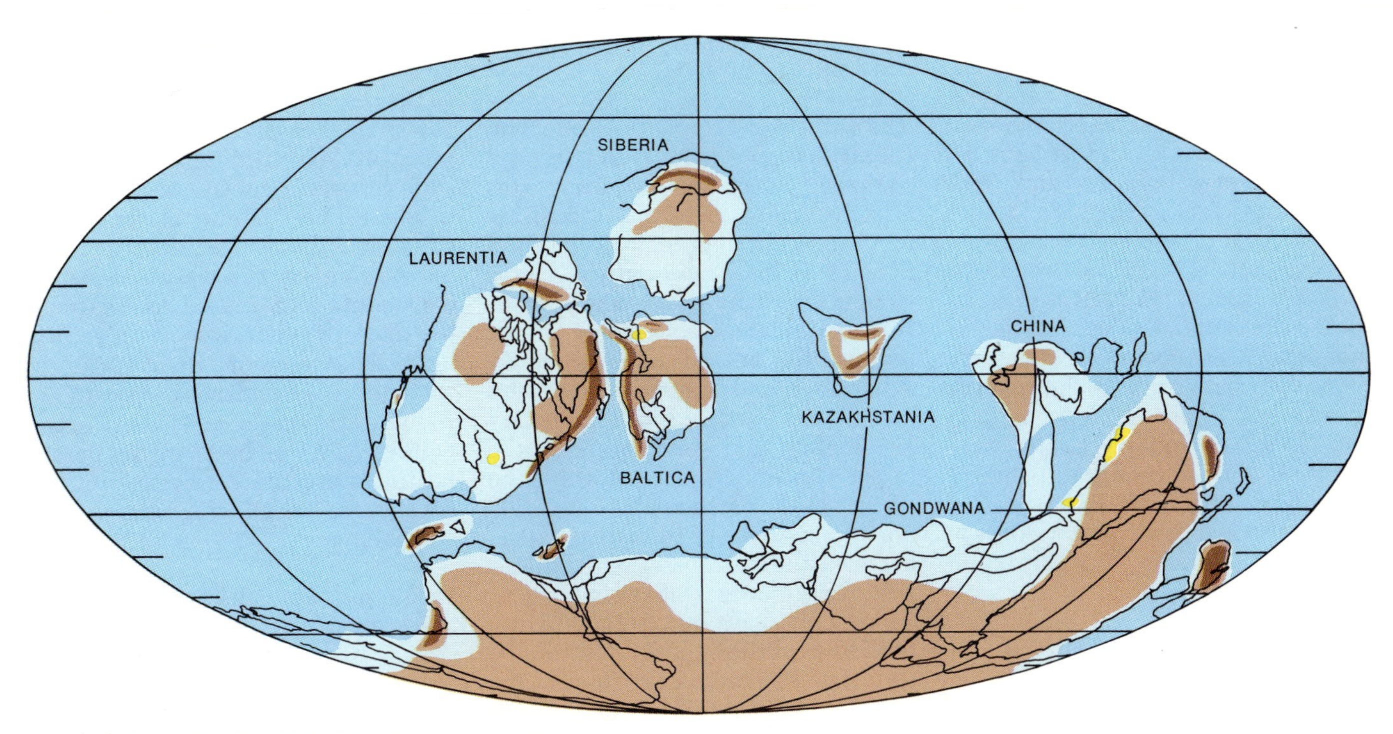

Figure 8. Middle Silurian (435–430 million years ago)

freshwater, and marine deposits, as do suites of sedimentary structures unique to particular depositional settings.

Some sediments reflect climatic conditions. Detrital sediments are indicative of humid environments. Coals formed in freshwater swamps where temperature and rainfall were both adequate for abundant plant growth. Hot, dry shorelines and protected basins along coasts in desert regions are the sites for deposition of salts and other evaporite minerals from brines left by evaporating seawater or playa lakes. In cold environments, glaciers leave distinctive deposits, called *tillites,* with features imposed by ice flow and the peculiar conditions of freezing and thawing at the melting edges of the ice. In areas where nutrients from deep water are supplied in abundance to surface waters, such as zones of coastal upwelling, the sediments may include cherts, phosphorites, or deposits rich in organic material, the result of high biologic productivity.

Geographic features can also be inferred from the type and structure of rock bodies. Delta complexes and

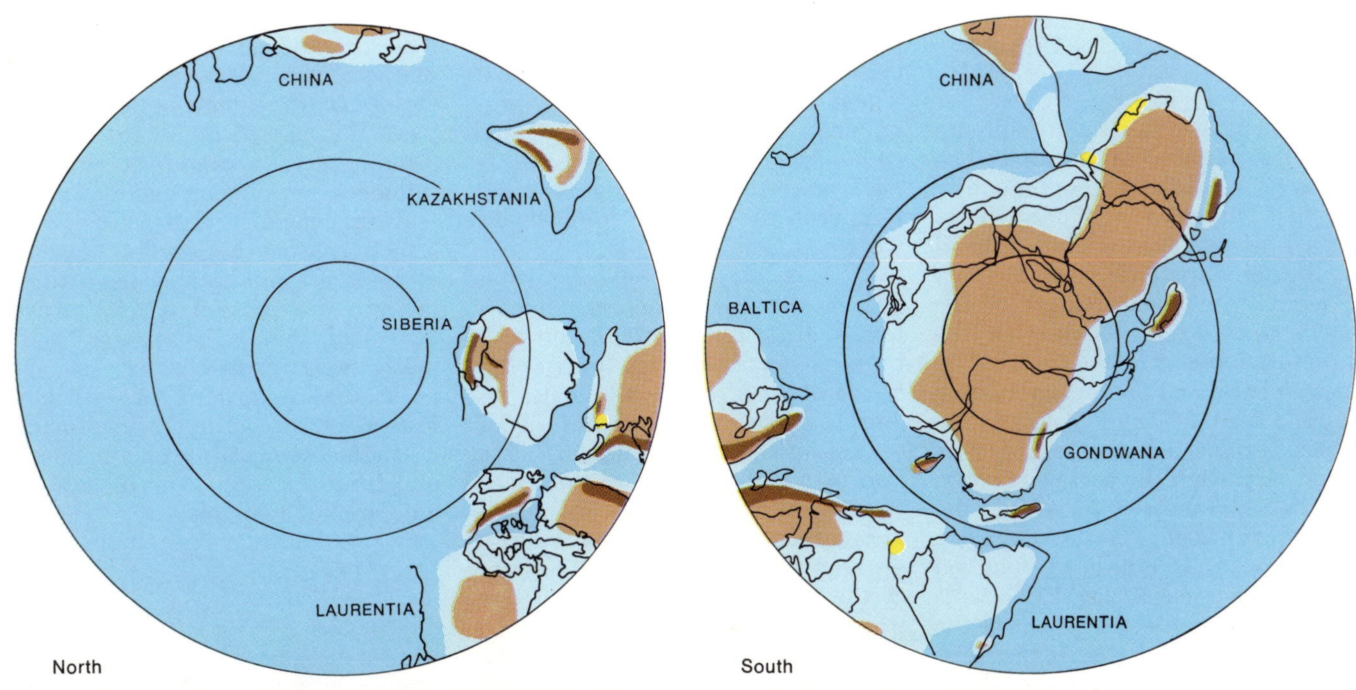

Figure 9. Middle Silurian, North and South Polar projections

deep-sea fans have typical internal features and distinctive three-dimensional forms. Belts of folded and thrust-faulted strata associated with metamorphic rocks and igneous intrusions represent areas that were mountain ranges at the time the deformation and metamorphism occurred, even if they are eroded to lowlands today. Andesitic igneous rocks were erupted along continental margins where subduction of oceanic crust occurred, as is happening in western South America and in the Aleutians today. Basalts were erupted and normal fault basins developed where rifting, associated with seafloor spreading, started, as in the modern African rift valleys or in Iceland.

The distribution of environments on the modern earth is related to the systematic arrangement of topographic features associated with plate-tectonic processes and to the climatic belts produced by atmospheric and oceanic circulation. These patterns are not random now, and they weren't in the past. The circulation of the atmosphere and oceans is an unchanging geophysical system produced by heat-transfer activities on a rotating earth as it is warmed by the sun. Despite the changes in geography over geologic time, some of which have produced local anomalies, the general climatic pattern of the earth is fixed over time. The present earth serves very well as the basis for a general model of global climate structure.

The modern world as a model

The sun is a stable main-sequence star and probably has not altered its intensity of radiation very much in the last billion years. The earth has always been spherical and has orbited at the same distance from the sun. Therefore the heat budget of the earth, with excess heating in the equatorial region, heat transfer by atmospheric and oceanic circulation toward the poles, and cooling in the polar regions, can be regarded as fixed. Although the earth's rotation has been slowed somewhat by tidal friction, it has not decreased by more than about 15% in the last half billion years. This means that the Coriolis effect, which deflects motions to the right in the Northern Hemisphere and to the left in the Southern Hemisphere, has remained nearly constant.

The influence of the Coriolis effect on the airflow generated by heating and cooling of the earth's surface creates a latitudinally zoned pattern of atmospheric circulation (Strahler and Strahler 1978). The equatorial belt is characterized by hot, humid conditions with irregular surface winds, because the heated air is primarily expanding and rising rather than moving in a particular direction. Dry air, cooling and contracting in the upper atmosphere, sinks toward the surface of the earth at about lats. 30° north and south of the equator (the horse latitudes), causing dry climates with irregular surface winds. The sinking air flows out from this belt both toward the equator and toward the poles. The surface flow toward the equator is deflected by the Coriolis effect into strong prevailing easterly winds (the trade winds). The surface flow of air poleward is also deflected by the Coriolis effect and becomes the prevailing westerly winds of the temperate belts at lats. 40°–50° north and south of the equator.

In the polar regions cooling causes the air to contract and sink. This cold polar air flows as surface winds toward the equator, and these winds are deflected by the Coriolis effect into easterly winds in high latitudes. The polar front, where the equatorially trending polar air intersects the poleward flow of temperate air, is a region of high precipitation. The 23.5% tilt of the earth's rotational axis generates the seasonal fluctuations in climate as the earth orbits the sun. These fluctuations are most pronounced in the high temperate latitudes, where the polar front shifts back and forth, causing seasonal temperature changes from below freezing to above.

The interaction of this regular pattern of atmospheric circulation with the land-sea distribution (geography) of the earth produces a climatic regime that is also regular but does not consist of simple latitudinal belts. Except for locations very near the equator or at extremely high latitudes, most continents have quite different climates on opposite coasts at the same latitude (Fig. 3). The tropical humid zone widens toward the eastern sides of continents, where moisture is brought from the ocean by prevailing easterly winds, but it is narrow on the western sides, confined to the narrow zone of intense heating where surface air is rising and losing moisture. The arid belts rise in latitude across the continents from west to east. They extend closer to the equator in the belts of easterly winds on the rain-starved western sides of continents and extend poleward in the midlatitude regions of prevailing westerly winds on their similarly rain-starved eastern sides. As with the wet eastern sides of continents in the tropics, the wet belts in temperate latitudes are much broader on the windward, west-facing margins of continents.

In the modern world these features are seen in the extensive arid belt which rises from low latitudes of the Sahara in eastern Africa to the high-latitude Gobi desert of western Asia. The humid regions of southeast Asia (Indochina, Burma, south China) are at the same latitude as the Sahara in the trade-wind belt (of easterlies), while France and Austria are at the same latitude as the Gobi Desert in the belt of prevailing westerlies. In the Southern Hemisphere, the Atacama Desert extends equatorward on the west coast of South America into latitudes occupied by the Amazon rain forest to the east, and the south Chile rainy zone is at the same latitude as the dry Argentinian pampas. The Andes create intense contrasts even across the narrow South American continent because of the rain shadow caused by their height.

The distribution of features in the modern world serves as a model for understanding climate patterns in the past. Because so many features of sedimentary rocks reflect climatic influence, we can map major climatic features on paleogeographic reconstructions. And the fact that climatic systems are predictable from their modern distribution means, as we have said, that we can use paleoclimatic data to cross-check paleomagnetic determinations of latitude. Paleoclimatic features also provide latitudinal information about regions and times for which paleomagnetic data are not available.

The Paleozoic world

Figures 5, 6, 7, 8, 10, 11, 12, and 14 are reconstructions of world paleogeography at seven times during the

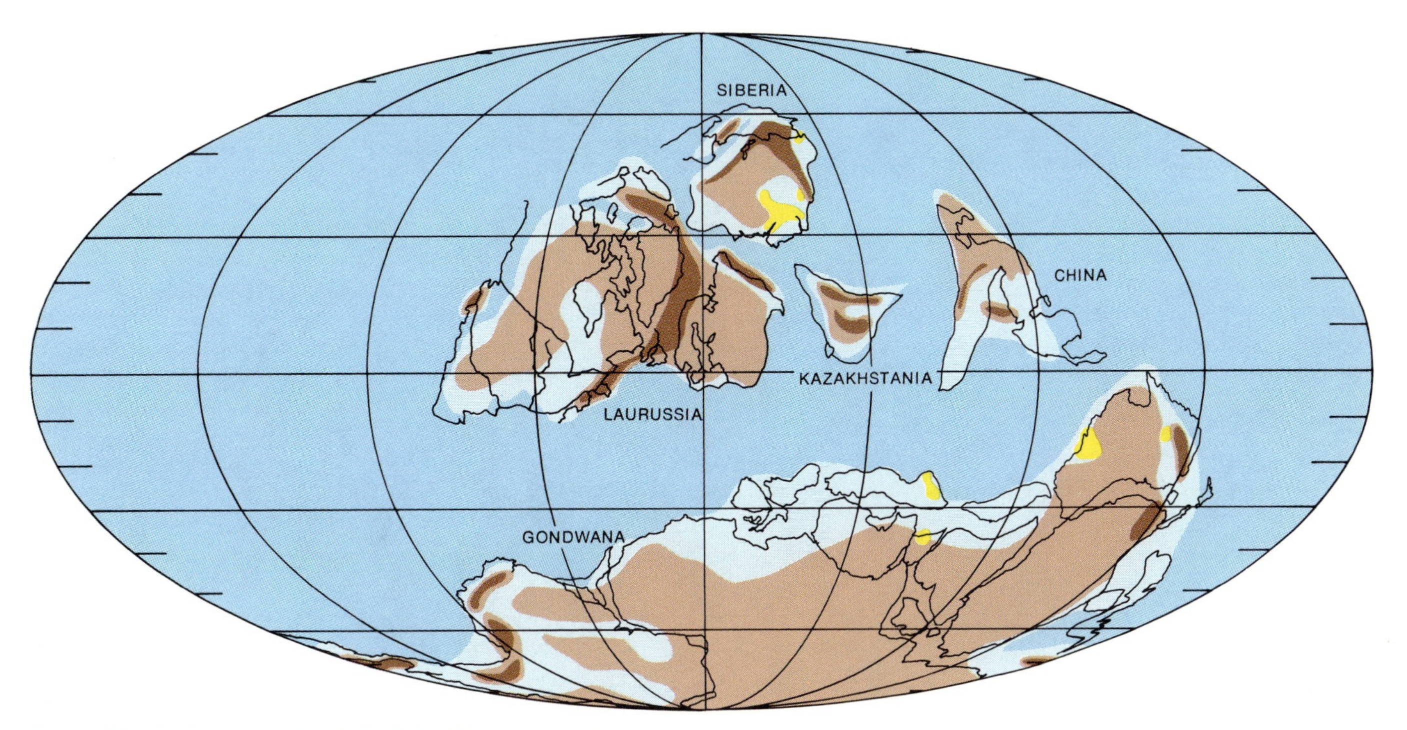

Figure 10. Late Early Devonian (410–405 million years ago)

Paleozoic. These reconstructions show the evolution of the Paleozoic world from the Late Cambrian through the Late Permian. We have used a Mollweide projection, which is an equal-area projection that avoids the excessive area expansion in higher latitudes of the familiar Mercator projection. Although there is angular distortion near the edge of the map at high latitudes, the Mollweide projection also shows the entire earth's surface, including the polar regions, which a Mercator projection cannot do.

The pageant of geographic change through the Paleozoic is profound. None of the Paleozoic geographies is similar to that of our modern earth. In the modern world (Fig. 4) the continents are grouped into three extensive north-south masses—the Americas, Europe/Africa, and eastern Asia through Indonesia to Australia—which partly isolate three equatorially centered oceans—the Pacific, Atlantic, and Indian oceans. The modern continents are mostly emergent, with only narrow continental shelves flooded by shallow seas. There are close connections between almost all the continents surrounding the North Pole, and the small Arctic Ocean is virtually landlocked by the belt of high-latitude land extending from Greenland across northern North America and Siberia to Scandinavia. Antarctica covers the South Pole. Extensive high mountain belts are characteristic, most prominently the Alps-Caucasus-Himalayas system, extending across the Eurasian supercontinent, and the Andes–Rocky Mountains system, extending from Tierra del Fuego to Alaska in the Americas. Areas forming climatically sensitive sediments today are shown on the reconstruction for correlation with modern climate distribution and for comparison with the distribution of ancient deposits.

The ancient Late Cambrian world (Fig. 5) contrasts sharply with the world of today. In the Late Cambrian the continents of the Paleozoic were isolated from each other and dispersed around the globe in low tropical latitudes. The ocean basins were extensively interconnected and the polar regions were occupied by broad open oceans. There was no land above lats. 60° north or south. Shallow seas had transgressed onto the low-lying continental platforms earlier in the Cambrian period and covered large areas of Laurentia, Baltica, Siberia, Kazakhstania, and China in the middle Late Cambrian. The major highlands were in northeastern Gondwana (Australia and Antarctica today), eastern Siberia, and central Kazakhstania. Erosion had reduced the topography of Laurentia and Baltica to low levels.

During the Ordovician (Fig. 6) and Silurian (Fig. 8) Gondwana moved southward from its Late Cambrian position on the equator halfway around the globe from Siberia to a position straddling the South Pole. This change can be followed by comparing the latitude of Gondwana in Figures 5, 7, and 9. The oldest record of glaciation in the Paleozoic era is of Late Ordovician age in what is now the Sahara Desert: tillites were deposited at the time that this part of Gondwana was actually crossing the South Pole.

During the movement of the plate containing Gondwana, Baltica also shifted position along a parallel path from the Cambrian to the Silurian. The two paleocontinents may have been located on one lithospheric plate. Siberia also shifted from equatorial to north temperate latitudes during the early Paleozoic.

These movements were marked by mountain building along the eastern margin of Laurentia (the Taconic orogeny) in the middle and late Ordovician and later mountain building in western Baltica during the Silurian (the Caledonian orogeny), as the ocean between Laurentia and Baltica closed. Shallow seas were widespread on the continents throughout most of the Ordovician and Silurian. Climatic zonation is detectable in the distri-

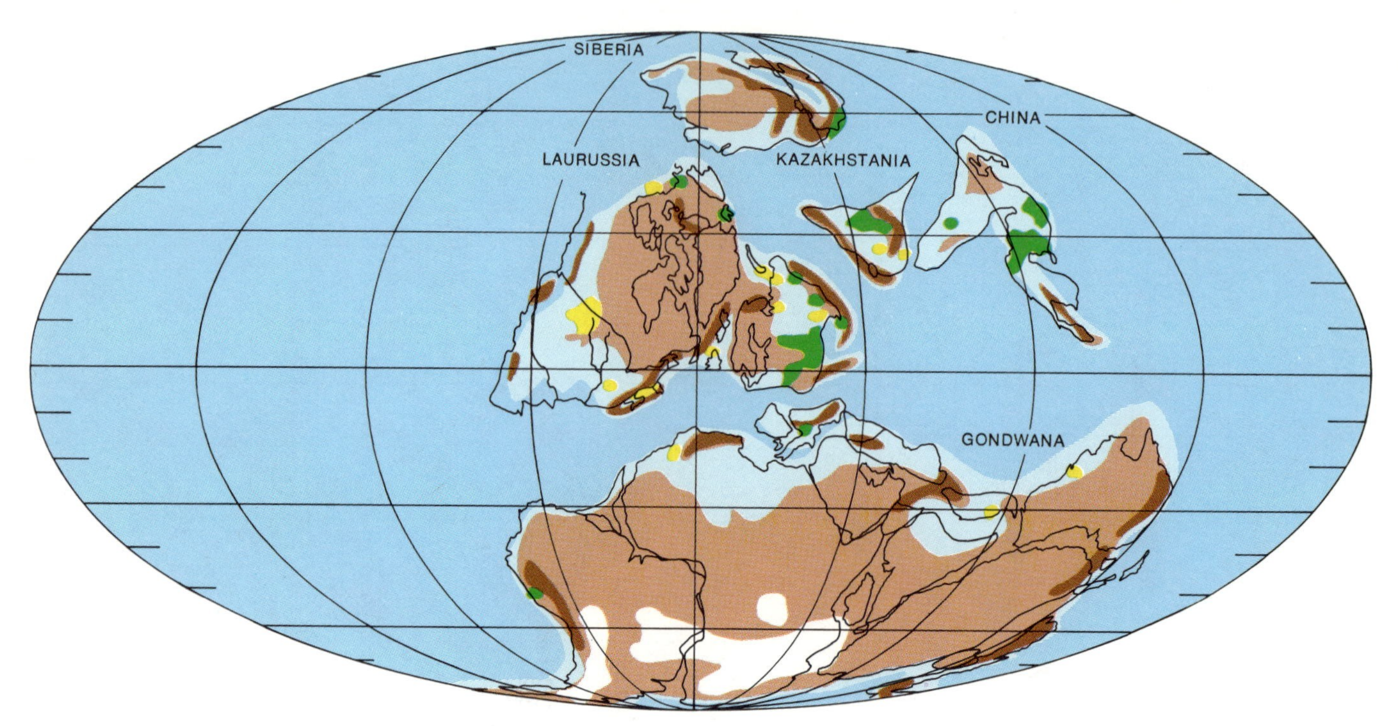

Figure 11. Middle Early Carboniferous (360–340 million years ago)

bution of evaporite deposits, which are concentrated between lats. 15° and 30° north and south, as in the modern world.

Frigid polar climates are not detectable in the Cambrian or most of the Ordovician. This may be because the open polar oceans of these times were always warmed by oceanic circulation, or it may simply be that the lack of polar land prevented preservation of a geologic record of polar climates. As mentioned above, the first land areas to enter polar latitudes, the North African portion of Gondwana, was glaciated in the Late Ordovician. Marine faunas of relatively low diversity—and therefore probably from temperate climates—are known from the higher south latitude areas of Gondwana in the Ordovician and Silurian, and a distinctive low-diversity fauna is found in the Silurian along the margin of Siberia, which first moved as far north as lat. 40° north.

Major reorganization of world geography is apparent by the Silurian. The shift southward of Gondwana from its Cambrian position had opened the former North Polar ocean basin until it not only circled the world at high northern latitudes but extended in an unbroken expanse southward across the equator to high southern latitudes. The former South Polar ocean basin of the Late Cambrian had been displaced to middle southern latitudes by the shifts of Gondwana, Baltica, and Siberia and had become a partly enclosed basin between Baltica, Kazakhstania, and Gondwana. This was the start of the development of the Tethys Sea, a region characterized by distinctive marine faunas throughout the Late Paleozoic and Mesozoic.

By the Early Devonian (Fig. 10), Laurentia and Baltica had collided to form a larger continent, Laurussia. The collision began in the Late Silurian with the Caledonian orogeny in northwestern Baltica. Mountain building continued into the Devonian as the Acadian orogeny in eastern Laurentia. These uplands along the suture between the formerly separate paleocontinents were located in the equatorial belt, and large volumes of detrital sediments were eroded from them. Nonmarine fluvial sediments covered large parts of eastern North America (the "Catskill Delta") and northern Europe; these sediments were deeply weathered and are stained red from iron oxides—hence the name Old Red Sandstone in Great Britain. Although land plants had begun to evolve in the Silurian, it is in these tropical nonmarine Devonian deposits that abundant larger fossil land plants first appear.

The Early and Late Carboniferous reconstructions (Figs. 11 and 12) show that Gondwana continued to move across the South Pole and entered the same hemisphere as Laurussia, closing the ocean between them. Their collision in the Late Carboniferous resulted in the Hercynian orogeny, which extended across central Europe, and the Alleghenian orogeny in eastern North America. Baltica had begun colliding with Laurentia in the Silurian and Devonian at a location relative to Laurentia far to the south of the position it occupied later. Repositioning took place in the Carboniferous, during the collision between Laurussia and Gondwana, when what had been Baltica was displaced northward along a series of faults extending from coastal New England, across Newfoundland, and through Scotland along the zone of weakness at the original suture.

From the Silurian through the Carboniferous, Siberia moved to high latitudes and Kazakhstania and China moved westward. By the Late Carboniferous, Kazakhstania and Siberia had collided and all the paleocontinents were clustered tightly as Pangea began to take shape (Fig. 13). Mountain belts extended along the suture between Laurussia and Gondwana (the southern Appalachian and Hercynian belts) and along the reactivated suture where

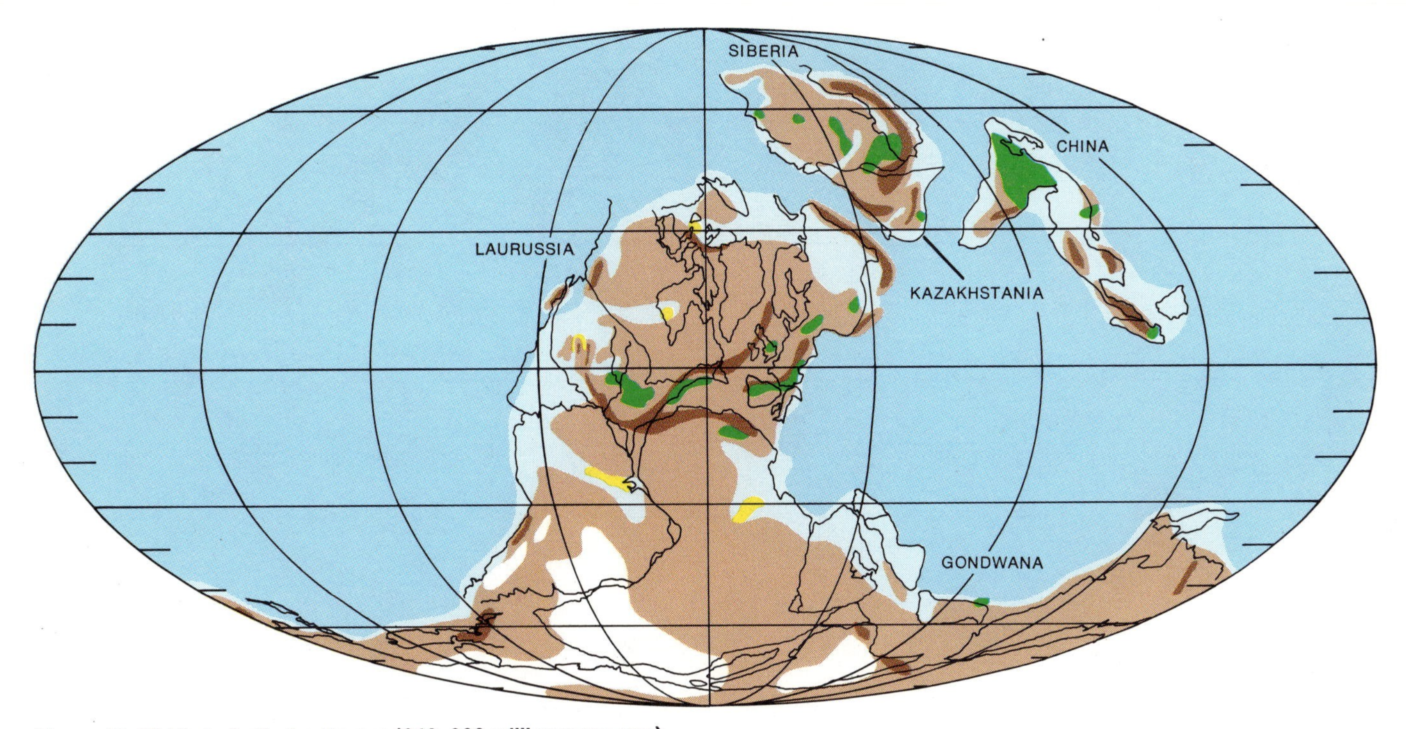

Figure 12. Middle Late Carboniferous (310–300 million years ago)

Baltica sheared northward relative to Laurentia (the northern Appalachian-Caledonide belt). Mountain systems marking subduction zones also developed during the Carboniferous on the eastern side of Baltica, as the ocean between Baltica and Kazakhstania-Siberia was closing, and in the Mongolian portion of Siberia, as the ocean between Siberia and China also closed.

The distribution of climatically indicative deposits in the Late Carboniferous (Fig. 12) reconstruction is particularly interesting. The great coal reserves of eastern North America, western Europe, and the Donetz Basin of the USSR lie in the equatorial zone. The coal swamps developed on marshy delta platforms built by rivers bringing detrital sediments from the adjacent mountain ranges. The large volume of both detrital sediments and plant remains testifies to the high rainfall in this tropical belt. Plant fossils in the coals do not show strong seasonal growth rings, which implies that they grew under constantly warm rather than temperate seasonal conditions. The belt of tropical coals is narrow in the west and broadens to the east, as predicted by the climate models (see Fig. 3).

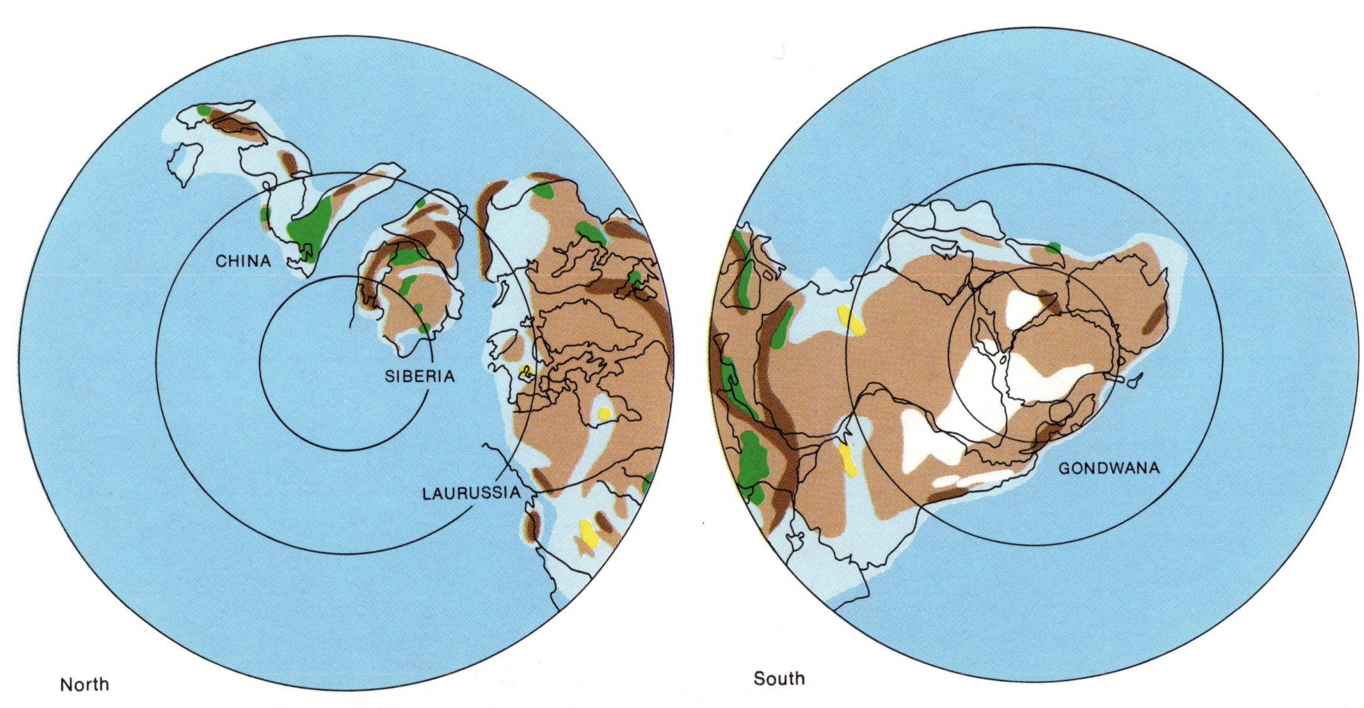

Figure 13. Middle Late Carboniferous, North and South Polar projections

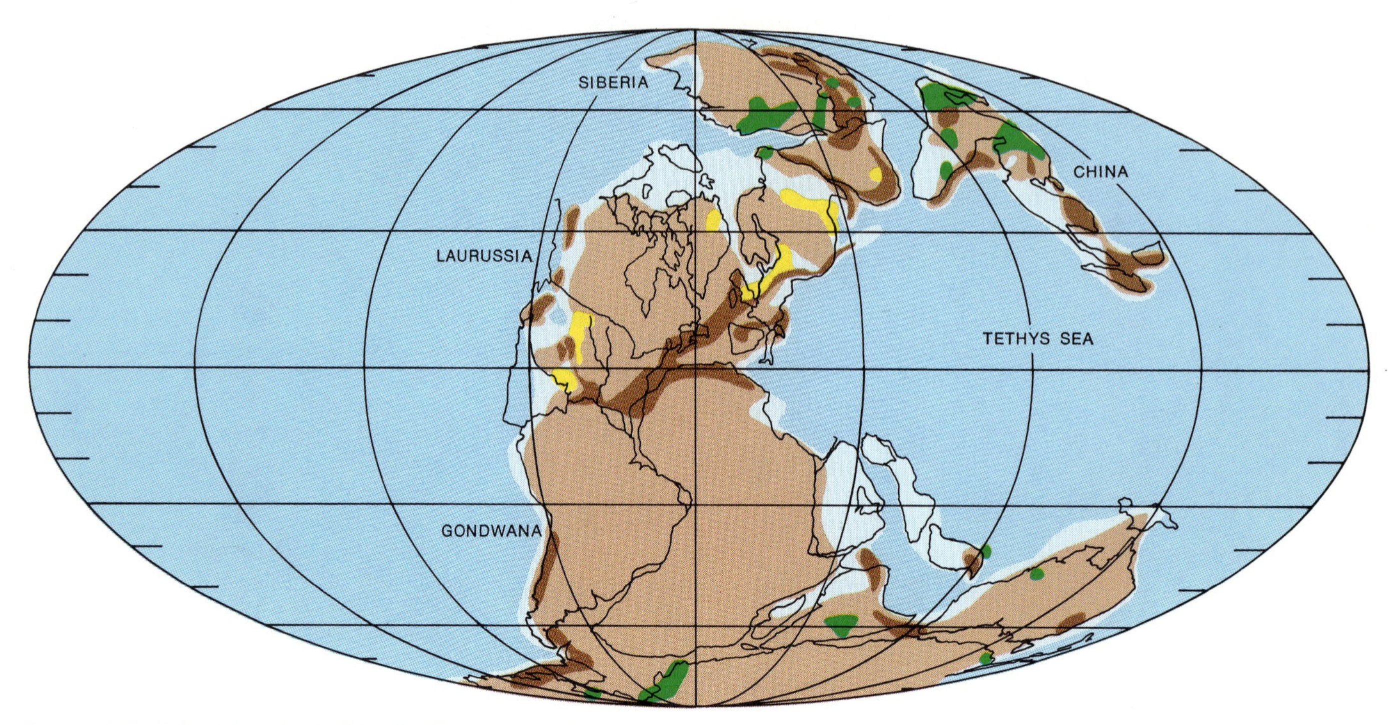

Figure 14. Early Late Permian (260–250 million years ago)

Evaporite deposits, indicative of low rainfall, occur in far western Laurussia at low latitudes and farther east between lats. 15° and 30° north and in a belt in Gondwana between lats. 20° and 30° south, again as expected from the climatic model. In the north temperate belt (lats. 40°–60° north) extensive detrital sediments and coals in Siberia and China indicate abundant rainfall. The seasonal nature of the climate of this belt is indicated by well-developed growth rings in plant fossils from Siberia. Seasonal growth rings, imposed by interruption of growth during cold winter months, are also found in plant fossils of both Carboniferous and Permian age from the south temperate latitudes of Gondwana.

Glaciation is recorded by widespread tillites in southern Gondwana. Ice sheets were present above lat. 60° south from the Early Carboniferous into the Early Permian. At their most extensive they flowed equatorward as far as the middle temperate latitudes, just as the Pleistocene glaciers flowed south from the Arctic regions of North America and Europe less than 100,000 years ago. The fact that the Carboniferous glacial deposits are

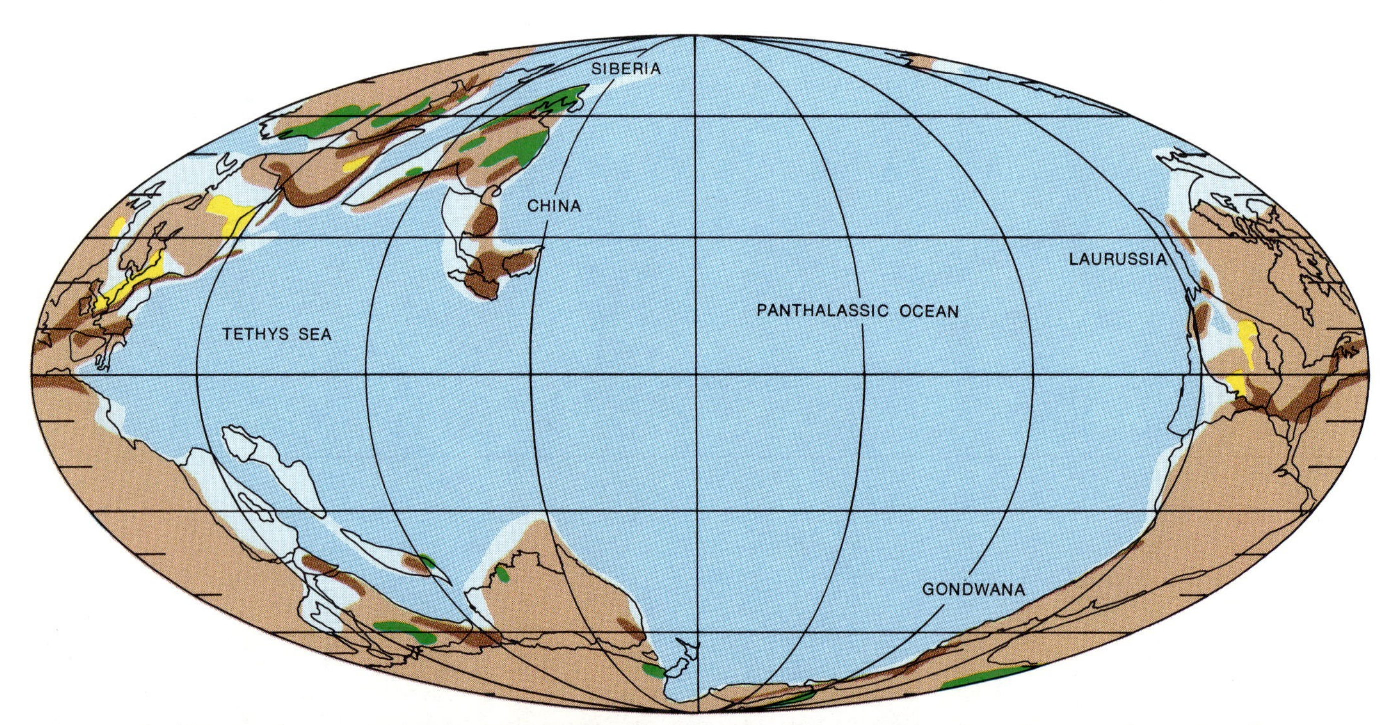

Figure 15. Early Late Permian, view of earth rotated 180° from the view in Figure 14

now found on continents scattered across half the earth's surface (from South America through southern Africa to Antarctica and India) was one of the strongest intimations of continental drift before the theory of plate tectonics.

Pangea was nearly assembled by the Late Permian (Fig. 14). It is worth noting, however, that a Pangea containing all continental blocks probably never formed completely. In eastern Gondwana during the Permian, rifting had begun that pulled Tibet, Iran, and Turkey away from the Tethyan margin of Gondwana as a separate, isolated block or blocks before China collided with the Mongolian region of the former Siberian block in the Triassic.

The effects of continental collision reach far beyond the simple suturing together of two formerly separate continental blocks. Collision deforms rocks through the entire thickness of the lithosphere along the colliding margins of the plates. The resulting folding and thrust faulting cause the rocks to "pile up" on themselves, thus thickening the continental crust in the zone of collision. Because lighter continental crust is buoyed up on the denser mantle, these belts of thickened, deformed continental crust form uplifted mountain belts.

And because the same mass of crustal material as formerly existed in undeformed, thinner crust is packed into these thickened belts, the area covered by continental crust is decreased during collision, just as the area covered by a rug is less if it is crumpled up against a wall rather than spread flat on a floor. The decrease in continental crustal area during collision is taken up by an increase in the area of the ocean basins, with a concomitant increase in their volume. This can cause a general lowering of sea level in relation to the continents, as shallow seas flooding continental platforms drain into the enlarging ocean basins.

The series of continental collisions through the late Paleozoic (Devonian–Permian) formed extensive mountain belts and decreased the total area of continental crust. As a result of this process and possibly also because of changes in the volume of mid-ocean ridges at the spreading centers, sea level was effectively lowered and most of the shallow seas that had flooded the low continental platforms through much of the Paleozoic were drained. The larger areas of exposed land contributed to increased climatic severity.

Permian terrestrial sediments indicative of arid and semiarid conditions were widespread. The mountains in the tightly sutured region between Laurussia and Gondwana were high enough to disrupt the subtropical easterly winds and create an intense rain shadow even in tropical latitudes, much as the Andes do in western South America today. Evaporites of Permian age extend to what was the equator in western Laurussia. Desert conditions also extended north of lat. 40° in eastern Laurussia on the side of the continent downwind from the westerly winds, just as deserts extend northward in eastern Asia today.

The clustering of the continents into Pangea had an extraordinary geographic consequence. An enormous single interconnected ocean developed (Fig. 15). This "world ocean," sometimes called Panthalassa, not only spanned the globe from pole to pole but extended for 300° of longitude at the equator, twice the distance from the Philippine Islands to South America across the modern Pacific. Circulation in this giant ocean had to have a major impact on Permian climates. For example, the equatorial currents driven by the trade winds flowed uninterrupted around five-sixths of the circumference of the earth. The east-facing (Tethyan) coast of Pangea, against which these currents impinged, must have been extremely warm. Ancient "Gulf Streams" would have circulated these warm waters into higher latitudes, causing an especially strong climatic asymmetry between eastern and western Pangea. The Permian was indeed a time of geographic extremes.

Paleogeography and geologic history

The old dictum "the present is the key to the past" has a rather specific meaning for geologic history. The processes of geology, including plate tectonics, have probably operated much as we observe them today over long spans of geologic time, and this is certainly the case if we accept reasonable variation in their rates. But although geographic features have always reflected the operation of processes and systems we observe today, the order of geographic change is a unique historical sequence. Thus the present *is* the key to understanding past processes, but it *is not* the key to describing past configurations.

Because plate-tectonic processes operate constantly, there has been no stable or even average geography of the earth during the past half billion years. Paleogeography changed continuously, passing from one extreme configuration through a series of intermediates to a different extreme, as illustrated by these maps, and then changing still further. The geographies of the Cambrian, Permian, and the present are simply single steps in this dynamic pattern.

The Cambrian, with its isolated, equatorially distributed continents and two polar but interconnected oceans, was totally unlike the Permian, with its single concentration of continents stretching from pole to pole and its immense, equatorially centered Panthalassic Ocean. The widespread shallow seas that flooded continental platforms in the Early Paleozoic contrast with the large proportion of exposed land in the Late Paleozoic. These two extremely different periods also differed totally from the modern world, which is the product of the breakup of Pangea into nearly interconnected north–south continental belts and large, semi-isolated ocean basins.

The modern world, of course, is just another transient stage in geologic history. Sound—though broadly outlined—predictions on the future course of geographic change indicate that the Atlantic Ocean, which has been opening for the past 150 million years as the Americas have been moving westward, will grow larger. The Atlantic coast of North America will probably develop into an Andean type of mountain system in the not-too-distant geologic future. There is very little subduction in the Atlantic today, but it is likely to begin in the next 50 million years, especially in the relatively old western North Atlantic.

On the other side of our continent, Baja California and the part of Southern California west of the San

Andreas Fault will continue on their present course of northwestward motion as part of the Pacific plate. California will not "fall into the sea," but the part west of the San Andreas fault may become a New Zealand-like small continental island moving away from mainland North America in the next 50 to 100 million years. Across the Pacific Ocean, Australia, which has been moving northward away from Antarctica for the last 40 million years, will move north past Indochina and may collide with China, Japan, or far eastern Russia. The world's largest ocean then will be the interconnected Indian-Antarctic-South Pacific.

The historical development of our earth has followed a nonrepetitive path through time. The world of the Permian was a world alien to the one in which we live; the world of the Cambrian was equally alien, to the Permian as well as the present; and the world 100 million years from now will be alien to our present one. It is within this framework of changing geographies, with markedly different extreme configurations and long intervening transitions, that we must cast our ideas of geologic history.

References

Burk, C. A., and C. L. Drake, eds. 1974. *The Geology of Continental Margins.* New York: Springer-Verlag.

Burke, K., J. F. Dewey, and W. S. F. Kidd. 1977. World distribution of sutures: The sites of former oceans. *Tectonophysics* 40: 69–99.

Dewey, J. F., and J. M. Bird. 1970. Mountain belts and the new global tectonics. *J. Geophys. Res.* 75:2625–47.

Dickenson, W. R. 1970. Relation of andesites, granites, and derivative sandstones to arc-trench tectonics. *Revs. of Geophysics* 8: 813–60.

Dietz, R. S., and J. C. Holden. 1970a. The breakup of Pangea. *Sci. Am.* 223(4):30–41.

———. 1970b. Reconstruction of Pangea: Breakup and dispersion of continents, Permian to present. *J. Geophys. Res.* 75: 4939–56.

McElhinny, M. W. 1973. *Paleomagnetism and Plate Tectonics.* Cambridge Univ. Press.

Mitchell, A. H., and H. T. Reading. 1969. Continental margins, geosynclines, and ocean floor spreading. *J. Geol.* 77:629–46.

Nierenberg, W. A. 1978. The deep sea drilling project after ten years. *Am. Sci.* 66:20–29.

Scotese, C. R., R. K. Bambach, C. Barton, R. Van der Voo, and A. H. Ziegler. 1979. Paleozoic base maps. *J. Geol.* 87:217–68.

Smith, A. G., and J. C. Briden. 1977. *Mesozoic and Cenozoic Paleocontinental Maps.* Cambridge Univ. Press.

Smith, A. G., J. C. Briden, and G. E. Drewry. 1973. Phanerozoic world maps. In *Organisms and Continents through Time,* ed N. F. Hughes, pp. 1–42. Paleontological Association Special Papers in Paleontology, no. 12.

Strahler, A. N., and A. H. Strahler. 1978. *Modern Physical Geography.* Wiley.

Van der Voo, R., and R. B. French. 1974. Apparent polar wandering for the Atlantic-bordering continents: Late Carboniferous to Eocene. *Earth Science Reviews* 10:99–119.

Ziegler, A. M., K. S. Hansen, M. E. Johnson, M. A. Kelly, C. R. Scotese, and R. Van der Voo. 1977. Silurian continental distributions, paleogeography, climatology, and biogeography. *Tectonophysics* 40:13–51.

Ziegler, A. M., C. R. Scotese, W. S. McKerrow, M. E. Johnson, and R. K. Bambach. 1977. Paleozoic biogeography of continents bordering the Iapetus (pre-Caledonian) and Rheic (pre-Hercynian) oceans. In *Paleontology and Plate Tectonics,* ed. R. M. West, pp. 1–22. Milwaukee Public Museum, Special Publications in Biology and Geology, no. 2.

———. 1979. Paleozoic paleogeography. *Ann. Revs. Earth and Planet. Sci.* 7:473–502.

Zonenshayn, L. P., and A. M. Gorodnitskiy. 1977a. Paleozoic and Mesozoic reconstructions of the continents and oceans, article 1: Early and Middle Paleozoic reconstructions. *Geotectonics* 11:83–94.

———. 1977b. Paleozoic and Mesozoic reconstructions of the continents and oceans, article 2: Late Paleozoic and Mesozoic reconstructions. *Geotectonics* 11:159–72.

Jack A. Wolfe

A Paleobotanical Interpretation of Tertiary Climates in the Northern Hemisphere

Data from fossil plants make it possible to reconstruct Tertiary climatic changes, which may be correlated with changes in the inclination of the earth's rotational axis

Anyone who has even a slight acquaintance with paleoclimatic literature is well aware that the last 1 to 1.5 million years of the Quaternary have been characterized by major episodic glaciations of the continents of the Northern Hemisphere, and hence the period is atypical of much of geologic time. A commonly accepted thesis on climates preceding Quaternary glaciation is that, from some time in the Late Cretaceous or early Tertiary (some 40–80 m.y. ago), when the earth's climate was characterized by generally higher temperatures and higher equability of temperature than now, both overall temperature and equability have gradually decreased, culminating in Quaternary glaciation. Further, some researchers have maintained that even as long ago as the early Tertiary, temperatures were only moderately higher than now, even at high latitudes. (See Table 1 for the geologic time span dealt with in this article.)

An increasing accumulation of data from a multitude of sources has, however, largely negated such once commonly accepted theses. A significant warm episode during the Miocene (see Fig. 1) was first documented in Europe by Mai (1964) and has subsequently been substantiated in other regions such as Japan (Tanai and Huzioka 1967), western North America (Wolfe and Hopkins 1967; Addicott 1969), and New Zealand (Devereux 1967). Alpine glaciation is known to have begun in Alaska during the Miocene (Denton and Armstrong 1969; Plafker and Addicott 1976), when at least part of the Antarctic ice sheet was also present (Kennett 1977).

The most dramatic climatic event, however, occurred during the middle of the Tertiary. MacGinitie (1953) recognized that, if certain floras in Oregon were as close in time as some stratigraphic evidence indicated, a rapid and major climatic change must have occurred, a decrease in temperature that was considered significant but gradual by Nemjč (1964) and Zhilin (1966). Utilizing newly available radiometric ages, Wolfe and Hopkins (1967) demonstrated that this major climatic deterioration had occurred within 1 or 2 m.y.

This temperature decrease has subsequently been recognized in many regions. I had previously (1971) termed it the "Oligocene deterioration," but since recent work in relating the marine and nonmarine chronologies indicates that, in the widely accepted chronology based on marine plankton, the event occurred at the end of the Eocene, I will refer to it as the "terminal Eocene event." In the Southern Hemisphere, the terminal Eocene event is closely associated with the initiation of cold bottom water in the oceans (Kennett 1977), while on the continents of the Northern Hemisphere the event is emphasized by a major decrease in equability of temperature (Wolfe 1971).

Foliar physiognomy

A thorough review of all pertinent paleoclimatic data for the Tertiary would be a lengthy and prodigious task. In this paper I will largely limit the discussion to the paleoclimatic data based on fossil plants from middle to high latitudes (<30°) of the Northern Hemisphere. For the Paleocene and Eocene, the North American data, which are based on leaf remains, are the most relevant. In Europe, the major Eocene floral sequence is based on fruit and seed floras. In eastern Asia, some Oligocene floras have been described (Tanai 1970), but most of the Paleocene and Eocene assemblages remain undescribed and unanalyzed (Tanai 1967).

There are several advantages in basing paleoclimatic interpretations on

Table 1. The subdivisions of the Cenozoic Era

		Million years ago
Quarternary		
Holocene		.012
Pleistocene		1.5
Tertiary		
Pliocene	Neogene	5
Miocene	Neogene	23
Oligocene	Paleogene	33
Eocene	Paleogene	53
Paleocene	Paleogene	65

In 1957, Erling Dorf delivered the Ermine Cowles Case Memorial Lecture at the University of Michigan, a lecture sponsored by Sigma Xi. His lecture, on Tertiary climates from a paleobotanist's viewpoint, which was later published in American Scientist, *shows some parallels in conclusions with the present article, but much of the basic information Dorf accepted has undergone major revision by subsequent paleobotanical and stratigraphic work. A version of the present paper was also delivered as a Case Lecture in October 1975.* *Jack A. Wolfe was educated at Harvard and Berkeley and is a geologist at the U.S. Geological Survey. His interests are in Cenozoic floras of western North America and systematics and phylogeny of angiosperms. Address: Paleontology and Stratigraphy Branch, U.S. Geological Survey, 345 Middlefield Road, Menlo Park, CA 94025.*

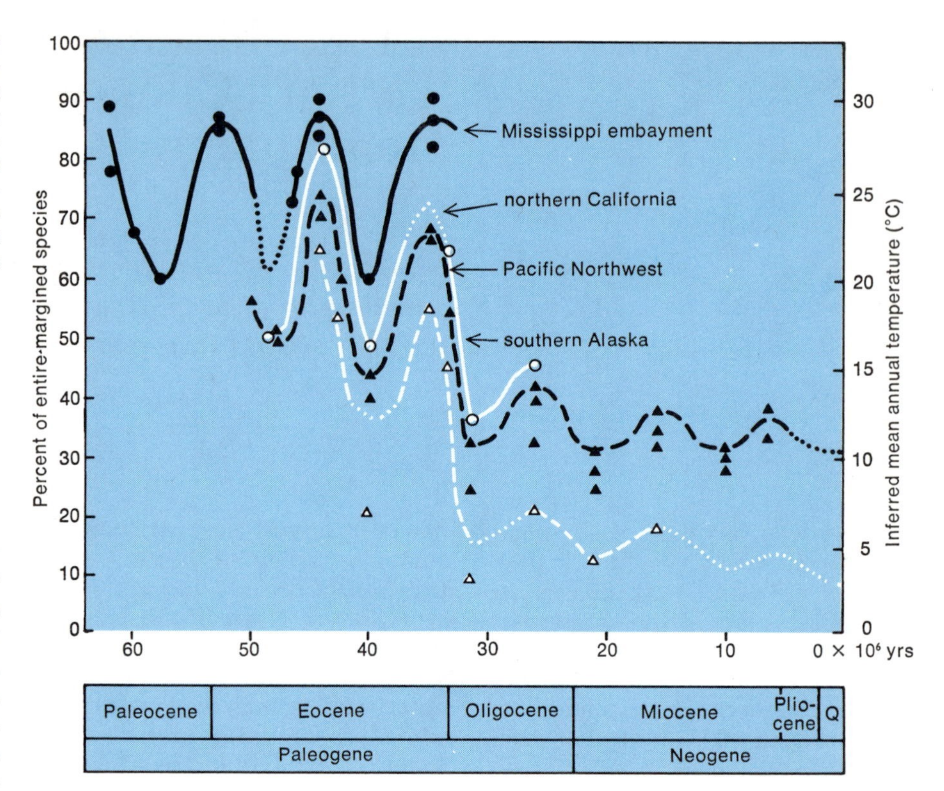

Figure 1. From the estimated percentages of species with entire-margined leaves in four locations in North America, it is inferred that a sharp drop in mean annual temperature took place in the early Oligocene and has continued—at least at high latitudes—to the present day. At middle latitudes mean annual temperature has, overall, not changed since the Oligocene. Dotted intervals indicate that leaf-margin data are either lacking or not considered reliable.

the physical aspects (or physiognomy) of fossil leaf assemblages. The physical characteristics of vegetation occupying similar climates in widely separated regions are highly similar, although the regions may have only a few taxa in common. Among the most conspicuous of such characteristics are the similarities in the appearance of foliage. On the other hand, vegetation occupying dissimilar climates in one region typically has different physical characteristics, although many taxa may occur throughout the region.

Thus the environment tends to select plants that have certain physical aspects for a given climatic type, whether this climatic type is separated by oceans or, presumably, by major periods of time. Just as we can expect the Tropical Rain forest in Africa to have the physical characteristics of Tropical Rain forest in other regions of the world, so we can also expect the present Tropical Rain forest to have the physical characteristics of the Tropical Rain forest of the Eocene.

On the other hand, the Tropical Rain forest of Indonesia has a floristic composition markedly different from the composition of the Tropical Rain forest of Brazil. Such differences have resulted from a variety of historical factors, both geographic and evolutionary. These historical factors will also result in floristic differences between the Tropical Rain forest of the Eocene in a given region and any part of the modern Tropical Rain forest. Considering that the floristic composition of any vegetational type is continually undergoing change, then the determination of the vegetational type (and hence climatic type) represented by a fossil assemblage is best accomplished by analyzing the physical characteristics of the assemblage rather than its floristic composition. And, the further back in time, the more dissimilar the floristic associations are to present associations and the more problematic become climatic inferences.

Among the most useful physiognomic characters of broad-leaved foliage are: type of margin, size, texture, type of apex, and type of base and petiole (see Fig. 2). In areas of high mean annual temperature and precipitation, for example, leaves typically have "entire"margins (i.e. lacking lobes or teeth), are large, are coriaceous ("leatherlike"—an indication of an evergreen habit), and have a high proportion of attenuated apices (i.e. "drip-tips," particularly common on lower-story plants); and a moderate number have cordate (heart-shaped) bases associated with palmate venation and joints ("pulvini") in the petiole—a combination of characters typically associated with the vine, or liana, habit.

The general correlation between type of leaf margin and climate was first

Figure 2. Physical characteristics of leaves largely represent adaptations to the environment and are thus good indicators of climate. The small (microphyllous) leaf (*top*), an alder leaf from southern Alaska, has an incised margin and is characteristic of cool climates. The vine leaf (*bottom*) from the Philippines has a swollen and jointed petiole, palmate venation, and a cordate (heart-shaped) base, as do the leaves of most vines. The attenuated tip (drip-tip) indicates a humid habitat, and large (mesophyllous) size and entire (smooth) margin are characteristic of most tropical plants. White bars represent 30 mm.

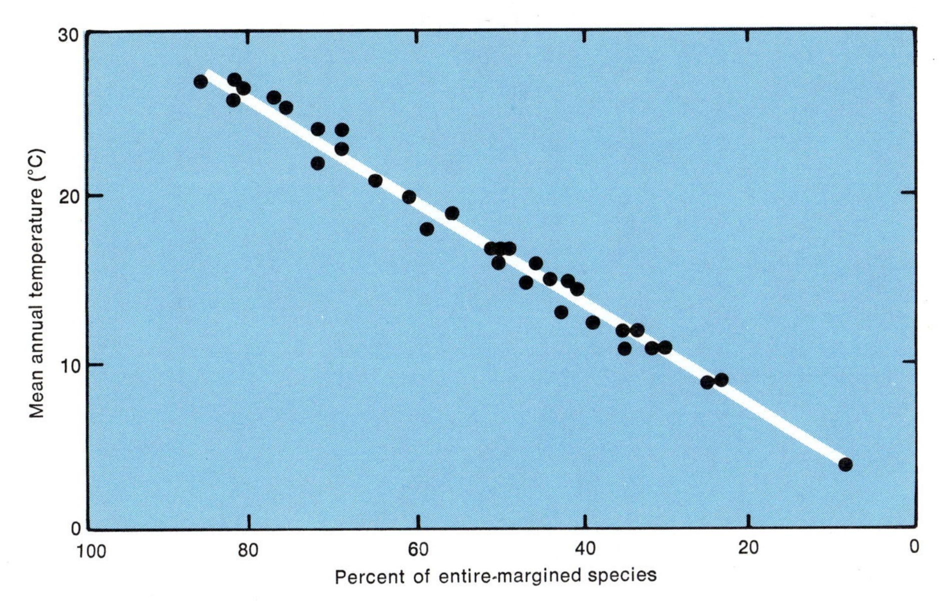

Figure 3. The percentage of species with entire-margined leaves in the humid to mesic broad-leaved forests of eastern Asia increases in direct proportion to the mean annual temperature of the particular forest.

documented by Bailey and Sinnott (1915) and has since been sporadically applied to interpretations of fossil assemblages. A recent compilation of analyses of woody vegetation in the humid to mesic (moderately humid) forests of eastern Asia has shown a strong correlation between the percentage of species with entire leaf margins and mean annual temperature (Fig. 3). Compilations of leaf-margin data of secondary vegetation—vegetation on disturbed sites that has not reached a climax stage—and of the broad-leaved element in coniferous forests do not display such a correlation.

Although leaf size is an important criterion in studies of extant vegetation, the application of this parameter to fossil assemblages is highly problematic, because leaves of different sizes may be differentially selected in the process of transport and preservation (Spicer 1975). Further, leaf-size changes can be related to precipitation and soils as well as to temperature. In the following discussion, I have used a generalized and modified version of the Raunkiaer (1934) system of leaf sizes: *mesophyll* for the larger mesophyll and larger classes, *notophyll* for the smaller mesophyll class (Webb 1959), and *microphyll* for the smaller classes.

The significance of physiognomy to the paleobotanist attempting paleoclimatic reconstructions is that the major physiognomic subdivisions of vegetation (which are partly based on foliar characters) have been found to correspond closely with certain major temperature parameters (Wolfe, in press). Figure 4 shows that mean annual temperature (an approximation of heat accumulation) is of major significance in determining what type of vegetation prevails, as are warm-month means. Only two cold-month means are of major significance. The 1°C mean separates dominantly broad-leaved evergreen (above 1°C) from broad-leaved deciduous (below 1°C) forests; in the areas that have cold-month means between 1°C and −2°C, notophyllous broad-leaved evergreens occur as an understory element, and in regions of even greater winter cold, notophyllous broad-leaved evergreens are lacking. The 18°C cold-month mean—a commonly accepted boundary between "tropical" and "subtropical"—has no relevance to the distribution of vegetation.

Estimates of mean annual temperature can be based on the percentage of entire-margined species in a given fossil assemblage. More difficult to infer is the mean annual *range* of temperature, which, in some cases, can be estimated only within broad parameters. In other cases, however, mean annual range of temperature can be accurately inferred. For example, if two succeeding assemblages have the same leaf-margin percentage of 50% (mean annual temperature ~17°C), and if the younger assemblage is dominantly microphyllous and the older assemblage is dominantly notophyllous, then reference to the framework of Figure 4 indicates that a mean annual range of temperature of 6°C was reached some time between the two assemblages. A second example is that of the Miocene Seldovia Point flora, which represents vegetation slightly inland from the coast of southern Alaska. Foliar physiognomic (as well as floristic) criteria indicate a mean annual temperature of 6–7°C (Wolfe and Tanai, in press). Other paleobotanical data indicate that the broad-leaved deciduous forest represented by the Seldovia Point flora merged with coniferous forest toward the coast. Again, reference to Figure 4 indicates a mean annual range of temperature of about 26–27°C.

Two major problems that have hampered many climatic inferences from paleobotanical data have been the lack of floras in even moderately close stratigraphic successions and the total misinterpretations of the age and climatic significance of high-latitude Tertiary floras. These misinterpretations arose from acceptance of the undocumented concept of an "Arcto-Tertiary Geoflora"—that the Eocene vegetation in Alaska represented temperate broad-leaved deciduous forest that, unchanged, gradually migrated southward to middle latitudes. In North America, both problems have, to a high degree, been overcome. In Alaska there are stratigraphic sequences of floras—many independently dated—that represent most of the Tertiary. In the Pacific Northwest, numerous floras—again, many in stratigraphic succession and/or independently dated—occur in early Eocene and younger rocks. In the Mississippi embayment region, an almost complete sequence of floras represents most of Paleocene and Eocene time. Almost all these floras represent coastal plain vegetation, and thus one major variable in interpreting the significance of paleoclimatic inferences—altitude—is held approximately constant. Certain floras from interior areas add other dimensions to paleoclimatic models, but the altitu-

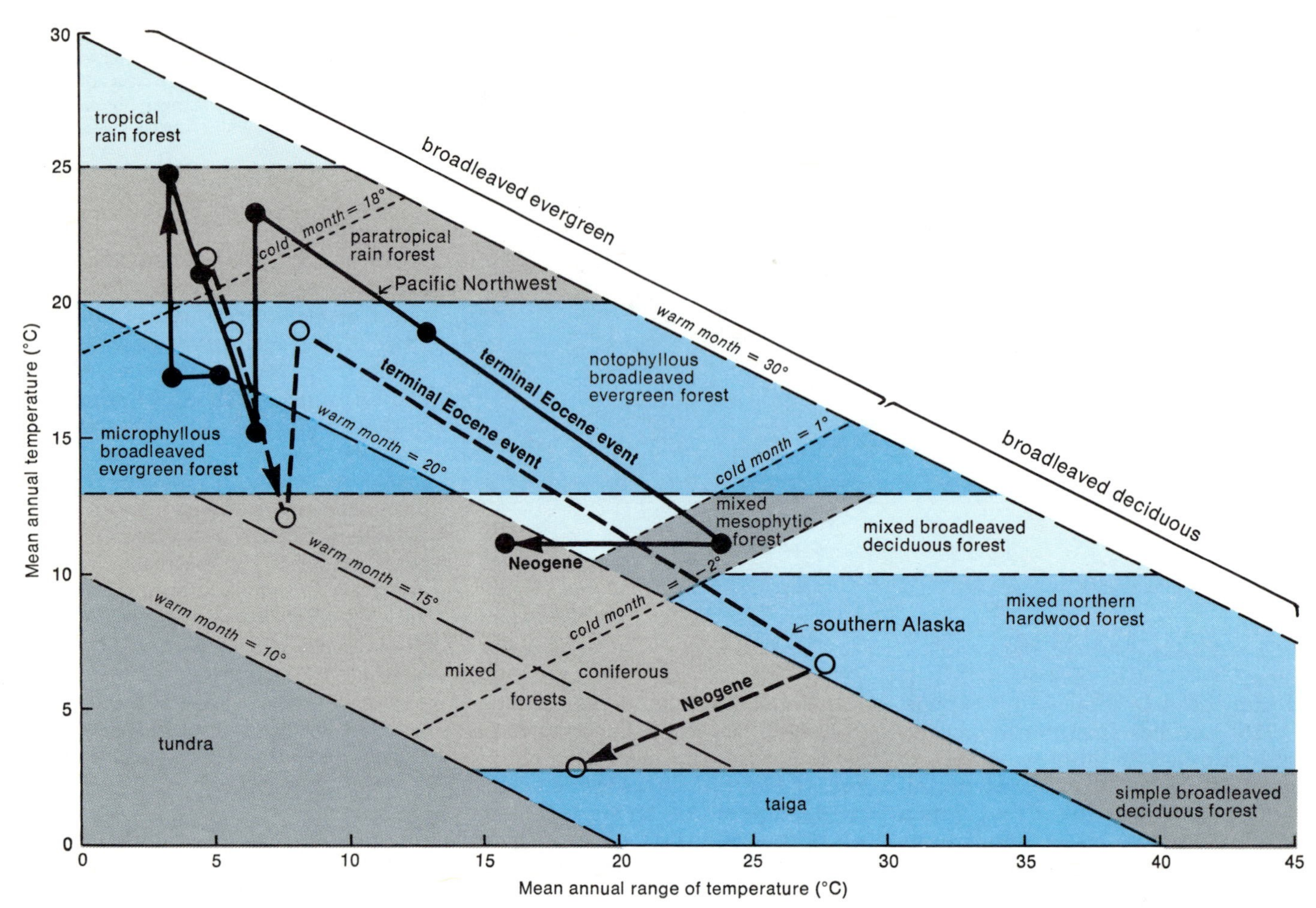

Figure 4. The humid to mesic forests of the Northern Hemisphere can be approximately circumscribed by various major temperature parameters. By comparing leaf assemblages in southern Alaska and the Pacific Northwest to the modern vegetation, we can infer the mean annual temperature and mean annual range of temperature for the assemblages. Major changes in temperature parameters are indicated for the time span between the middle Eocene and the Quaternary—showing a dramatic increase in mean annual temperature and an increase in mean annual range of temperature during the terminal Eocene event.

dinal factor introduces a problematic variable.

Paleocene and Eocene climates

The most complete sequence of Paleogene leaf floras in a small area is that of the Puget Group in western Washington (Wolfe 1968). The Puget assemblages extend from an estimated 50 m.y. ago (late early Eocene) up to about 34 m.y. ago (latest Eocene). In this sequence, the floras all contain numerous leaf species that have drip-tips and/or probable liana leaf physiognomy, and coriaceous (i.e. ≅ evergreen) texture dominates. Thus, all the assemblages represent vegetation that apparently grew under abundant year-round precipitation and would be classed as broad-leaved evergreen rain forests. Major changes in margin and size of the leaf assemblages occurred, however, during deposition of the Puget Group (Fig. 5). The leaf-margin data alone indicate major (perhaps 7–9°C) fluctuations in mean annual temperature, and, in a general manner, the leaf-size data also indicate climatic fluctuations. Sequences of floras in western Oregon, eastern Oregon, and northeastern California parallel the Puget sequence (Wolfe 1971).

The Puget leaf-size data are, however, possibly significant in a context other than mean annual temperature. In the lower part of the sequence, although the leaf-margin data do not significantly change, there is a pronounced movement from a notophyllous to a microphyllous forest. If mean annual temperature was approximately constant during this interval, then mean annual range of temperature must have decreased (i.e. equability of temperature increased). Indeed, the combination of physiognomic data indicates a mean annual range of temperature about half that at present in coastal western Washington, which is highly equable now in comparison to most other mid-latitude areas of the Northern Hemisphere. Although other workers have on questionable floristic interpretations (e.g. Berry 1914, p. 66–67) suggested that the Eocene was characterized by high equability, the Puget physiognomic data provide strong evidence that the Eocene was in fact highly equable.

One of the major corollaries of high equability during the Eocene has been generally overlooked, particularly by the proponents of the concept of an "Arcto-Tertiary Geoflora," who long argued that the Eocene vegeta-

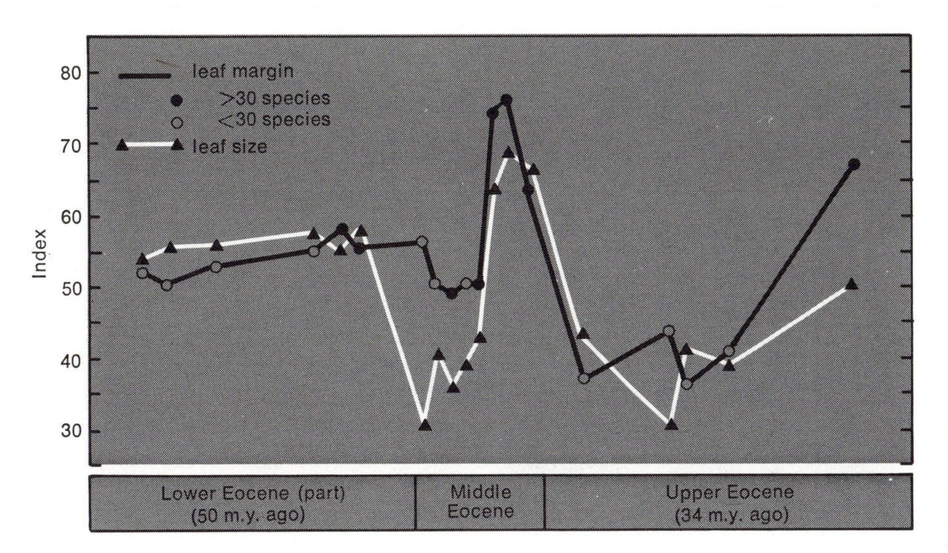

Figure 5. Foliar physiognomic data for Eocene assemblages in the Puget Group of western Washington show that both leaf margin and size underwent great changes between the early and late Eocene, indicating major fluctuations in mean annual temperature. The leaf size index is % microphyllous species + % notophyllous species × 2 + % mesophyllous species × 3 − 100 × 0.5.

tion in regions such as Alaska represented temperate broad-leaved deciduous forest. As in the now highly equable areas of the Southern Hemisphere or on tropical mountains, temperate and mesic broad-leaved deciduous forests could *not* have existed in the Northern Hemisphere during the Eocene. That is, the latitudinal temperature gradient would fall far to the left side of Figure 4—far from temperatures that would support temperate broad-leaved deciduous forest.

Notable also in the Puget analyses is that, during the late middle Eocene (ca. 45 m.y.), the vegetation was marginally Tropical Rain forest. A contemporaneous assemblage in northern California—the Susanville flora (lat 40°)—has a leaf-margin percentage of 82, concomitant with a leaf-size index of 79. The other physiognomic data are consistent with inferring the Susanville flora to be Tropical Rain forest. This indicates that Tropical Rain forest (and the 25°C isotherm) occurred at least 20° and possibly 30° poleward of the present northern limit.

Many assemblages that those who argued for an "Arcto-Tertiary Geoflora" interpreted as temperate broad-leaved deciduous forest are indeed that type; however, these assemblages are typically of Neogene age (cf. the radiometric data of Triplehorn et al. 1977), rather than Eocene, as they had supposed. In fact, only two small Eocene assemblages were known from Alaska until the last decade.

Collections from the Eocene at 60–61° latitude in the Gulf of Alaska region (Wolfe 1977) represent the latter half of the epoch. The late middle Eocene floras—correlative with the Susanville flora—represent Paratropical Rain forest and indicate the warmest climate (ca. 22°C mean annual temperature) of the Tertiary in Alaska (Wolfe 1972). The warmth indicated by the foliar physiognomic data is fully substantiated by the floristic evidence: included are feather and fan palms, mangroves, and members of other families now dominantly or entirely tropical (Wolfe 1977).

Recent geologic data (e.g. Jones et al. 1977) indicate that parts of southern Alaska were once at low latitudes and have drifted northward. The drift and accretion of these plate fragments to Alaska were, however, accomplished by the beginning of the Tertiary. The various major models of plate tectonics are unanimous in suggesting that, in general, western North America rotated southward during the Tertiary. That is, the paleolatitudes of these western North American floras were probably higher than the present latitudes of the fossil localities.

The latitudinal temperature gradient along the Pacific Coast of North America is today very moderate—about 0.5°C/1° latitude. During the late middle Eocene, however, the gradient was even lower. The Susanville (ca. 27°C mean annual temperature at 40° N.), the Puget (ca. 25°C mean annual temperature at 48° N.), and the Gulf of Alaska (ca. 22°C mean annual temperature at 60° N.) data indicate a temperature gradient of about 0.25°C/1° latitude.

The Alaskan Paleocene assemblages are exceedingly difficult to interpret climatically. In southeastern Alaska on Kupreanof Island (lat 57°) there are large assemblages that in all aspects of physiognomy represent Notophyllous Broad-leaved Evergreen forest (mean annual temperature approximately 18°C). Yet, only 2–4° latitude northward, the bulk of the broad-leaved evergreen element is unrepresented in the even larger assemblages of the Chickaloon and West Foreland formations. In features of foliar physiognomy such as margin and size, the Chickaloon assemblages would appear to correspond to Notophyllous Broad-leaved Evergreen forest (Wolfe 1972), yet the Chickaloon assemblages are dominantly broad-leaved deciduous, and even the broad-leaved deciduous element is not as diverse as in present temperate broad-leaved deciduous forests. The minor broad-leaved evergreen element includes palms and certain dicotyledonous families that are not to be expected in temperate broad-leaved deciduous forests. Thus, although almost all physiognomic characters and limited floristic data point to temperatures that should support dominantly broad-leaved evergreen vegetation, the vegetation was dominantly deciduous. Such assemblages occur throughout much of Alaska and Siberia north of latitude 60° during the Paleocene.

In the southeastern United States a large number of Paleogene leaf assemblages that represent coastal plain vegetation occur. Detailed work by many stratigraphers allows an accurate placement of these assemblages in stratigraphic sequence (Fig. 6). The oldest assemblages—early Paleocene (Midway Group)—are those from Naborton and Mansfield, Louisiana. The physiognomy of these assemblages is clearly indicative of Tropical Rain forest (mean annual temperature about 27°C). The succeeding late Paleocene (lower part of Wilcox Group) assemblages represent Paratropical Rain forest—an indicated cooling consonant with data from the continental interior (Wolfe and Hopkins 1967). In the earliest Eocene assemblages (upper part of Wilcox Group), however, leaf size is reduced, the probable liana type of leaf is not as common as earlier, and drip-tips are uncommon; at the same time, the leaf-margin data suggest a warm interval.

How much of a hiatus exists between the Wilcox and Claiborne assemblages is uncertain, but I am assuming that most of the late early and early middle Eocene is missing, at least in the floral sequence. The large assemblages from Puryear, Tennessee, and Granada, Mississippi (the bulk of the "Wilcox flora" of various authors, but actually Claiborne in age; cf. Dilcher 1973a), are of late middle Eocene age. Dilcher (1973b) considered the climatic inferences based on such assemblages to be puzzling; however, the scarcity of probable lianas, the scarcity of drip-tips, and the small leaf size concomitant with a high leaf-margin percentage are characteristic of dry tropical vegetation (cf. Rzedowski and McVaugh 1966).

It is perhaps also significant that these Claiborne assemblages contain a diversity of Leguminosae, a family common in dry tropical vegetation today. Apparently a cooling occurred near the Claiborne-Jackson boundary, but the one cool assemblage (interestingly, once interpreted by Berry, 1916, to be of Pleistocene age) is unfortunately small. In any case, the Mississippi embayment sequence indicates a pronounced drying trend from the Paleocene into at least the middle Eocene.

In the continental interior, the Paleocene floral sequence also shows a definite cooling from the early into the late part of the epoch (Wolfe and Hopkins 1967; Wolfe, in press). Hickey (1977) suggests a renewed warming trend near the Paleocene-Eocene boundary, which would parallel the Mississippi embayment trend. As in that area, the interior

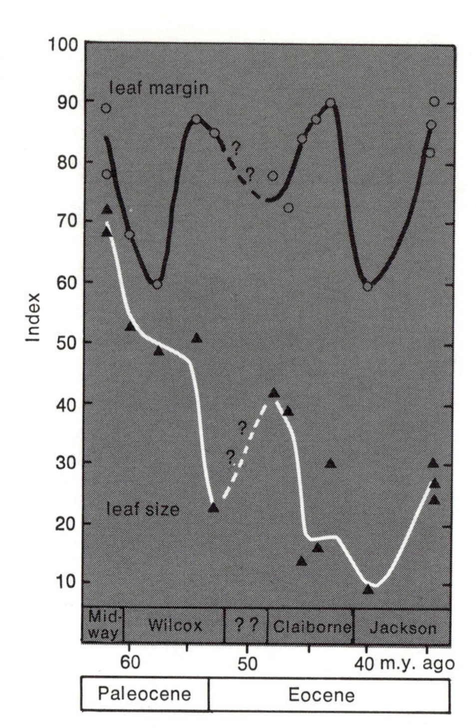

Figure 6. Indexes from foliar physiognomic data for Paleocene and Eocene assemblages in the Mississippi embayment region indicate a pronounced drying trend from the Paleocene into at least the middle Eocene.

Paleocene assemblages all indicate humid to mesic vegetation.

In the Eocene, however, the pattern in the interior becomes greatly complicated. The late early and early middle Eocene assemblages (the earliest Eocene assemblages are unstudied) from central and northern Wyoming represent definite mesic conditions, but the assemblages from southern Wyoming and adjacent Colorado and Utah represent pronounced dry conditions (MacGinitie 1969, 1974). How this situation is related to the presence of mountains and consequent rain shadows is uncertain. Today, the predominant sources of moisture for this region are southerly; one would expect the more southern area (southern Wyoming) to be moister if the Eocene circulation pattern were similar. Later Eocene leaf assemblages from this region are poorly known, except for the latest Eocene Florissant flora of central Colorado (MacGinitie 1953). The climatic significance of this flora is problematic because the altitude at which the Florissant beds were deposited is unknown.

To the west, a number of floras are known from an ancient uplifted area that stretched from Nevada north into British Columbia. The known assemblages represent mesic coniferous forest. Two—the Princeton, British Columbia (Arnold 1955), and the Republic, Washington (Berry 1929)—are of early middle Eocene age and represent the same cool interval as documented in western Washington. The Copper Basin and Bull Run floras from northern Nevada (Axelrod 1966) are correlative with the late Eocene cool interval.

The Paleocene and Eocene floras from North America thus provide the basis for a number of climatic inferences. (1) An overall gradual warming took place from the Paleocene into the middle Eocene, with gradual cooling until the terminal Eocene event. (2) Cool intervals occurred during the late Paleocene, the late early to early middle Eocene, and the early late Eocene. The difference between the intervening warm intervals was, in mean annual temperature, about 7°C. (3) The cool intervals were about 4 to 5°C (mean annual temperature) warmer than the present. (4) Mean annual range of temperature during the middle Eocene was about half that of the present. (5) Mean annual range of temperature decreased from the early into the middle Eocene and possibly increased slightly until the end of the Eocene. (6) The latitudinal temperature gradient during the middle Eocene along the west coast of North America was about half that of the present. (7) The west coast of North America received abundant precipitation during that period. (8) The southeastern United States experienced a pronounced drying trend from the Paleocene into at least the middle Eocene.

Oligocene and Neogene climates

The most profound climatic event of the Tertiary took place at the end of the Eocene. In middle to high latitudes of the Northern Hemisphere, the vegetation changed drastically. Within a geologically short period of time, areas that had been occupied by broad-leaved evergreen forest became occupied by temperate broad-leaved deciduous forest. A major decline in mean annual temperature occurred—about 12–13°C at latitude 60° in Alaska and about 10–11°C at latitude 45° in the Pacific Northwest. Just as profound, however, was the shift in temperature equability: in the Pacific Northwest, for example, mean annual range of temperature, which had been at least as low as 3–5°C in the middle Eocene, must have been at least 21°C and probably as high as 25°C in the Oligocene (Fig. 4; Wolfe 1971).

One of the major aspects of early Oligocene floras at middle to high latitudes is their lack of diversity, which was followed by enrichment during the remainder of the Oligocene (Wolfe 1972, 1977). The lack of diversity would be expected following a major and rapid climatic change such as the one that characterized the terminal Eocene event—that is, few lineages were preadapted or could rapidly adapt to the new temperature extremes.

Although the late early to early middle Miocene warming has been recognized throughout the world, some evidence indicates a warm interval during the late Oligocene (see references cited by Wolfe 1971) and perhaps, to a lesser extent, a warming during the latest Miocene (Wolfe 1969; Barron 1973). These warm intervals, however, were not as warm in comparison to adjoining cool intervals as were the Paleocene-Eocene warm intervals.

The climatic trends following the terminal Eocene event, aside from the minor fluctuations, are of great significance. One trend that can be demonstrated in areas north of latitude 30° is an increase in equability, a trend that runs counter to putative models of Neogene climatic change (cf. Axelrod and Bailey 1969).

Mean annual range of temperature was, during the Oligocene in western Oregon, as great as 21–25°C, but the present value is 12–16°C. At latitude 60° in Alaska, the mean annual range of temperature during the Miocene warm interval was at least as high as 26–27° C, in contrast to the current value of 18°C in the same area (Fig. 4). Similar declines in mean annual range of temperature can be demonstrated in other areas of the Northern Hemisphere, for example, in eastern Asia (Wolfe and Tanai, in press).

Overall trends in mean annual temperature since the terminal Eocene event are dependent on latitude. In southern Alaska (lat 60°), a decline of about 4°C can be documented since the early to middle Miocene. The salient feature of this high-latitude trend is that almost all the change appears to be the result of a decline in summer temperature, which would greatly enhance the "over-summering" of snow fields and, in turn, the initiation of widespread glaciation.

In the Pacific Northwest (lat 42–46° N.), no overall change in mean annual temperature appears to have occurred since the terminal Eocene event. In California and Nevada, climatic inferences from Neogene floras are so greatly complicated by altitudinal and rain shadow factors that extension of these inferences to other areas would at present be unjustified. The few Neogene floras based on leaf remains from eastern North America are too small to be of value in this context.

In Europe, the Neogene floras are found at about the same or higher latitudes as those in the Pacific Northwest; correcting for plate tectonic movements, the Pacific Northwest and European Miocene floras would have been at about equivalent latitudes. The European Neogene sequences typically display—as in the Pacific Northwest—an overall change from broad-leaved deciduous (with a broad-leaved evergreen element, particularly in the Miocene warm interval) to coniferous forest. This implies predominantly a decrease in mean annual range of temperature, possibly along with some decline in warm-month and consequently in mean annual temperatures.

In eastern Asia, the assemblages from Sakhalin (lat 50°) and Kamchatka (lat 55°) show much the same temperature trend as those in Alaska, whereas in Hokkaido (lat 42–45°) only a decrease in mean annual range of temperature occurred (Wolfe and Tanai, in press). South of Hokkaido the floras of Oligocene and Neogene age apparently indicate a contradictory trend—at least in part. In the early Miocene, for example, broad-leaved deciduous forest occupied lowland Kyushu (lat 32°) and southern Honshu (lat 35°; Tanai 1961)—areas now occupied by broad-leaved evergreen forest. Although it can be inferred that mean annual range of temperature has decreased by about 2–4°C, the major point is that mean annual temperature has increased by 3–4°C (Fig. 7). In the middle Miocene of northern Taiwan (lat 25°)—an area now occupied by Paratropical Rain forest—the lowland vegetation was Notophyllous Broad-leaved Evergreen forest (Chaney and Chuang 1968), indicating an increase of at least 2°C in mean annual temperature.

It is significant in this context that Muller (1966) has recorded pollen of elements such as alder and spruce from Borneo (lat 5°), while Graham and Jarzen (1969) have recorded similar cool-climate indicators from the Oligocene to the Miocene in Puerto Rico (lat 18°). In these instances the authors explained the presence of the cool element by suggesting the existence of mountains even higher than those now in the respective areas—although there are no geologic data to support such inferences. Graham (1976), however, explained the presence of cool-climate indicators in the Miocene of Veracruz (lat 19°) by suggesting that temperatures were cooler than now. I suggest that the presence of cool-climate indicators is consistent with the data from Taiwan and Kyushu and implies that, following the terminal Eocene event, low latitudes were cooler than at present.

It is noteworthy that the amount of change in mean annual range of temperature increases as latitude gets higher. At lower latitudes, the major change was apparently an increase in winter temperature that resulted in an overall increase in mean annual temperature and a slight decrease in mean annual range. At about 45° latitude, winter temperature increased by about the same amount as summer temperature decreased, with the result that mean annual temperature remained constant while mean annual range decreased moderately. At high latitudes, summer temperature decreased significantly, leading to a moderate decrease in mean annual range.

One of the obvious consequences of the above trends is an increase in the latitudinal temperature gradient. Such an increase would necessarily increase the intensity of the subtropical high-pressure cells (Willett and Sanders 1959), which, in turn, would bring increasing drought—particularly in summers—to the west coasts of the continents. Such an increase in

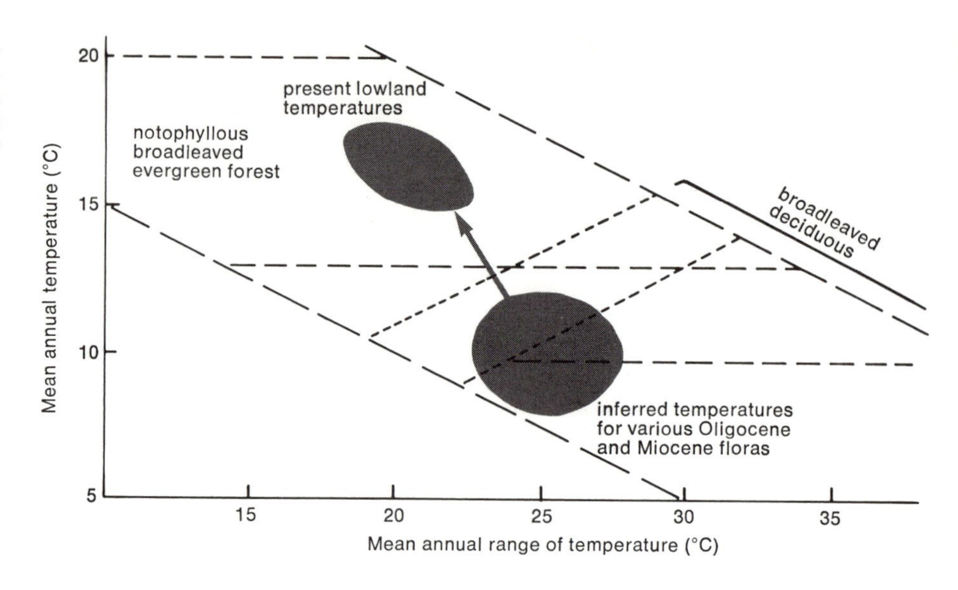

Figure 7. It can be inferred from the leaf assemblages that, in southern Japan, broad-leaved deciduous forest in the early Miocene gave way to broad-leaved evergreen forest and that mean annual temperature increased by 3–4°C.

summer drought during the Neogene has been well documented in western North America (Chaney 1944).

Causes of major climatic changes

Milankovitch (1938) proposed that the episodic glaciations of the Quaternary resulted from changes in the inclination of the earth's rotational axis. The changes Milankovitch considered were those due to perturbations, precession, and other known phenomena that result in minor changes in the inclination. In turn, inclinational changes would cause changes in insolation (the amount of radiation received from the sun) that would vary according to latitude. These astronomical explanations of Quaternary climatic change have received considerable support from many researchers, who consider that the paleoclimatic data fit well the timing based on calculations of minor inclinational changes.

Is it entirely coincidental that the divergent latitudinal patterns following the terminal Eocene event fit very well a model resulting from a significant decrease in the inclination of the earth's rotational axis? According to Milankovitch's hypothesis, under conditions of decreasing inclination (and assuming an atmospheric circulation pattern similar to that of the present), the pattern of climatic change would be (1) an increase in winter temperatures at lower latitudes (resulting in an increase in mean annual temperature), (2) an increase in winter temperatures about equal to a decrease in summer temperatures at latitude 43° (resulting in no increase in mean annual temperature), (3) a decrease in summer temperatures at higher latitudes (resulting in a decrease in mean annual temperature), and (4) a decrease in mean annual range of temperature proportional to the latitudes.

These are precisely the changes that are inferred from paleobotanical data for the Oligocene and Neogene and would indicate that a significant decrease in the earth's inclination has occurred during the last 30 million years.

Conditions during the Paleocene-Eocene were strikingly different from those during the Oligocene and younger epochs. During the thermal and equability maximum of the middle Eocene, the west coast of North America was wet and the southeastern United States was comparatively dry. The presence of humid broad-leaved evergreen forests in Alaska would, concomitant with the other data, argue for a circulation that involved the poleward flow of warm, moist air along the west coast; as this flow returned equatorward over the eastern part of the continent, the heating air would, of course, become drier. That is, the Eocene pattern may have been dominated by north-south (meridional) flow rather than being dominated by regional cells, as at present. What factor(s) could bring about such a pattern?

Readers who are plant physiologists may have been startled to learn that during the Eocene, broad-leaved evergreen forest extended north of latitude 60°. The principles of plant physiology would argue against such vegetation under the prolonged dark winters at such latitudes (Mason 1947; van Steenis 1962). The present distribution of broad-leaved evergreens is consistent with such principles—that is, notophyllous evergreens occur on mountains in California that have the same fundamental temperature parameters as lowland areas at more northern latitudes, where notophyllous broad-leaved evergreens are absent (Wolfe, in press). Notophyllous broad-leaved evergreens—except for a very few conspicuous taxa—today do not occur north of 50° latitude, and most are equatorward of 40–45°. Such considerations provide strong evidence that the middle Eocene Alaskan assemblages, with their diverse and dominant notophyllous to mesophyllous broad-leaved evergreen element, could not have existed under the present light conditions at the latitude of the fossil localities. Light was, in fact, considered a negative factor in the possibility of explaining trans-Pacific disjunctions of tropical broad-leaved evergreen groups via the land bridge that is now the Bering Straits (e.g. van Steenis 1962), and yet such groups are now known to have occurred in the Beringian region (van Buesekom 1971; Wolfe 1972).

Under conditions of high temperatures and prolonged winter darkness, the predicted vegetation would be broad-leaved deciduous with a minor broad-leaved evergreen element (van Steenis 1962), composed of those evergreens tolerant of low light levels (as a limited number of broad-leaved evergreen taxa are today). This is apparently the type of vegetation that existed in Alaska north of latitude 60°, during the Paleocene.

I am suggesting that the data thus far indicate that, during the middle Eocene, the light conditions at latitude 60° were more favorable to the growth of broad-leaved evergreens than at present. During the Paleocene, in fact, latitude 57° (but not 60°) supported a diverse broad-leaved evergreen forest. The only factor that could produce more light at these northern latitudes is a significantly smaller (at least 15° less during the middle Eocene than now) inclination of the earth's rotational axis. A smaller inclination would certainly be consistent with the low mean annual range of temperature during the Paleocene and Eocene: today this temperature parameter is primarily (although not entirely) a function of latitudinal position because of the inclination.

A suggestion that the inclination was, in comparison to today, considerably smaller is, if the obvious warmth of the Alaskan middle Eocene is accepted, contradicted by the Milankovitch calculations, i.e. a smaller inclination would yield lower insolation at high latitudes and hence lower temperatures. These calculations were, however, based on the assumption that the insolational values could be directly translated into temperature values—i.e. that the present atmospheric circulation pattern was, in general, constant throughout geologic time. But we have seen that the Paleocene and Eocene circulation pattern could not have been like that of the present. Could an increased insolational gradient under a low inclination have been the major driving force for a dominantly meridional circulation, a circulation that would have more than compensated for decreased annual insolational values at high latitudes? Perhaps at some critical value of inclination, the atmospheric circulation changes from one that is dominantly cellular (as it is today and was during the Oligocene and Neogene) to one that is dominantly meridional.

If the major climatic trends during the Tertiary were largely the result of inclinational changes, then from the Paleocene to the middle Eocene, inclination decreased gradually from a value of perhaps 10° to a value approaching 5°. The inclination then began to increase slightly until the end of the Eocene, when the inclination increased rapidly to 25–30°. Since then, the inclination has gradually decreased to the present average value of 23.5°.

The drastic change in inclination suggested as the cause of the terminal Eocene event would have had a profound effect on the earth's crust. It is significant that a number of researchers have suggested major tectonic changes at the end of the Eocene. For example, Molnar et al. (1975) suggest that the tectonic patterns of the South Pacific were different in the pre-Oligocene than now and that the current patterns were achieved at the end of the Eocene. Menard (1978) similarly indicates major changes in the northeastern Pacific at the end of the Eocene.

Yet, even assuming the validity of this model of inclinational change, the several fluctuations in mean annual temperature are not explained. From available radiometric data, the fluctuations appear to represent a cycle about 9.5 m.y. in duration (Wolfe 1971). Presumably such regular fluctuations would result from fluctuations in the amount of solar radiation reaching the earth; certainly no model of plate tectonic movements could explain such fluctuations.

Much additional information is needed, particularly from the continental interiors, to develop accurate models of temperature and precipitation distribution during the Paleocene and Eocene. Low-latitude leaf floras of Oligocene and Neogene age are needed to determine whether the low-latitude data thus far accumulated are anomalous or typical. More studies of modern depositional environments are needed to understand fully the significance of what is actually found in fossil assemblages. More information from rigidly controlled experiments is needed to determine the physiological response of broad-leaved woody plants to low light levels.

This review has shown the type of data that can be obtained from fossil plants. Brooks (1949) noted that the evidence available to him could not be explained solely by geographic factors, and even attempts to explain Paleocene and Eocene climates by the changing positions of the continental plates are only partially satisfactory (Frakes and Kemp 1973). The evidence now available indicates even more radical differences between present climates and those of the past than were recognized by Brooks. Attempting to fit such data into a "steady-state" hypothesis would be doing an injustice to the data similar to that done to geologic data prior to the general acceptance of plate tectonics.

References

Addicott, W. O. 1969. Tertiary climatic change in the marginal northeast Pacific Ocean. *Science* 165:583–86.

Arnold, C. A. 1955. Tertiary conifers from the Princeton coal field of British Columbia. *Michigan Univ. Mus. Paleontol. Contr.* 12:245–58.

Axelrod, D. I. 1966. The Eocene Copper Basin flora of northeastern Nevada. *Calif. Univ. Pubs. Geol. Sci.* 59:1–125.

Axelrod, D. I., and H. P. Bailey. 1969. Paleotemperature analysis of Tertiary floras. *Palaeogeography, Palaeoclimatology; Palaeoecology* 6:163–95.

Bailey, I. W., and E. W. Sinnott. 1915. A botanical index of Cretaceous and Tertiary climates. *Science* 41:831–34.

Barron, J. A. 1973. Late Miocene-early Pliocene paleotemperatures for California from marine diatom evidence. *Palaeogeography, Palaeoclimatology, Palaeoecology* 14: 277–91.

Berry, E. W. 1914. The Upper Cretaceous and Eocene floras of South Carolina and Georgia. U.S.G.S. Prof. Paper 84.

———. 1916. The Mississippi River bluffs at Columbus and Hickman, Kentucky, and their fossil flora. *U.S. Natl. Mus. Proc.* 48: 293–303.

———. 1929. A revision of the flora of the Latah formation. U.S.G.S. Prof. Paper 154-H, pp. 225–65.

Brooks, C. E. P. 1949. *Climate through the Ages,* 2nd ed. London: Benn.

Chaney, R. W. 1944. Summary and conclusions. In *Pliocene Floras of California and Oregon,* ed. R. W. Chaney, pp. 353–83. Carnegie Inst. Washington publ. 553.

Chaney, R. W., and G. C. Chuang. 1968. An oak-laurel forest in the Miocene of Taiwan (Part 1). *Geol. Soc. China* 11:3–18.

Denton, G., and R. L. Armstrong. 1969. Miocene-Pliocene glaciations in southern Alaska. *Am. J. Sci.* 267:1121–42.

Devereux, I. 1967. Oxygen isotope paleotemperature measurements on New Zealand Tertiary fossils. *New Zealand J. Sci.* 10: 988–1011.

Dilcher, D. L. 1973a. Revision of the Eocene flora of southeastern North America. *Palaeobotanist* 20:7–18.

———. 1973b. A paleoclimatic interpretation of the Eocene floras of southeastern North America. In *Vegetation and Vegetational History of Northern Latin America,* ed. A. Graham, pp. 39–59. Amsterdam: Elsevier.

Frakes, L. A., and E. M. Kemp. 1973. Paleogene continental positions and evolution of climate. In *Implications of Continental Drift to the Earth Sciences,* ed. D. H. Tarling and S. K. Runcorn, pp. 535–58. Academic Press.

Graham, A. 1976. Studies in neotropical paleobotany. II: The Miocene communities of Veracruz, Mexico. *Missouri Bot. Garden Annals* 63:787–842.

Graham, A., and D. M. Jarzen. 1969. Studies in neotropical paleobotany. I: The Oligocene communities of Puerto Rico. *Missouri Bot. Garden Annals* 56:308–57.

Hickey, L. J. 1977. Stratigraphy and paleobotany of the Golden Valley Formation (early Tertiary) of western North Dakota. *Geol. Soc. Am. Mem.* 150.

Jones, D. L., N. J. Silberling, and J. Hillhouse. 1977. Wrangellia—a displaced terrane in northwestern North America. *Can. J. Earth Sci.* 14:2565–77.

Kennett, J. P. 1977. Cenozoic evolution of Antarctic glaciation, the circum-Antarctic ocean, and their impact on global paleoceanography. *J. Geophys. Res.* 82:3843–60.

MacGinitie, H. D. 1953. *Fossil Plants of the Florissant Beds, Colorado.* Carnegie Inst. Washington publ. 599.

———. 1969. The Eocene Green River flora of northwestern Colorado and northeastern Utah. *Calif. Univ. Pubs. Geol. Sci.* 83:1–140.

———. 1974. An early middle Eocene flora from the Yellowstone–Absaroka volcanic province, northwestern Wind River basin, Wyoming. *Calif. Univ. Pubs. Geol. Sci.* 108:1–103.

Mai, D. H. 1964. Die Maxtixioideen-Floren im Tertiär der Oberlausitz. *Palaontolog. Abh.,* Abt. B, 2:1–92.

Mason, H. L. 1947. Evolution of certain floristic associations in western North America. *Ecol. Monographs* 17:201–10.

Menard, H. W. 1978. Fragmentation of the Farallon plate by pivoting subduction. *J. Geol.* 86:99–110.

Milankovitch, M. 1938. Astronomische Mittel zur Erforschung der erdgeschichtlichen Klimate. *Handbuch der Geophysik* 9: 593–698.

Molnar, P., T. Atwater, J. Mammerick, and S. M. Smith. 1975. Magnetic anomalies, bathymetry, and the tectonic evolution of the South Pacific since the late Cretaceous. *Royal Astron. Soc. Geophys. J.* 40:383–420.

Muller, J. 1966. Montane pollen from the Tertiary of northwestern Borneo. *Blumea* 14:231–35.

Nemjč, F. 1964. Biostratigraphic sequence of floras in the Tertiary of Czechoslovakia. *Časopis pro mineralogii a geologii, Rocnik* 9:107–9.

Plafker, G., and W. O. Addicott. 1976. Glaciomarine deposits of Miocene through Holocene age in the Yakataga Formation along the Gulf of Alaska margin, Alaska. In *Symposium on Recent and Ancient Sedimentary Environments in Alaska,* ed. T. P. Miller, pp. Q1–Q23. Alaska Geological Society.

Raunkiaer, C. 1934. *The Life Forms of Plants and Statistical Plant Geography.* Oxford: Clarendon Press.

Rzedowski, J., and R. McVaugh. 1966. La vegetacion de Nueva Galicia. *Michigan Univ. Herbarium Contr.* 9:1–123.

Spicer, R. A. The sorting of plant remains in a Recent depositional environment. 1975 diss., London Univ.

Tanai, T. 1961. Neogene floral change in Japan. *Hokkaido Univ. Fac. Sci. J.,* ser. 4, 11: 119–298.

———. 1967. On the Hamamelidaceae from the Paleogene of Hokkaido, Japan. *Palaeont. Soc. Japan Trans. Proc.,* N.S., 66:56–62.

———. 1970. The Oligocene floras from the Kushiro coal field, Hokkaido, Japan. *Hokkaido Univ. Fac. Sci. J.,* ser. 4, 14:383–514.

Tanai, T., and K. Huzioka. 1967. Climatic implications of Tertiary floras in Japan. In *Tertiary Correlation and Climatic Changes in the Pacific,* ed. K. Hatai, pp. 89–94. Sendai: Sasaki Printing and Publishing Co.

Triplehorn, D. M., D. L. Turner, and C. W. Naeser. 1977. K-Ar and fission-track dating of ash partings in coal beds from the Kenai Peninsula, Alaska: A revised age for the Homerian Stage-Clamgulchian Stage boundary. *Geol. Soc. Am. Bull.* 88:1156–60.

van Buesekom, C. F. 1971. Revision of *Meliosma* (Sabiaceae) section, *Lorenzanea* excepted, living and fossil, geography and phylogeny. *Blumea* 19:355–529.

van Steenis, C. G. G. J. 1962. The land-bridge theory in botany. *Blumea* 11:235–372.

Webb, L. J. 1959. Physiognomic classification of Australian rain forests. *J. Ecol.* 47:551–70.

Willett, H. C., and F. Sanders. 1959. *Descriptive Meteorology.* Academic Press.

Wolfe, J. A. 1968. Paleogene biostratigraphy of nonmarine rocks in King County, Washington. U.S.G.S. Prof. Paper 571.

———. 1969. Neogene floristic and vegetational history of the Pacific Northwest. *Madroño* 20:83–110.

———. 1971. Tertiary climatic fluctuations and methods of analysis of Tertiary floras. *Palaeogeography, Palaeoclimatology, Palaeoecology* 9:27–57.

———. 1972. An interpretation of Alaskan Tertiary floras. In *Floristics and Paleofloristics of Asia and Eastern North America,* ed. A. Graham, pp. 201–33. Amsterdam: Elsevier.

———. 1977. Paleogene floras from the Gulf of Alaska region. U.S.G.S. Prof. Paper 997.

———. In press. Temperature parameters of humid to mesic forests of eastern Asia and relation to forests of other regions of the Northern Hemisphere and Australasia. U.S.G.S. Prof. Paper 1106.

Wolfe, J. A., and D. M. Hopkins. 1967. Climatic changes recorded by Tertiary land floras in northwestern North America. In *Tertiary Correlation and Climatic Changes in the Pacific,* ed. K. Hatai, pp. 67–76. Pacific Sci. Cong., 11th, Tokyo, Aug. 1966, Symp. 25.

Wolfe, J. A., and T. Tanai. In press. The Miocene Seldovia Point flora from the Kenai Group, Alaska. U.S.G.S. Prof. Paper 1105.

Zhilin, S. G. 1966. A new species of *Carya* from the late Oligocene. *Paleont. Zhur.* 4:104–8 (in Russian). English translation available from Telberg Book Co., New York.

". . . and the record low for this date is 147° below zero, which occurred 28,000 years ago during the Great Ice Age."

Peter H. Raven
Daniel I. Axelrod

History of the Flora and Fauna of Latin America

The theory of plate tectonics provides a basis for reinterpreting the origins and distribution of the biota

Until the late 1960s, most attempts to interpret the global distribution of plants and animals were based upon the assumption that the relative positions of continents and islands had remained unchanged since their formation. Within the past few years, however, the now generally accepted theory of plate tectonics has provided a reasonable explanation for the movement of continents hypothesized by Alfred Wegener more than 50 years ago but generally discredited in this country until recently.

In simple terms, plate tectonics is a unified theory that envisions the earth's crust (lithosphere) as composed of a limited number of fairly rigid plates—as few as six in some reconstructions. The plates move in relation to one another, some at rates as high as 10 centimeters per year. They are 50 to 100 kilometers thick and include both ocean basins and continents. They are generally aseismic except near or at their boundaries, where moving plates jostle one another, resulting in intense earthquake activity and volcanism.

Three different kinds of movement occur at plate boundaries. First, during major rifting, lavas well up, solidify, and accumulate on each side of the rift. This process is continuous along all the active mid-ocean ridges, and as new crust is added to the adjacent plates, they grow laterally and move apart, causing the older sea floor, continents, and islands to be rafted to new positions. Second, the moving plates are thrust back into the mantle along subduction zones at the site of ocean trenches, such as the Aleutian trench. Third, two plates may slide past one another without plate modification, with the zone of movement marked by a major strike-slip fault. The San Andreas fault of California is an example of this type of movement.

Biological results of continental movement

When land masses move, they may be rafted across latitudinal belts of climate that may then lead to new opportunities for evolutionary change, widespread impoverishment of the biota, or the total decimation of a rich biota.

An example of new opportunities for evolutionary change is provided by the history of Australia, which has moved north during the past 55 million years from a cool-temperate southern position adjacent to Antarctica to its present position, lying directly across one of the great zones of aridity that flank the tropics. There is no evidence for the existence of arid zones in Australia prior to about 20 million years ago; the vast arid to subhumid stretches that now characterize most of the continent are of relatively recent origin. The ancient plants and animals have survived with little change in the cool, humid climate of Australia's southeastern corner and in Tasmania, as well as locally northward along the mountains to New Guinea, where the ancient southern biota has been joined recently by plants and animals of tropical requirements that migrated across the tropical Malaysian-Indonesian lowlands (Raven and Axelrod 1972).

The history of India exemplifies biotic impoverishment. As the subcontinent moved north from temperate to southern arid, to inner tropical, to northern dry latitudes, its flora and fauna changed greatly. Austral gymnosperms, now mainly restricted to the Southern Hemisphere where they take the place of the pines, firs, and spruces familiar in the north, once inhabited India. They became extinct there as the land mass moved northward across the equator, as did many other groups of plants and animals.

Antarctica, formerly joined with Australia, moved southward during the past 55 million years to occupy high polar regions, with the consequent almost complete elimination of its plants and animals. In the Triassic, a whole fauna of large reptiles, including the famous *Lystro-*

Peter H. Raven is Director of the Missouri Botanical Garden and Engelmann Professor of Botany at Washington University, St. Louis. Dr. Raven received his Ph.D. from the University of California, Los Angeles, in 1960. He taught at the Claremont Graduate School and at Stanford University before going to St. Louis in 1971. His interest in plate tectonics was kindled during a sabbatical leave in New Zealand, at the Department of Scientific and Industrial Research, in 1969–70. Dr. Raven is a specialist on the evening primrose family (Onagraceae), on taxonomic theory, plant-animal interactions, and biogeography.
Daniel I. Axelrod, Professor of Botany at the University of California, Davis, obtained his doctorate from the University of California, Berkeley, in 1938. He has been a National Research Fellow at the U.S. National Museum, and held a joint appointment in geology and botany at UCLA before transferring to the Davis campus in 1968. His publications deal with the early history of flowering plants, the paleoecology and evolution of Tertiary vegetation, and environmental factors of evolution. Reprint requests: Peter H. Raven, Missouri Botanical Garden, 2315 Tower Grove Ave., St. Louis, MO 63110.

saurus, inhabited both Antarctica and Africa as well as India and China; this fauna has not yet been discovered in South America. Fossil remains of the Triassic *Lystrosaurus* fauna have now been found in rocks at 85° S latitude, where poikilothermic vertebrates could not possibly survive the long Antarctic nights. These fossils suggest that the continent must have moved southward to its present position (Axelrod and Raven 1972). No fossil vertebrates that can be dated later than the Triassic have been found in Antarctica, but then, in Africa, no fossil mammals have been found dating from the period from the Triassic to the Eocene (Fig. 1), a gap of some 150 million years! It would certainly be premature to conclude that no vertebrates lived in Antarctica after the Triassic, or that Antarctica did not form a pathway for their migration at a later date.

Theories of the origin of South American biota

What do the new geological syntheses have to tell us about the origin of plants and animals of South America? Twenty-five years ago, the illustrious American paleontologist and systematic biologist George Gaylord Simpson (1950) addressed himself to this question in the pages of *American Scientist.* He provided a clear and cogent outline of the history of the Latin American fauna based upon his own scholarly investigations and the geological information available at the time. Like Philip J. Darlington, whose invaluable *Zoogeography* (1957) significantly advanced our understanding of the distribution of animals a few years later, Simpson based his conclusions on the assumption of an unchanging world map, with a few exceptions. One of the most notable exceptions was the rise of the Central American land bridge, an event now dated at 5.7 million years ago (Emiliani, Gaertner, and Lidz 1972).

Both Simpson and Darlington, on the assumption of continental stability, accepted as an important dictum the notion first clearly articulated by W. D. Matthew that the dominant groups of animals evolved mainly in the Northern Hemisphere landmasses of Eurasia and North America and then radiated to other parts of the world. According to this theory, survivors of earlier evolutionary lines would be concentrated at the southern tips of South America, southern Africa, and Australasia. Now that we understand that the continents have been moving, we know that this principle cannot be accepted completely. Instead, the changing positions of the continents and the connections or proximity between them during the history of a particular group of organisms must now be taken into account.

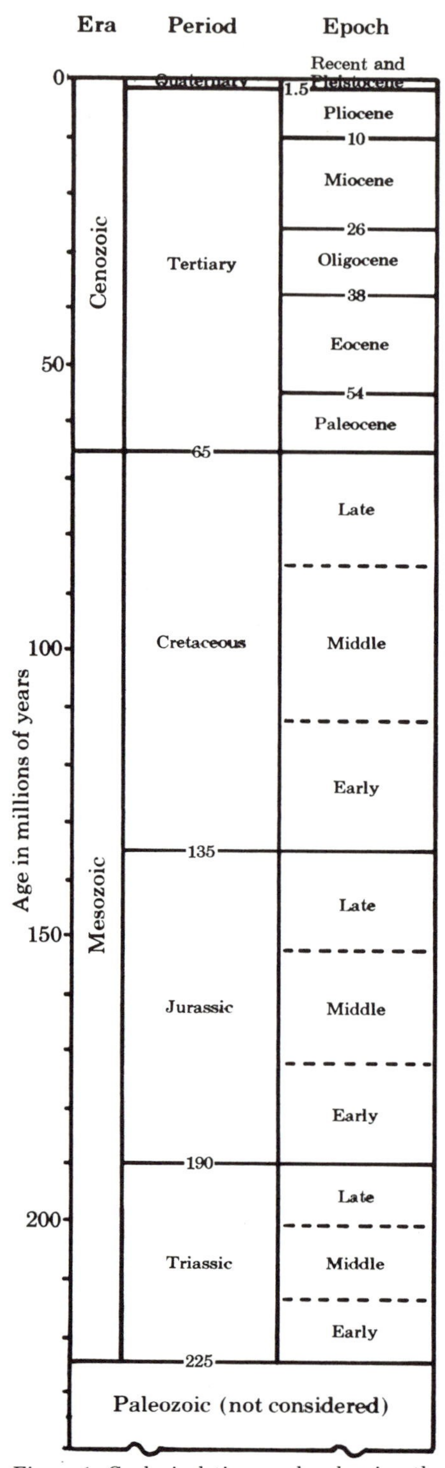

Figure 1. Geological time scale, showing the principal time divisions referred to in the text.

One major conclusion developed by Simpson was that the mammal faunas of North and South America had evolved largely in isolation. With the emergence of the Central American land bridge, the two faunas began to mix, starting with the mid-Pliocene arrival in South America of a group of small arboreal, raccoonlike carnivores of the North American genus *Cyonasua.* In all, 19 families of North American mammals spread into South America, 15 reaching temperate areas; in the opposite direction, 24 southern families of mammals have reached Central America, of which 12, including the porcupine and the armadillo, reached temperate North America (Savage 1974). Many families later became extinct. Eventually, about the same number of families inhabited each continent as formerly, but with many changes in the list for each area.

This fascinating picture is still regarded as scientifically sound. What is in need of reevaluation, in view of our current understanding of the geological history of South America, is the origin of the older mammals and the other animals and plants living in South America before the formation of the Panamanian land bridge. Where did the ancestors of old, extinct groups of mammals such as the condylarths, litopterns, notoungulates, astrapotheres, and pyrotheres come from, and what was the source of those later arrivals, the New World primates and the hystricomorph rodents? Were they really derived from North America, as Simpson thought probable?

Plate tectonic history of South America

Geological evidence, universally accepted in its broad outline, indicates that in Middle Cretaceous time (Fig. 2) South America was directly connected with Africa and, via Antarctica, with Madagascar, India, and Australia. Separation in the South Atlantic commenced 125–130 million years ago, in the

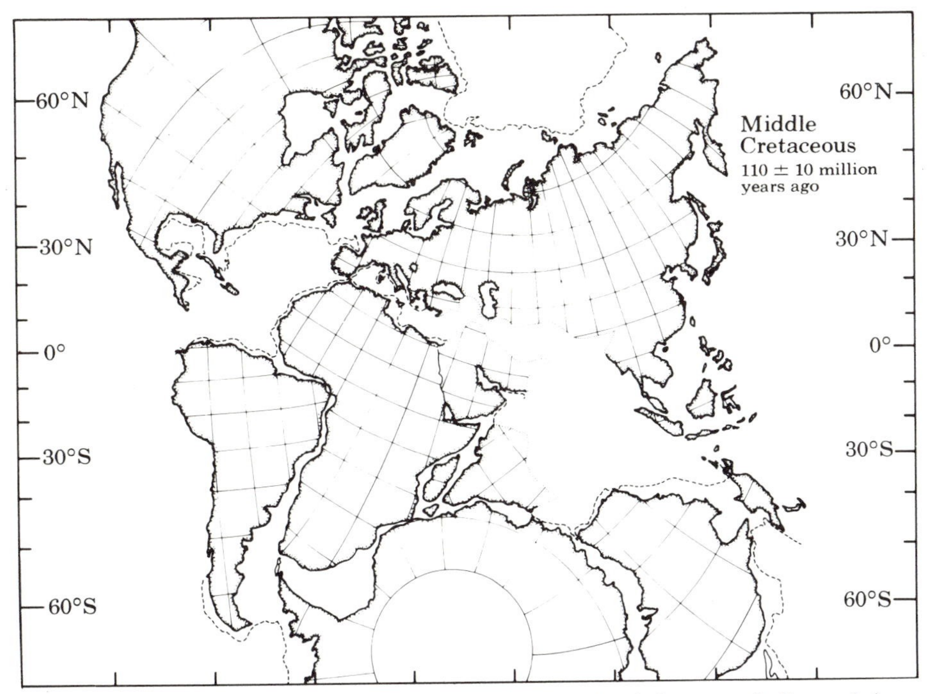

Figure 2. In the Middle Cretaceous, 110 ± 10 million years ago, South America was directly connected with Africa, and via Antarctica with Madagascar, India, and Australia. (After McKenzie and Sclater 1973).

Lower Cretaceous (summary in Raven and Axelrod 1974), but the final marine connection associated with the spreading apart of Africa and South America took place in equatorial latitudes slightly less than 90 million years ago (Reyment 1969, 1972; Reyment and Tait 1972; Douglas, Moullade, and Nairn 1973; Raven and Axelrod 1974). At the close of the Cretaceous, some 65 million years ago, about 600 km probably separated Africa and South America at their closest points. They were still linked, however, by numerous islands along the Mid-Atlantic Ridge and its flanks. Fossils indicating the presence of the dinosaur genus *Laplatasaurus* in the Upper Cretaceous in South America, Madagascar, and India offer biological proof of these connections, but strata of appropriate age are very uncommon on the mainland of Africa, where the genus has not yet been discovered.

Connections between South America and Australasia, available only to organisms able to spread into temperate climates, persisted even longer (Fig. 3). The biological relationships between South America and New Zealand–New Caledonia are not as close as those between South America and Australia, chiefly because New Zealand and New Caledonia parted company with Australia–Antarctica some 30 million years earlier, in the Late Cretaceous (80 million years ago). This resulted in the preservation of a unique relict fauna and flora on New Zealand and New Caledonia, which became progressively more isolated during the past 80 million years, remaining in an equable moist climate while conditions in Australia deteriorated throughout the Tertiary. Although rifting between Australia and Antarctica commenced about 55 million years ago, the continental margins did not separate until 49 million years ago (McGowran 1973). More or less direct migration to Antarctica through the Tasmanian area may have been possible for perhaps another 10 million years via the continental South Tasman Rise. The initial break between the Antarctic Peninsula and Tierra del Fuego probably occurred before the end of the Cretaceous, but they remained in proximity. Cool water apparently began to flow through the seaway between Australia and Antarctica 36–41 million years ago, and this probably dates their definitive separation (Foster 1974). Thus more or less direct migration, probably in part by island-hopping, between Australia and South America seems to have been possible until the Early Oligocene.

Figure 3. The silver beech (*Nothofagus menziesii*), photographed in Upper Caples Valley, Southland, New Zealand, is a relict of the Antarcto-Tertiary forest that extended across now cold, unforested southern lands (Antarctica, Palmer Peninsula, Fuegia). These areas remained in proximity into Middle Eocene time, linking the forests of Australia-New Zealand with those of temperate Argentina-Chile. (Photo by J. H. Johns, A.R.P.S., courtesy of the New Zealand Forest Service.)

In short, there were fairly close connections between tropical South America and Africa until the close of the Cretaceous, and excellent opportunities for migration across islands between South America and Australia existed until the Early Oligocene. What then was the relationship between South America and North America at these times?

During most of the Cretaceous, about 3,000 km separated the southern continental margin of North America (in Oaxaca, Mexico) from the continental margin of South America, the Guiana Shield (Fig. 4). This exceeds the present distance (about 2,500 km) between Freetown, Sierra Leone, and Recife, Brazil. The entire Atlantic and Gulf Coastal Plain, including Florida, was submerged until the Oligocene, which increased the distance between North and South America. A general period of uplift near the close of the Cretaceous resulted in northern Central America and the northern Andes beginning to approach their modern configuration. From that time until a direct land connection was established some 5.7 million years ago, plants and animals would have had to migrate across some 1,300 km of water between northern Nicaragua and northern Colombia by means of volcanic islands that became increasingly more numerous during the Tertiary.

In the Caribbean region, very complicated geologically, there were small islands in the area of Cuba by the Late Jurassic, but most existing islands east or south of Cuba are no older than Late Cretaceous. At that time, a chain of small to medium-sized islands had appeared, first in the west and then spreading eastward along the Antillean chain, with the Lesser Antilles formed most recently. In general, all the West Indies have increased in size and altitude throughout the Tertiary and subsequently, and, now larger and higher than they have ever been, they constitute a more important pathway for the migration of plants and animals between North and South America than at any time in the past. At no time have they ever formed a direct pathway for migration between North and South America, as does Central America: they have never been joined by continuous land, though Cuba may have been connected to Central America in the Jurassic, before the other islands came into existence.

For tropical to warm-temperate organisms, opportunities for interchange were probably greater between South America and Africa than between South and North America until approximately 45 million years ago, in the Late Eocene. This inference is based upon the assumption of a relatively constant rate of divergence since the initial separation of South America and Africa about 90 million years ago, plus a steady rate of convergence between South and North America since that time, with increasingly direct land connections. Pathways for the migration of cool-temperate organisms between southern South America and Australia were relatively direct until about 40 million years ago.

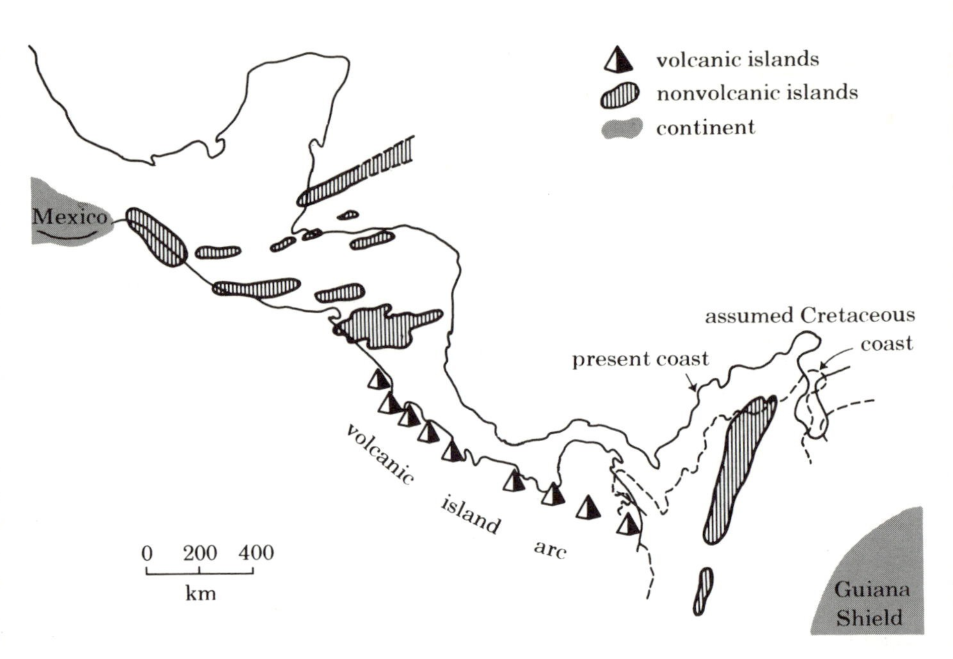

Figure 4. During most of the Cretaceous the continental margins of North and South America *(grey)* were separated by almost 3,000 km of water. In the late Cretaceous an arc of volcanic islands apparently permitted some migration of plants and animals between the two areas. (After Dengo, 1973.)

History of South American vertebrates

In view of these considerations, we may infer that Africa and South America shared a common flora and fauna of tropical to warm-temperate organisms until the Middle Cretaceous, some 90 million years ago; that the South American plants and animals evolved in isolation until they were joined by numerous migrants coming directly from the north during the past 5.7 million years; and that additions to the flora and fauna of South America came most likely from Africa until about Late Eocene time, and then increasingly from North America until the present. Are these assumptions reflected in present patterns of distribution and in the known fossil record?

For the mammals, the gap in the fossil record in Africa from the Triassic to the Eocene—about 150 million years—makes it impossible to prove the existence of the original South American groups on that continent. In addition, Africa was more or less directly connected with Europe at various times up to the Early Paleocene (63 million years ago), subsequently becoming more widely separated from it (Dewey et al. 1973). North America and Europe were in direct overland contact via the Greenland route until about 49 million years ago (McKenna 1972). Consequently, overland migration between Africa (which would have shared a common mammal fauna with South America until about 90 million years ago) and North America via Europe was possible at various times until about 63 million years ago, in the Paleocene. Connections such as these might account for similarities between the faunas of North and South America up to that time.

The near-union of Africa and Europe about 18 million years ago (Cooke 1972; Hallam 1973) ended Africa's approximately 45 million years of isolation from Eurasia, interrupted, perhaps, for a brief period in the Eo-Oligocene time, ?40–?35 million years ago (Berggren and Van Couvering 1974). Their proximity resulted in a renewed interchange of plants and animals between Africa and Eurasia, particularly tropical Asia, which is reflected in the present-day faunas and floras of these regions. South America, on the other hand, lacked overland connections with other areas for more than 80 million years, from the mid-Cretaceous to the Pliocene—which may

Figure 5. The yapock, or water opossum (*Chironectes minimus*), is one of the characteristic marsupials of South America. After becoming extinct in North America about 20 million years ago, the marsupials recolonized the area by way of a land bridge in the Late Pleistocene. (New York Zoological Society photo.)

partly explain the greater distinctiveness of its fauna and flora as compared with those of Africa.

Marsupials are very well represented in Australia and in South America (Fig. 5); the route by which they achieved this disjunct distribution has puzzled biogeographers for years. They are known from 100 million to about 20 million years ago in North America, from about 50–20 million years ago in Europe, and recently they have been reported from the Late Cretaceous of Peru. No fossil marsupials have been found in Africa. Whether they spread between South America (–Africa) and Australia via India and Antarctica in the mid-Cretaceous (see Fig. 2), or by a cool-temperate route across Antarctica subsequently, it is impossible to determine. They assuredly did not reach Australia from Asia, as experts formerly believed. No evidence exists for an exchange of marsupials between North and South America until the familiar opossum, *Didelphis,* eventually recolonized North America by a direct land connection in the Late Pleistocene. At that time, marsupials had been extinct in North America for about 20 million years. The distribution of an extinct group of mammals, the condylarths, presents problems similar to those for the marsupials.

One of the most distinctively South American groups of mammals is the notoungulates (Fig. 6), whose fossil record extends from the Paleocene to the Pleistocene. A single family of this group, Arctostylopidae, however, does not occur in South America. It consists of two genera, one from the Upper Paleocene of Mongolia and the other from the Upper Paleocene and Lower Eocene of North America. It is logical to assume that the ancestors of the Arctostylopidae did somehow disperse between North America and South America in earliest Tertiary or uppermost Cretaceous time, although even for them a less likely route via Africa and Eurasia cannot be completely ruled out.

Another distinctive South American group of mammals is the edentates. Various representatives of this group, including the anteaters, armadillos, and tree sloths, reached North America after the establishment of a direct land connection. Simpson (1945) and Darlington (1957) considered the Early Tertiary North American Metacheiromyidae to be edentates, and hence indicative both of migration from North America to South America at that time and of a North American origin for the edentates as a group. Emry (1970) later showed that the Metacheiromyidae are not edentates but are related to the Pholidota, a mainly Old World group that includes the scaly anteaters and pangolins. Edentates therefore afford no evidence for mi-

Figure 6. A generalized notoungulate (*Thomashuxleya*), from Patagonia. In the Early Tertiary, one family reached North America and Asia; all other known members are South American. (From Simpson 1936.)

gration between North and South America before the Pliocene, and they were more probably derived from a Cretaceous mammal fauna shared with Africa.

Both primates and caviomorph rodents appear in the fossil record of South America in the Early Oligocene (Patterson and Pascual 1972). These South American groups reached North America in the Pliocene and later. The caviomorph rodents, which include guinea pigs, chinchillas, agoutis, and New World porcupines, are broadly related to various African groups, such as the Old World porcupines, and both groups are considered hystricomorph rodents. The recent discovery in North America of fossils resembling these groups (e.g. Wood 1972) suggests that rodents with these characteristics evolved independently in North America, or that a Eurasian stock arrived before immigration from South America was possible. Thus the fossils do not provide evidence for migration between North and South America. Rafting from Africa to South America, which was about equally distant from North America and from Africa in the Oligocene, appears the most likely explanation for the arrival of both hystricomorph rodents and monkeys in South America at that time.

To summarize the record for mammals, on the basis of present evidence only the notoungulates appear to have passed between North and South America prior to the establishment of a direct land connection. There is no evidence for an early "intense exchange of land mammals" between the two continents, as suggested by Darlington (1957, p. 365): precisely the opposite is indicated by present distributions and by a critical examination of the fossil record in the light of plate tectonics.

For other groups of vertebrates, the patterns are similar and reflect their relative powers of dispersal (e.g. for birds, see Cracraft 1973b). Among the amphibians, for example, caecilians evidently dispersed overland between Africa and South America, as did frogs in general and probably Pipidae, Bufonidae, and perhaps Leptodactylidae in particular (Savage 1973; Cracraft 1973a). The familiar treefrogs, *Hyla,* originated in South America but appear in the fossil record in the Early Oligocene of Saskatchewan, suggesting sweepstakes dispersal from South to North America at least that early. The toads, *Bufo,* appear in North America in the Lower Miocene and in Europe in the Eocene, which, in view of their evident origin in South America, suggests they also crossed water barriers in reaching the north. *Eleutherodactylus* and related genera may have reached Central America from South America in the Miocene, judging chiefly from their extensive evolutionary radiation there. In summary, the amphibians, by virtue of their greater flexibility in crossing water barriers (Darlington 1957), seem to have begun to pass between North and South America in some numbers in the Miocene, but only *Bufo* and *Hyla* appear to provide examples of earlier migration.

Reptiles also cross water barriers with some facility. Fossils of dinosaurs that seemed to link North and South America have been cited as evidence of early connections between the two land masses. As currently interpreted, however, the dinosaurs seem instead to indicate connections between South America, Africa, and the Old World, and perhaps a roundabout connection with North America via Europe. Several groups are nearly cosmopolitan. Two families of lizards, the iguanids and teiids, were probably dispersed by island-hopping between North and South America by the Eocene and Late Cretaceous, respectively. The teiids subsequently became extinct in North America, where they are now represented solely by genera that arrived in the Miocene and later. Like the amphibians, the reptiles appear to have begun to migrate in numbers between South and North America in the Miocene, across relatively short distances, and only the teiid and iguanid lizards appear to have passed between these continents earlier.

One of the most critical groups for an interpretation of the history of South America's fauna is that of the primary freshwater fishes, which never enter salt water at any time and are incapable of surviving there. They are more conservative in crossing water barriers than any group except the mammals. The dominant freshwater fishes of Africa and South America are characoids and siluroids, both derived from ancestral forms that dispersed directly between these continents when they were united. In South America, the characoids gave rise to the endemic gymnotids, and in Africa cyprinoids entered from the north after separation of these continents. Thus there are no gymnotids in Africa and no cyprinoids in South America.

The three other primary freshwater fish families of South America are all shared with Africa. No primary freshwater fishes of North American origin extend south of Nicaragua, and none has reached South America, contrary to the elaborate scheme proposed by Darlington (1957). A few genera of characoids of South American origin have extended their range north into Mexico after the establishment of a direct overland connection, and one, *Astyanax,* has reached the Rio Grande. The cichlids of Central America are secondary freshwater fishes that reached Central America from South America across saltwater gaps, whereas the poeciliids of Central America were probably derived from North American cyprinoids.

In summary, there is evidence for only five instances of dispersal of vertebrates between North and South America prior to the Miocene: (1) teiid lizards, Late Cretaceous; (2) arctostylopid mammals, Late Paleocene; (3 and 4) toads of the genus *Bufo* and iguanid lizards, Eocene; (5) the frog genus *Hyla,* Early Oligocene. In view of the richness and variety of the vertebrate faunas of North and South America, the small amount of demonstrated interchange certainly accords well with geological evidence suggesting a very wide separation of the Americas in Cretaceous through Oligocene time.

The flowering plants of South America

South America is the richest in species of flowering plants (angiosperms) and most poorly known of the three major tropical areas of

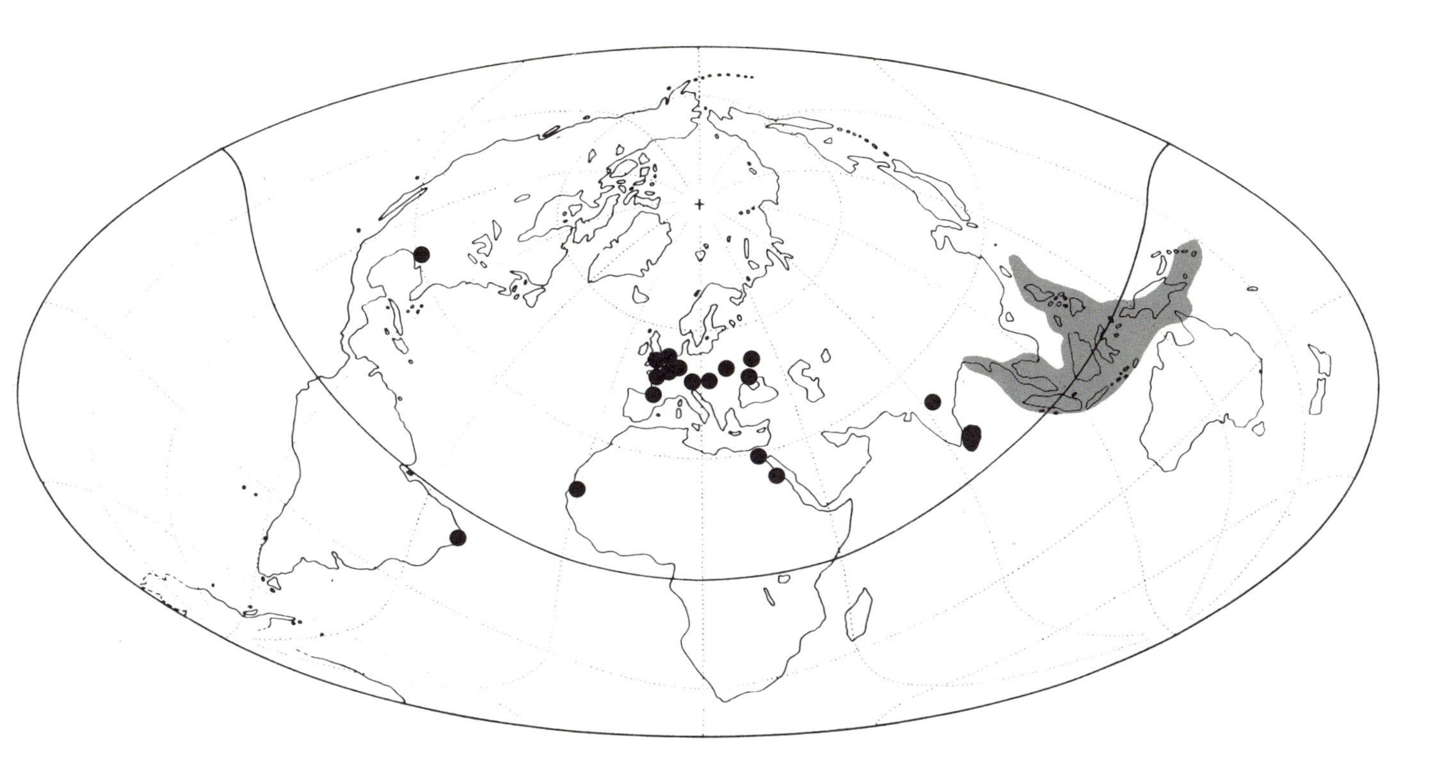

Figure 7. The genus *Nypa* (see Fig. 8) was distributed widely *(dots)* throughout the warmer parts of the globe in the Paleocene and Eocene. Its range is now confined to the area from Southeast Asia out into the tropical Pacific (shaded area). Data from Tralau 1964 and Moore 1972.

the world. In order to understand the origins of this rich flora, one must first consider the age of the angiosperms (summary in Raven and Axelrod 1974). The oldest fossil pollen certainly determined as angiospermous is from the Early Cretaceous, some 125 million years ago. Even if, as seems likely, angiosperms with pollen indistinguishable from that of the gymnosperms are older, there is no evidence that any of the existing orders of angiosperms appeared until the mid-Cretaceous, about 110 million years ago. By the close of the Cretaceous, 65–70 million years ago, a number of modern genera and families were in existence, including Proteaceae, Myrtaceae, Arecaceae (Palmae), and Sapotaceae, and the hollies (*Ilex*), southern beeches (*Nothofagus*), alders (*Alnus*), and sycamores (*Platanus*).

Critical for understanding the biogeography of flowering plants is the relationship between South America and Africa in the Paleocene and Eocene (38–65 million years ago). At the start of this time, approximately 600 km separated the two continents, and the shallow seaway between them was dotted with volcanic islands that provided way stations for sweepstakes dispersal. By the end of the Eocene they were some 1,400 km apart; there probably were fewer islands; and South America was now closer to North America than to Africa.

In Paleocene time South America was about as far from Africa as Yucatán is from Jamaica. Jamaica was completely submerged 15 million years ago (Robinson and Lewis 1971), but has since acquired a flora of some 3,000 species of angiosperms, as well as a fauna comprising a number of species of the frog *Eleutherodactylus*, four species of *Hyla*, many lizards, some snakes, an extinct primate, a rodent, and various other animals. If an amount of exchange similar to that which populated Jamaica had taken place over water between South America and Africa in Paleocene and Early Eocene time, we would be unable at present to distinguish its effects from those of direct overland migration.

Figure 8. The palm *Nypa*, shown growing along the edges of a tidal swamp on Truk in Micronesia, with coconut palms (*Cocos nucifera*) in the background. (Photo by F. R. Fosberg.)

Africa and South America shared a common flora of angiosperms into Eocene time (Figs. 7, 8). The opportunities for dispersal between them decreased progressively from their initial separation just under 90 million years ago. Conversely, the opportunities for migration between North and South America progressively increased from the Eocene onward, as South America gradually stocked North America with the original flora and fauna of West Gondwanaland (Africa plus South America), first by sweepstakes dispersal during the Eocene and the Oligocene, and then more rapidly and effectively as the land connection was finally established.

The distinctiveness of the African and South American floras has often been stressed, and there seem to be three basic reasons for it. First, a very large segment of the evolution of the flowering plants took place after the two continents were no longer in relatively direct contact; progressive change doubtless led to considerable differentiation. Second, Africa has been directly accessible to immigration from tropical Asia for more than 18 million years, and the migration of subtropical plants and animals between them has been interrupted only by the spread of arid climate in the region during the past several million years. South and North America have been, at least from the standpoint of a readily dispersible group such as the flowering plants, in close contact for about the same period of time, and there has been extensive exchange between them. Most important of all, however, is the third factor: massive extinction resulting from climatic change in continental Africa. From the Miocene onward, the altitude of eastern Africa has increased some 8,000 feet, and arid climates have spread over the continent.

Also beginning in the Miocene, the Benguela Current has brought cold water, associated with the onset of glaciation in Antarctica, to flow along the west coast of Africa, subjecting an area previously covered with rainforest to only seasonal precipitation. In contrast, the rising Andes blocked the effects of spreading aridity over the lowlands of South America, and Southeast Asia–Malaysia owes its isothermal climate to its high degree of insularity. Finally, the trend toward aridity was exacerbated by the Pleistocene arid cycles, which reduced the extent of tropical rainforest and led to the spread of savannas everywhere.

Figure 9. The closely related families Musaceae, Strelitziaceae, and Heliconiaceae link the Old and New World tropics. Their ancestors dispersed between Africa and South America when those continents were much closer together. Shown here are (*top*) *Ravenala* (Strelitziaceae), the traveler's palm, from Madagascar (photo by A. H. Gentry); (*middle*) *Phenakospermum* (Strelitziaceae) from northern South America (photo by A. H. Gentry); and (*bottom*) *Heliconia pendula* (Heliconiaceae), from the Central and South American tropics (photo by G. S. Daniels). *Heliconia* is mainly American but ranges across the Pacific to the Solomon Islands and New Guinea, which it probably reached by long-distance dispersal.

In view of these relationships, it is not surprising that Africa has the most impoverished of all tropical floras, with probably less than a third of the number of species of angiosperms of tropical South America, despite its greater area. Madagascar, on the other hand, which has been separate from the mainland of Africa since at least the Late Cretaceous, has a richer and more nearly balanced flora. Groups such as palms, ferns, orchids, and bamboos, which are well represented in all tropical parts of the world except continental Africa, are also well represented on Madagascar, presumably because its insularity has helped to maintain a more equable climate and because it is removed from the arid effects of cold currents. Madagascar probably retains a sample of the original balanced flora of Africa (Fig. 9), in the same sense that New Zealand and New Caledonia retain a sample derived from the original cool-temperate flora of Australia.

Origins of the Latin American biota

In conclusion, it is appropriate to consider the same problem that fascinated Simpson (1950): Why does Latin America—Mexico to Argentina—seem to have a common biota? Before the separation of Africa and South America, the two areas constituted a supercontinent that was by far the largest block of land in the tropics. Extensive arid areas, both climatic and edaphic, would have existed in the interior and afforded different kinds of evolutionary opportunities that may have been conducive to rapid evolution. If we consider the world of the Early Cretaceous, it is clear that this supercontinent, called West Gondwanaland, would have constituted an evolutionary site of the same magnitude that Eurasia plus North America does at present, with the same sorts of opportunities for evolution and subsequent migration to the four corners of the earth. Both now and in the Early Cretaceous, Australasia is one of the far corners; but in the Early Cretaceous, North America and perhaps northeastern Asia, if it was

Figure 10. The creosote bush (*Larrea divaricata*) is a member of a primarily South American genus which probably reached North America by long-distance dispersal within the past several million years. (Photo by Dade W. Thornton, from National Audubon Society.)

separated by the sea from the main part of Eurasia, were far corners also.

It should not therefore be surprising that the dominant flora and fauna of the lowland tropics of the New World originated in West Gondwanaland, or that they spread north to occupy similar sites in Central America, the West Indies, and southern North America when South America moved into range for sweepstakes dispersal. In contrast, the contributions from tropical North America to South America appear very few for any group of plants or animals. Exactly the same process took place in Australia: the tropical lowlands were taken over by Asian and Malaysian plants and animals when the two areas moved within range, with the plants moving earlier and across much wider water gaps.

The plants and animals of temperate North America, on the other hand, have spread only slowly into South America. Such genera as fir (*Abies*), alder (*Alnus*), sweetgum (*Liquidambar*), beech (*Fagus*), walnut (*Juglans*), and elm (*Ulmus*) had already reached the mountains of southern Mexico 16 million years ago. In northern South America, walnuts were present about 8 million years ago, alders about 700,000 years ago, and oaks (*Quercus*) about 150,000 years ago. For the many herbaceous montane plants that have extended their range into South America along the mountains, conditions have been favorable from the Late Pliocene onward.

Appropriate cool climates first appeared in tropical latitudes in the Pleistocene, and the mountains and volcanoes that provide way stations through Central America were fully uplifted during the Late Pliocene and more recently. Opportunities for the migration of cool-temperate and montane plants and animals between North and South America were never greater than in Pleistocene and Recent times, and such groups as the gooseberries and currants (*Ribes*), the locoweeds (*Astragalus*), willows (*Salix*), and evening primroses (*Oenothera*), now represented by many species in South America, probably arrived on that continent only during the past million years or so and evolved rapidly under the influence of the expanding and fluctuating climates characteristic of Pleistocene cycles.

The arid areas of North and South America possess similar vegetation types and even have a number of plant genera in common, although relatively few genera of animals. For example, the vegetation type termed the *monte* in Argentina is very similar to that of the Sonoran and Chihuahuan deserts of North America and, like them, is dominated by creosote bush (*Larrea;* Fig. 10) and mesquite (*Prosopis*). In the past, authors have at times considered the possibility of an arid Cretaceous corridor connecting these regions, along which plants and animals could have migrated directly. The existence of such a corridor is ruled out not only by the geological realities but by the actual representation of plants and animals in the two regions. Not only are the animals very different, even when they are associated with the same sorts of plants in both continents, but the plants common to both constitute only a small percentage of the respective floras. If there had been direct migration, we would expect much more extensive similarities between the floras. We would also expect the animals to have migrated with the plants.

During the Pleistocene, the pluvial periods of the temperate regions were periods of aridity in the tropics (summary in Raven and Axelrod 1974, pp. 608–11; Williams 1975). Areas of grassland and thorn scrub were enlarged, and the opportunities for stepwise exchange of plants and animals between the arid areas of North and South America were maximal. At any earlier period, arid areas were restricted in the region of Central America and the West Indies, because the islands and mountains were too small to create effective rain shadows. The arid glacial-age periods in the tropics would have been suitable times for the exchange of many grazing mammals such as horses, glyptodonts, ground sloths, and camels between North and South America.

The theory of plate tectonics has helped us to view the biogeography of South America in a new light. The extreme importance of West Gondwanaland (South America plus Africa) as a site for evolution in the Cretaceous and the impact of its fragmentation on the evolution of its original biota are now apparent. The subsequent relationships between these fragments and the lands with which they came into contact have contributed in an important way to the present patterns of distribution of organisms. An increasingly accurate and complete analysis of the fossil record will gradually enable biologists to use the facts of plate tectonics more extensively in their interpretations of the evolution and biogeography of many groups of plants and animals.

References

Axelrod, D. I., and P. H. Raven. 1972. Evolutionary biogeography viewed from plate tectonic theory. In J. A. Behnke, ed., *Challenging Biological Problems: Directions toward Their Solution.* New York: Oxford University Press. pp. 218–36.

Berggren, W. A., and J. A. Van Couvering. 1974. Neogene geochronobioclimatopaleomagnetostratigraphy: A Mediterranean synthesis. *Abstr. Pap. 1974 Annual Meet. Geol. Soc. Amer.*, pp. 1022–24.

Cooke, H. B. S. 1972. The fossil mammal fauna of Africa. In A. Keast, F. C. Erk, and B. Glass, eds., *Evolution, Mammals, and Southern Continents.* Albany: State Univ. New York Press. pp. 89–139.

Cracraft, J. 1973a. Vertebrate evolution and biogeography in the Old World tropics: Implications of continental drift and paleoclimatology. In D. H. Tarling and S. K. Runcorn, eds., *Implications of Continental Drift to the Earth Sciences.* London and New York: Academic Press. Vol. 1: 373–93.

Cracraft, J. 1973b. Continental drift, paleoclimatology, and evolution and biogeography of birds. *J. Zool.* (London) 179:455–545.

Darlington, P. J., Jr. 1957. *Zoogeography: The Geographical Distribution of Animals.* N.Y.: Wiley.

Dengo, G. 1973. *Estructura geológica, historia tectónica y morfológia de América Central.* 2nd ed. Central Regional de Ayuda Técnica, A.I.D., Mexico.

Dewey, J. F., W. C. Pitman, III, W. B. F. Ryan, and J. Bonin. 1973. Plate tectonics and the evolution of the Alpine system. *Geol. Soc. Amer. Bull.* 84:3137–80.

Douglas, R. G., M. Moullade, and A. E. M. Nairn. 1973. Causes and consequences of drift in the South Atlantic. In D. H. Tarling and S. K. Runcorn, eds., *Implications of Continental Drift to the Earth Sciences.* London and N.Y.: Academic Press. Vol. 1: 517–37.

Emiliani, C., S. Gaertner, and B. Lidz. 1972. Neogene sedimentation on the Blake Plateau and the emergence of the Central American isthmus. *Palaeogeogr., Palaeoclimat., Palaeoecol.* 11:1–10.

Emry, R. J. 1970. A North American Oligocene pangolin and other additions to the Pholidota. *Bull. Amer. Mus. Nat. Hist.* 142:455–510.

Foster, F. J. 1974. Eocene echinoids and the Drake Passage. *Nature* 249:751.

Hallam, A. 1973. Distributional patterns in contemporary terrestrial and marine animals. In N. F. Hughes, ed., *Organisms and Continents through Time.* Palaeontol. Assoc. London, Spec. Pap. Palaeontol., 12. pp. 93–105.

McGowran, G. 1973. Rifting and drift of Australia and the migration of mammals. *Science* 180:759–61.

McKenna, M. C. 1972. Was Europe connected directly to North America prior to the Middle Eocene? *Evol. Biol.* 6:179–89.

McKenzie, D. P., and J. G. Sclater. 1973. The evolution of the Indian Ocean. *Sci. Am.* 228(5):62–72.

Moore, J. E., Jr. 1972. Palms in the tropical forest ecosystems of Africa and South America. In B. J. Meggers, E. S. Ayensu, and W. D. Duckworth, eds., *Tropical Forest Ecosystems in Africa and South America: A Comparative Review.* Washington, D.C.: Smithsonian Inst. Press. pp. 63–88.

Patterson, B., and R. Pascual. 1972. The fossil mammal fauna of South America. In A. Keast, F. C. Erk, and B. Glass, eds., *Evolution, Mammals, and Southern Continents.* Albany: State Univ. New York Press. pp. 247–309.

Raven, P. H., and D. I. Axelrod. 1972. Plate tectonics and Australasian paleobiogeography. *Science* 186:1379–86.

Raven, P. H., and D. I. Axelrod. 1974. Angiosperm biogeography and past continental movements. *Ann. Missouri Bot. Gard.* 61:539–673.

Reyment, R. A. 1969. Ammonite biostratigraphy, continental drift and oscillatory transgressions. *Nature* 224:137–40.

Reyment, R. A. 1972. The age of the Niger Delta (West Africa). *24th Internat. Geol. Congr.*, sect. 6:11–13.

Reyment, R. A., and E. A. Tait. 1972. Biostratigraphical data of the early history of the South Atlantic Ocean. *Roy. Soc. London Trans.* 264:55–95.

Robinson, E., and J. F. Lewis. 1971. Field guide to aspects of the geology of Jamaica. In *Internat. Field Institute Guidebook to the Caribbean Island-Arc System,* pp. 2–39 (Jamaica section). Amer. Geol. Inst.

Savage, J. M. 1973. The geographic distribution of frogs: Patterns and predictions. In J. L. Vial, ed., *Evolutionary Biology of the Anurans.* Columbia: Univ. Missouri Press. pp. 351–445.

Savage, J. M. 1974. The Isthmian link and the evolution of Neotropical mammals. *Contr. Sci. Nat. Hist. Mus. Los Angeles Co.* 260:1–51.

Simpson, G. G. 1936. Skeletal remains and restoration of Eocene Entelonychia from Patagonia. *Amer. Mus. Novitates* 82:1–12.

Simpson, G. G. 1945. The principles of classification and a classification of the mammals. *Bull. Amer. Mus. Nat. Hist.* 85:1–350.

Simpson, G. G. 1950. History of the fauna of Latin America. *Am. Sci.* 38:361–89.

Tralau, H. 1964. The genus *Nypa* Van Wurmb. *Kungl. Svenska Vetenskap. Handl.* 10(1):5–29.

Williams, M. A. J. 1975. Late Pleistocene tropical aridity synchronous in both hemispheres? *Nature* 253:617–18.

Wood, A. E. 1972. An Eocene hystricognathous rodent from Texas: Its significance in interpretations of continental drift. *Science* 175:1250–51.

"Some of these youngsters have come up with a terrific new idea—feathers."

PART 4 *Vertebrate Paleontology*

John H. Ostrom

Bird Flight: How Did It Begin?

Did birds begin to fly "from the trees down" or "from the ground up"? Reexamination of Archaeopteryx *adds plausibility to an "up from the ground" origin of avian flight*

Paleontologic data have provided unique clues about various events in the history of life that we could not otherwise have known about—for example, the question of human origins, the ancestry of mammals, and the origins of land vertebrates. It may be true that the exact lines of descent in these particular examples (and other lineages) are not known precisely, but few would disagree that the following general evolutionary histories are widely accepted as true: mankind evolved from a late Tertiary apelike creature, probably by way of the australopithecines; mammals originated from advanced cynodonts of the long-extinct mammal-like reptilian order, Therapsida; and land vertebrates arose somewhat more than 350 million years ago from a now-extinct group of fishes, the rhipidistian crossopterygians. These are but a few of many evolutionary histories that have been inferred from the fossil record.

One such historical narrative deduced from the fossil record is the origin of birds and the intriguing question of how bird flight might have evolved. Flight is one of the most remarkable of all animal adaptations, yet it has evolved numerous times quite independently—more than half a dozen times in vertebrate animals alone. Flight in vertebrates ranges from passive (non-self-powered) parachuting (flying squirrels, flying frogs) and gliding (the colugo and flying lizards) to fully self-powered flight of birds, bats, and probably the extinct flying reptiles known as pterosaurs. The most improbable flight is that of the flying snake, *Chrysopelea,* but it is the varied forms of avian flight that have fascinated mankind since ancient times, as is evidenced by Egyptian and Greek mythology. The diverse ways in which birds fly have been revealed in the studies by Brown (1948, 1951, 1953, 1961, 1963), Greenewalt (1960a, 1960b, 1962, 1975), Hartman (1961), Pennycuick (1960, 1968, 1971, 1975), Tucker (1969, 1973), and others, to the extent that the aerodynamics of bird flight is fairly well understood. But how it all began is not so clear.

A century ago, it was supposed by some (e.g. Mivart 1871) that bird flight came into being by a sudden spontaneous appearance of fully developed wings, where previously there had been only normal forelegs for walking. Today, most of us are confident that such a major and complex innovation as flight developed through a sequence of many small changes that occurred (however slowly or rapidly) over many generations, with every stage being well adapted for its own particular setting and way of life. Assuming this latter mode to be correct, what can we say about the origins of bird flight? And what is the evidence?

Previous speculations on this question have produced two quite different scenarios. Stated very simply, these are that birds began to fly "from the trees down"—or "from the ground up." The first is the widely favored and very logical "arboreal theory," first suggested by O. C. Marsh in 1880 and skillfully elaborated recently by Walter Bock (1965, 1969). The second is the often ridiculed and seemingly less probable "cursorial theory," commonly attributed to Baron Francis Nopcsa (1907, 1923), but originally outlined in 1879 by Samuel Wendell Williston.

The arboreal theory

In brief, the arboreal theory postulates that the primitive ancestors of birds must have been tree-dwelling animals which developed increasing need and skill to leap from branch to branch, and then from tree to tree. The selective advantage of any anatomical change that would increase body surface area relative to body mass is obvious—first, to slow the rate of fall from a mis-leap; later, to provide lift and directional descent in longer and longer glides; and finally, to permit longer and longer gliding leaps between trees. As Marsh expressed it in 1880:

> The power of flight probably originated among small arboreal forms of reptilian birds. How this may have commenced, we have an indication in the flight of *Galeopithecus* [colugo], the flying squirrels (*Pteromys*), the flying lizard (*Draco*) and the flying tree frog (*Rhacophorus*). In the early arboreal birds, which jumped from branch to branch, even rudimentary feathers on the forelimbs would be an advantage as they would tend to lengthen a downward leap or break the force of a fall.

John H. Ostrom is Professor of Geology at Yale University and Curator of Vertebrate Paleontology at Yale's Peabody Museum of Natural History. After receiving his Ph.D. from Columbia University, he taught first at Brooklyn College and then at Beloit College before joining the faculty at Yale in 1961. His research has centered on the history and evolution of Mesozoic reptiles, chiefly dinosaurs and flying reptiles, and on the origins of birds and bird flight. *The author is indebted to many students and colleagues for critical and lively discussions on the subject of this article, but only the author should be held responsible for the speculations set forth. Photographs are by the author unless indicated otherwise.* *Address: Department of Geology and Geophysics, Yale University, New Haven, CT 06520.*

The flaw in this picture lies with the analogs chosen by Marsh. They are all *quadrupedal* rather than bipedal arborealists (see Fig. 1), and with the exception of *Draco* (where a fold of skin, the patagium, is supported by elongated ribs), the flight membrane is stretched between the fore and hind limbs. I will return to this point later.

In 1965, Walter Bock presented a more detailed version of the arboreal theory as a model to show that macroevolutionary differences between members of a phyletic lineage were not necessarily incompatible with known principles of microevolutionary change. Listing only the "major" hypothetical adaptive stages in their order of occurrence—(1) ancestral ground-dwelling quadrupedal reptile, (2) bipedal ground-dweller, (3) bipedal and arboreal life, (4) leaping between trees, (5) parachuting, (6) gliding, (7) active, powered flight—Bock made it clear that he visualized many intermediate gradational, but fully adapted, stages between each of these. This sequence culminated in the macroevolutionary transition from a reptilian nonflier to powered avian flight.

The major significance of Bock's paper was to show that at each stage there would be different, perhaps unique, selective pressures inducing still further changes subject to yet other selective forces. In his model, flight would be selected for only in the final stages. The critical point of this model, though, is that, in order to fly, the animal first had to be able to climb. However, considering the design of modern birds, together with that of the oldest-known bird, *Archaeopteryx,* that skill may not have been part of the repertoire of primitive birds, or even of bird ancestors. I will return to this point also.

Figure 1. Artist Zdenek Burian has reconstructed a pre-*Archaeopteryx* stage (commonly termed "pro-avis") in the evolution of avian flight. No fossil evidence exists of any pro-avis. It is a purely hypothetical pre-bird, but one that must have existed—whether or not it resembled the creature shown here. Notice that Burian visualized this transitional stage as quadrupedal, rather than bipedal, following the prevailing view at the time that bipedality in birds arose as an adaptation to arboreal life. (From Augusta and Burian 1961, with the permission of Artia Publishers, Prague.)

The cursorial theory

The cursorial theory postulates a sequence of stages from a primitive quadrupedal reptile, to a facultative biped (like the lizards *Basiliscus* and *Crotaphytus,* which at high speeds run on their hindlegs, but at slower speeds move on all fours), to an obligatory cursorial biped, followed by stages of elongation of the forelimbs and enlargement of "scales" on the arms to increase their surface area, thereby forming ever larger "thrust" surfaces. Flapping action of these "proto-wings" supposedly added thrust to that provided by the hindlimbs, resulting in greater acceleration and faster running speed. Ultimately, this is presumed to have led to flight velocities and to at least partial conversion of the forelimbs from "propellers" into wings.

One of the key criticisms that has been leveled at this hypothesis is that, once the animal is airborne, the main thrust source (i.e. traction of the hind feet against the ground) would be lost and velocity would diminish. However, at that primitive "flight" stage there might well have been selective advantage in lifting up off the ground, at least momentarily—either for escape, or in pursuit of prey—that outweighed any disadvantage caused by loss of traction. A more critical flaw, however, is the miniscule amount of additional thrust that could have been generated by those earliest enlarged "scales" on the incipient "wing" pushing against the air. Certainly this could not have been anywhere near enough additional thrust to produce a measurable increase in running speed, and thus be selected for.

The cursorial theory of bird flight origins has received virtually no acceptance, apparently for several very good reasons: Nopcsa's seemingly absurd flailing-arm propellers; the seemingly impossible "bootstrap" effort required for the animal to lift itself by means of flapping protowings; the simple logic of cheap gravity-powered flight from an ele-

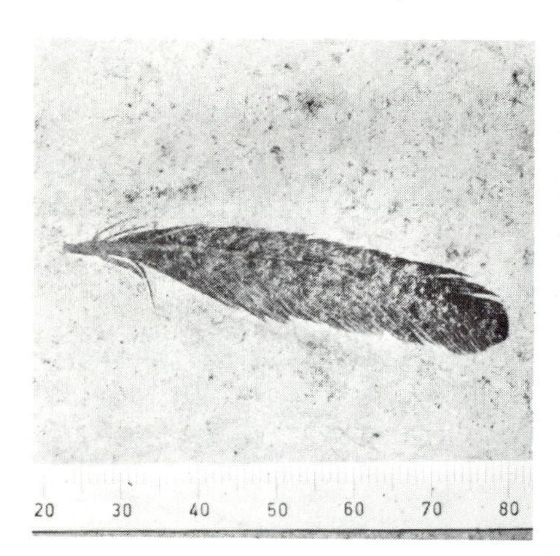

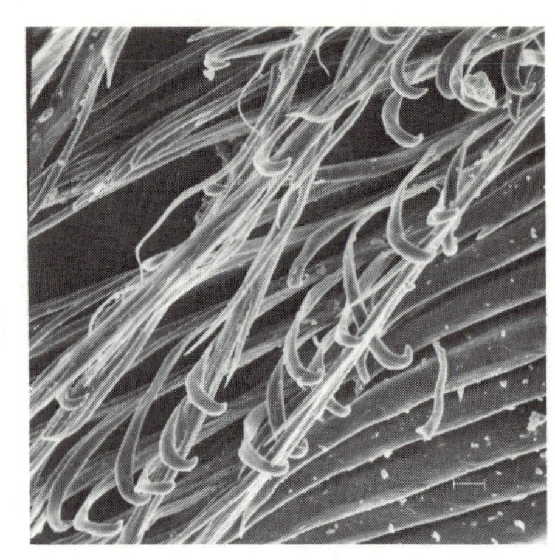

Figure 2. The fossil imprint (*left*) of a solitary feather that has been referred to *Archaeopteryx,* and represents the original discovery of Jurassic birds, is compared with a modern contour (vaned) feather (*center*). This fossil imprint was found on a slab of the famed Solnhofen lithographic limestone in a Bavarian stone quarry in 1861. It clearly shows the uniformly parallel alignment (except at two places near the tip) of numerous barbs, which indicates that these barbs bore minute barbules with microscopic hooklets, just as is shown in the scanning electron micrograph of a modern contour feather (*right*). The fossil imprint shown here is in the collections of the Humboldt Museum für Naturkunde in East Berlin, and its counterpart is in the Bayerische Staatssammlung für Paläontologie und historische Geologie in Munich. (Electron micrograph courtesy of A. Pooley; scale bar equals 10 microns. Modern feather photographed by W. Sacco; scale units are in millimeters.)

vated perch; and the fact that nearly all present-day vertebrate fliers (except flying fish, of course) are arboreal. Needless to say, some bats and many birds are not now arboreal, but these may be presumed to be relatively recent adaptations for habits and habitats quite different from those of earlier arboreal ancestors.

I noted above that Marsh's analogs were all quadrupeds, and implied that they might not be appropriate models upon which to base speculations about bipedal avian flight origins. It is important to note that *all* nonavian vertebrate fliers, except flying snakes and fish (and possibly pterosaurs), are quadrupedal walkers in which all four limbs are involved in the flight apparatus. Birds are unique. They are the only vertebrate fliers in which the flight apparatus is constructed entirely of the forequarters (and the tail) and does not involve the hindlimbs at all. By a unique sequence of circumstances, birds evolved *independent* dual locomotory systems—wings for flight and legs for running, walking, or swimming—and this may be the chief reason for their remarkable success.

It is possible that the extinct flying reptiles (pterosaurs) also had wings constructed only of the forelimbs, but their hind legs were not designed for bipedal walking. It appears that, as with nearly all bats, excursions by most pterosaurs across the ground surface must have been awkward and required the use of the wings as legs. Birds, on the other hand, are skilled bipeds—some of them extraordinarily so (take the roadrunner, for example). It seems to me that only one conclusion is possible: avian flight evolved by a very different sequence of adaptive stages than did that of all other fliers. An arboreal origin of flight appears to be the only possible explanation for quadrupedal fliers, and for *Draco* and the flying snakes, but it may not have been the way bird flight came about.

What is the evidence?

Archaeopteryx

The anatomical differences among modern vertebrate fliers point up the probably very different scenarios that led to the different forms of vertebrate flight, but they do not provide a clear picture of how bird flight might have begun. The answer may be indicated, though, in the fossil remains of the oldest-known bird—*Archaeopteryx*—from the 150-million-year-old Jurassic Solnhofen limestones of Bavaria, in West Germany. Although possibly not directly ancestral to any modern birds, these specimens constitute the only direct evidence available pertaining to the earliest stages of bird evolution. At the present time a total of only six specimens are known: five partial to virtually complete skeletons and an isolated imprint of a solitary feather that is presumed to have come from the same kind of animal. That feather imprint (Fig. 2) plus the details preserved in the feather impressions surrounding two of the skeletons indicate the existence of microscopic feather construction essentially identical to that of modern feathers. Add to this the arrangement of large contour-type feathers along the arms of those two specimens (the London and Berlin specimens) in a pattern very much like that of modern birds' wings. These suggest at least some degree of flight capability—perhaps.

Unlike the feathers, however, the skeletal anatomy of *Archaeopteryx* is not so birdlike as we might expect. In fact, there is just one skeletal feature of *Archaeopteryx* that is *exclusively* avian—the wishbone, or furcula (fused clavicles). All other features of the skeleton are also found in various small carnivorous dinosaurs (coelurosaurian theropods). It is this dinosaurian nature of *Archaeopteryx* that led me to theorize that modern birds, perhaps by way of *Archaeopteryx*, are the direct lineal descendants of the small carnivorous dinosaurs (Ostrom 1973, 1975, 1976a).

This is not the main point of this article, however. These striking anatomical similarities perhaps indicate much more than a close evolutionary relationship between coelurosaurs and *Archaeopteryx*. They also suggest very similar adaptations, and thus similar modes of life. This may be the most important clue about the early stages of bird evolution and the origins of avian flight.

Figure 3. The Berlin specimen of *Archaeopteryx lithographica* (some would assign it to a separate species, even genus, *Archaeornis siemensi*), from the Solnhofen limestone, shows the modernlike arrangement of the "flight" feathers in an elliptical-shaped "wing." This individual was about the size of a common pigeon. The most remarkable of the six known specimens of *Archaeopterxy,* it was found in 1877 and now is in the collections of the Humboldt Museum.

Upon first examination of the specimens of *Archaeopteryx*—especially of the famed Berlin specimen (Fig. 3)—we see what appears to be a birdlike creature with long arms and clear imprints of feathered "wings" remarkably like those of modern birds. Surely these "wings" must have been for flight, as nearly all previous authors have concluded. But closer inspection reveals some puzzling—even paradoxical—conditions which throw that conclusion into doubt, at least in terms of powered flight.

First of all, there is no sternum (breast bone) preserved in any of the specimens. Modern flying birds have a large well-ossified sternum, which usually has a prominant deep midline keel that serves as the main site of attachment of the large flight muscles. Presumably, a sternum was present in *Archaeopteryx,* at least in a cartilaginous state, but there is no evidence of that. What does this indicate about the size and power of the flight muscles? Perhaps nothing. After all, in bats the sternum is quite small and often lacks a keel. But the ventral flight muscle (M. pectoralis) of bats is small, comprising less than 10% of total body weight, as compared with birds, in which it typically equals 15% to 20% of body weight and ranges to an astonishing 33% in the dove *Leptotila* (Hartman 1961). Thus, the absence of a sternum in *Archaeopteryx* seems important—and perhaps indicates weak pectoral muscles.

Another surprising feature of the skeletal "flight apparatus" of *Archaeopteryx* is the shape and thin sheetlike construction of the coracoids—the robust strutlike bones in modern birds that brace the shoulder against the breast bone. This coracoid brace is critical in flying birds, because it immobilizes the shoulder socket so that the full contractile force of the flight muscles is applied to the power stroke of the wing, and there is no loss of muscular force by downward displacement of the shoulder. In *Archaeopteryx,* the coracoids are thin, half-moon-shaped sheets of bone—short in length and not at all robust—fragile "braces" (?) between the shoulder and the sternum. This too suggests that the pectoral region was subject to relatively weak stresses resulting from relatively small "flight" muscles.

Even more interesting is the absence of the triosseal canal (Ostrom 1976b), the pulleylike structure of the shoulder skeleton in modern birds that reverses the action of the supracoracoideus muscle so that it powers the wing *recovery* stroke, rather than adding to the downward power stroke. Without that canal, this muscle could not possibly have functioned to elevate the wing in *Archaeopteryx,* and the recovery stroke must have been powered by the relatively weak and mechanically less efficient dorsal muscles, such as the deltoids.

Examination of the "wing" skeleton of *Archaeopteryx* reveals further

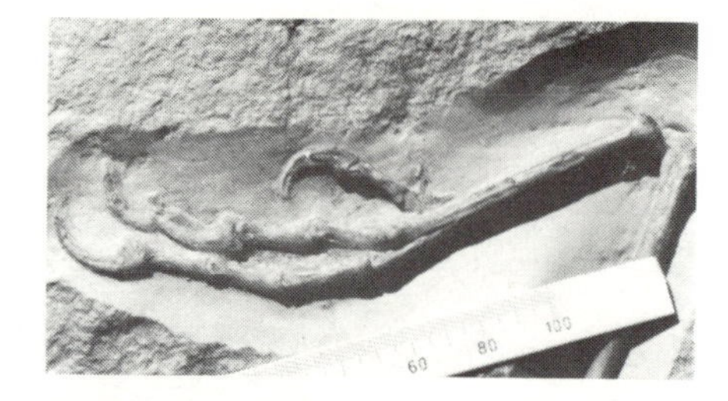

Figure 4. The left feet of the London (*top*) and Berlin (*center*) specimens of *Archaeopteryx* show the reversed hallux and the birdlike form similar to that of a modern pigeon (*bottom*). The same foot design is also characteristic of some theropod dinosaurs. (Bottom photo by W. Sacco.)

anomalies. The three fingers are not fused together, as they are in all modern birds—a condition that is generally thought to be an adaptation related to the need for a solid skeletal platform for the firm attachment of the primary flight feathers. Also, the wrist of *Archaeopteryx* consists of several separate small bones and clearly had much more articular mobility than is found in modern birds, whose wrists permit flexion and extension only in the plane of the wing, perpendicular to the wing stroke. That would seem to be a major mechanical deficiency in the "wing" of *Archaeopteryx.*

The upper arm bone (humerus) of *Archaeopteryx* has a very large deltopectoral crest, the site of attachment of the pectoralis muscle. That would seem to indicate a fairly large muscle mass—possibly for flight. But this same crest, of comparable relative size, exists in all theropod dinosaurs, which also lack ossified breast bones and obviously were not fliers. On the other hand, the humerus lacks all the other processes and tubercles that are prominently developed in modern flying birds and that are the sites of attachment of the special muscles that fold the wing compactly against the back and flanks. This suggests that *Archaeopteryx* probably was not able to fold the wing tightly against the body—a presumably important action for protecting the "wing" (if it actually *was* an airfoil) when not in use.

All the skeletal features just described play critical roles in the action of avian wings, both during flight and when not in use. Their absence in *Archaeopteryx* is clear evidence that despite the modern "winglike" appearance of the feather arrangement, the "flight" apparatus of this ancient bird was far from fully evolved. In fact, all these features as developed in *Archaeopteryx* are essentially the same as in certain small carnivorous dinosaurs, such as *Ornitholestes, Velociraptor,* and *Deinonychus.*

There is one curious exception, though. Two of the specimens of *Archaeopteryx* (the London specimen and the Maxberg specimen) feature a robust boomerang-shaped bone that almost certainly is a furcula, or wishbone. Believed to be formed by fusion of the paired clavicles, the furcula is unique to birds and generally is interpreted as a specialized feature of the avian flight apparatus, possibly acting as a springlike spacer to counteract the transverse compressive component of the force produced by contraction of the flight muscles, which otherwise would tend to squeeze the shoulder sockets toward each other. If *Archaeopteryx* possessed a furcula, then one is tempted to conclude that it had powerful pectoral muscles and therefore it surely must have flown. The "wing" outline seems to add strong support to that conclusion, as Saville (1957), de Beer (1975), and others have noted. But what about all the above-mentioned deficiencies?

The general consensus seems to be that *Archaeopteryx* probably was at least a glider, but no more than a feeble, flapping flier—*if* it flapped at all (Bock 1965, 1969; Yalden 1970). In any case, it is clear that *Archaeopteryx* was a long way from modern birds as far as its flight equipment was concerned.

Another very interesting aspect of

Archaeopteryx is in the construction of the hindquarters. Although the pelvis was not co-ossified into a large synsacrum as in modern birds (does that indicate that it lacked the shock-absorbing capacity of the modern avian landing gear?), the hind legs and feet are much more similar to those of living birds than is the forelimb or shoulder. There can be no question that this animal was an obligatory biped with a posture and gait very much like that of today's birds, but the hindlimb was not yet fully modern.

The foot consisted of three main toes in a near-symmetrical arrangement, with the median toe (the third) the longest and directed forward, almost exactly as in modern birds. The first toe, or hallux, was short and placed behind the second in a fully opposable position, again as in modern birds. But the metatarsals consisted of three main bones, rather than a single co-ossified tarsometatarsus. However, these three metatarsals were fused together at their upper ends in one specimen (the Maxberg specimen) and appear to have been similarly fused in all the other specimens since these bones are not separated from each other in *any* specimen of *Archaeopteryx* (Fig. 4). That would seem to be a clear indication of an early trend, as far back as *Archaeopteryx,* toward the modern avian state of metatarsal co-ossification.

The ankle itself was mechanically quite close to that of living birds, with a simple hingelike joint. Flexion and extension took place at a single level, between the distal tarsal bones, which were fixed against the upper ends of the metatarsals, and the proximal tarsals (astragalus and calcaneum), which fit tightly against the lower ends of the shin (tibia and fibula) bones.

The importance of these conditions is clear: as early as the *Archaeopteryx* stage, evidence exists showing that the functions of the hindquarters and forequarters had evolved entirely independently. In fact, those specimens show that at that early stage, the running apparatus was far more advanced (toward the modern avian condition) than was the flight apparatus. The respective limb proportions (femur vs. tibia vs. metatarsus) also strongly suggest that *Archaeopteryx* was a highly developed cursorial animal, fleet of foot, and probably a ground-dweller.

Figure 5. This reconstruction of *Archaeopteryx* in an arboreal setting was painted by Rudolph Freund in 1965 to illustrate an article on the origin of feathers by K. C. Parkes (1966). Notice the nongrasping hallux—a deliberate rendering by the artist, who stated: "I have tried to show all the controversial aspects of *Archaeopteryx* as clearly as possible (the number of primaries [he shows eight], the use of the claws on the wing, the method of feeding, and the method of flying/gliding) so that subsequent investigators and illustrators will have a target at which to shoot." (Courtesy of K. C. Parkes, with permission of Carnegie Museum of Natural History and *The Living Bird.*)

This last item is one of the major points of contention among students of *Archaeopteryx*. Contrary to almost everyone else's conclusion that *Archaeopteryx* must have been arboreal (Fig. 5), I find no compelling anatomical evidence in any of the specimens that that was so. I agree that it *could* have been arboreal, but the specimens themselves indicate otherwise. This point may seem unimportant, but in fact it is critical for choosing between the two hypotheses of flight origins.

It is widely acknowledged that *Archaeopteryx* lacked virtually all of

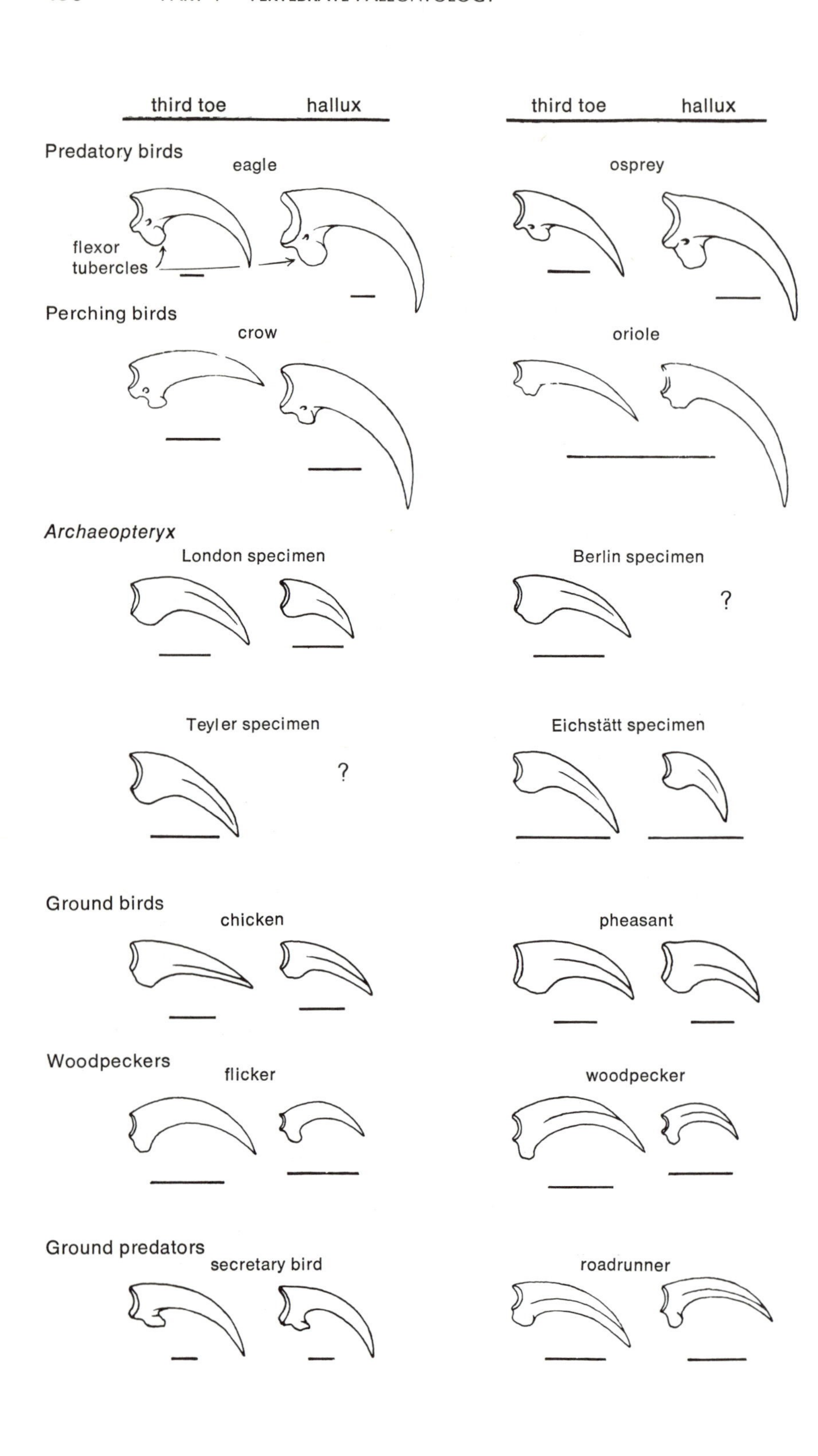

Figure 6. The bony claws, or unguals, from the feet of four of the specimens of *Archaeopteryx* are compared here with the corresponding elements of several types of modern birds. Each example includes the ungual of the median (third) toe and of the hallux. All third-toe unguals are drawn to the same unit length for convenient comparison, and each companion sketch of the hallux ungual is drawn to the same scale to show relative size differences between these two claws; the degree of enlargement or reduction of each claw is indicated by the horizontal lines, all of which equal 5 mm. Orientations are standardized by showing the articular facets in vertical position. In nearly all respects, the unguals of *Archaeopteryx* resemble those of ground-dwelling birds more closely than any of the others: they feature slight curvature, robust form, shallow, poorly defined flexor tubercles, and a hallux claw shorter than the other claw. (From Ostrom 1974.)

the skeletal specializations that are essential parts of the flight apparatus of modern *powered* avian fliers. Yet, the general consensus is that *Archaeopteryx was* at least a glider, which leads to the seemingly logical conclusion that *Archaeopteryx must* have been arboreal, since it obviously could not "glide" from the ground up. In support of this arboreal habit, advocates point to the reversed hallux of *Archaeopteryx,* which, according to them, must have been for grasping, as in all living perching birds. That evidence, however, is not as conclusive as it might seem.

First of all, the hallux of *Archaeopteryx* is situated higher on the metatarsus (see Fig. 4) than is to be expected in a perching or grasping foot. Second, the hallux is relatively short, as compared with the three main toes *and* with the hallux in passerines or predatory birds, where grasping is critical. Third, the bony claw, or ungual, of the hallux is distinctly shorter than that of the median, or third, toe, whereas in modern perching birds and birds of prey it usually is considerably longer than the ungual of the third toe. Finally, and perhaps most important, the shapes of the bony claws of all four toes in *Archaeopteryx* are only moderately curved and have *very weakly developed* flexor tubercles. In all living perching birds and birds of prey all claws are strongly curved and have very prominent flexor tubercles. The relatively low tubercles in *Archaeopteryx* (compared with those of passerines) clearly indicate that the claws could not have been tightly flexed with comparable force. It is significant, I think, that the claws of *Archaeopteryx* most closely resemble the relatively straight, low-tubercled claws of modern ground birds, such as the pheasant, partridge, and quail (Fig. 6). These anatomical facts point to a ground-dwelling existence for *Archaeopteryx,* rather than to life in the trees.

Of course, one can argue that *Archaeopteryx* might have reverted to a ground habit from an arboreal ancestry in which the earliest phases of flight were achieved. But if that were true, then how, or for what, were the "wings" of *Archaeopteryx* used, since the skeletal features seemingly essential to powered flight are lacking?

A hypothesis of preflight evolution

As I mentioned above, the overall nature of the skeleton of *Archaeopteryx* is remarkably similar to that of certain small theropod dinosaurs (coelurosaurs). The similarities include almost identical three-fingered hands, metacarpals and carpals, shoulder region, foot and ankle, parts of the pelvis, and the vertebral column, including a long unbirdlike tail. Some critics claim that these anatomical similarities do not necessarily mean a close phyletic relationship between *Archaeopteryx* and the theropods, arguing that they are the result of convergent or parallel evolution. That is a possibility, but a very remote one, as I have tried to show elsewhere (Ostrom 1976a). There simply are too many detailed similarities. However, regardless of which explanation is accepted, these striking anatomical similarities also mean that these two kinds of animals were similar in many aspects of their life styles (Fig. 7). They must have been doing the same kinds of things.

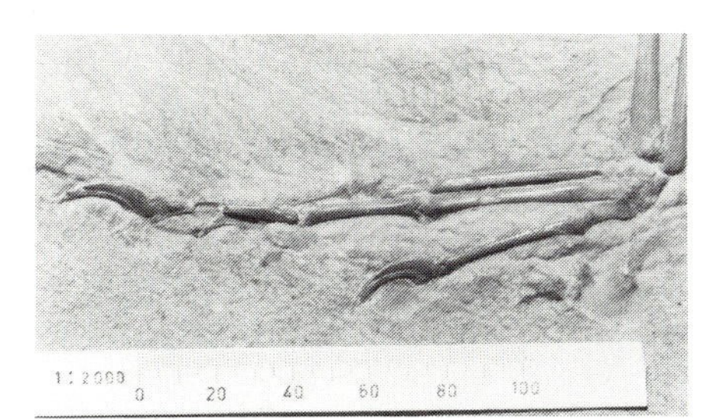

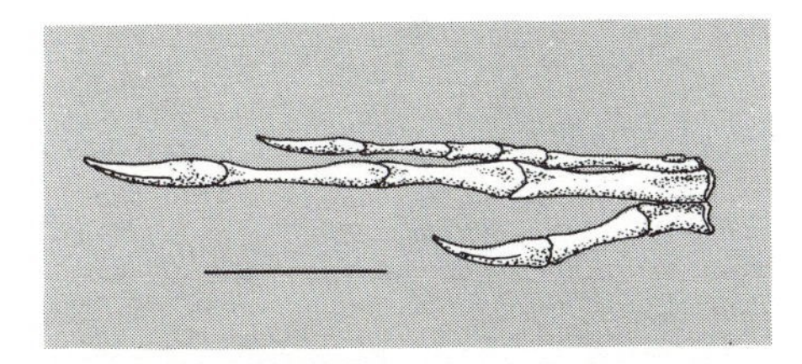

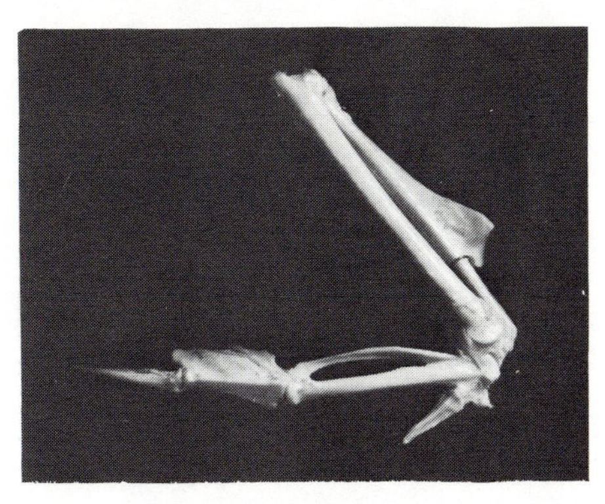

Figure 7. Comparison of the left hand of the Berlin specimen of *Archaeopteryx* (*top*), that of the small theropod dinosaur *Ornitholestes* (*center*; scale bar equals 5 cm), and that of a modern pigeon (*bottom*) shows the striking similarity between *Archaeopteryx* and the dinosaur. (Bottom photo by W. Sacco.)

Small theropods such as *Compsognathus, Ornitholestes,* and *Velociraptor* clearly were fleet-footed bipedal predators. The foot in some (perhaps all) bore a fully or partly reversed hallux, but that certainly was not for arboreal activity. The long arms and three-fingered, strongly clawed hands obviously were for grasping and, in view of their sharp piercing teeth, probably for seizing prey, not branches. For theropods as a whole, prey size probably ranged from cat-size up. As for *Archaeopteryx,* the tiny sharp teeth in its jaws make it clear that it too was a predator. But its small size, about that of a common pigeon, indicates that it lived on much smaller prey, most probably insect-size organisms. The image that comes to mind is that of a fast-running, bipedal predator flushing small animals or insects and running them down. In short, *Archaeopteryx* appears to have been a miniature theropod—except that it had feathers.

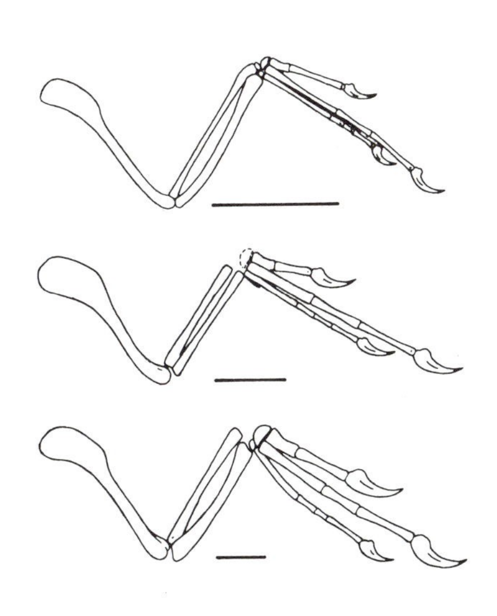

Figure 8. Sketches of the right arm and hand of *Archaeopteryx* (*top*) and of the theropods *Ornitholestes* (*center*) and *Deinonychus* (*bottom*) show the striking anatomical similarities. The humeri are drawn to the same length to minimize size-related differences; true sizes are indicated by the scale bars, which equal 5 cm. (From Ostrom 1976a.)

The most critical components of the avian flight apparatus are the flight feathers, which are greatly enlarged versions of the contour-type feathers that cover the rest of the body in modern birds. In today's birds, these body contour feathers provide extremely efficient insulation, minimizing heat loss when ambient temperatures are lower than body temperatures. But on the wings and tail, they form airfoils producing lift and propulsion.

We don't know what the original selective value of primordial feathers might have been. It has been suggested that they originated in conjunction with flight (Parkes 1966), as heat shields (Regal 1975), for coloration (Mayr 1960), and as insulation to retain body heat (Ostrom 1974). Whatever the original function, it seems most likely that feathers appeared and evolved as a general body covering, rather than in isolated patches, and most probably were related to epidermal functions or properties. If this presumption is correct, then our quest for the original factors behind the origin of bird flight can be rephrased, in part, to: What were the reasons for modification of small contour feathers on the arms and tail into greatly enlarged contour "flight" feathers? Was it to increase the surface area to slow the rate of fall? Or could there be some other explanation?

The arboreal theory, as described by Bock, depicts a bipedal ancestor moving from the ground into the trees, becoming increasingly arboreal and perfecting the ability to leap from branch to branch and tree to tree. The specimens of *Archaeopteryx* clearly indicate that bipedal carriage preceded evolution of the flight machinery. Indeed, it is that prior development of bipedality that released the forelimbs from the normal business of support and locomotion, making them available for other activities. In theropods, especially the small coelurosaurs, the ultimate (and perhaps the original) activity was for grasping prey. The specimens of *Archaeopteryx* seem to indicate that that probably was true of the ancestors of *Archaeopteryx*—and perhaps of *Archaeopteryx* as well, since its forelimbs are virtually identical to those of some coelurosaurian theropods (Fig. 8).

The arboreal theory is a very logical model; otherwise it would not have been so widely accepted. But it does seem to have one major weakness: it is predicated on the invasion of an arboreal niche by an obligatory bipedal ancestor (Tucker 1938). That

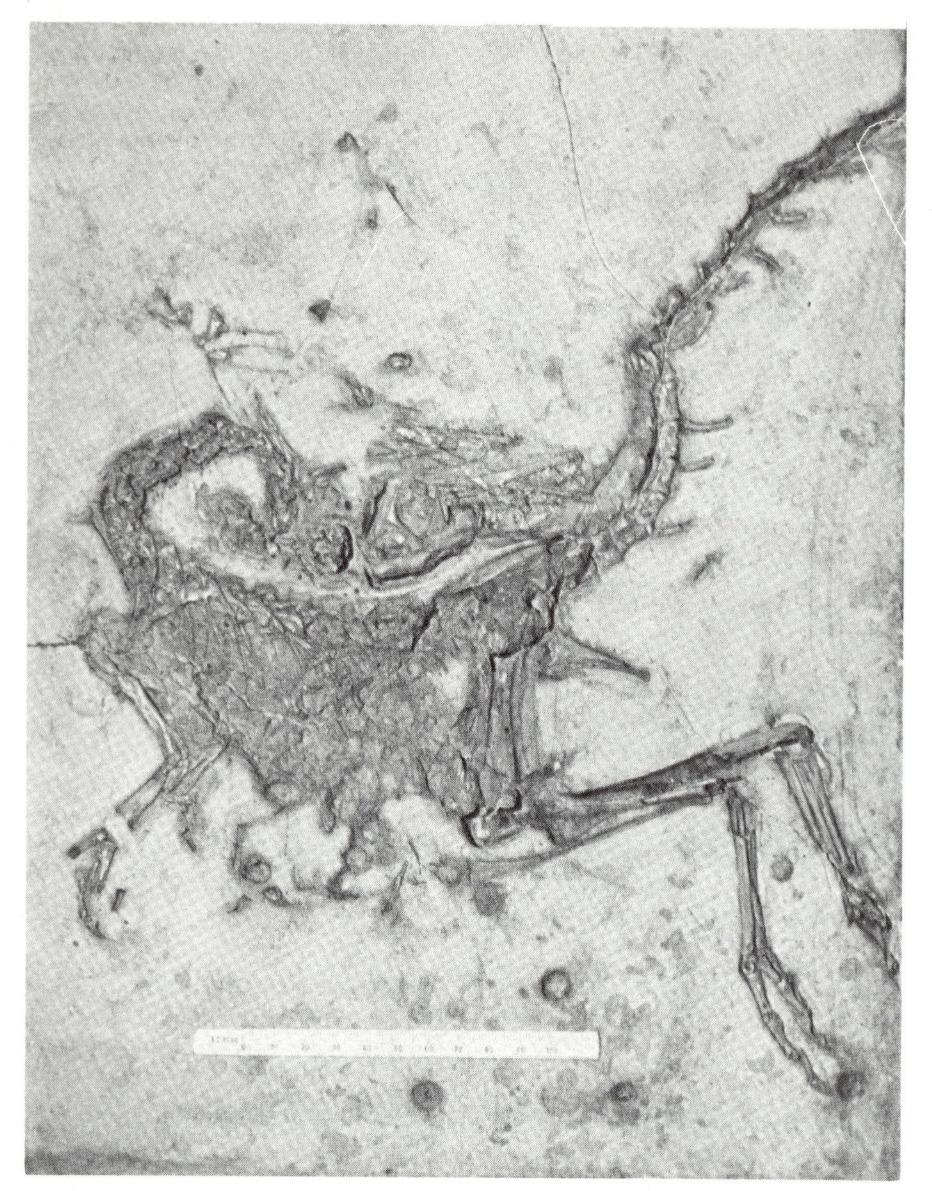

Figure 9. This pigeon-sized specimen of *Compsognathus,* the smallest known carnivorous (theropod) dinosaur, came from the same Solnhofen limestone that produced all the specimens of *Archaeopteryx.* The head has been twisted back over the body. Except for the shorter arms and hands, it is very similar to *Archaeopteryx* (Fig. 3).

may not seem odd at first, because so many modern birds are tree-dwellers and they are bipedal. In fact, birds are the most common of all arboreal vertebrates. But in modern birds, the wing is fully perfected for flight, and the feet are all that remain for grasping *any* perch. When we look around for other arboreal creatures, almost without exception they are quadrupedal—with all four limbs adapted to provide maximum clinging capacity. Furthermore, several kinds have even modified the tail into an additional clinging appendage. The tree kangaroos (*Dendrolagus*) would seem to be an exception as "bipedal" tree inhabitants, but, unlike other kangaroos, they have forelimbs and hindlimbs that are nearly equal in size, and all four feet are modified for clinging (Rothschild and Dollman 1936). Apparently all four limbs are used in moving about in the trees.

In short, it seems highly unlikely to me that a bipedal (and presumably cursorial) animal would shift to an arboreal habitat and still retain obligatory bipedal stance—unless the forelimbs were *already* highly adapted for some critical, nonsupport, nonclinging action. Even given the unlikely invasion of the tree world by an obligate biped, if this lineage evolved into more and more proficient leapers (as the model specifies), I would expect the forelimbs to have been perfected more and more to the task of grasping the next landing spot, and not to have evolved into a parachute, or a wing. Or, if the forelimbs of this obligatory bipedal pro-avis had been used to assist in climbing, why did the lineage not revert to quadrupedal clinging?

The distinctly independent actions and functions of the hind and fore quarters of modern birds (and of *Archaeopteryx* as well) indicate complete functional separation of the two ever since bipedal locomotion was first achieved in bird ancestry. I suggest that, unlike other vertebrate fliers (which remained quadrupedal), birds became arboreal *after* they had developed the power of flight—*not before.* I further suggest that the unique and highly diverse nature of avian flight was possible *only* because it was preceded by uncompromised, obligatory bipedal posture and locomotion that allowed the forequarters in a proto-*Archaeopteryx* stage to be adapted for other entirely separate and unrelated activities. The question remains, though: How did this lead to flight? What was the original selective advantage of enlarging the initially small contour feathers that covered the arms and hands?

It seems to me that two important clues point to how, or why, this feather enlargement came about. First, the small size of *Archaeopteryx* and the presence of tiny *sharp* teeth in its jaws are surely solid evidence that *Archaeopteryx* was insectivorous. Second, the coelurosaurian dinosaurlike design of the forelimb and shoulder, and especially the hand and wrist, suggests a coelurosaurlike use—at least in its ancestors (although it is so nearly identical in *Archaeopteryx* that I find it difficult not to believe a coelurosaurlike action there too, despite the presence of feathers). These lead to the thought that there may have been a behavioral connection between this coelurosaurlike predatory design of the forelimb and the development of enlarged "flight" feathers.

If the "proto-*Archaeopteryx*" lineage evolved from a somewhat larger predaceous coelurosaurian stock into a *smaller* animal insectivore niche (see Figs. 9 and 10), and if these animals used the three-fingered hands to seize ever smaller and smaller prey, any modifications that would improve insect-catching ability would be highly advantageous. Is it possible that the initial (pre-*Archaeopteryx*) enlargement of feathers on those narrow hands might have been to increase the hand surface area, thereby making it more effective in catching insects? Continued selection for larger feather size could have converted the entire forelimb into a large, light-weight "insect net." It is not difficult to visualize how advantageous these paired "insect nets" would be in snaring leaping insects, or even in batting down escaping flying insects. The "proto-wing" at first may have been used in sweeping movements to flush insects concealed in ground cover or in low vegetation, but subsequent enlargement of the forelimb surface area seems best explained as increasing the snaring range—reaching those insects that were flushed beyond the range of the mouth. From simple sweeping arm movements, to flapping swipes at leaping or flying insects, to flapping leaps up after escaping insects seems to be a logical sequence within the insectivorous precursors of *Archaeopteryx*.

From this beginning, it would seem to require relatively small evolutionary steps leading to brief flapping flights after fleeing insects, with selection continuing to improve the adductive powers of the pectoral muscles, to firm up the shoulder skeleton, and to align and firmly implant the "wing" feathers, with the final achievement of an *Archaeopteryx* grade and the beginnings of powered flight. There is no question in my mind that *Archaeopteryx* was well beyond the simple leaping, insect-catching stage and perhaps was at the threshold of powered flight. Its wing form, so similar to the elliptical wing form (Saville 1957) widespread among modern birds, is strong evidence that *Archaeopteryx* probably was at least a weak, flapping flier. But it seems unlikely that its flight proficiency was enough to catch in flight the various large flying insects so plentiful during Solnhofen time (Late Jurassic). The structural changes necessary to make it a skillful powered flier, even comparable to the contemporaneous pterosaurs, were still great.

When I first proposed the "cursorial predator" theory several years ago (1974), it generated quite a bit of interest but apparently very little acceptance. As one friendly critic put it, "The enlarged 'insect-nets' would produce so much increased drag that they would work against the cursorial running skills." I don't find this a serious problem because, more than likely, the "nets" would have been held in streamlined position during the chase and not employed as snares until within range of the prey (see Fig. 10). It is also possible that the expanded snaring range would have more than compensated for any loss of running speed resulting from increased drag. Possible too is increased leg muscle power to overcome what-

Figure 10. The author's attempt at reconstruction of a hypothetical stage in the early evolution of birds shows a pre-*Archaeopteryx* stage (*top*) and *Archaeopteryx* (*bottom*). Proto-*Archaeopteryx* illustrates an early stage in the enlargement of feathers on the hands and arms as aids in catching insects. The enlarged tail feathers are hypothesized as aerodynamic stabilizers enhancing agility and quick manuevering during the chase after prey. *Archaeopteryx* presumably was at or just past the threshold of powered flight, but is drawn here in a similar predaceous pose. Body, limb, and feather proportions on the wings and tail of *Archaeopteryx* are based on the Berlin specimen. Proportions of proto-*Archaeopteryx* are based on *Archaeopteryx* specimens and the theropod *Ornitholestes*. Scale bars = 5 cm.

ever added resistances were encountered.

More serious, perhaps, is the widely held view that the excessive energetic cost for a proto-*Archaeopteryx* to flap itself up off the ground makes any cursorial theory most unlikely. That view may not be correct, though. Clark (1977) has commented:

> If birds are derived from bipedal archosaurs, and the forelimbs developed strong powers of adduction in their capacity as insect nets, then it is conceivable that the first fliers may have had muscles capable of the power outputs required for hovering or low speed flight.

We can add to this that the earliest ascents were not powered by the forelimbs alone, but almost certainly were initiated by powerful leaps generated by the strong hindquarters. Clark goes on to say:

> However, if *Archaeopteryx* was capable of brief periods of hovering or of ascending flight (for which power requirements are also quite high) then it should immediately have been capable of sustained level flight at moderate speeds, as the power requirements for this are minimal.

My cursorial predator scenario leading up to *Archaeopteryx* is speculative to be sure, but no more so than the arboreal theory, given the anatomy of *Archaeopteryx* and modern birds. Enough points of evidence contradict, or raise serious questions about, a preflight arboreal stage in the history of birds to give us cause to reconsider carefully whether the "up from the ground" origin of bird flight is so improbable after all. Make no mistake: I emphatically reject the Williston–Nopcsa hypothesis that has the proto-wings developing as propellers to increase running speed. But I do believe that the predatory design of the wing skeleton in *Archaeopteryx* is strong evidence of a prior predatory function of the proto-wing in a cursorial proto-*Archaeopteryx*, which turned out to be preadaptive for flight. It is conceivable that a predatory function of the forelimb still existed in *Archaeopteryx* and that it was this adaptation that led to the unique nature of avian flight.

References

Augusta, J., and Z. Burian. 1961. *Prehistoric Reptiles and Birds.* Prague: Artia Publishers.

Bock, W. J. 1965. The role of adaptive mechanisms in the origin of higher levels of organization. *Syst. Zool.* 14:272–87.

———. 1969. The origin and radiation of birds. *Ann. New York Acad. Sci.* 167:147–55.

Brown, R. H. J. 1948. The flight of birds. I. The flapping cycle of the pigeon. *J. Exp. Biol.* 25:322–33.

———. 1951. Flapping flight. *Ibis* 93:333–59.

———. 1953. The flight of birds. II. Wing function in relation to flight speed. *J. Exp. Biol.* 30:90–103.

———. 1961. The power requirements of birds in flight: Vertebrate locomotion. *Zool. Soc. London Symp. 5:95–99.*

———. 1963. The flight of birds. *Biol. Rev.* 38:460–89.

Clark, B. D. 1977. Energetics of hovering flight and the origin of bats. In *Major Patterns in Vertebrate Evolution,* ed. M. K. Hecht, P. C. Goody, and B. M. Hecht, pp. 423–25. NATO Advanced Study Institute Series, Series A: Life Sciences, Vol. 14. Plenum Press.

de Beer, G. 1975. The evolution of flying and flightless birds. *Oxford Biology Readers* 68:1–16. Oxford Univ. Press.

Greenewalt, C. H. 1960a. The wings of insects and birds as mechanical oscillators. *Proc. Amer. Philos. Soc.* 104:605–11.

———. 1960b. *Hummingbirds.* Doubleday.

———. 1962. Dimensional relationships for flying animals. *Smithson. Misc. Coll.* 144 (2):1–46.

———. 1975. The flight of birds. *Trans. Amer. Philos. Soc.* 65(4):1–67.

Hartman, F. A. 1961. Locomotor mechanisms of birds. *Smithson. Misc. Coll.* 143 (1):1–91.

Marsh, O. C. 1880. *Odontornithes: A Monograph on the Extinct Toothed Birds of North America.* Vol. 7, Report of the geological exploration of the fortieth parallel. Prof. Papers of the Engineer, Dept. U.S. Army, No. 18, Washington, DC. 201 pp.

Mayr, E. 1960. The emergence of evolutionary novelties. In *The Evolution of Life* I, ed. S. Tax, pp. 349–80.

Mivart, St. G. 1871. *Genesis of Species.* Macmillan.

Nopcsa, F. 1907. Ideas on the origin of flight. *Proc. Zool. Soc. London.* pp. 223–36.

———. 1923. On the origin of flight in birds. *Proc. Zool. Soc. London.* pp. 463–77.

Ostrom, J. H. 1973. The ancestry of birds. *Nature* 242:136.

———. 1974. *Archaeopteryx* and the origin of flight. *Quart. Rev. Biol.* 49:27–47.

———. 1975. The origin of birds. *Ann. Rev. Earth Planet. Sci.* 3:55–77.

———. 1976a. *Archaeopteryx* and the origin of birds. *Biol. J. Linn. Soc. London* 8:91–182.

———. 1976b. Some hypothetical anatomical stages in the evolution of avian flight. *Smithson. Contrib. Paleobio.* 27:1–21.

Parkes, K. C. 1966. Speculations on the origin of feathers. *The Living Bird* V:77–86.

Pennycuick, C. J. 1960. Gliding flight of the fulmar petrel. *J. Exp. Biol.* 37:330–38.

———. 1968. Power requirements for horizontal flight in the pigeon, *Columba livia. J. Exp. Biol.* 49:527–55.

———. 1971. Gliding flight of the white-backed vulture, *Gyps africanus. J. Exp. Biol.* 55:13–38.

———. 1975. Mechanics of flight. In *Avian Biology,* ed. D. S. Farner and J. R. King, vol. 5, pp. 1–75. Academic Press.

Regal, P. J. 1975. The evolutionary origin of feathers. *Quart. Rev. Biol.* 50:35–66.

Rothschild, L. W., and G. Dollman. 1936. The genus *Dendrolagus. Trans. Zool. Soc. London* 21(6):477–551.

Saville, D. B. O. 1957. Adaptive evolution in the avian wing. *Evol.* 11:212–24.

Tucker, B. W. 1938. Functional evolutionary morphology: The origin of birds. In *Evolution,* ed. Sir G. de Beer, pp. 321–36. Oxford: Clarendon Press.

Tucker, V. A. 1969. The energetics of bird flight. *Sci. Amer.* 220:70–78.

———. 1973. Bird metabolism during flight: Evaluation of a theory. *J. Exp. Biol.* 58: 689–709.

Williston, S. W. 1879. Are birds derived from dinosaurs? *Kansas City Rev. Sci.* 3:457–60.

Yalden, D. W. 1970. The flying ability of *Archaeopteryx. Ibis* 113:349–56.

J. W. Osborn

The Evolution of Dentitions

The study of evolution suggests how the development of mammalian dentitions may be controlled

Because teeth contain the most durable of biological tissues they have left the best, albeit patchily distributed, record of vertebrate evolution. Even for the history of man, who is a newborn infant in 500 million years of vertebrate evolution, teeth provide the most complete record, with much of human paleontology being based on detailed measurements of the sizes and shapes of teeth. Sometimes, as in the famous Piltdown forgery, the sequence in which teeth erupt becomes important. The wear on the Piltdown canine tooth indicated a relatively early eruption time compared with the molars. This supported the theory that the Piltdown skull had human affinities because, in apes, the permanent canine erupts relatively later.

J. W. Osborn is a Reader in Anatomy in relation to Dentistry at London University. Following his dental degree he obtained the Fellowship in Dental Surgery of The Royal College of Surgeons while building a part-time dental practice. However, in 1964 he became a full-time anatomist, his major research interest being the structure and development of tooth enamel. He has recently published a lengthy review of his work in this field. In 1969 he began investigating the evolution of enamel but changed his interest when F. R. Parrington, of Cambridge University, and A. W. Crompton, of Harvard University, (with whom he now works "across the pond") explained to him the intriguing problems involved in changing from a reptilian to a mammalian dentition. In 1971 he was appointed Visiting Agassiz Professor in Vertebrate Paleontology at Harvard University. He is currently engaged in the problem of the controls required to develop the shapes of teeth—once more leaning on paleontology. The author wishes to thank Andrew Sita-Lumsden for many fruitful discussions. Address: Anatomy Department, Guy's Hospital Medical School, London Bridge SE1 9RT, England.

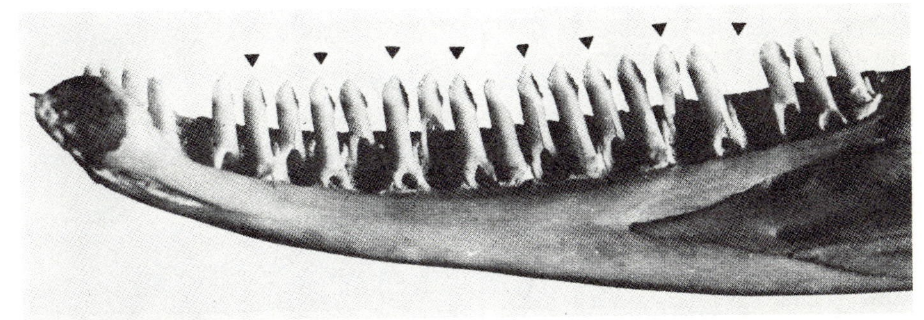

Figure 1. The dentition of a lizard, in which teeth continue being replaced throughout life. All the replacement teeth have been lost from this dried jaw, but the extent to which each was developed matches the size of the hole it occupied beneath its predecessor. Teeth are replaced in waves which sweep from the back to the front of the jaw through alternate tooth positions. In this jaw an erupting tooth at position 21 has been lost, but the forward-running replacement wave is obvious (*triangles*). The other replacement waves can readily be traced.

Like man and apes, every species of mammal is characterized by a sequence in which teeth erupt into the mouth. Compared with those of reptiles, these sequences are usually very irregular: for example, in man, if the permanent teeth within a jaw quadrant are numbered 1 to 8 from the front, they erupt in the sequence 6, 1, 2, 4, 5, 3, 7, 8, with minor variations. Because this sequence is repeated in all humans it is evidently genetically controlled, but its particular significance in terms of development, function, and evolution is not known. To unravel this problem it is necessary to trace the evolution of recent mammals from their reptilian ancestors of the Permian, about 300 million years ago.

Unlike mammals, which are heterodont (the teeth have different shapes) and diphyodont (they have only 2 dentitions—milk, or deciduous, and permanent), reptiles are usually homodont (all the teeth are alike) and polyphyodont (the teeth are constantly being replaced throughout life). It has been calculated (*1*) that an elderly crocodile may have replaced its front teeth about 50 times. It is intriguing that the teeth of most submammalian vertebrates, including reptiles, are replaced in the following complex but regular sequence. Suppose that one side of a jaw contains 20 teeth and that a new tooth now erupts at the back of the jaw to become the first tooth to occupy the 21st tooth position. Soon after this the odd-numbered teeth are replaced in the sequence 19, 17, 15, 13, and so on. At about this time a tooth appears in the 22nd position, to be followed by replacement of teeth 20, 18, 16, 14, and so on. These replacement waves continue to be initiated and to sweep from the back to the front of the jaw through alternate positions throughout life (Fig. 1). However, in some animals (for example most snakes) the replacement waves sweep through alternate positions in the opposite direction, from the front to the back of the jaw.

Because mammals evolved from reptiles it is evident that, somewhere along the line leading toward recent mammals, the eruption sequence changed. By studying the fossil record between reptiles and mammals it should be possible to plot the way in which mammalian eruption sequences evolved. And if we can understand how the regular reptilian eruption sequences are controlled, it should then be possible to understand how the irregular mammalian eruption sequences are controlled.

Three theories have been put forward to explain the general relationship between the dentitions of reptiles and mammals: the *Odontostichi* (*2*), *Zahnreihe* (*3*), and Tooth Family (*4*) theories. These theories will now be considered.

Three dentition theories

The *Odontostichi* theory was originally concerned with tooth replacement in the elasmobranchs. Bolk in 1922 (*2*) noted that in these cartilaginous fish, the even-numbered teeth are all replaced at about the same time, later to be followed by replacement of the odd-numbered teeth and so on (Fig. 2). He recognized two types of dental units, an *odontostichos*, which is a "horizontal" unit consisting of alternate teeth, and a tooth family, which is a "vertical" unit, consisting of all the teeth which successively occupy a tooth position (Fig. 3A). Bolk concluded that the pattern of tooth replacement in elasmobranchs is due to the existence of two types of *odontostichi*, the even and the odd-numbered *odontostichi*, which he incorrectly concluded had significantly different developmental sites. All the tooth families within an *odontostichos* contribute their replacement teeth to the dentition at the same time, thereby accounting for the strict alternation with which teeth are replaced in elasmobranchs.

Bolk now concluded (without evidence) that reptiles replace teeth in the same pattern as elasmobranchs. And he further concluded (again without evidence) that diphyodont mammalian dentitions have retained the two *odontostichi* present in reptiles (and fish and amphibia): one *odontostichos* has been pushed together to produce the deciduous teeth and the permanent molars, and the other *odontostichos* has been pushed together to produce the replacement teeth (Fig. 3B). This theory now seems most improbable because it does not account for the sequences in which teeth are initiated and erupt in either reptiles or mammals.

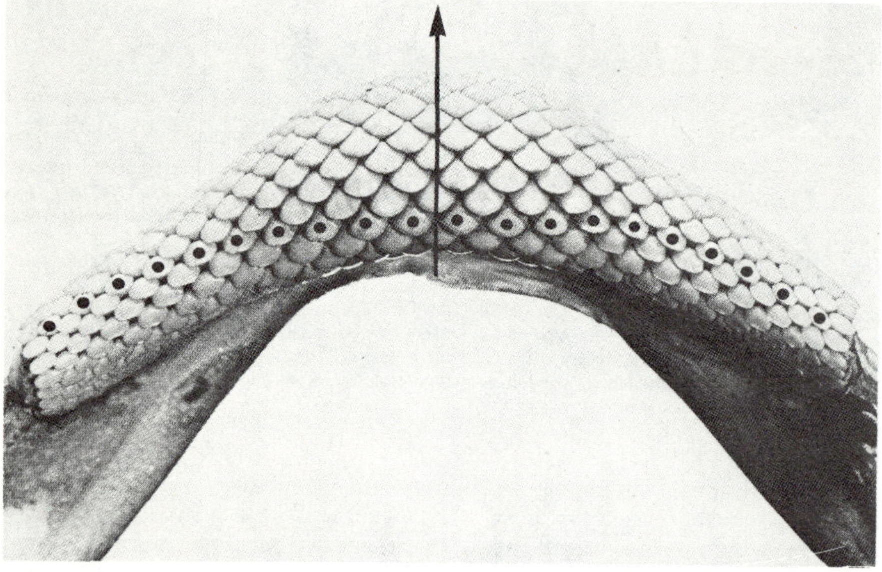

Figure 2. In the lower jaw of an elasmobranch fish, each tooth rolls over the jaw in the direction of the arrow and is finally shed. The alternation between rows of teeth (*odontostichi*) produces a regular pattern. One *odontostichos* is indicated by dots.

We now turn to the *Zahnreihe* theory. Wave replacement of alternate teeth, as opposed to the strict alternation observed by Bolk, was originally described in a mammal-like reptile of the early Triassic (*5*). It was later described in Permian pelycosaurs (*6*) and subsequently extensively documented for fossil and recent submammalian vertebrates by Edmund (*3*).

It must be noted that Edmund's own observations were confined to replacement waves in adult dentitions. However, Woederman (*7*) had already described the development of the dentition in crocodile embryos. Basing his conclusions on Woederman's data, together with some previous observations on tooth development in fish, Edmund constructed the *Zahnreihe* theory (in 1960) to account for the wave replacement of alternate teeth.

He supposed that there exists at the front of a reptilian jaw a region in which impulses are generated at regular intervals of time (Fig. 4A). Each impulse leads to the propagation of a stimulus which sweeps from the front to the back of the jaw at a regular speed. When the stimulus reaches a tooth position it causes a new tooth to be initiated. Surprising though it may seem, despite the fact that rows of teeth are initiated in sequence from front to back, such a system can readily lead to the wave replacement of alternate teeth from the back to the front of the jaw (Fig. 4A). Since its formulation, Edmund's *Zahnreihe* theory (each row of teeth developed in response to a stimulus is called a *Zahnreihe*) has dominated research into the control of tooth replacement.

Edmund (*3*) now suggested that mammalian dentitions are equivalent to two reptilian *Zahnreihen* (Fig. 4B). This seemed to provide the answer to the relationship between reptilian and mammalian dentitions until it was pointed out (*8, 9*) that mammals neither develop nor erupt teeth in the front to back sequence predicted by the *Zahnreihe* theory. And later (*10*) it was also shown that no study of tooth development in any vertebrate embryo (including reptiles) supports the theory either.

The Tooth Family theory will only be briefly outlined at this point. It supposes that the tooth family rather than the tooth row or *Zahnreihe* is the important unit of dentitions—in other words, that the rate of tooth replacement is "vertically" controlled within each tooth family rather than "horizontally" along tooth rows. In order to understand the theory it is necessary to present the data on dental development in reptiles.

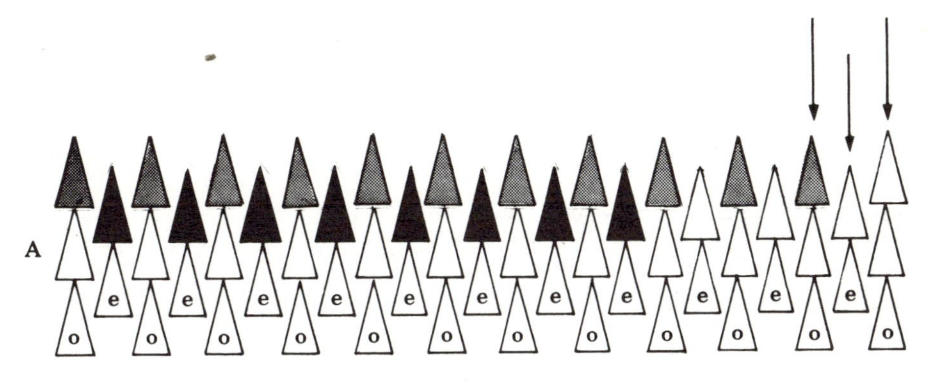

Figure 3. Bolk's *Odontostichi* theory (2) attempted to explain the evolution of mammalian dentitions. (A) The jaws of elasmobranchs contain alternating rows of teeth called *odontostichi* (see Fig. 2); an even-numbered (e) and odd-numbered (o) *odontostichos* are shown. The arrows point to 3 tooth families. **(B) Bolk supposed that one *odontostichos (shaded)* was pushed together in the evolution of the deciduous (replaced) teeth and the 3 permanent (unreplaced) molars, while the next *odontostichos (black)* was pushed together to produce the replacement teeth.**

Dental development in reptiles

The embryogenesis of reptilian dentitions has been described for *Sphenodon* (*11*), the crocodile (*7*), and the common lizard, *Lacerta vivipara* (*10*). Of these studies the last is the most comprehensive because every developing dentition (totaling 25 lower jaw quadrants) was reconstructed in full; in neither of the other studies were the dentitions reconstructed. From my own data (*10*) I concluded that it is very difficult, if not impossible, to recognize equivalent tooth germs in different embryos unless the whole dentition is reconstructed.

The jaws of *Lacerta vivipara* hatchlings are about 2 mm long, each quadrant containing about 30 teeth at different stages of development (Fig. 5D). It appeared certain that the precision of the tooth replacement waves in the hatchling was achieved by an equivalent precision during development. But such precision makes it difficult to account for the tiny rudimentary teeth which are frequently produced in the apparently random positions 3, 5, 8, 10, and 13 (Fig. 5D). The solution to this problem became apparent only when the teeth in all the reconstructions had been correctly numbered. The probable sequence of development is shown in Figure 6.

The first tooth bud to be initiated is at the back of the embryonic jaw in what ultimately becomes position 11. This first bud will be called the "dental determinant." From the dental determinant, buds are initiated in sequence toward the front and the back of the jaw. However, those buds anterior to the dental determinant are progressively separated by interstitial growth of the embryonic jaws, with the result that space is created between them for the later development of an intervening row (Fig. 5A, B). Thus, the first row becomes separated and ultimately occupies positions 11, 9, 7, 5, and 3 (buds a in

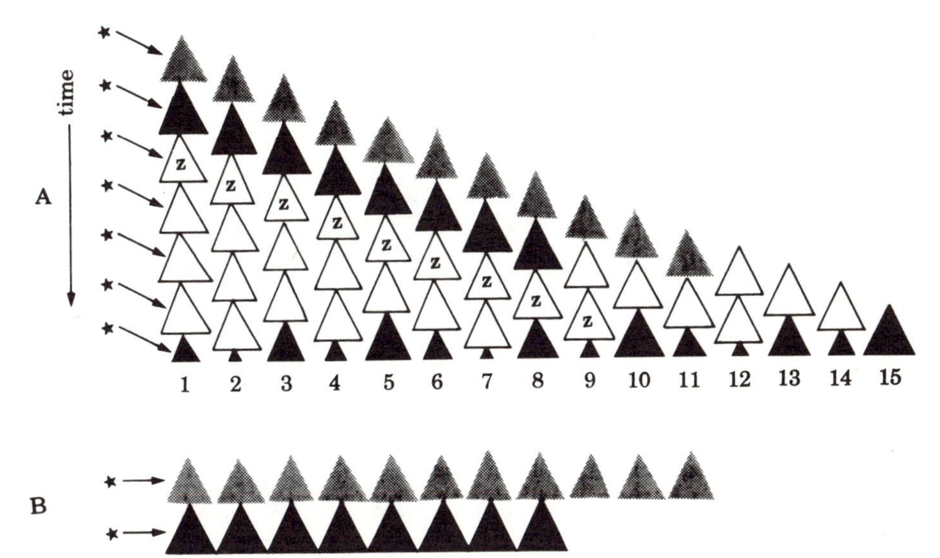

Figure 4. Edmund's *Zahnreihe* theory (*3*) to explain the evolution of mammalian dentitions. (A) In submammalian vertebrates impulses (*starred*) are regularly initiated at the front of the jaw. Each impulse propagates a stimulus which travels back through the jaw initiating a new tooth (triangle) at each tooth position in sequence from the front to the back of the jaw. (Time is represented by a vertical axis.) Such a row is referred to as a *Zahnreihe* (the 3rd *Zahnreihe* is indicated by the letter z). Consider the bottom row of apparently randomly sized black teeth, each size representing the age of a tooth at the time when the 7th *Zahnreihe* has just been initiated. The tooth at position 15 forms a graded series with those at positions 13, 11, 9, and 7. Another wave starts at position 5, and another at position 10. The model demonstrates how the *Zahnreihe* theory explains the origin of the wave replacement of alternate teeth from the back to the front of the jaw. (B) Edmund suggested that the dentitions of mammals represent **the remains of 2 reptilian *Zahnreihen (shaded and black teeth in A)*. This appears to predict that the teeth of mammals should be initiated in sequence from the front to the back of the jaw.**

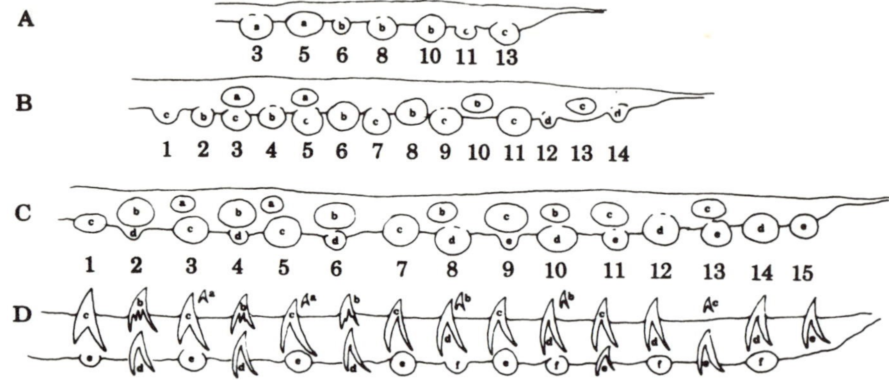

Figure 5. Scale diagrams of 4 stages of dental development in the common lizard, *Lacerta vivipara* (after Osborn, *10*). D represents a hatchling. The small teeth at positions 3, 5, 8, 10, and 13 in D are rudimentary teeth. The letters on each tooth indicate the alternating row to which the tooth belongs (c.f. Fig. 6). Thus the teeth in dentition A belong to 3 different rows—a, b, and c. The rudimentary buds which regress before any mineralized tissues have been deposited (the black teeth in Fig. 6) have already been lost from positions 7, 9, 11, and 12 of dentition A (the bud at position 11 is a replacement belonging to the 3rd row, c). The buds at positions 3, 5, 8, 10, and 13 of dentition A become the rudimentary teeth of the hatchling (D). Room is made for the intervening buds 4, 7, 9, and 12 of dentition B by interstitial jaw growth separating the buds of dentition A. In the hatchling (D), teeth are being replaced in waves which sweep from the back to the front of the jaw through alternate tooth positions: rows c, d, and e are complete, while only 4 buds from row f have as yet been initiated. The 3 anterior teeth of row b are already being resorbed prior to replacement by teeth from row d.

Figs. 5 and 6), while the second intervening row occupies positions 12, 10, 8, 6, 4, and 2 (buds b in Figs. 5 and 6). Interstitial growth of the jaw now ceases, but anteriorly a little further room is created for a final tooth to develop with the 3rd row; this is position 1 (Fig. 5B).

It is now possible to explain why rudimentary teeth are produced in positions 3, 5, 8, 10, and 13. This distribution seems to be related to the time required for the tooth-forming tissues to achieve the competence (in an embryological sense) to contribute to tooth development. During the formation of the earliest buds, none of the tissues is yet capable of laying down mineral; therefore the earliest buds (11a, 9a, 7a, and 12b in Fig. 6) are rudimentary and regress (they are not seen in Fig. 5A). The mesodermal cells of later initiated buds become competent to lay down dentine, but the associated enamel organs cannot yet lay down enamel. The teeth initiated during this time are rudimentary teeth: those which ultimately occupy positions 3, 5, 8, 10, and 13 (3a, 5a, 8b, 10b, 13c in Fig. 6). However, positions 3, 5, and 10 sometimes contain rudimentary buds, and occasionally position 6 contains a rudimentary tooth. In all later developed buds both the dental papilla and the enamel organ are competent to lay down dentine and enamel respectively: all these buds become fully developed teeth.

It remains to explain how the above sequence is controlled. The back-to-front sequence in which the buds of the first tooth row (buds 11, 9, 7, 5, and 3) are initiated is related to the direction in which neural crest cells migrate into the developing jaws. These neural crest cells, which have been shown (*12* and *13* among others) to be responsible for initiating tooth buds, migrate forward from the developing neural tube through the jaws.

Finally, it is necessary to explain why tooth buds are evenly separated. It seems probable (*10*) that a tooth bud generates around it a zone of tissue in which the initiation of another tooth bud is temporarily inhibited; were this not the case, one massive fused tooth would be produced which would extend the whole length of the jaw, be-

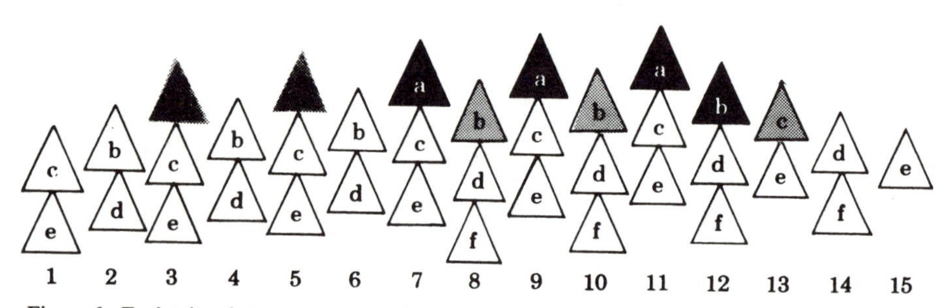

Figure 6. Each triangle represents a tooth germ in the developing dentition of *L. vivipara* (c.f. Fig. 5). Time passes from above downwards. The earliest initiated tooth germs *(black)* are rudimentary buds, those initiated a little later *(shaded)* are rudimentary teeth. From the dental determinant (tooth a at position 11) tooth families are initiated in alternation forward (9a, 7a, 5a, 3a) and in sequence backward (12b, 13c, 14d, 15e).

cause all the relevant jaw tissues (even the lip ectoderm, *14*) are potentially competent to take part in the initiation of teeth. The separation of the first row of tooth buds due to interstitial jaw growth (Fig. 5A, B) allows the intervening tissues to escape from the inhibitory zones around each of the buds in the first tooth row. The sequential (as opposed to alternate) initiation of buds behind the dental determinant (11a, 12b, 13c, 14d, 15e in Fig. 6) is due to the presence of tip rather than interstitial growth of potential tooth-forming tissues in this region.

We have now accounted for the sequence in which tooth families are initiated in a homodont reptile. It is evident that the observed sequence (Fig. 6) cannot be explained by either the *Odontostichi* (Fig. 3) or *Zahnreihe* (Fig. 4) theories. However, the Tooth Family theory, with which they have been compared, was originally proposed (*8*) to account for the manner in which the wave replacement of alternate teeth is controlled, not the sequence in which tooth families are initiated.

Tooth replacement

In order to understand the control of tooth replacement it is necessary to understand why teeth are replaced. The most commonly accepted reason for the multiple tooth replacements in reptiles is that they compensate for the wear and accidental loss of teeth. However, although the probably careless and imprecise eating habits of reptiles do often lead to accidental tooth loss, this is unlikely to be the reason for mammalian tooth replacement because the firmly rooted teeth of mammals are very rarely lost accidentally.

What is more likely is that tooth replacement is part of a growth process, at least in mammals. The small teeth that can be accommodated in the small jaws of a child would be quite inadequate in the large jaws of the adult. Unfortunately, with few exceptions such as the rodent incisor and some canines, teeth cannot grow once they have erupted, and so in order to increase tooth size it is necessary to replace a smaller tooth by a larger tooth. I suggest that the same explanation is true for the multiple tooth replacements in submammalian vertebrates. A young crocodile, for example, requires small teeth. Throughout life its lengthening jaws require progressively larger teeth if the two are to be efficiently matched. It must therefore replace its teeth throughout life.

If the above explanation is true of all submammalian vertebrates (and it is supported by the fact that the teeth which they shed with such apparent waste are often remarkably little worn), then why do mammals not replace their teeth more frequently? This question is particularly relevant when it is realized that the dentition is one of the weakest "links" in the "chain" of most mammals' lives: failure of the dentition rapidly leads to death. To answer the above question it is necessary to look at the differences between tooth function in mammals and reptiles.

The upper and lower teeth of most reptiles cannot be brought close together when the jaws are closed because the lower teeth are set inside the upper teeth and the lower jaw cannot be moved sideways sufficiently to bring them into contact. The teeth of reptiles function merely to hold food before it is swallowed. But the teeth of mammals shear across each other to grind food before it is swallowed. The grinding mechanism is efficient because the upper and lower sets of shearing planes developed on each complex tooth mutually wear against each other to produce a precise fit. If teeth were always being replaced, the precisely matched shearing planes would constantly be disrupted (*15*). Therefore, in order to maintain the advantages of an efficient shearing dentition, the permanent teeth of mammals are not replaced. To offset the wear problem the teeth are firmly rooted and are covered with a thick layer of very hard enamel. Tooth replacement in all animals is probably a growth phenomenon, and its control must be looked for in terms of a growth control.

Replacement control

Many growth processes can be expressed by a power equation of the form $T = xe^{yL}$, where T is time, L is length, and x and y are constants. If the addition of new tooth families at the back of the dentition is a growth phenomenon (equivalent to growth in length of the dentition), it seems probable that it could be expressed by an equation of the form $T_{(n)} = pe^{qn}$, where $T_{(n)}$ is the time at which the nth family appears at the back of a jaw quadrant, n is the number of the family, and p and q are constants.

If tooth replacement is a growth phenomenon (equivalent to growth in height of the dentition) and it is controlled within the tooth family, then it might be expressed by the equation $T_{(r)} = ve^{wr}$, where $T_{(r)}$ is the time at which the rth replacement erupts in a particular family and v and w are constants related to the family.

The above two equations can be combined to produce the equation

$$T_{(n)r} = ke^{ar+bn} \quad (1)$$

where $T_{(n)r}$ is the time at which the rth replacement erupts in the nth family ($T_{(n)o}$ is the time at which the nth family appears in the mouth), and k, a, and b are constants. Theoretically this equation can fully describe tooth replacement in a reptilian dentition (*16*). Provided teeth are replaced at less than 4 times the rate at which new families are added to the back of the dentition, there is a greater than 4:1 chance that teeth will be replaced in waves that sweep through alternate tooth positions even if no other form of control were present (*16*).

From equation (1) the time taken to add the nth tooth family to the dentition is given by

$$T_{(n)o} - T_{(n-1)o} = ke^{bn} - ke^{b(n-1)} = ke^{b(n-1)}(e^b - 1) = T_{(n-1)o}(e^b - 1)$$

In other words the time taken to add the nth family is a constant $(e^b - 1)$ times the age of the animal at which the adjacent anterior family was added to the dentition ($T_{(n-1)o}$). This result can be explained in terms of the inhibitory zone which it was earlier suggested temporarily surrounds a newly initiated tooth; it takes progressively

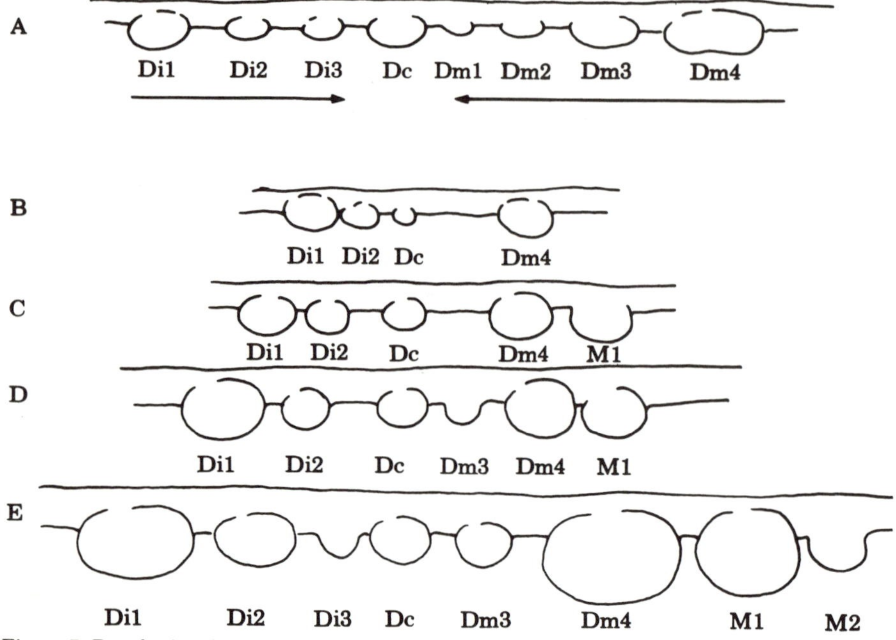

Figure 7. Developing dentitions of the American mole *Scapanus latimanus* (A) (after Ziegler, *19*) and the golden mole *Eremitalpa* (B to E) (after Kindahl, *22*). The sizes of the buds in A indicate that the deciduous incisors develop in sequence from front to back and the deciduous molars in sequence from back to front (*arrows*). In *Eremitalpa* it can be seen that room for Di3 and Dm3 is created by interstitial jaw growth.

longer to add new tooth families because, in conformity with the progressive decrease in the rate at which the jaws grow in length, it takes progressively longer for the potential tooth-forming tissues to escape from the inhibition of the anteriorly adjacent tooth family.

Also from equation (1), if N is a particular family, the time taken to replace the rth tooth is given by

$$T_{(N)r+1} - T_{(N)r} = ke^{a(r+1)+bN} - ke^{ar+bN} = ke^{ar+bN}(e^a - 1) = T_{(N)r}(e^a - 1)$$

In other words, the time taken to replace a tooth is a constant ($e^a - 1$) times the age of the animal when its predecessor erupted. And in terms of inhibition, each tooth inhibits the initiation of its replacement for a constant times the age of the animal when the tooth itself was initiated. It will also be noted from the above result that every family initiates replacement teeth at a rate which differs from all other families because the value of $T_{(N)r}$ in the expression $T_{(N)r}$ ($e^a - 1$) is different for every family.

The Tooth Family theory is best summarized by the original equation $T_{(n)r} = ke^{ar + bn}$. Tooth replacement involves two controls: an overall control represented by e^{ar} which is the same for every family (it is independent of n), and a family specific control, e^{bn}, which differs for every family (because n is different for every family).

Finally, if the addition of new tooth families represents growth in length of the dentition and tooth replacement represents growth in height of the dentition, then the pattern of unerupted, erupting, and functioning teeth represents the two-dimensional shape of the dentition—a shape which is expressed as the wave replacement of alternate teeth. The maintenance of this shape by two control constants (a and b) is as easy (or as difficult) to understand as the maintenance of the shape of a bone during the growth of an animal.

Tooth initiation in recent mammals

Now that we have explained the polyphyodonty of submammalian vertebrates as the expression of a growth phenomenon controlled by a tooth family specific and an overall control, it is necessary to look at the evidence of tooth initiation in recent mammals. The ancestral eutherian mammal is generally considered to have possessed 3 incisors, a canine, 4 premolars, and 3 permanent molars in each jaw quadrant. With very few exceptions (for example, the aquatic mammals), recent eutherian mammals either possess this dental formula (expressed as I3, C1, Pm4, M3) or have lost one or more teeth. What was the sequence in which teeth were initiated in the ancestral eutherian mammal? To answer this question it is necessary to look at the data for recent mammals, in particular those which still possess the full eutherian complement.

The most accurate method of assessing the sequence in which teeth are initiated is to reconstruct the developing dentitions of serially sectioned, closely timed embryos. However, the earliest mammalian embryos in which tooth buds are found nearly always seem to contain a newly initiated incisor, canine, and deciduous molar. This is true even of a very closely timed series of human embryos (*17*).

In the absence of early embryos it is probably safe to assume that where two teeth are at different stages of development, the more developed was the earlier initiated, particularly if the relevant teeth are each at a stage before the onset of mineralization. Using this criterion, data for the mole (*18, 19*) seem to indicate that the 3 deciduous incisors and 3 permanent molars are each initiated in sequence from the front to the back of the jaw, whereas the 4 deciduous molars are initiated from the back to the front of the jaw (Fig. 7A). It will be noted that the mole dentition is one of the few which has maintained the full eutherian dental complement.

However, most studies (including the above study of the mole) have not been specifically concerned with the sequences of tooth initiation. For example, despite his own data for the mole (*19*) Ziegler constructed a theory to explain the evolution of mammalian dentitions which assumes that the teeth of all reptiles and mammals are initiated in sequence from the front to the back of the jaw (*20*). In fact nearly all the data for insectivores (*21*) show the same sequences as those of the mole (Fig. 7B). I am preparing a paper in which these data will be discussed, but for the moment it will be assumed that in the ancestral eutherian mammal, just as in the mole and most other recent eutherian mammals, the deciduous incisors were developed in front-to-back sequence, the deciduous molars developed from back to front, and the permanent molars developed from front to back.

It has already been stated that the earliest embryos in which tooth formation is evident usually contain a deciduos incisor, canine, and molar. These teeth will be called the incisor, canine, and molar determinants, respectively. It will be noted that in the ancestral eutherian mammal, Dm4 (the 4th deciduous molar) was probably the molar determinant. In the reduced postcanine dentitions of most eutherian mammals the only difference is that the penultimate rather than the most posterior deciduous molar may sometimes be the molar determinant.

Reconstructions of developing insectivore dentitions (*21*) seem to show that room for the deciduous incisors and molars is created in the embryonic jaws by interstitial growth between the incisor and canine determinant, and the canine and molar determinants, respectively (Fig. 7B to E). This accounts for the sequences in which teeth develop in these regions: just as in reptiles, a new bud is initiated when the presumptive tooth-forming tissues have escaped from the inhibitory regions generated around a developing tooth. For the same reason the permanent molars develop in front-to-back sequence as space becomes available posteriorly in the growing jaws. However, it will be noted that no explanation has yet been offered for the early appearance of the three determinants. In terms of development I have no data or solution to this problem; any sequence appears possible, including all three arising at the same time. But if ontogeny can be compared with phylogeny, the canine is the most likely to be initiated first.

Dentition in mammal-like reptiles

Before suggesting how mammalian dentitions may have evolved it is necessary to look at data for the mammal-like reptiles which bridged the gap between reptiles and mammals. Obviously we can only surmise the sequences in which teeth were initiated in embryos of these animals, but given a complete enough growth sequence it is possible to deduce the sequences in which teeth were replaced. And it so happens that there is a good growth sequence of the early Triassic mammal-like reptile *Thrinaxodon*, which is generally believed to have been close to the line of mammal evolution.

Figure 8. A lower jaw of the early Triassic mammal-like reptile, *Thrinaxodon*. Behind the canine are 2 A-type teeth: the more posterior is erupting and its tip is broken. This is followed by 2 M types: the more posterior has recently erupted. Three P types terminate the dentition. The hole beneath the most anterior of these contains the tip of an erupting tooth, while the most posterior P type was probably still beneath the soft tissues lining the oral cavity. The postcanines were replaced in waves which passed from the back to the front of the jaw through alternate tooth positions.

Both the pre- and postcanines in the small, almost certainly carnivorous *Thrinaxodon* were replaced in waves which swept through alternate tooth positions: the large canine was regularly replaced. To this extent *Thrinaxodon* had a dentition like that of a homodont reptile, with the exception that it had evolved a canine tooth. However, the postcanine dentition was graded anteriorly from rather simple conical teeth (but more complex than those of the earlier reptiles) to complexly cusped posterior teeth (Fig. 8). *Thrinaxodon*'s heterodont postcanine dentition makes it an important link between reptiles and mammals.

No matter what size of *Thrinaxodon* jaw is studied it nearly always contains a developing or erupting tooth at the back of the dentition. This also is like a primitive reptile. But despite the constant addition of new tooth families, no dentition seems to contain more than about 7 postcanines (6 or 8 may be present). The answer to this problem is that when a tooth was added to the back of the dentition, one was generally lost from the front of the postcanine dentition (*23*).

The 7 postcanine teeth of *Thrinaxodon* consist of 3 different tooth types referred to as P, M, and A types (*4*) (Fig. 8). These symbols represent posterior, middle, and anterior, the positions in which the different tooth types are invariably situated. Because new P types were always erupting at the back of the dentition, while A types were being lost at the front, it might seem that an old animal would have possessed a postcanine dentition consisting entirely of P types. But in fact all jaws contain about 3 A types followed by 2 M types, 2 P types, and an erupting P type. The way in which this dentition was probably maintained is shown in Figure 9.

Every postcanine tooth position was successively occupied by P, M, and A types. This result provides excellent support for the concept that the tooth family is the unit of dentitions, the postcanine unit being a P, M, A sequence. Furthermore it is also clear that tooth replacement in *Thrinaxodon* served to increase the size of the teeth in the dentition so that they would match the size of the growing jaw and the requirements of the growing animal.

Diademodon, a gomphodont cynodont of the Triassic, also had a heterodont postcanine dentition in which teeth were probably constantly being replaced in response to the requirements of the growing animal. Each of 20 or more postcanine tooth families sequentially added to the back of the dentition contained 6 different tooth types replacing each other in a carefully timed sequence which led to a gradual increase in the surface area for crushing food, matched by the increase in jaw length. Just as for the maintenance of wave replacement of alternate teeth, so also for *Diademodon's* postcanine dentition it can be shown mathematically that the rate of tooth replacement was probably related to a tooth family specific and an overall control (*24*).

In *Diademodon*, the 6th tooth in each postcanine family appears to have been "intentionally" removed from the jaw and replaced by bony tissue called a plugged socket. Increasingly large replacement canines encroached on the potential space created at the front of the postcanine dentition. However, in *Thrinaxodon* it seems possible either that anterior postcanine families were finally suppressed owing to the proximity of developing canines (*25*) (Fig. 9) or that the families were more intentionally lost, as in *Diademodon*.

There remains the evidence of the Upper Triassic (Rhaetic) *Eozostrodon*, the earliest known animal possessing a sufficient number of mammalian characteristics to be called a mammal. For example, its jaw joint was partly mammalian, and it was almost certainly diphyodont (*26*). It appears to have had 5 upper and 4 lower premolars together with 4 and sometimes 5 permanent molars. The premolars (which replaced deciduous molars) were themselves lost and replaced by plugged sockets like those of *Diademodon*. Rather surprisingly, some molars also appear to have been replaced by plugged sockets in older (larger) animals (*26*).

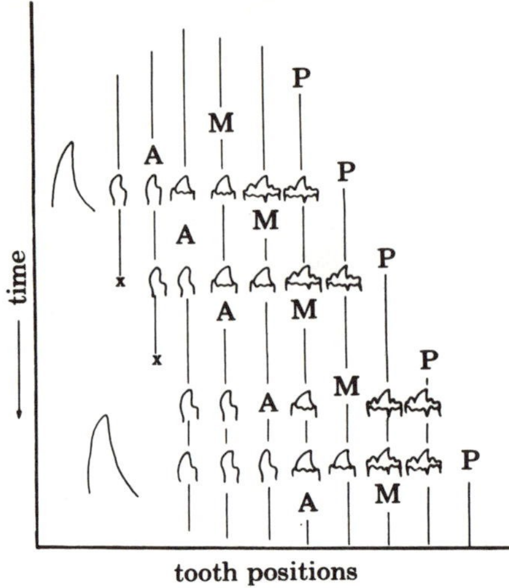

Figure 9. Postcanine replacement in *Thrinaxodon*. The letters P, M, and A indicate the time at which the respective tooth types erupt at each tooth position. The dentitions of 4 different aged animals have been drawn. All contain either 2 or 3 A types, 2 M types, and 2 or 3 P types (c.f. Fig. 8). In one dentition a recently erupted M type and an erupting A type have not been included. The anterior postcanine families were suppressed (x), perhaps owing to the development of increasingly large canines (*drawn on the left*).

From the Upper Triassic through to the Tertiary there is remarkably little data on tooth replacement in mammals. However, as will now be shown, we have sufficient data from the above animals, together with the data on recent reptiles and mammals, to be able to outline a plausible sequence in which the dentitions of mammals evolved.

Tooth initiation in submammalian vertebrates

In the earliest vertebrate dentitions which have so far been studied, teeth appear to have been replaced in waves which passed from the back to the front of the jaw. It seems probable that these teeth were initiated in the same sequence as those of *L. vivipara* (*10*). The sequence is shown in Figure 6, and the adult pattern of tooth replacement is the same as that of the hatchling (Fig. 5D). Although the regularity of the replacement waves shows little trace of the sequence in which tooth families were initiated in the embryo, it is evident that there is an important difference between the alternating sequence in which families were initiated in front of the dental determinant and the sequential initiation of families behind the dental determinant (Fig. 6).

Wave replacement of alternate teeth in front of the dental determinant was initiated before any teeth were replaced (rows a and b in Fig. 6): it was the outcome of the alternating sequence in which families were initiated in this region. But the rate of tooth replacement needed to be controlled in order for wave replacement of alternate teeth to develop between the sequentially initiated families behind the dental determinant (11a, 12b, 13c, 14d, 15e in Fig. 6). And a more widespread overall control of tooth replacement was required for the waves passing through the families behind the dental determinant to blend into the waves already present in front of the dental determinant (*16*).

The earliest heterodont dentitions often contain one or two enlarged teeth in the middle, or anterior to the middle, of the maxillary bone: for example, the Carboniferous reptile *Hylonomus* and the upper Carboniferous captorhinomorphs. The same is true of the Permian pelycosaurs. What was the most probable sequence of tooth initiation?

In terms of function the jaws were by now separated into three distinctly different regions. The precanines captured food, the canine was a weapon, and the postcanines immobilized prey or food prior to swallowing. It seems likely that teeth were replaced at a different rate in each region; in particular, the large canine would have been more slowly replaced because it took longer to develop. It is surely more probable that these differences in function and the rate of tooth replacement would evolve from, rather than be separately superimposed on, the pre-existing dental asymetry present in embryos of the homodont ancestors. In other words, the canine evolved at the site of the dental determinant, the precanines were initiated in alternation toward the front of the jaw, and the postcanines were initiated in sequence toward the back of the jaw. This may not have been the only site at which a caniniform tooth evolved, but it is certainly the most satisfactory functionally and one which probably had the greatest potential for subsequent evolutionary change. To understand this it is necessary to turn again to *Thrinaxodon*.

In order to understand tooth replacement in *Thrinaxodon* we must first seek the selective advantage conferred by its dentition. It seems likely that, together with the canine, the simple A types found toward the front of the postcanine dentition either punctured or held prey. The shape of the P types indicates that their primary function was not to hold or puncture struggling prey; A types would have been more efficient for this purpose. What seems possible is that these teeth were used to crack bones. For this purpose a moderately serrated surface was required in order to prevent the bone's slipping (large spaces between slender cusps or teeth, such as between the A types, would have been much less efficient), and the bone-splitting teeth needed to be sited where maximum jaw power could be developed. In order to break bones it did not matter that the upper and lower cheek teeth could not be brought together (as in most reptiles). However, it was necessary that the P types should be firmly rooted (as are all the teeth in *Thrinaxodon*). The intermediate M types could function either with the A or P types.

It is not difficult to visualize how the dentition of *Thrinaxodon* might have been initiated. The canine remained the dental determinant separating two developmentally different jaw regions. The families within the precanine region were still initiated in alternation toward the front of the jaw, and those in the postcanine region might still have been initiated in sequence toward the back of the jaw (but see below). The major innovation was the distinct change in shape of successive teeth in each postcanine family. This evolutionary step will now be considered.

Changes in shape

In many recent Lacertilia the earlier members of each more posterior family (perhaps those behind the dental determinant) have rather small anterior and posterior accessory cusps situated on the flanks of the major cusp (*27*). The posterior and then the anterior cusps are gradually lost by successive replacement teeth, with the result that the tooth families now produce unadorned conical teeth. It can be visualized that, owing to the presence of incipiently tricuspid teeth, the short dentitions of the young are relatively more serrated than the longer dentitions of the adult. Presumably the 3-cusped teeth of the young have the advantage that fewer are required to produce the same number of grasping serrations as simple conical teeth.

The above type of dentition in which the accessory cusps of hatchling teeth are gradually lost by successive replacement teeth is exactly what might be predicted for an ancestor of *Thrinaxodon*. Those animals with larger accessory cusps were selected for in successive generations, with the result that the complex P type evolved. However, in conformity with its ancestors, successive replacement teeth were less complex. Indeed the maintenance of this characteristic had a selective advantage because it resulted in the anterior, and therefore older, postcanine families containing puncturing conical teeth, while the newly initiated posterior families contained complex "crushing" teeth.

It was now an advantage to reduce the rate of tooth replacement in the postcanine region. Too rapid a rate of tooth replacement would have resulted in the rapid loss of crushing P types and a postcanine dentition which contained nearly all puncturing A types. However, too slow a rate of tooth replacement would have resulted in too few A types anteriorly.

It will be noted that the above evolutionary step implies that the tooth family with its P, M, A sequence was a genetic unit of *Thrinaxodon*'s dentition. If one postcanine family was changed then all postcanine families were changed. In terms of morphogenetic "coding" it was possible to evolve differences between precanine, canine, and postcanine families by taking advantage of the fact that each of these major units was contained in a developmentally different part of the jaw. However, the following analysis suggests that the anterior postcanine families in *Thrinaxodon* may have differed from the posterior postcanine families. The data to support this suggestion come from *Diademodon*.

The evolution of a molar determinant

It will be recalled that most of the 20 or so postcanine families in *Diademodon* each contained 6 teeth. However, a very small (very young) *Diademodon* jaw contains 5 postcanine teeth, which are the 5th, 5th, 5th, 5th (gomphodonts), and 4th (intermediate gomphodont) teeth, respectively, of the 6-tooth sequence. Due to the very small size of this jaw, the implication is that, unlike the remaining postcanine families, the anterior postcanine families did not develop the first 2 or 3 replacement teeth (*28*). The absence of these earlier teeth from the anterior postcanine positions would have been an advantage because the 5th tooth in the 6-tooth sequence, the large grinding gomphodont, appears to have been the most useful postcanine tooth: a newly hatched (probably herbivorous) *Diademodon* whose short postcanine dentition contained only the first 3 teeth in the 6-tooth sequence, which were slicing (sectorial) teeth, would have been unable to grind food.

Suppose that a newly hatched *Thrinaxodon* also required a miniaturized version of the adult dentition—say, a P type in postcanine position 3, M type in position 2, and A type in position 1. If all postcanine families were alike, this would have required P and M types from position 1 and an M type from position 2 to have been developed and lost before hatching. One way to achieve this would be to initiate the anterior P types at a stage before the dental tissues were competent to lay down mineral. In other words, the relevant buds were coded for P types but regressed like the earliest developed buds in *L. vivipara*. However, it seems possible that another method may have been evolved.

Although this is not the place for a detailed discussion of the determination of tooth shape, it is necessary to draw attention to the fact

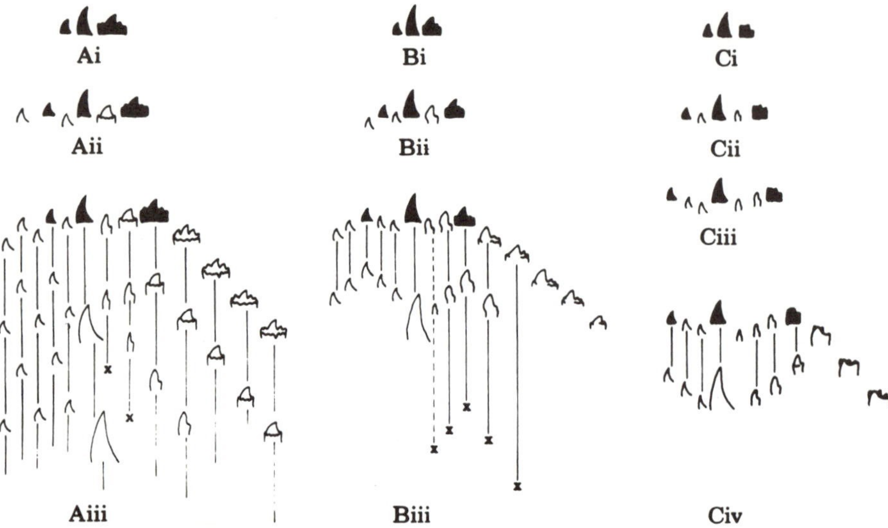

Figure 10. The diagram indicates the way in which the probable sequence of tooth initiation in the earliest eutherian mammals (C) may have evolved from that of a mammal-like reptile (A). In *Thrinaxodon,* incisor, canine, and postcanine determinants (*black teeth*) were initiated at about the same time (Ai). By interstitial jaw growth (Aii and Aiii), room was created for an M and then an A type at postcanine positions 2 and 1, respectively, the postcanine determinant (a P type) ultimately occupying postcanine position 3 (Aiii). The eruption times of each tooth type are represented in Aiii. The 2 most anterior postcanine families were suppressed early in life (x in Aiii). In *Eozostrodon* (B) the postcanine families were probably initiated in the same sequence as those of *Thrinaxodon.* Anterior postcanine families were usually terminated by a plugged socket (x in Biii). Generally, the first and second postcanine sockets were plugged in sequence from back to front (*26*). The 1st deciduous molar may not have been replaced (*dotted line*), as in *Thrinaxodon* and recent eutherian mammals having 4 deciduous molars. In the earliest eutherian mammal, Dm4 was probably the molar determinant, room being made for the remaining deciduous molars by interstitial growth between the molar and canine determinants (Ci, ii, iii, iv). In *Thrinaxodon* the incisors were probably initiated forward in alternation (A). It is suggested that in the *Eozostrodon*-like animal Di3 was the incisor determinant (B). During the next step in evolution the most anterior incisor was lost (leaving 4 incisors as in *Eozostrodon*) (*26*), and then the remaining incisor anterior to the incisor determinant was lost in the eutherian mammals (C). It will be noted that the evolution of eutherian mammals was characterized by a gradual increase in the interstitial growth of tissues on both sides of the canine tooth at the expense of tip growth of the incisor and molar regions.

that the (mesodermal) dental papilla rather than the (ectodermal) enamel organ of tooth germs is probably responsible for determining tooth shape in mice (and presumably all mammals) (*14*). For example, an incisor enamel organ grafted onto a molar dental papilla produces a molar tooth. Even small fragments of enamel organ or lip ectoderm grafted onto a papilla result in the formation of a tooth whose shape is the presumptive shape of the tooth germ from which the papilla was taken. Therefore tissue derived from the jaw mesoderm probably determines the shape of a tooth. This suggests that the changes in jaw mesoderm responsible for sequentially producing the P, M, and A types in each postcanine family of *Thrinaxodon* were triggered by the initiation or development of the preceding tooth type.

Suppose that *Thrinaxodon,* like recent mammals, possessed a molar determinant—in other words, that the 3rd postcanine family was initiated adjacent to the canine and that the 1st and 2nd tooth families were initiated in succession, following interstitial growth between the canine and molar determinants. Just posterior to the canine, jaw mesoderm generated the P-type shape of the molar determinant (Fig. 10Ai). This newly differentiated jaw mesoderm was not only responsible for generating the papilla of the replacement for this P type but also, by interstitial growth of the jaw, for generating the papilla of the tooth anterior to the molar determinant, which was therefore also an M type (Fig. 10Aii).

Jaw mesoderm which had generated an M type was triggered to continue by generating an A type. Therefore, further interstitial jaw growth resulted in the formation of an A type in the newly initiated family posteriorly adjacent to the canine. Thus the 3 anterior postcanine positions of the hatchling would be occupied by A, M, and P types, respectively, without any tooth having been replaced. Each family continued by developing the next tooth in the P, M, A sequence (Fig. 10Aiii). However, the anterior postcanine families sometimes replaced A types with even more diminutive teeth (*4*). The above explanation carries the implication that tooth families are the genetic units of dentitions. It should be noted that, behind the molar determinant, P types started molar families because the associated mesoderm was newly differentiated.

The above technique may or may not have been evolved by *Thrinaxodon;* juvenile *Thrinaxodons* were so small that it is unlikely sufficient numbers will ever be found to test the theory. It is even less likely that sufficient early mammals will be found, because although *Thrinaxodon* was not large it was a giant compared with *Eozostrodon* (*26*), the earliest undoubtedly diphyodont animal. It took over three years of sifting through a truckload of rubble to find the 4 or 5 jaw fragments (the largest being about ½ cm long) that firmly established the diphyodont nature of *Eozostrodon* (*26*). It is therefore unlikely that we will ever find much evidence for the sequences in which the teeth of early mammals were replaced, let alone initiated. But it seems improbable that postcanines were initiated in sequence from the front to the back of the jaw, as Ziegler has suggested (*20*), not only for Mesozoic but also (contrary to the data which are available) for recent mammals.

This is not the place for a detailed discussion of fossil data, which are so limited that they could probably be used to support several theories. But if the concept of a molar determinant is accepted, it is possible to explain most of the available data for recent and extinct mammals. Reduction in the rate of tooth replacement can account for the diphyodont nature of *Eozostrodon*'s dentition (Fig. 10B). Just as in *Thrinaxodon,* the anterior postcanines were initiated in sequence anteriorly in front of the molar determinant, and the morphological gradient was similar. The premolar sockets were plugged, usually in sequence from the front of the jaw, but occasionally more irregularly (*26*). It may be that the 1st deciduous molar was not replaced, as in recent mammals, accounting for its early plugging (c.f. the 1st postcanine in *Thrinaxodon*). Further irregularities could be accounted for if Dm4, rather than Dm3, was occasionally the molar determinant, implying variation in interstitial jaw growth. The decrease in number of molar families as compared with *Thrinaxodon* was related to a reduction in posterior jaw growth.

Further increase in interstitial jaw growth between the canine and molar determinants, together with the loss of one or two molar families from the back of the dentition, can account for the sequence of tooth initiation in recent eutherian mammals (Fig. 10C). The eutherian postcanine dentition may have evolved by a progressive increase in interstitial jaw growth together with loss of molar families.

The precanines

We now return to the sequences in which precanines were initiated. Because in *Thrinaxodon* the precanines were alternately replaced, like those of its ancestors, it seems probable that they were also initiated in alternate sequence from back to front, also like those of its ancestors (Fig. 10A).

The incisors of recent eutherian mammals are generally initiated in sequence from the front to the back of the jaw. However, in a recent marsupial it appears that, of the 5 upper incisors, I1 and I5 are the last to be developed (*29*). For convenience I have incorporated this latter sequence in the dentition of *Eozostrodon* (Fig. 10B), although this animal had 4 lower incisors (*26*). Thus I3 was the incisor determinant; I4 and I5 were initiated in a region of interstitial jaw growth and I1 and I2 in a region of anterior tip growth. In terms of what is biologically possible, this is no more difficult to accept than the equivalent evolution of a molar determinant. But just as for the molar determinant, so also for the incisor determinant it is necessary to find a selective advantage for the new sequence in which incisors may have been initiated.

Although there can never be any supporting evidence, it seems possible that the earliest newborn mammals would have become firmly attached to the mother's nipple, rather like recent metatheria, but that when released from the nipple they would immediately have had to forage for themselves. Intermittent suckling on demand would probably have evolved separately and later. If this is true, just prior to being released from their mother's nipple the earliest infant mammals, like newly hatched reptiles, would have required a functioning dentition. Apart from the anterior incisors, all the teeth could have erupted before the infant was released because they formed the side walls of a tunnel which surrounded the nipple. But eruption of the anterior incisors must have been delayed in order to provide the space through which the nipple entered the mouth.

It is of interest here that Mills (*30*) states there is a particularly large midline diastema between the upper incisors of *Morganucodon* (= *Eozostrodon*). If this is true, then the requisite space for the mother's nipple may have been provided by inserting it between the position of the lost upper anterior incisors (suppressed during evolution) and the procumbent lower anterior incisors.

It is evident that the sequence of incisor eruption implied for *Eozostrodon* (Fig. 10B) would have allowed the anterior incisors to erupt after the remainder of a functioning dentition. The evolution of this sequence is no more difficult to accept than that given for the evolution of the molar determinant. It will be noted that in both cases the teeth initiated following interstitial jaw growth were initiated and erupted earlier than those initiated following tip growth.

There would now have been a selective advantage for evolving different suckling periods which were matched to the physical development of different species of infant. Too short a time would have led to the wasteful death of many infants that were incapable of foraging for themselves. This led to the evolution of a behavior pattern which permitted suckling on demand by an infant that had earlier been released from the nipple. However, too long a suckling time would have involved the mother in wasting energy on suckling infants that were physically capable of foraging for themselves; in terms of selective advantage the energy would have been better employed in breeding a new generation.

This latter argument can account for the evolution of the eutherian incisor formula and pattern of tooth eruption. By prolonging the time *in utero*, the eutherian infant was better protected than the metatherian infant. Since all suckling is intermittent in eutherian mammals, it was an advantage to erupt the incisors as early as possible in order that the infant could supplement its milk diet. Provided the mother continued to permit suckling until the infant was capable of foraging enough food by itself, those infants (presumably insectivorous) possessing an uninterrupted span of incisors at the front of the jaw would have had a selective advantage when foraging. This ultimately led to the loss of the 2 late-developed anterior incisors, with the result that the earliest initiated (and erupted) incisor, the incisor determinant, became sited towards the midline of the jaw (Fig. 10C) and the ancestral 2 anterior incisors were lost.

"Irregular" sequences

Now that we have arrived at the earliest eutherian dentition consisting of I3, C1, Pm4, M3, it is worth summarizing how the initiation of such a dentition may be controlled. The embryonic jaws of mammals contain three major developmental segments: incisor, canine, and molar segments. Each segment is provided with genetic coding for a paradigm tooth shape. A first tooth, the determinant, appears in each of the three developmental segments at about the same time during development. Each generates around it a zone of tissue in which further initiation of teeth is inhibited. The incisor and molar jaw segments (not the canine segment) are potentially capable of expanding anteriorly and posteriorly, thereby generating new jaw tissue for the initiation of further teeth. However, tip growth of the incisor region has been suppressed in most eutherian mammals, with the result that the most anterior incisor is the incisor determinant and I2 and I3 are developed following interstitial jaw growth between I1 and the canine.

The molar region grows both by expansion between the canine and molar determinants, which provides space for deciduous molars to be initiated in sequence forward, and by expansion posteriorly, which provides space for the permanent molars in sequence posteriorly. Successive dental papillae derived from a single colony of jaw mesoderm retain successively less ability to generate the paradigm molar shape, with the result that they are less complex. Thus Dm3, Dm2, and Dm1 are progressively less complex because they are derived in ("horizontal") sequence from jaw mesoderm which has already generated Dm4. And for the same reason their ("vertical") replacements, the premolars, are even less complex. The permanent molars are sequentially initiated in newly differentiated jaw mesoderm behind the molar determinant. Therefore each has a shape which is closer to that of the paradigm molar than the deciduous molars are.

We can now give an explanation for the irregular sequences in which the teeth of eutherian mammals erupt. The irregularity is due to the existence of four developmental regions: the incisor, canine, and anterior and posterior molar regions. Within each region the teeth develop in sequence. For example, if we number the 8 deciduous teeth and 3 permanent molars of the mole in sequence from the front to the back of the jaw (i.e. the teeth in the first mammalian *Zahnreihe* of Edmund), they are initiated approximately in the sequence 1, 4, 8, 7, 2, 6, 3, 9, 10, 5, 11 (*31*). When separated into developmental regions, the sequences are 1, 2, 3 (the deciduous incisors); 4 (the canine); 8, 7, 6, 5 (the deciduous molars); 9, 10, 11 (the permanent molars). It is only because all four regions are developing at the same time that the integrated sequence appears to be so irregular.

In a later paper (in preparation) many of the irregularities in mammalian dentitions will be discussed. The present model suggests that several currently accepted tooth homologies may not only be inaccurate but also very misleading in evolutionary terms. For example, if the above explanations are accepted, the molar determinant is presumably homologous between different dentitions. In many primates the so-called Dm3 (the penultimate deciduous molar) is the molar determinant. If this tooth is homologous with the ancestral molar determinant, it should be designated Dm4, in which case the ancestral M1 is now replaced by a Pm5. However, this heresy will be shelved for the present.

References

1. D. F. G. Poole. 1961. Notes on tooth replacement in the Nile crocodile, *Crocodilus Niloticus*. *Proc. Zoo. Soc.* 136:131–40.
2. L. Bolk. 1922. Odontological essays. 5. On the relation between reptilian and mammalian dentitions. *J. Anat.* 57:55–75.
3. A. G. Edmund. 1960. Tooth replacement phenomena in the Lower Vertebrates. *Contr. Life Sci. Div. R. Ont. Mus.* 56:1–42.
4. J. W. Osborn and A. W. Crompton. 1973. The evolution of mammalian from reptilian dentitions. *Breviora* 399.
5. F. R. Parrington. 1936. On tooth replacement in the theriodont reptiles. *Phil. Trans. R. Soc. Lond.* B. 226:121–42.
6. A. S. Romer and L. I. Price. 1940. Review of the Pelycosauria. *Spec. Pap. Geol. Soc. Amer.* 28.
7. M. W. Woederman. 1921. Beitrage zur Entwicklungsgeschichte von Zahnen und Gebiss der Reptilian. IV. Uber die Anlage und Entwicklung der Zahne. *Arch. mikvosk. Anat. EntwMech* 95:265–395.
8. J. W. Osborn. 1970. New approach to *Zahnreihen*. *Nature* Lond. 225:343–46.
9. J. W. Osborn. 1972. On the biological improbability of *Zahnreihen* as embryological units. *Evolution* 26:601–7.
10. J. W. Osborn. 1971. The ontogeny of tooth succession in *Lacerta vivipara* Jacquin (1787). *Proc. R. Soc. Lond.* B. 179:261–89.
11. H. S. Harrison. 1901. The development and succession of teeth in *Hatteria punctata*. *Q. Jl. Microsc. Sci.* 44:161–219.
12. S. Sellman. 1946. Some experiments on the determination of larval teeth in *Ambystoma mexicanum*. *Odont. Tidskr.* 54:1–28.
13. G. R. de Beer. 1947. The differentiation of neural crest cells into visceral cartilages and odontoblasts in *Amblystoma*, and a re-examination of the germ layer theory. *Proc. R. Soc. Lond.* B. 134:377–98.
14. E. J. Kollar. 1972. In H. C. Slavkin and L. A. Bavetta, eds., *Developmental Aspects of Oral Biology*. New York: Academic Press, pp. 126–49.
15. A. W. Crompton. In press. Postcanine occulsion in cynodonts and Tritylodontids. *Bull. Brit. Mus. (N. H.) Geology*.
16. J. W. Osborn. Submitted. A mathematical model describing tooth replacement in primitive reptiles. *Evolution*.
17. T. Ooe. 1957. On the early development of the human dental lamina. *Okaj. Fol. Anat. Jap.* 32:97–108.
18. H. Sicher. 1916. Die Entwicklung des Gebisses von *Talpa europaea*. *Arb. Anat. Inst. Wiesbaden* 54:31.
19. A. C. Ziegler. 1972. Processes of mammalian tooth development as illustrated by dental ontogeny in the mole *Scapanus latimanus* (Talpidae: Insectivora). *Archs. Oral Biol.* 17:61–76.
20. A. C. Ziegler. 1971. A theory of the evolution of therian dental formulas and replacement patterns. *Q. Rev. Biol.* 46:226–49.
21. M. E. Kindahl. 1967. Some comparative aspects of the reduction of the premolars in the Insectivora. *J. Dent. Res.* 46:805–8.
22. M. E. Kindahl. 1963. On the embryonic development of the teeth in the Golden Mole, *Eremitalpa (Chrysochloris) Granti* Broom in South Africa. *Arkiv. Zoologi.* 16:97–115.
23. A. W. Crompton. 1963. Tooth replacement in the cynodont *Thrinaxodon liorhinus* Seeley. *Am. S. Afr. Mus.* 46:479–521.
24. J. W. Osborn. Submitted. On tooth succession in *Diademodon*. *Evolution*.
25. J. A. Hopson. 1964. Tooth replacement in cynodont, dicynodont and therocephalian reptiles. *Proc. Zool. Soc. Lond.* 142:625–54.
26. F. R. Parrington. 1971. On the Upper Triassic mammals. *Phil. Trans. Roy. Soc. Lond.* B. 261:231–72.
27. J. S. Cooper. The dental anatomy of the genus *Lacerta*. Ph.D. thesis, 1963, Bristol, England.
28. J. A. Hopson. 1971. Postcanine replacement in the gomphodont cynodont *Diademodon*. In D. M. and K. A. Kermack, eds., *Early Mammals*. London: Academic Press.
29. B. K. B. Berkowitz. 1967. The dentition of a 25-day pouch-young specimen of *Didelphis virginiana* (Didelphidae: Marsupiala). *Archs. Oral Biol.* 12:1211–12.
30. J. R. E. Mills. 1971. The dentition of *Morganucodon*. In D. M. and K. A. Kermack, eds., *Early Mammals*. London: Academic Press.
31. M. E. Kindahl. 1957. Notes on tooth development in *Talpa europaea*. *Arkiv. Zoologi.* 11:187–91.

A. W. Crompton
Pamela Parker

Evolution of the Mammalian Masticatory Apparatus

The fossil record shows how mammals evolved both complex chewing mechanisms and an effective middle ear, two structures that distinguish them from reptiles

The problem of how, why, and when the mammalian ear and jaw configuration arose from that of reptiles has intrigued paleontologists and anatomists for more than a century. One hundred years ago comparative anatomists and embryologists (e.g. Reichert 1837) determined that most of the bones of the reptilian and mammalian skulls were homologous, or comparable. They could show, for example, that the bones forming the jaw joint in reptiles—the articular and the quadrate—were the homologues of two of the bones in the mammalian middle ear—the malleus and the incus. They did not, however, explain why in the evolution of mammals from reptiles this dramatic change had taken place.

Documentation of the gradual evolution of mammals from reptiles required the discovery of many well-preserved fossils. It has taken about one hundred years to accumulate the evidence, and only in recent years were the first examples of the oldest mammals found (see Parrington 1971 and Crompton and Jenkins 1973 for general reviews), adding the final and most important link in the transition from primitive reptiles to early mammals. The transition covers about 120 million years. These discoveries, together with the advent of new techniques such as cinefluoroscopy and electromyography to study the functioning of the chewing mechanisms of mammals (see Hiiemae, in press, for a review) and reptiles (Frazetta 1962; Impey 1967; Throckmorton 1976), have enabled us to begin to understand the evolution of the mammalian middle ear and masticatory apparatus from those of reptiles.

A. W. Crompton is currently director and Alexander Agassiz Professor of Zoology at the Museum of Comparative Zoology. He has previously served as director of the Peabody Museum of Natural History at Yale and director of the South African Museum in Cape Town. His principal research interests are in vertebrate paleontology and functional anatomy. Pamela Parker is at present the curator of mammals at the Brookfield Zoo. After graduating from Yale, she was a postdoctoral fellow at the Museum of Comparative Zoology for two years before moving to Chicago. Her research interests are in population biology and comparative studies of mammals. *The authors wish to thank L. Meszoly, who did the drawings, and A. Coleman, who took the photographs on which some of the illustrations are based. This work was made possible by grants from NSF and NIH.* *Address: Museum of Comparative Zoology, Harvard University, Cambridge, MA 02138.*

Jaw structure

Mammals have a three-boned middle ear and a lower jaw, or mandible, consisting of a single bone, the dentary, whereas reptiles have only one bone in the middle ear, the stapes, and a lower jaw made up of at least seven bones (Figs. 1 and 2). The reptilian dentary supports the teeth and is followed by a series of postdentary bones. In mammals the jaw joint is formed between the temporal bone and the single bone of the lower jaw, the dentary. In reptiles the jaw joint lies between the last postdentary bone, the articular, and a skull bone, the quadrate. The quadrate is usually joined to the temporal bone and the braincase. This comparison implies that during the evolution of mammals from reptiles the postdentary bones and the quadrate decreased in size and the dentary progressively increased until it reached the temporal bone, when a new (mammalian) jaw articulation developed. The small bones of the reptilian jaw joint were then incorporated into the middle ear as the malleus and incus.

The evolutionary history of the mammalian jaw is in part recapitulated during the embryological development of mammals. The malleus (=articular) originates as part of the lower jaw, and only in later stages of foetal development is the separation between these elements achieved. In young marsupials, which start to suckle at a very immature stage, the initial functioning jaw joint lies between the articular and the incus; i.e. the jaw joint at these stages is like that of reptiles. Only in more advanced stages of development, when the dentary establishes contact with the temporal bone, do the articular and incus separate from the lower jaw and move into the middle ear.

Before discussing whether the fossil evidence in fact shows that these changes took place as suggested or attempting to explain why, it is necessary to review briefly several other differences between the masticatory apparatuses of reptiles and mammals.

Tooth replacement. Figure 3 compares the tooth-replacement patterns in mammals and reptiles. In most mammals the dentition is divided into four distinct regions: incisor, canine, premolar, and molar. The teeth in the first three have two distinct generations. The molars are never replaced and are added to the jaw from the front backward during growth. In marked contrast, the teeth of reptiles are replaced continuously through life by alternate tooth replacement or simply adding teeth at the end of the row. Thus newly hatched or recently born reptiles start life with a full set of functional teeth

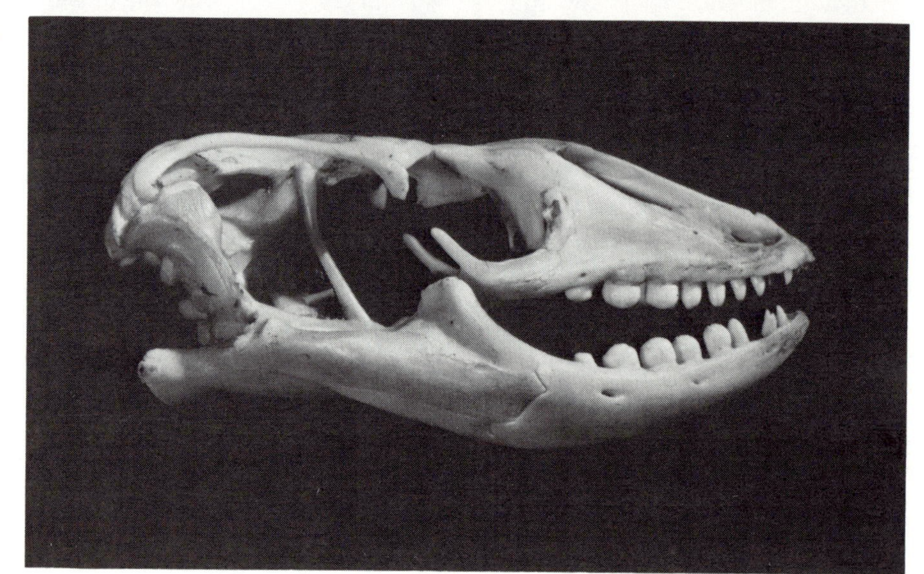

Figure 1. These lateral views of the skull of the American opossum (*Didelphis virginiana*) and the savannah monitor (*Varanus exanthermaticus*) show the distinguishing features of mammalian and reptilian jaw structures, which are labeled in the diagram below. (Photographs by A. Coleman.)

that enables them to process their own food. Growing reptiles may increase their skull length by a factor as great as ten as teeth are added (Edmund 1962).

Since mammals suckle their young, eruption of the milk teeth can be delayed, and up to 80% of the postnatal cranial growth takes place before the first set of functional teeth is fully erupted; only a slight further increase in the size of the jaws is necessary to accommodate the milk-teeth replacements and the new, permanent molars. In man this new set of teeth requires only a 15% increase in the length of the lower jaw. Ewer (1963) and Hopson (1973) have suggested that the single-replacement characteristic of mammalian dentitions is directly related to the advent of lactation and parental care. Not until a mammal has erupted several of its milk teeth is it usually large or strong enough to eat much the same kinds of food as the adult.

Jaw muscles. In mammals, the crowns of the molars and sometimes also the premolars support several shearing surfaces, which accurately match similar surfaces on the occluding teeth in the opposite jaw. These surfaces break down food when the lower teeth are forced against the upper. To bring a series of matching shearing surfaces on upper and lower teeth into precise contact during occlusion, the lower jaw must be capable of moving not only upward and downward but also forward and backward and from side to side, or in a combination of these movements. The size and shape of the shearing

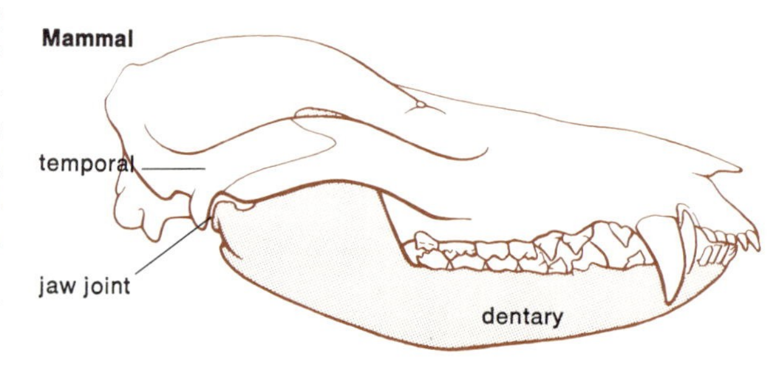

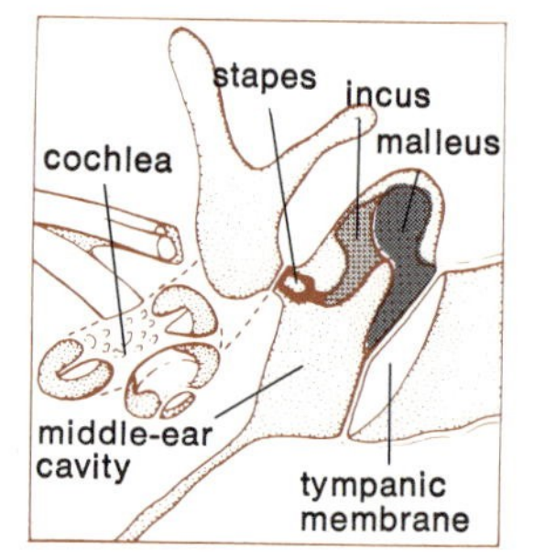

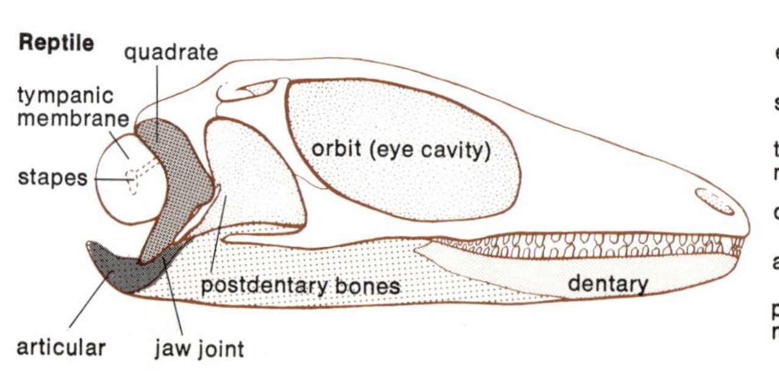

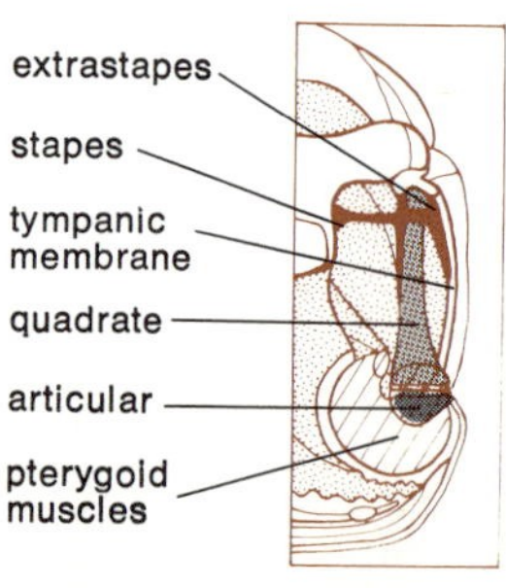

Figure 2. In mammals such as the American opossum the lower jaw is a single bone, the dentary, and there are three bones in the middle ear, the malleus, incus, and stapes. In reptiles such as the early Triassic *Prolacerta* the lower jaw is made up of several bones, and there is only a single bone in the middle ear, the stapes. The jaw-joint bones of the reptile, the articular and quadrate, are homologous with the malleus and incus of the mammalian middle ear. (Mammalian ear after Bekesy 1957; reptilian ear after Allin 1975.)

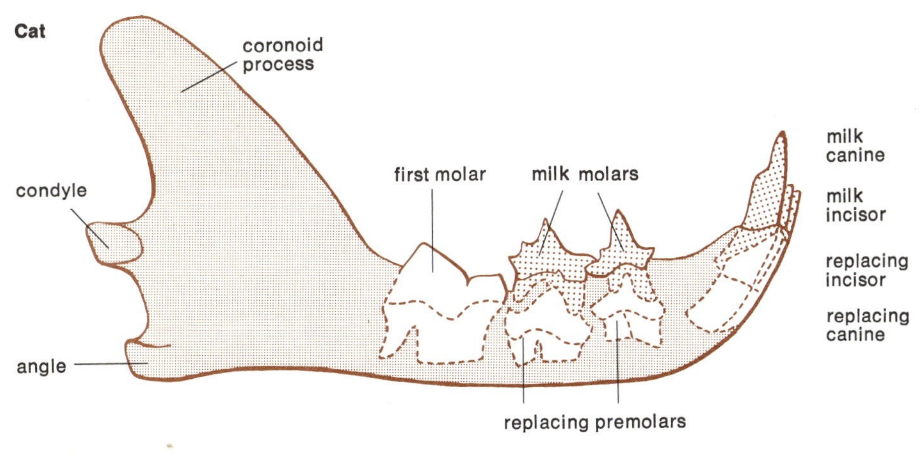

Triassic mammal-like reptile (Thrinaxodon)

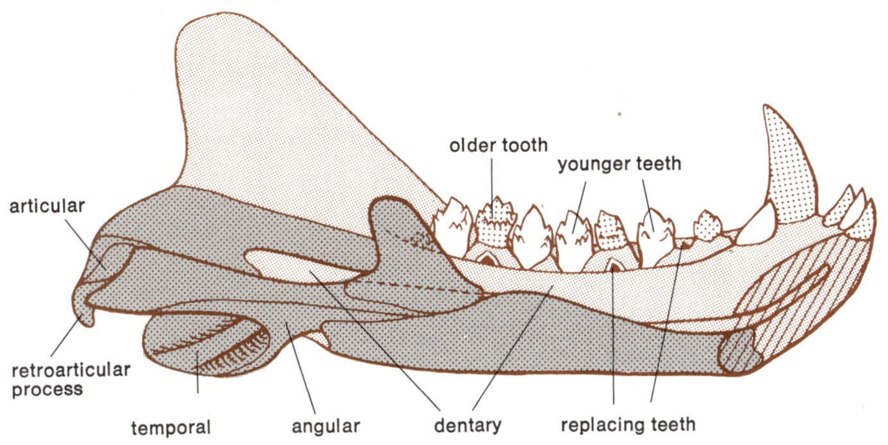

Figure 3. Tooth replacement in mammals and reptiles differs. In a cat, the milk teeth are replaced by permanent incisors, canines, and premolars; there are no predecessors to the molars. In reptiles and mammal-like reptiles, such as *Thrinaxodon,* new teeth (white) tend to erupt between older teeth (stippled), and replacement usually continues throughout the life of the individual. Replacing teeth can be seen below the older teeth.

surfaces and the pattern of these movements vary considerably among mammals and reflect the types of food chewed. For instance, in many mammals the cusps, or high points, of the teeth are fairly slender, and if the cusps supporting the shearing surfaces were to be forced against one another rather than to slide past one another, they could very easily be fractured.

The ability to move the jaw in a complex pattern is made possible in part by the orientation and position of the jaw-closing muscles (see Fig. 4). Essentially, each side of the lower jaw is held in a sling of muscles (Stein 1939). A large masseter muscle inserts on the outer surface of each side of the lower jaw and pulls it upward, forward, and outward. The counterpart of the masseter muscle, the pterygoideus muscle, inserts on the inside and pulls it upward, forward, and inward. Additional force increasing the power of bite is provided by the temporalis muscle, which inserts mainly on the coronoid process and pulls upward, backward, and inward. Differential contractions of parts or all of these muscles make it possible for the jaw to move from side to side or from front to back. Side-to-side movement, for example, is especially evident in herbivores, which must grind rather than pulp food.

Recent work using cinefluorographic and electromyographic techniques suggest that, in mammals, the sequence and strength of contractions of the musculature operating the jaws and tongue are in part determined by sensory feedback from the oral cavity, jaws, teeth, muscles, and the jaw joint (Thexton 1976; Crompton et al., 1977; Hiiemae, in press). Factors such as the consistency of particle size of food determine jaw gap, the frequency of jaw movements, and tongue movement.

The mechanics of the jaw joint of reptiles contrast markedly to those of the mammalian jaw (see Fig. 4). For example, in a herbivorous Late Triassic dinosaur such as *Heterodontosaurus,* the lower jaw is essentially a third-class lever, with the jaw joint acting as the fulcrum, the jaw-closing muscles as the force, and the bite force as the load. Since the jaw-closing muscles are inserted closer to the jaw joint than to the point of bite, the vertical forces passing through the jaw joint or fulcrum were probably greater than those generated between the teeth at the point of bite. Reptiles lack a masseter muscle inserting on the outside of the jaw. Furthermore, as can be seen in the crocodile in Figure 4, the jaw-closing muscles of a reptile insert on the inner, upper, and lower surfaces of the lower jaw and do not form muscular slings. These muscles pull the jaw upward and inward and forward and inward. When the muscles on one side contract, therefore, they tend to deflect that jaw ramus toward the midline of the skull. It appears that the pterygoid bone of the palate is armed with powerful flanges to limit this movement and also to serve as areas of attachment for the large pterygoideus muscles. Although these flanges prevent medial deflection of either side of the jaw, it must be remembered that the muscle arrangement also requires that the jaw joint be strong enough to resist powerful medially directed forces. As would be expected, in crocodiles and other typical reptiles the jaw joint is fairly massive in relation to skull size and can withstand these forces.

Jaw opening. Figure 4 illustrates the different arrangement of muscles in mammals and reptiles. In all birds and reptiles, a depressor muscle running from a backward projection of the lower jaw, the retroarticular process, to the back end of the skull plays a major role in opening the jaw. In primitive mammals, such as the American opossum, the jaw is opened by two groups of muscles, one that runs from the front and lower surface of the lower jaw to the hyoid bone (the suprahyoid muscles) and another that runs from the back of the skull, behind the jaw joint, from the scapula and from the sternum to the hyoid bone (the infrahyoid muscles).

Opening of the jaw and the movements of the base of the tongue are controlled by differential contraction of these muscles.

The hyoid muscles and the action of the tongue are very important for the effective breakdown of food in the mouth. Mammalian mastication usually does not consist of a simple bite followed by a swallow, but rather, food is chewed. After each bite the food is collected by the tongue and the cheeks and placed between the teeth as the jaws close. The organization of the tongue and hyoid muscles also makes suckling possible. A partial vacuum is created by the tongue around the maternal teat, helping to withdraw the milk.

Why are there marked differences between the teeth and jaws of reptiles and those of mammals? The answer to this question hinges upon a feature that at first would seem to have little to do with jaws and teeth, namely, the structure of the middle ear (see Fig. 2).

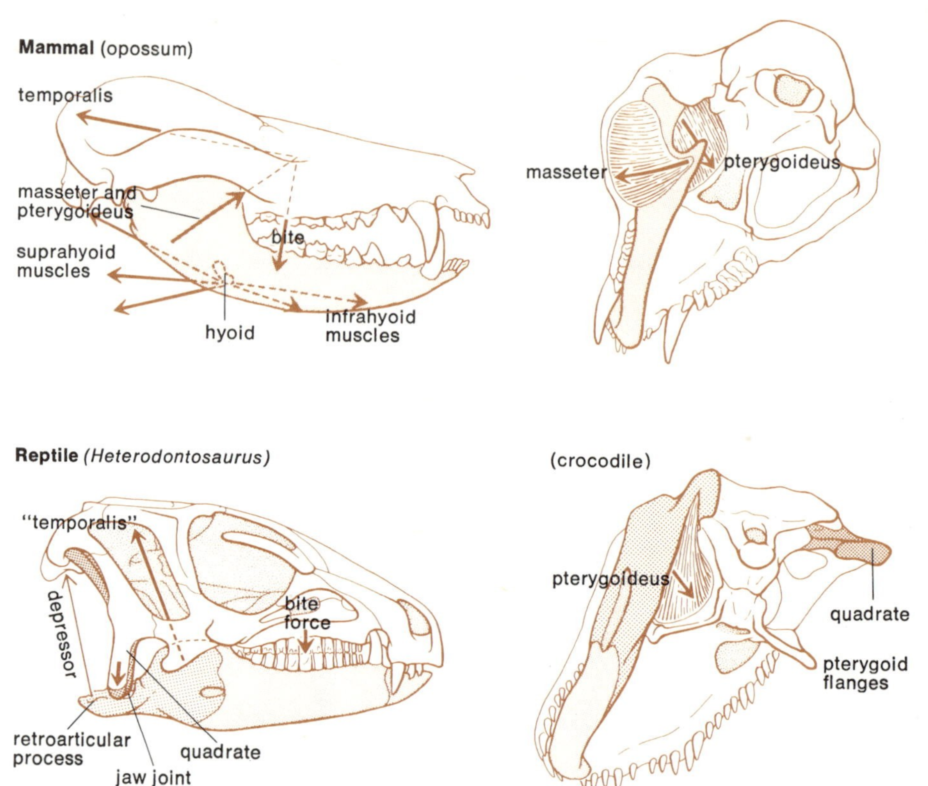

Figure 4. The muscles that open and close the jaws of a mammal are arranged differently from those in a reptile. In the American opossum, the muscles (indicated by arrows) are arranged so that the force of the muscles and the force generated between the molar teeth tend to meet and establish a stable triangle. The jaw is opened by the synchronous contraction of the infrahyoid and suprahyoid muscles, which are all attached to the hyoid bone. In reptiles, such as the late Triassic dinosaur *Heterodontosaurus,* the lower jaw acts as a third-class lever, and the vertical force acting through the jaw joint or fulcrum is greater than that acting through the point of bite. The jaw is opened by a depressor muscle running from the back of the skull to the retroarticular process, although infrahyoid and suprahyoid muscles may play a minor role. In reptiles, such as crocodiles, the jaw-closing muscles are inserted on the inner surfaces of the lower jaw and tend to pull the sides of the lower jaw toward the midline. This is prevented by a strong jaw joint and the flanges on the pterygoid bone.

In reptiles and birds, sound is conducted by a single small bone, the stapes, and a cartilaginous extension of this bone, the extrastapes. In mammals, vibrations are conducted via three small bones—the malleus, the incus, and the stapes—across an air-filled space called the middle-ear cavity. The eardrum or tympanic membrane forms part of the outer wall of this cavity; the fenestra ovalis, a passageway to the inner ear, is covered by another membrane that forms part of the medial surface of the cavity; and the small bones are arranged between the two membranes. Fully terrestrial animals such as the modern "higher" vertebrates have a large external tympanic membrane to pick up weak airborne sounds (Bekesy 1957). Because the vibrations collected from the tympanic membrane are transferred to the much smaller membrane covering the fenestra ovalis, the vibrations are amplified. This amplification system helps to overcome the impedance between air and the aqueous environment of the inner ear. Although the middle-ear structure of reptiles and mammals is different, the principle is essentially the same, and there is no obvious reason to believe that a three-boned middle ear is superior to a single-boned middle ear in conducting vibrations of the tympanic membrane. Hearing ability is well developed in both mammals and birds, and the latter have an essentially reptilian middle ear. (The improved ability in mammals to analyze complex sounds and to determine the direction from which they come is probably related to the enlargement of the cochlea region of the inner ear to house a longer basilar membrane, and to the addition of thousands of additional neurons to the brain to process information from the inner ear.)

The first steps leading to these two different types of middle-ear and jaw organization probably took place within the earliest reptiles (Carroll 1974), sometime during middle to late Carboniferous times, roughly 300 million years ago. When the ancestors of terrestrial vertebrates moved from an aquatic to a terrestrial environment, one of the problems they confronted was hearing, since detection and amplification of waterborne and airborne sounds require different anatomical structures.

The aquatic ancestors (Rhipidistian fishes) of amphibians and reptiles had an upper jaw partially supported by a bone called the hyomandibular (Jarvik 1975; Thomson 1966). The early "stem" reptiles (see Fig. 5) probably retained the hyomandibular as a support for the back of the palate and the upper jaw bones, which included the quadrate. Recent work by Manley (1972), Allin (1975), and Carroll (in press) indicates that these early fossil reptiles had probably not developed the typical reptilian tympanic membrane or middle ear, and they were probably incapable of detecting weak airborne sounds. These authors have suggested that these early reptiles placed the lower jaw flat against the substrate; groundborne vibrations were then picked up and transmitted from the lower jaw via

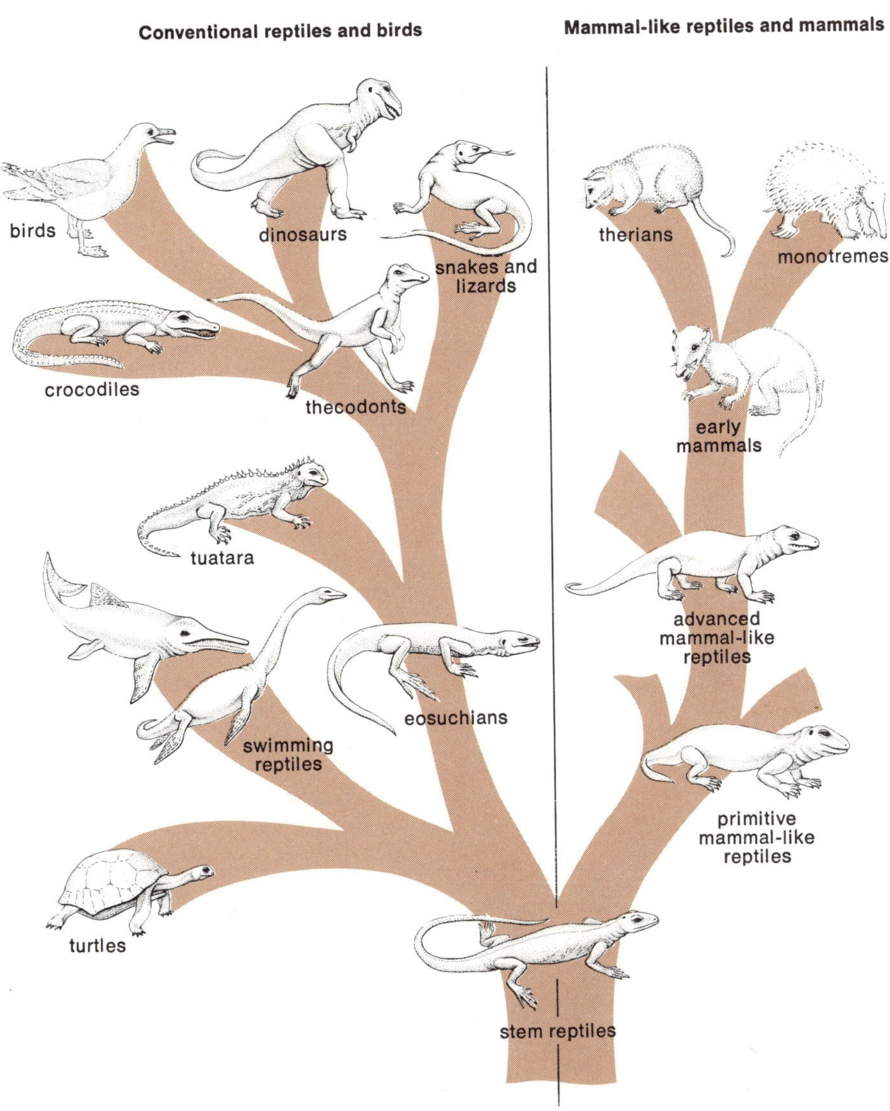

Figure 5. A simple family tree of higher vertebrates illustrates the basic dichotomy between "conventional" reptiles and birds on the one hand and mammal-like reptiles and mammals on the other, both branching out from the same ancestral stem reptiles

the jaw-joint bones and the hyomandibular, called the stapes in amphibians and higher vertebrates, to the inner ear.

All fish, including those which gave rise to the first amphibians and reptiles, have a specialized skeleton, called the hyobranchial skeleton, which helps support the gills and gill slits. Although the gills are lost in terrestrial forms, the front part of this skeleton, the hyoid skeleton, is retained in a highly modified form that in reptiles plays a major role in supporting the tongue and floor of the mouth. The stapes is also a modified part of this hyoid skeleton. In mammals, the hyoid skeleton is generally reduced to a small bone, the hyoid (see Fig. 4), which lies in front of the larynx and serves as an attachment for the principal muscles of the tongue. Barry (1968) and Allin (1975) have suggested that in the earliest stem reptiles the hyoid and tongue skeleton may have retained a connection with the stapes, as it does in reptiles such as that living fossil the tuatara, from New Zealand. If such was the case in early reptiles, this skeleton too could have picked up vibrations from the ground and transferred them directly to the inner ear via the stapes.

Whether or not early reptiles actually used bone conduction in this way we will probably never know for certain. But active terrestrial animals would obviously benefit from being able to perceive airborne sounds rather than relying exclusively on ground sounds as a means of detecting movement in their vicinity. The two main groups that descended from the earliest reptiles solved this problem in two completely different ways, and here began the fundamental difference in the ears and jaws of reptiles and mammals.

The reptile group that was destined to give rise to turtles, lizards, snakes, crocodiles, and dinosaurs, and that for the sake of convenience we will refer to as "conventional" reptiles, ceased to use the hyomandibular (the future stapes) to support the jaw articulation (the quadrate). In this group the stapes lightened and became attached to an external tympanic membrane via a cartilaginous rod called the extrastapes (see Fig. 2). It is important to note that in these reptiles and their descendants, the birds, the tympanic membrane lies behind and is partially supported by the quadrate bone. In the phylogenetic line leading from the earliest stem reptiles to the earliest of the conventional reptiles a very rapid decrease in the diameter and mass of the stapes took place. An explanation for this change might be that once a tympanic membrane developed in early reptiles, a reduction in the mass and size of the stapes greatly increased the stapes' efficiency in transmitting vibrations picked up by the tympanic membrane. This type of middle ear has been retained with only minor changes in all living terrestrial reptiles and birds.

The earliest stem reptiles gave origin not only to the conventional reptiles, but also to a group known as the mammal-like reptiles. These forms dominated the terrestrial scene for more than 100 million years. They became extinct shortly after they gave rise to the earliest mammals, about 180 million years ago. In the early mammal-like reptiles the stapes was large and, as in early reptiles, served as a support for the back of the skull and palate. Several workers (e.g. Parrington 1946 and Hopson 1966) have suggested that the mammal-like reptiles had a conventional-reptilian single-bone middle ear with a tympanic membrane situated behind the quadrate. They based this conclusion partially on the fact that in advanced mammal-like reptiles such as *Thrinaxodon,* a groove in the temporal bone terminates in a lip near the stapes/quadrate contact (Fig. 9). They suggested that the lip supported a tympanic membrane and the groove a cartilaginous ear trumpet or external auditory meatus similar to that

Figure 6. The principal jaw-closing muscles in mammal-like reptiles developed and changed in orientation. In early stem reptiles the vertical forces generated at the jaw joint were as large as or larger than those generated at the point of bite. The progressive decrease in the size of the postdentary bones was accompanied by the development of a coronoid eminence and, later, a coronoid process. This permitted a change in the direction of pull but not in the leverage of the temporalis muscle. The insertion of the pterygoideus muscle shifted progressively forward and eventually moved from the postdentary bones onto the angle of the dentary. As the intersection of the lines of force of these muscles migrated forward, the force generated at the jaw joint during chewing would have decreased in magnitude, permitting a reduction in the size of the postdentary bones and jaw joint. When the extensions of the main components of force of the jaw-closing muscles and the bite force meet, only minimal vertical forces are generated at the jaw joint.

Early Permian stem reptile

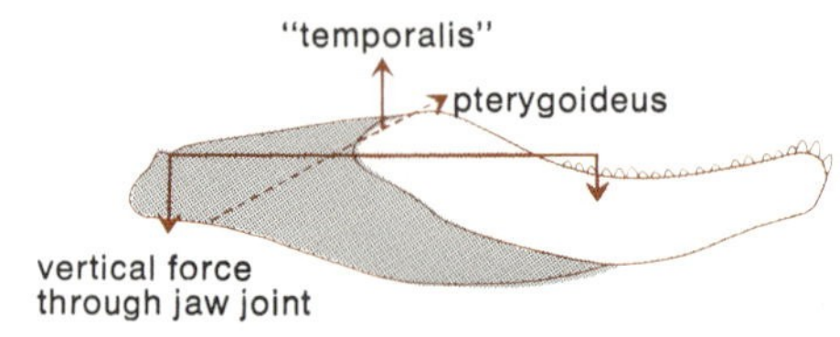

Early Permian mammal-like reptile *(Dimetrodon)*

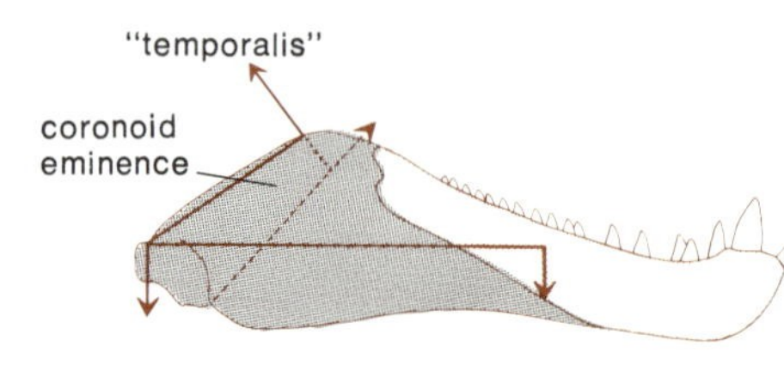

Late Permian therocephalian *(Pristerognathoides)*

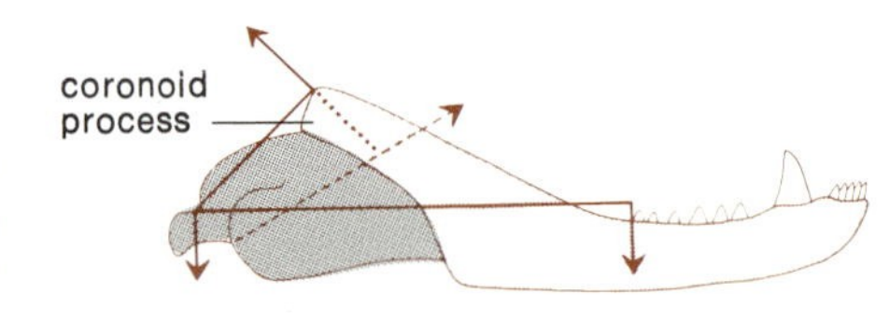

Early Triassic cynodont *(Thrinaxodon)*

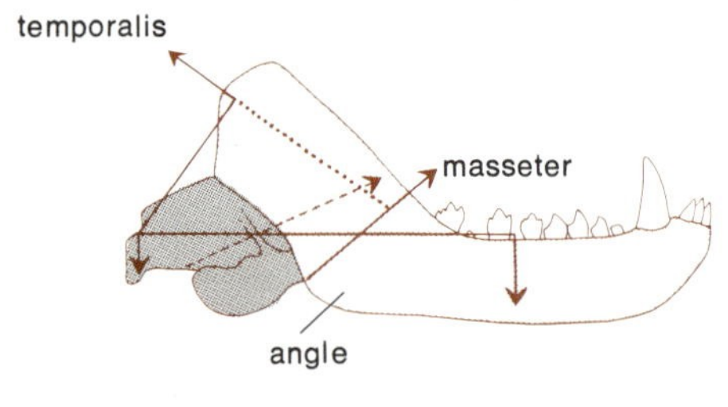

Early Triassic cynodont *(Trirachodon)*

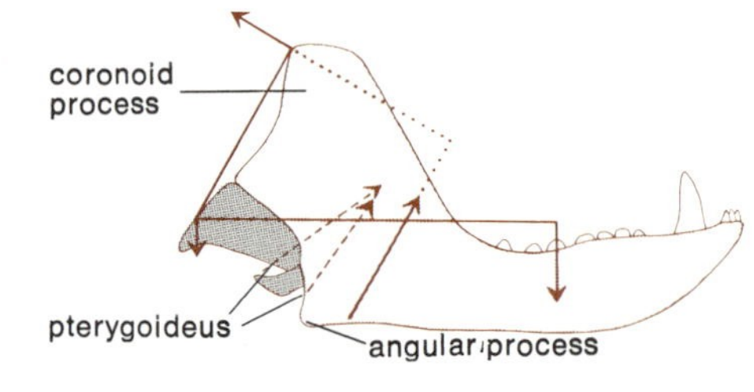

Late Triassic ictidosaur *(Diarthrognathus)*

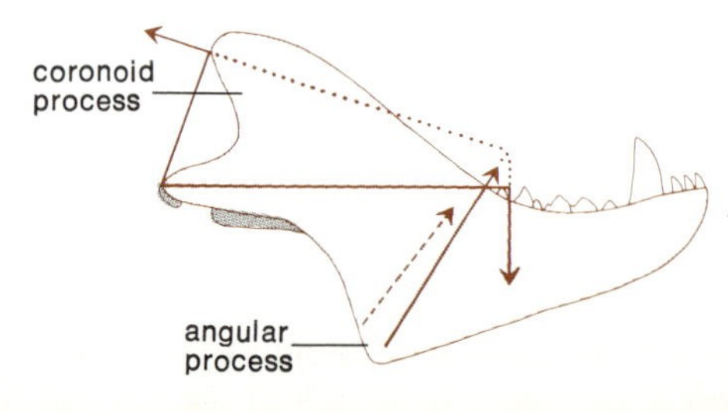

found in the living egg-laying mammals, the monotremes.

But the middle ear can only amplify weak airborne sounds if the tympanic membrane is considerably larger than the area of the footplate of the stapes where it fits into the fenestra ovalis. It is doubtful whether the arrangement suggested by Parrington and Hopson could have been functional, because a tympanic membrane placed in the position suggested by them would necessarily have been very small, not, in fact, much larger than the area of the footplate of the stapes. Allin (1975) has therefore suggested that the early mammal-like reptiles continued to use the lower-jaw joint bones to transmit groundborne vibrations from the lower jaw to the stapes, which then transmitted them to the inner ear.

He has further suggested that in the more advanced mammal-like reptiles an outpocketing or diverticulum extended from the pharynx toward the exterior below the lower jaw and the pterygoid musculature, in front of and below the jaw joint, rather than above and behind it as in conventional reptiles. The diverticulum appears to have extended upward into an area bordered medially by the body of the lower jaw behind the dentary bone and externally by a thin flange of bone extending backward from the angular bone and called the reflected lamina (Fig. 8). (The angular is one of the several bones of the lower jaw located behind the dentary.) A membrane extending from the dorsal edge of the reflected lamina to the body of the lower jaw may have formed a dorsal wall to this diverticulum. Allin claims that this membrane and the reflected lamina together formed a large surface area that functioned as a mandibular tympanic membrane to pick up airborne vibrations, transmitting these to the inner ear via the jaw-joint bones and stapes. Unlike a small tympanic membrane situated behind the quadrate, the surface area of a mandibular tympanic membrane would have been much larger than the area of the footplate of the stapes—about thirty times larger. It is possible that the groove on the temporal bone did contain an ear trumpet as suggested by Parrington (1946). This may not have terminated near the quadrate but could have looped forward to reach the mandibular tympanic membrane and reflected lamina of the angular.

In conventional living and fossil reptiles and birds, therefore, a tympanic membrane is partially supported by a jaw-joint bone, the quadrate, behind the joint, whereas in mammal-like reptiles a tympanic membrane may have been supported by a reflected lamina of the angular bone in front of the jaw articulation. We may never know why a tympanic membrane developed in different positions in mammal-like reptiles and conventional reptiles, but it is clear that a mandibular tympanic membrane would have permitted the chain of bones consisting of angular, quadrate, articular, and stapes to be used to conduct airborne vibrations picked up by the mandibular tympanic membrane, in addition to, or rather than, vibrations picked up by the lower jaw from the ground. Consequently, no basic change in the function of this chain of bones was needed when the shift from transmission of groundborne to transmission of airborne vibrations took place.

The phylogenetic history of the mammal-like reptiles is characterized by a progressive decrease in the size of the postdentary bones and accompanying increase in the size of the dentary. It has previously been claimed (Crompton 1963) that this change in jaw structure improved masticatory function by inserting all the jaw muscles on the bone that holds the teeth. If Allin's view that the jaw-joint bones were involved in

Late Permian therocephalian (*Pristerognathoides*)

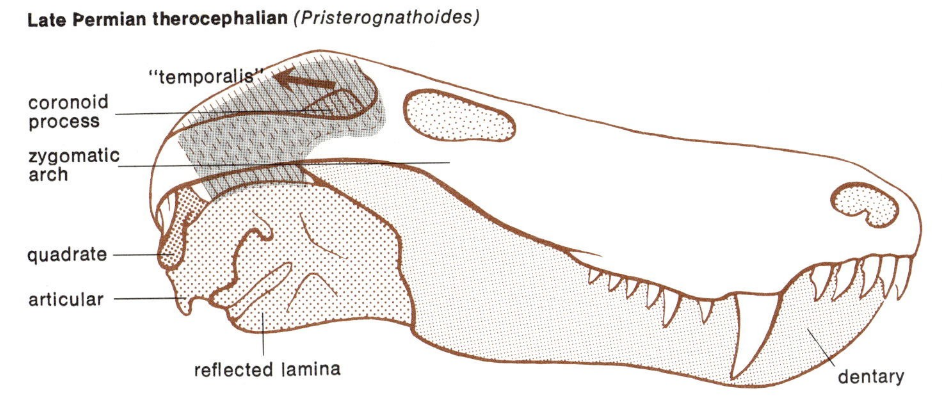

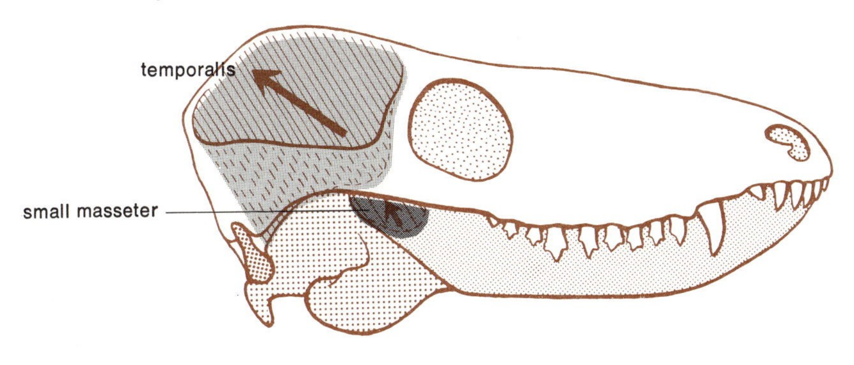

Early Triassic cynodont (*Thrinaxodon*)

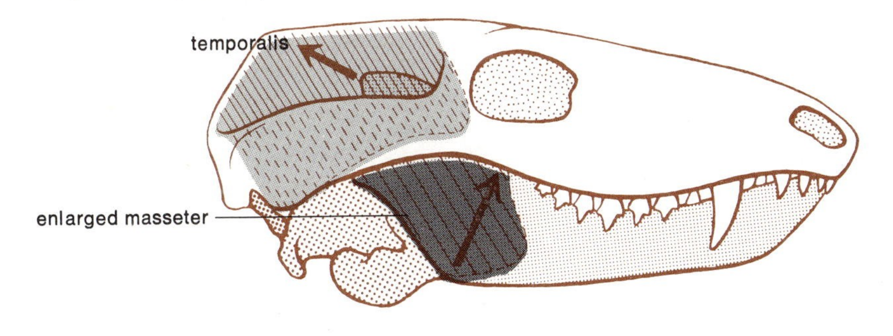

Figure 7. In a primitive mammal-like reptile such as *Pristerognathoides,* no jaw-closing musculature inserted on the outer surface of the dentary and the zygomatic arch fitted snugly against the coronoid process. In cynodonts (mammal-like reptiles from which mammals originated), the postdentary bones further reduced in size, and part of the "temporal" muscle mass migrated onto the outer surface of the dentary. In order for this musculature to reach the outer surface, it was necessary for the zygomatic arch to bow away from the dentary. A small masseter muscle was present in *Protocynodon,* an early cynodont, but in more advanced cynodonts, such as *Thrinaxodon,* which had even smaller postdentary bones, the mass of the masseter musculature had increased and its insertion area had invaded the whole back end of the dentary including the angle. Each side of the lower jaw in this form, as in all later mammals, was thus held in a muscle sling, and horizontal forces acting through the jaw joint could be reduced and controlled. (*Protocynodon* after Mendrez 1972).

hearing is correct, then there would also have been a selective advantage to decreasing the size of the postdentary bone, the quadrate, and the stapes, and removing them from participation in the jaw joint.

Evolution of the lower jaw

Mammal-like reptiles were probably capable of maintaining a fairly constant body temperature, which was probably higher than the average ambient temperature (Bakker 1975). Homeothermy, like house heating and cooling, is expensive in terms of energy. One way the energy from food can be made rapidly available to the animal is through the mechanical breakdown of food (mammals use teeth and birds use a gizzard), which speeds up the chemical process of digestion. Mechanical breakdown is especially important if fairly tough or firm foods are used as a nutrient source. In the line leading to mammals and advanced mammal-like reptiles, it is possible to observe a relative increase in the space available for jaw-closing muscles and in the areas available for the origin and insertion of these muscles (see Fig. 7) (Crompton 1963; Barghusen 1972). This increase presumably led to a relatively greater bite force and more complex jaw movements.

Mammal-like reptiles seem, therefore, to have been caught in what appear to have been two conflicting trends: a decrease in the size of the jaw-joint bones because these bones were involved in hearing; and an increase in the forces to which the jaw was subject because of the relative increase in the jaw musculature. If the lower jaw of the mammal-like reptile functioned like a third-class lever, an enlargement, not a shrinking, of the jaw joint would be expected. The fossil record, however, shows the simultaneous accommodation of these two trends: jaw muscles increasing in relative size, the dentary enlarging in relation to skull size, and the jaw joint and postdentary bones becoming progressively smaller and smaller (see Fig. 7).

There is a simple way to reduce the forces acting through the jaw joint: rearrange and reorient the jaw muscles (Crompton 1963; DeMar and Barghusen 1972). The gradual decrease in the size of the jaw joint is accompanied by (1) the acquisition and progressive development of a coronoid process, an upward vertical projection at the back of the lower jaw, so that the temporalis muscle pulls backward and upward rather than directly upward, and (2) the development of an angle or angular process, a downward projection at the back of the lower jaw, to serve for the insertion of the masseter and pterygoideus muscles, both of which pull upward and forward. Figure 6 illustrates the gradual change of direction of the principal components of these muscles that permitted a progressive unloading of the jaw joint during mastication. The main components of the forces produced by the jaw muscles of mammals now can meet above the molars and can be balanced around this meeting point so that the jaw joint is not load-bearing. Mammals are thus unique in that they can generate enormous forces between occluding molars without generating large vertically oriented forces on the jaw joint (Maynard Smith and Savage 1959).

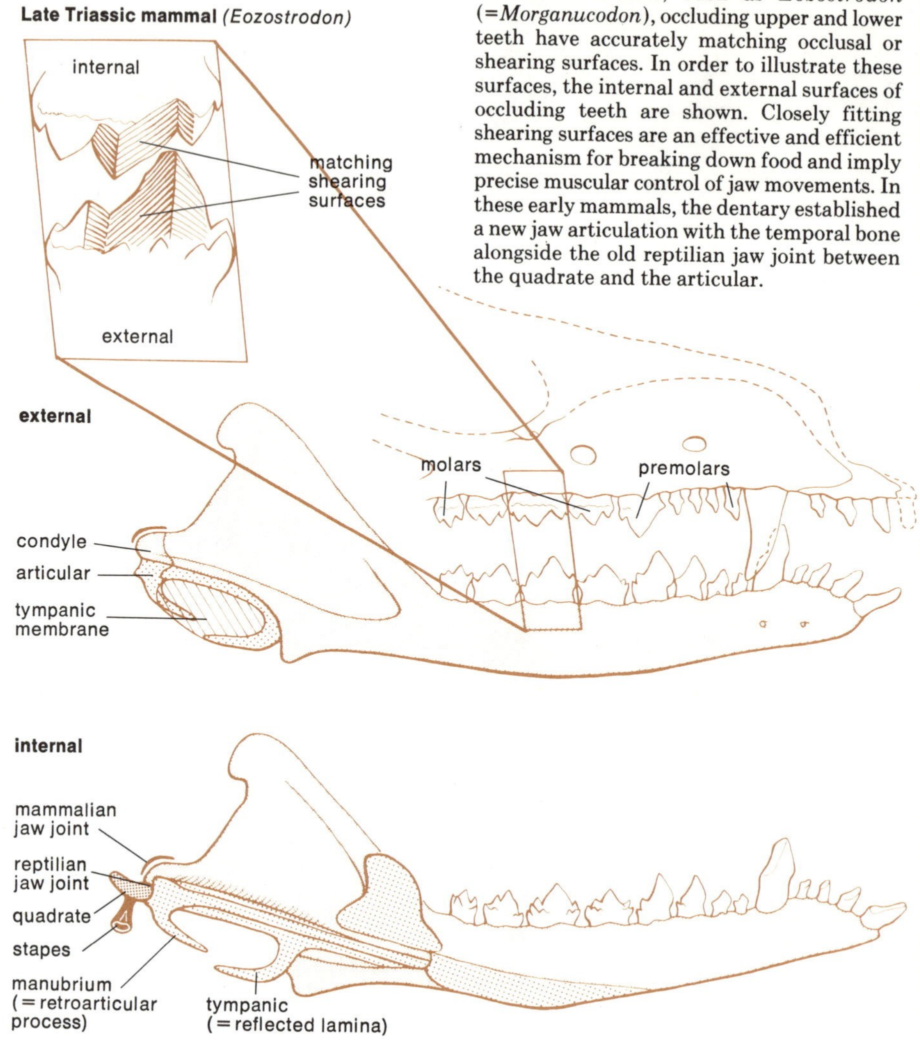

If, however, the jaw-joint bones are to become relatively smaller than those of a fairly primitive mammal-like reptile such as a therocephalian (Fig. 7), reducing the vertical forces acting through the jaw joint is only part of the problem. In the discussion of the structure of the reptile jaw it was shown that transverse flanges on the pterygoid bones (see Fig. 4) and a strong jaw joint were necessary to withstand the medially directed forces caused by the jaw-closing muscles. In mammal-like reptiles, compensation had to be made for reduced jaw-joint bones and the therefore weaker joint by reducing the horizontal forces acting through the joint as well, since if these forces were large they could easily have dislocated a small and fragile jaw joint.

A simple way to control medially directed forces acting through the jaw joint is to counterbalance the forces on the sides of the jaw. In mammals the temporalis and pterygoideus muscles both tend to deflect each side of the jaw medially, but the powerful masseters inserting on the outer sides of the jaw, which are not found in conventional reptiles, help to control this medially directed movement because they have a strong outwardly directed component. The migration of jaw-closing muscles onto the outer surfaces of the lower jaw to form the masseter complex in advanced mammal-like reptiles (Barghusen 1968; Crompton 1972; DeMar and Barghusen 1972) is well documented in the fossil record (see Fig. 7) and is accompanied by a further reduction in the size of the postdentary bones.

An important spin-off from having the sides of the lower jaw held in muscular slings is the controlled jaw movements discussed earlier. In the mammal-like reptiles such as *Thrinaxodon* (see Fig. 7), which had tearing postcanine teeth and also large pterygoid flanges to withstand medially directed forces acting on the jaw, jaw movements during feeding were essentially in a vertical plane. Upper and lower molariform (or postcanine) teeth did not come into contact with one another and therefore did not possess matching shearing surfaces. The teeth punctured food rather than cutting or grinding it. With the advent of a muscular sling and the possibility of controlled transverse jaw movements, upper and lower teeth could for the first time be brought into close contact as they sheared past one another during mastication, because the transverse flanges of the pterygoid bones were no longer a structural necessity. Shearing would have greatly improved the ability to break down food: the upper and lower teeth could function like a pair of scissors rather than a meat tenderizer. Once a system of finely controlled jaw movements was acquired, a critical threshold in the evolution of mammals was achieved, and the stage was set for the rapid evolution of many different types of shearing and grinding teeth (Crompton 1971).

Figure 8. In the oldest known mammals from the Late Triassic, such as *Eozostrodon* (=*Morganucodon*), occluding upper and lower teeth have accurately matching occlusal or shearing surfaces. In order to illustrate these surfaces, the internal and external surfaces of occluding teeth are shown. Closely fitting shearing surfaces are an effective and efficient mechanism for breaking down food and imply precise muscular control of jaw movements. In these early mammals, the dentary established a new jaw articulation with the temporal bone alongside the old reptilian jaw joint between the quadrate and the articular.

In the earliest mammals, one molar or premolar in the upper jaw occluded with two teeth in the lower jaw and vice versa (Fig. 8). Shearing surfaces with cutting edges are visible. It was essential that occluding upper and lower teeth continue to fit one another as the teeth were worn. If a new large unworn tooth erupted between two smaller worn teeth, as is the case in most reptiles, the established occlusal pattern would have been disrupted and the efficiency of the shearing function reduced. The time base of tooth development and eruption therefore changed: upper and lower molars, for example, are added sequentially from front to back (Osborn 1970, 1973) as the jaw increases in size.

The advanced mammal-like reptiles that were the ancestors of mammals probably possessed many of the characteristic features of mammalian masticatory apparatus—sequential addition of molars, matching shearing surfaces on molariform teeth, and controlled jaw movements—but they retained a small reptilian jaw joint (which transmitted vibrations from the mandibular tympanic membrane to the stapes). Continued reduction of the quadrate and the postdentary bones and enlargement of the dentary bone eventually brought the dentary close to and finally into contact with the temporal bone, establishing a new mammalian jaw joint alongside the reptilian jaw joint. Several forms (see Fig. 8, for example) document this transitional state (Kermack, Kermack, and Mussett 1968; Kermack, Mussett, and Rigney 1973).

Once the new articulation was established, the need to retain the old reptilian jaw joint as a jaw hinge was eliminated. The angular bone, which supported the mandibular tympanic membrane and some of the reduced postdentary bones and the quadrate, could separate from the jaw and become isolated in a middle-ear cavity (Fig. 9). And the development of a new mammalian jaw joint that was not involved in sound conduction once again allowed the jaw joint to bear both vertically and medially directed forces. The relatively rapid development of different types of jaws and dental configurations during the Jurassic may have been in part related to the acquisition of this new jaw articulation.

In conventional reptiles the jaw-opening muscle is attached to the backwardly directed retroarticular process. In advanced mammal-like reptiles and the earliest mammals (Figs. 8, 9), this process points down and forward rather than backward (as in the *Heterodontosaurus* in Fig. 4), and in early mammals it appears to have borne the tympanic membrane. This relationship is retained in modern mammals, in which the structure is called the manubrium (see the opossum in Fig. 9). Consequently, the reptilian depressor muscle, which inserted on the retroarticular process in early mammal-like reptiles, must have been progressively reduced and eventually lost as the retroarticular process gradually became smaller. Quite possibly the mammalian jaw-opening mechanism was present, at least in the earliest mammals and advanced mammal-like reptiles. The muscles opening the jaw in mammals are also involved in the control of the tongue, perhaps suggesting that these early mammals were capable of manipulating food in the mouth and suckling. That the earliest mammals also had two sets of teeth—milk and permanent (Parrington 1971) further supports the probability that lacta-

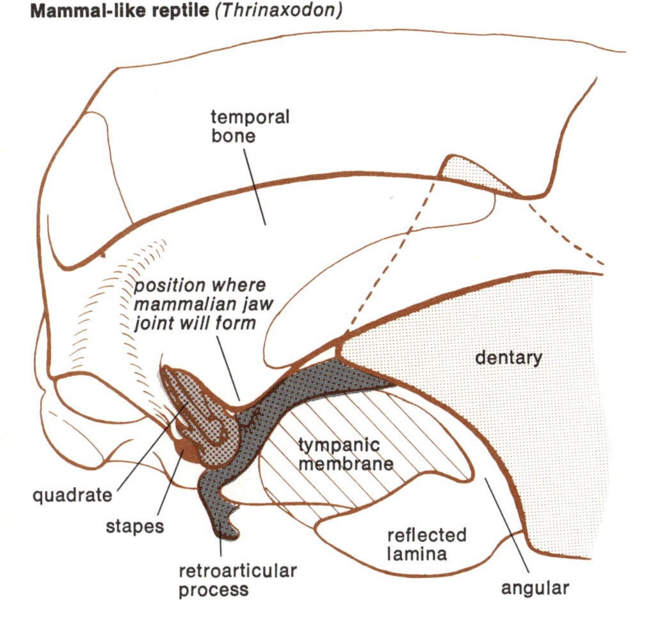

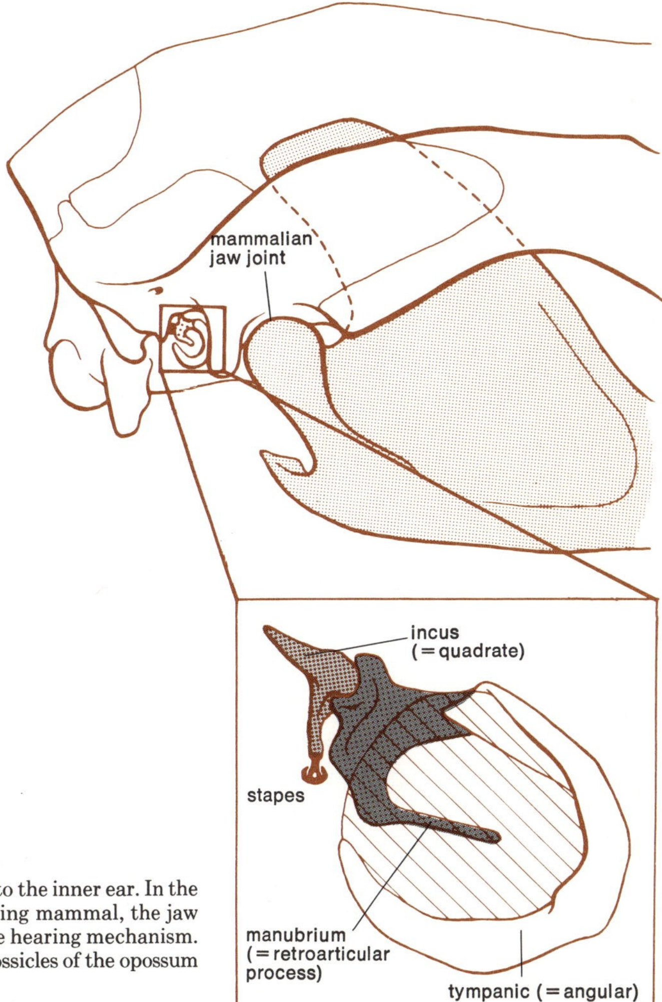

Figure 9. In the Triassic mammal-like reptile *Thrinaxodon*, the jaw-joint bones not only formed the jaw hinge but also conducted vibrations from the mandibular tympanic membrane, supported in the reflected lamina of the angular bone, to the stapes, which in turn conveyed the vibrations to the inner ear. In the opossum, a primitive living mammal, the jaw joint is separate from the hearing mechanism. (Enlargement of the ear ossicles of the opossum after Ihle et al. 1927).

tion and parental care were characteristic of the earliest mammals and possibly also the advanced mammal-like reptiles.

In mammalian ancestors, the masticatory system and the middle ear formed an integrated system or functional unit. With the help of the fossil record, it is possible to demonstrate how the selective forces that led to the improvement of both mastication and hearing have resulted in a separation of this unit into two distinct morphological parts: one for hearing and one for mastication. Almost ironically, though, the characteristic features of the mammalian masticatory system—namely, finely controlled and complex jaw movements, which are basic to the incredible diversity of mammalian dental types, were in large part made possible by a retention and specialization of features that in ancestors of mammals evolved primarily to improve auditory acuity.

References

Allin, Edgar F. 1975. Evolution of the mammalian middle ear. *J. Morph* 147:403–38.

Bakker, R. T. 1975. Experimental and fossil evidence for the evolution of tetrapod bioenergetics. In *Perspectives in Biophysical Ecology,* ed. David Gates and Rudolf Schmerl, pp. 365–99. Springer Verlag.

Barghusen, H. R. 1968. The lower jaw of cynodonts (Reptilia, Therapsida) and the evolutionary origin of mammal-like adductor jaw musculature. *Postilla* 116:1–49.

———. 1972. The origin of the mammalian jaw apparatus. In *Morphology of the Maxillo-Mandibular Apparatus,* ed. G. H. Schumacher. pp. 26–32. VEB Georg Thieme.

Barry, T. H. 1968. Sound conduction in the fossil anomodont *Lystrosaurus. Ann. S. Afr. Mus.* 1950:275–81.

Bekesy, Georg. 1957. The ear. *Sci. Am.* 97: 66–78.

Carroll, R. 1974. Problems of the origin of reptiles. *Biol. Revs.* 44:393–432.

———. In press. The origin of reptiles. In *Problems of Vertebrate Evolution,* ed. Andrews, Niles & Walker. Academic Press.

Crompton, A. W. 1963. On the lower jaw of *Diarthrognathus* and the origin of the mammalian lower jaw. *Proc. Zool. Soc. Lond.* 140:697–750.

———. 1971. The origin of the tribosphenic molar. In *Early Mammals,* ed. D. M. Kermack and K. A. Kermack. Suppl. no. 1, *J. Linn. Soc. (zool.)* 50:65–87.

———. 1972. The evolution of the jaw articulation of cynodonts. In *Studies in Vertebrate Evolution,* ed. K. A. Joysey and T. S. Kemp, pp. 231–52. Oliver and Boyd.

———. 1974. The dentitions and relationships of the southern African Triassic mammals, *Erythrotherium parringtoni and Megazostrodon rudnerae. Bull. British Mus. (Nat. His.) Geol.* 24:399–444.

Crompton, A. W., and F. A. Jenkins. 1973. Mammals from reptiles: A review of mammalian origins. *Ann. Rev. Earth and Plan. Sci.* 1:131–53.

Crompton, A. W., A. Thexton, P. Parker, and K. M. Hiiemae. 1977. The activity of the jaw and hyoid musculature in the opossum. In *Biology of Marsupials,* ed. B. Stonehouse and D. Gilmore, pp. 287–305. MacMillan, London.

DeMar, R., and H. R. Barghusen. 1972. Mechanics and evolution of the synapsid jaw. *Evolution* 26:622–37.

Edmund, A. G. 1962. Sequence and fate of tooth replacement in the Crocodilia. *Roy. Ont. Mus., Life Sci. Div. Contribution* 56: 1–41.

Ewer, R. F. 1963. Reptilian tooth replacement. *News Bull. Zool. Soc. So. Afr.* 4:4–9.

Frazetta, T. H. 1962. A functional consideration of cranial kinesis in lizards. *J. Morph.* 3:287–320.

Hiiemae, K. M. In press. Mammalian mastication: A review. In *Development, Function and Evolution of Teeth,* ed. P. M. Butler and K. A. Joysey. Academic Press.

Hopson, J. A. 1966. The origin of the mammalian middle ear. *Am. Zool.* 6:437–50.

———. 1973. Endothermy, small size and the origin of mammalian reproduction. *Am. Nat.* 107:446–52.

Ihle, J. E. W., P. N. van Kampen, H. F. Nierstraz, and J. Versluys. 1927. *Vergleichende Anatomie der Wirbeltiere.* Julius Springer.

Impey, O. R. Functional aspects of cranial kinetism in the Lacertilia. Ph.D. dissertation, 1967, University of Oxford.

Jarvik, E. 1975. On the saccus endolymphaticus and adjacent structures in osteolepiforms, anurans and urodeles. In *Problemes Actuels de Paleontologie: Evolution des Vertebres. Colloque international C.N.R.S.* 218:191–211.

Kermack, D., K. A. Kermack, and F. Mussett. 1968. The Welsh pantothere *Kuehneotherium praecursoris. J. Linn. Soc. (zool.)* 47: 407–23.

Kermack, K. A., F. Mussett, and H. W. Rigney. 1973. The lower jaw of *Morganucodon. J. Linn. Soc. (zool.)* 53:87–175.

Manley, G. A. 1972. A review of some current concepts of the functional evolution of the ear in terrestrial vertebrates. *Evolution* 26:608–21.

Maynard Smith, J., and R. J. G. Savage. 1959. The mechanics of mammalian jaws. *School Sci. Rev.* 141:289–301.

Mendrez, C. H. 1972. Revision du genre *Protocynodon* Broom 1949 et discussion de sa position taxonomique. *Paleont. Afr.* 14: 19–50.

Osborn, J. W. 1970. New approach to Zahnreihen. *Nature* 225:343–46.

———. 1973. The evolution of dentitions. *Am. Sci.* 61:548–59.

Parrington, F. R. 1946. On the cranial anatomy of cynodonts. *Proc. Zool. Soc. Lond.* 116: 181–97.

———. 1958. The problem of the classification of reptiles. *J. Linn. Soc. (zool.)* 44:99–115.

———. 1971. On the upper Triassic mammals. *Phil. Trans. R. Soc. Lond. (B)* 261:231–72.

Reichert, C. 1837. Über die Visceralbogen der Wirbeltiere im Allgemeinen und deren Metamorphosen bei den Vögeln und Säugetieren. *Archiv fur Anatomie, Physiologie und wissenschaftl. Medecin,* pp. 120–222.

Romer, A. S. 1956. *Osteology of the Reptiles.* Univ. Chicago Press.

Romer, A. S., and L. W. Price. 1940. Review of the Pelycosauria. *Geol. Soc. of Amer. Spec. Papers* 28:1–538.

Stein, M. R. 1939. The "mandibular sling." *Dental Survey* 1939:883–87.

Thexton, A. J. 1976. To what extent is mastication programmed and independent of peripheral feedback? In *Mastication: Proceedings of a Symposium on Clinical and Physiological Aspects of Mastication,* ed. D. J. Anderson and B. Matthews, pp. 213–20 John Wright & Sons.

Thomson, K. S. 1966. The evolution of the tetrapod middle ear in the rhipidistian-amphibian transition. *Am. Zool.* 6:379–97.

Throckmorton, G. S. 1976. Oral food processing in two herbivorous lizards, *Iguana iguana* (Iguanidae) and *Uromastix aegyptius* (Agamidae). *J. Morph.* 148:363–90.

Leonard Radinsky

Primate Brain Evolution

Comparative studies of brains of living mammal species reveal major trends in the evolutionary development of primate brains, and analysis of endocasts from fossil primate braincases suggests when these specializations occurred

Our knowledge of the evolution of primate brains comes from two sources: the fossil record and comparisons between the brains of the various kinds of living primates. The fossil record of primate brains consists of casts of the insides of fossil braincases, called endocranial casts, or endocasts. These casts reveal the size and shape of the brains of extinct primates and, in all but the largest-brained species, such as the great apes and humans, also reproduce details of the surface of the brain, including the patterns of cerebral convolutions. Fossil endocasts provide the only direct evidence of brain evolution, but the information they provide is limited to external morphology. Comparative studies of the brains of living primates, on the other hand, while they can provide an enormous amount of information, can only be considered indirect evidence of brain evolution. Although it is tempting to arrange living species in a series of presumed primitive to advanced stages, series of this sort may be misleading if they are taken to represent evolutionary lineages, because each species is specialized for its own ecological niche and most species are mosaics of primitive and advanced features. Since living species are the only source of information on most aspects of brain evolution, however, we must do the best we can with information obtained from studies of living brains, keeping in mind the pitfalls of inferring evolutionary sequences from this indirect evidence.

Leonard Radinsky is a vertebrate paleontologist who became interested in brains when he realized that the cortical mapping studies of neurophysiologists could be used to interpret sensory specializations of extinct mammals. He has written several papers on the phylogenetic and functional implications of the neuroanatomy of various living and fossil carnivores and primates, and, with the support of a grant from the National Science Foundation, he is currently studying the fossil record of ungulate brains, to see if there are correlations between changes in the brains of competing groups of ungulates and of their predators, the large carnivores. Dr. Radinsky earned his Ph.D. at Yale University in 1962 and taught biology at Boston University and Brooklyn College before going to the University of Chicago, where he is now an Associate Professor of Anatomy and Evolutionary Biology. Address: Department of Anatomy, University of Chicago, Chicago, IL 60637.

Primates are divided into two groups: the prosimians, or lower primates, and the anthropoids, or higher primates. The prosimians include the earliest primates, which first appeared about 70 million years ago, and are represented today by the lemurs, lorises, and galagos. The anthropoids first appeared about 35 million years ago, and include the New World and Old World monkeys, the apes, and humans. The relationships of *Tarsius*, a small, large-eyed jumping form from southeast Asia, are currently controversial. Primates evolved from insectivorans, and while living insectivorans (such as shrews, moles, and hedgehogs) are different in many ways from the ancient insectivorans that gave rise to primates, they nevertheless provide the best living evidence of what the brains of the primates' ancestors were like.

Comparative studies of living organisms

Despite the great amount of interest in primate evolution in general and in the evolution of primate brains in particular, until recent years there were surprisingly few broad comparative studies of primate brains. Now, however, thanks to studies such as those of H. Stephan, R. Bauchot, and their colleagues, there is available a sizable body of data on the brains of a large series of living primates and insectivorans.

The relationship between brain weight and body weight in the smallest-brained living insectivorans (the basal insectivorans of Bauchot and Stephan) is described by the equation: log brain weight (grams) $= -1.37 + 0.63$ log body weight (grams). That equation probably represents relative brain size in the insectivoran ancestors of primates as well. The 33 species of living insectivorans measured by Bauchot and Stephan have brains ranging from 0.80 times to about 3.0 times the basal insectivoran level described by that equation. Relative brain size in living prosimian primates ranges from 2.4 times to 7.0 times what would be expected in a basal insectivoran of comparable body weight, with the lemurs *Lepilemur* and *Hapalemur* at the low end of the range and the bizarre aye-aye, *Daubentonia*, at the high end. Most living anthropoid primates range in relative brain size from about 5.3 to 11.7 times the basal insectivoran level, with extensive overlap between New World and Old World monkeys and apes. Outliers are *Alouatta*, the howler monkey, with an Encephalization Index of only 4.76, and humans, with an E.I. of 28.76! Figure 1 illustrates the trend toward increased relative brain size in the progression from insectivoran to anthropoid.

Most of the parts of the brain measured by Stephan and Bauchot show increase in size relative to body weight in the series insectivorans→prosimians→anthropoids. (Exceptions are structures associated with olfaction, which decrease in relative size in that series.) Take for example the cerebellum, the part of the brain concerned with muscle coordination and the maintenance of bodily equilibrium. The cerebellum of prosimians ranges from about 2.9 to 8.1 times what would be expected in a basal insectivoran of comparable body weight, while the cerebellum of anthropoids ranges from about 4.3 to 11.5 times the basal insectivoran standard, except for that of humans, which is 20.9 times larger than would be expected in a basal insectivoran of human body weight.

Since total relative brain size also increased in the evolution from insectivorans to prosimians to anthropoids, the question arises as to whether the increase in relative size of various parts of the brain can be accounted for by the increase in total relative brain size. If we plot cerebellar volume versus total brain volume, the resulting graph shows both prosimians and anthropoids, including humans, to have a cerebellum of the size one would expect in a basal insectivoran of comparable brain size (Fig. 2). Thus, high progression indices, such as 20.9 for the human cerebellum, do not necessarily indicate a disproportionate increase in the relative size of specific parts of the brain.

Of the various parts of the brain measured by Stephan and Bauchot, the cerebral neocortex (the outer layer covering much of the forebrain, which includes visual, auditory, tactile, and motor centers as well as association areas) shows the greatest increase in size relative to body weight in comparisons between insectivorans, prosimians, and anthropoids. Modern prosimians have 8.3 to 26.5 times the amount of neocortex we would expect in a basal insectivoran of comparable body weight, with *Lepilemur* and *Hapalemur* at the low end and *Daubentonia* at the high end of the range. Anthropoids have about 21 to 61 times as much neocortex as basal insectivorans of comparable size. An exception is humans, who have 156 times as much! If we take into account the differences in brain size (Fig. 3), it turns out that prosimian brains have about 3⅓ to 3½ times as much neocortex, and anthropoid brains about 4 times as much neocortex, as would be expected in a basal insectivoran brain of comparable size. Human brains have the expected amount of neocortex for an anthropoid brain of their size.

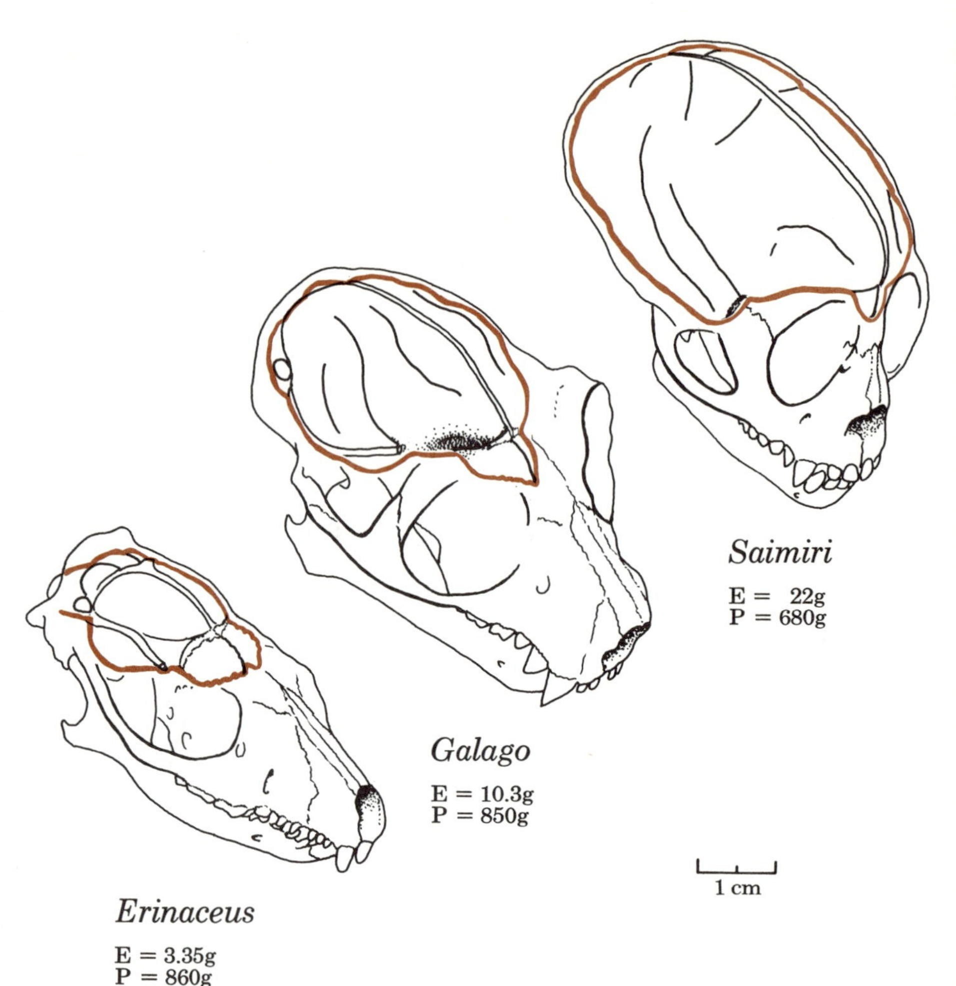

Figure 1. Endocasts (*color*) superimposed on skulls of *Erinaceus europaeus* (hedgehog), *Galago crassicaudatus* (galago), and *Saimiri sciureus* (squirrel monkey), representing basal insectivoran, prosimian primate, and anthropoid primate, respectively, illustrate the increase in relative brain size in these groups. All are drawn to the same scale. E = brain weight; P = body weight. (Weights are species means, from Stephan et al. 1970.)

The size of the olfactory bulb (relative to body weight) shows a reverse trend. This feature provides a rough indicator of the importance of the sense of smell in a given species. Among living prosimians, olfactory bulbs range from 1.3 to 0.13 times the size expected for the olfactory bulb of a basal insectivoran of comparable body weight, with most species falling below 0.90. *Galago demidovii* is at the high end of the range, and the lemur *Indri* at the low end. Among living anthropoids, olfactory bulbs range in relative size from 0.2 to 0.02 times the basal insectivoran level, with *Aotes*, the owl monkey, and some marmosets at the high end and humans at the low end. Most anthropoids fall below 0.10 times the basal insectivoran level in relative olfactory bulb size.

Stephan (1969) has provided comparative data on the relative volumes of the primary visual cortex (= striate cortex) and of the lateral geniculate nucleus, a major visual center. The relative sizes of these structures are probably a rough indicator of the relative importance of visual information for a given species. Data from basal insectivorans are not available for these structures, and for the purpose of

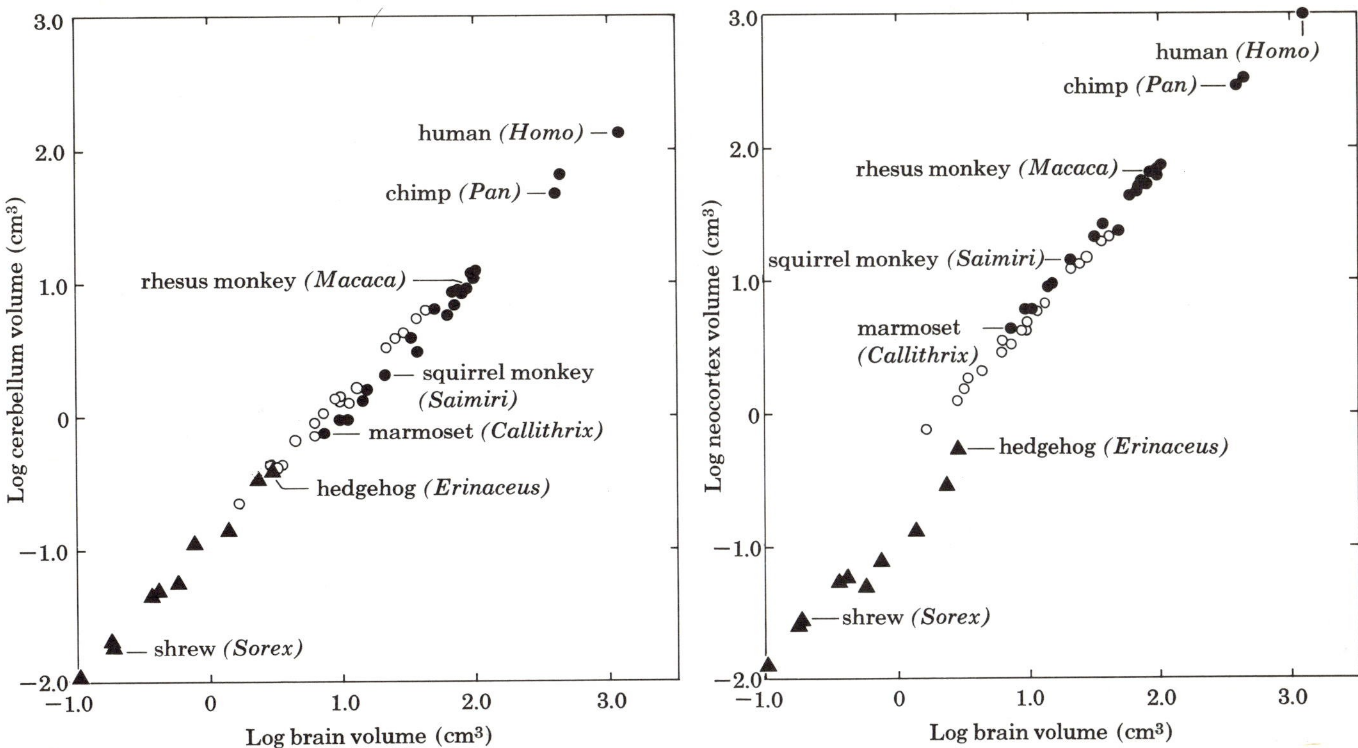

Figure 2. The relationship between cerebellum size and brain size is shown for 10 basal insectivorans (*triangles*), 18 prosimians (*open circles*), and 21 anthropoids (*solid circles*). Although the human cerebellum is 20.9 times larger than that expected for a basal insectivoran of comparable body weight, the graph shows that both prosimians and anthropoids, including humans, have a cerebellum of the size that would be expected in a basal insectivoran of comparable brain size. Apparently the increase in relative size of some portions of the brain is related to the increase in total relative brain size. (Data from Stephan et al. 1970.)

Figure 3. The relationship between neocortex volume and brain size is shown for 10 basal insectivorans (*triangles*), 18 prosimians (*open circles*), and 21 anthropoids (*solid circles*). Note that primates have relatively more neocortex for their brain size than do basal insectivorans. (Data from Stephan et al. 1970.)

comparison, the condition in *Lepilemur* was taken as a standard. In relative volume of primary visual cortex, most living prosimians range from 1.0 (in *Lepilemur*) to 3.2, except for *Tarsius,* which has 4.6 times as much primary visual cortex as would be expected in a *Lepilemur* of comparable body weight. In most living anthropoids, relative volume of primary visual cortex falls within a range of 3.0 to 7.5 times the *Lepilemur* standard, with *Alouatta* an outlier at 2.0. Humans fall within the anthropoid range at 5.0. Comparisons of the relative size of the lateral geniculate nucleus yield similar results, with anthropoids overlapping but averaging higher than prosimians. Although the published data on relative size of striate cortex and lateral geniculate nucleus do not permit direct comparison with brain volume, Stephan's (1969) graph of striate cortex versus body weight and Passingham and Ettlinger's (1974) graph of striate cortex versus neocortex suggest that the apparent trend toward increased relative size of striate cortex in anthropoids as compared to prosimians is accounted for by the increase in overall relative brain size.

Comprehensive surveys of the relative size or complexity of parts of the brain that might reveal major evolutionary trends in auditory, tactile, or motor systems have not been published. Verhaart (1970) reviewed the anatomy of the pyramidal tract (the main motor pathway) of about a dozen primate species and concluded that there was an increase in nerve fiber size and further caudal extent of the pyramidal tract into the spinal cord in the series prosimians→monkeys→apes and humans. That work did not consider the possible effects of body size on those features, however.

An example of a specialization of the somatic sensory (tactile) cortex has been demonstrated in *Ateles,* the prehensile-tailed spider monkey. The area of somatic sensory cortex receiving tactile information from the tail in *Ateles* is about five times as large as in *Macaca mulatta,* the rhesus monkey, a species of similar size that lacks a prehensile tail (Hirsch and Coxe 1958; corrections in Pubols and Pubols 1971), and the motor cortex for the tail in *Ateles* is similarly enlarged. That finding correlates with the observed use of the tail in *Ateles* (and other prehensile-tailed monkeys) for tactile exploration, manipulation, and gripping.

Insectivoran brains are relatively small, have little neocortex, and show few if any neocortical folds. With the evolution of relatively larger brains and more neocortex, folds appear in the neocortex of the brains of primates as well as in those of almost all other orders of mammals. The causes and mechanisms of cortical folding are not well understood. Comparisons of brains of different-sized prosimians, New World monkeys, apes, and humans reveal an increased amount of folding with increased brain size (see figures in Radinsky 1968 and 1974b; Hershkovitz 1970; Connolly 1950);

but this trend is not so obvious in Old World monkey brains. Cortical mapping studies (see e.g. Sanides 1970; Sanides and Krishnamurti 1967; Welker and Campos 1963) suggest that folds tend to develop between different functional areas of the cortex.

A major difference in the arrangement of cortical folds distinguishes most prosimian brains from those of anthropoids. In prosimians, as in most mammals with a comparable degree of neocortical folding, a longitudinally oriented fissure, the coronal sulcus, separates the representation of the head from that of the forelimb in the primary motor and somatic sensory cortex. In anthropoids, no major fissure exists in that position, but instead the primary motor cortex is separated from the primary somatic sensory cortex by a major transverse fissure, the central sulcus (Fig. 4). Pottos (*Perodicticus* and *Arctocebus*) are exceptions among prosimians, having a central sulcus similar to that seen in anthropoids. Many anthropoid brains have small sulci marking the boundary between head and forelimb representations in motor and somatic sensory cortical areas, but none have a complete coronal sulcus like that of the prosimians.

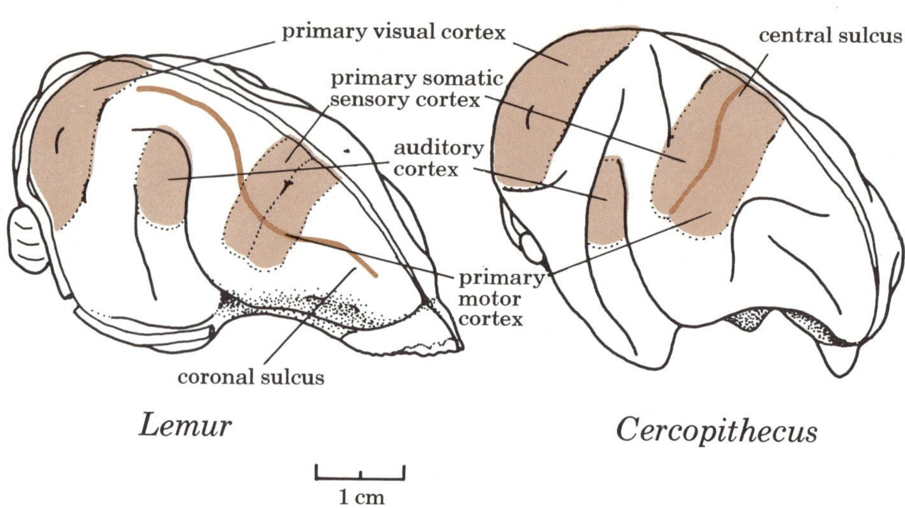

Figure 4. The endocasts of *Lemur variegatus* and *Cercopithecus talapoin* illustrate differences in cortical folding in prosimians and anthropoids. Areas in light color are major sensory and motor cortical areas, inferred from mapping studies of related species. The coronal sulcus, running lengthwise along the brain, is characteristic of prosimians as well as many other mammals. This feature is not seen in anthropoids, which have instead a central sulcus, running across the brain. The functional significance of that difference in cortical folding is not known.

The functional significance of the difference in cortical folding between prosimians and anthropoids is not obvious. It has been suggested that the transversely oriented anthropoid central sulcus reflects the requirements of packing a relatively large brain into a braincase impinged on by relatively large orbits in front (most anthropoid brains show prominent orbital impressions), and with expanded visual cortex at the back (the occipital lobe). However, that suggestion leaves unexplained the occurrence of a central sulcus in the prosimian pottos.

To summarize, evolutionary trends suggested by comparative studies of the brains of living insectivorans and primates include: increase in size of brain relative to body weight; increase in amount of neocortex (beyond what one would expect from the increase in brain size); decrease in relative size of olfactory bulbs; increase in amount of visual cortex and size of lateral geniculate body (possibly accounted for by the overall increase in brain size); development of a central sulcus in anthropoids rather than the coronal sulcus seen in prosimians. Now let us turn to the fossil record to see when those specializations occurred in the evolutionary history of primates.

Fossil endocasts

The fossil record of primate brains is scanty, owing to the scarcity of complete fossil primate skulls. Only about 13 genera of primates that lived before the Pleistocene Ice Age are represented by endocasts (see Fig. 5). I recently reviewed the record of fossil prosimian and anthropoid endocasts, excluding hominids (Radinsky 1970, 1974a), and Holloway (1972, 1975) has reviewed fossil hominid endocasts.

Contrary to statements in the literature, there is no good evidence relating to the brains of the earliest primates, members of the Paleocene evolutionary radiation (Radinsky, in press). That radiation is known mainly from fragments of jaws and teeth, and the few skulls discovered to date are not well enough preserved to yield endocasts. The oldest good record of primate brain morphology is an early Eocene (about 55 million years old) endocast of *Tetonius*, a small prosimian that appeared early in the second major evolutionary radiation of primates (Fig. 6). The *Tetonius* endocast is preserved in a skull that was discovered almost a hundred years ago in Wyoming, and it is still the only skull known for that genus. The shape of the *Tetonius* endocast suggests expansion of the occipital and temporal lobes, areas composed of visual cortex, and reduction in the size of the olfactory bulbs, compared to the condition in basal insectivorans. Compared to modern prosimians, *Tetonius* appears to have had a relatively smaller frontal lobe. Estimates of brain size and body weight (the latter not very reliable due to the lack of postcranial skeletal elements) suggest that relative brain size in *Tetonius* fell below that of modern prosimians but above that of basal insectivorans.

Endocasts of other early prosimians—*Smilodectes* (middle Eocene, about 50 million years old), *Adapis* and *Necrolemur* (late Eocene, about 40 million years old), and *Rooneyia* (early Oligocene, about 35 million years old)—agree with that of *Tetonius* in suggesting reduction of olfactory bulbs and expansion of visual cortex compared to the basal insectivoran condition, and relatively smaller frontal lobes as compared to modern prosimians. Brain weight–body weight estimates

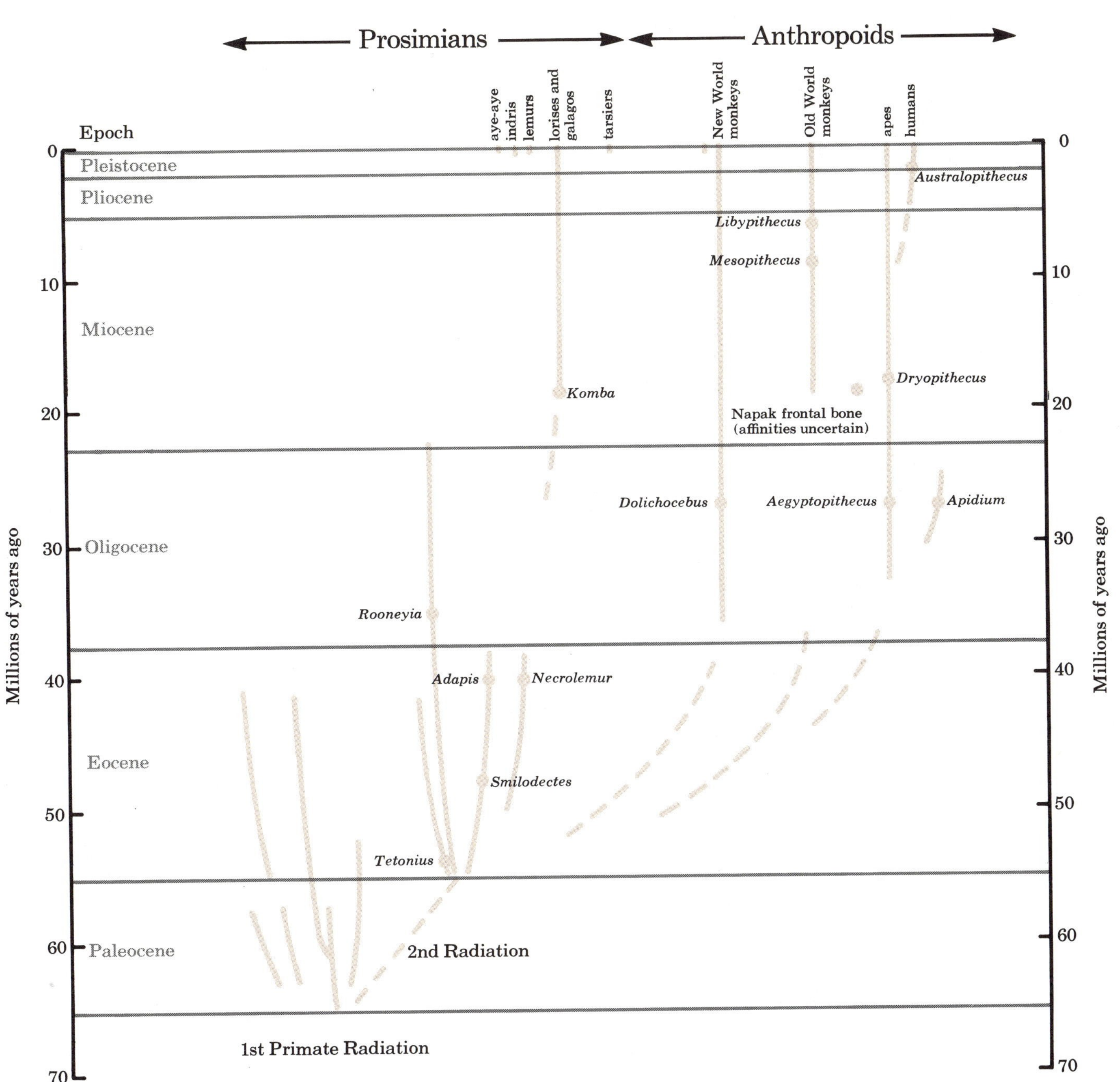

Figure 5. The primate family tree shows both living and extinct families. The origins of the major living groups are still uncertain. The sparse fossil record of primate brain evolution is represented by the specimens labeled at their approximate position in the geological time scheme (several Pleistocene prosimian and Old World monkey endocasts have been omitted). No good evidence exists regarding brains of species in the first primate radiation.

suggest that *Necrolemur* and *Rooneyia* may have attained the modern prosimian range of relative brain size. The oldest record of a prosimian brain of modern appearance is an endocast of *Komba*, a Miocene galago (about 18 to 20 million years old).

The oldest record of anthropoid brains is from the late Oligocene, about 25 to 30 million years ago, shortly after the first appearance of higher primates in the fossil record. The evidence consists of partial endocasts of *Aegyptopithecus* (Fig. 7), one of the oldest apes (and possibly ancestral to humans as well as to later apes), *Apidium*, a genus perhaps close to the ancestry of Old World monkeys, and *Dolichocebus*, one of the oldest New World monkeys. The *Aegyptopithecus* endocasts (Fig. 8) display primary visual cortex (bounded by the lunate sulcus) apparently expanded as in modern anthropoids and larger than would be expected in a prosimian endocast of comparable size.

The olfactory bulbs are reduced as in modern anthropoids and appear smaller than in modern prosimians. In addition, *Aegyptopithecus* had a central sulcus, a feature that distinguishes the brains of modern anthropoids from those of most prosimians. The frontal lobe of *Aegyptopithecus* appears to have been relatively smaller than that of modern anthropoids.

The *Apidium* endocast, which preserves only the frontal portion, suggests reduced olfactory bulbs compared to the prosimian condition. The *Dolichocebus* endocast preserves little surface detail and lacks olfactory bulbs, but it does show an occipital lobe (composed of visual cortex) expanded as in anthropoids and larger than in prosimians. Because of the problems of estimating body weights, it is impossible to determine whether the Oligocene anthropoids had brains that fell within the modern anthropoid range in relative size. The oldest record of an anthropoid brain of modern appearance is that of *Dryopithecus* (*Proconsul*), an 18-million-year-old ape.

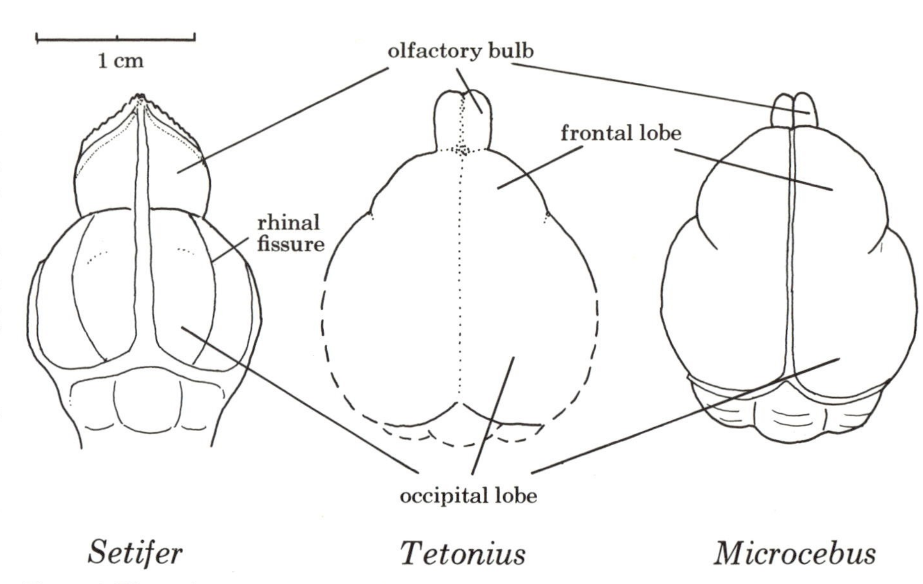

Figure 6. The endocast of *Tetonius homunculus*, a 55-million-year-old prosimian, is compared with endocasts of *Setifer setosus*, a living hedgehog-like basal insectivoran, and *Microcebus murinus*, the mouse lemur, a living prosimian primate. The frontal lobe is not delimited in *Setifer*. The rhinal fissure (visible in dorsal view only in *Setifer*) marks the lateral boundary of neocortex. All are drawn to the same scale. *Tetonius* was probably a larger animal than *Microcebus* but smaller than *Setifer*.

The oldest fossil record of hominid brains, that is, of species on or close to our ancestral line, consists of a small number of endocasts about 2 to 3 million years old. Because of their large size, they preserve few surface details. Nevertheless, in proportions and in the apparently expanded parietal association cortex (a feature whose exact function is unknown), they suggest a brain more similar to that of modern humans than to that of the apes. In relative size these early fossil hominid endocasts appear to have been between humans and other anthropoids (Pilbeam and Gould 1974).

In summary, the fossil evidence of primate brains suggests that the specializations of expanded visual cortex and reduced olfactory bulbs that distinguish modern prosimians from basal insectivorans had appeared by 55 million years ago, at the beginning of the second major evolutionary radiation of primates. At that time frontal lobes were still relatively small, and in at least one genus of fossil prosimian, relative brain size appears to have been between that of basal insectivorans and modern prosimians. By about 45 million years ago, at least some fossil prosimians appear to have exceeded the lower end of the modern prosimian range of relative brain size.

The anthropoid record suggests that 25 to 30 million years ago, shortly after the first appearance of anthropoids in the fossil record, the visual cortex had expanded and olfactory bulbs were reduced below the prosimian condition and appeared as in modern anthropoids. A main difference in cortical folding that distinguishes modern anthropoids from most prosimians—the transversely oriented central sulcus—was also present in at least some of the earliest anthropoids. Early anthropoids had relatively small frontal lobes, as did early prosimians. By about 18 million years ago, both prosimians and anthropoids with brains of modern aspect had appeared.

Uniqueness of primate and human brains

Introductory anthropology textbooks and other secondary sources frequently characterize primate brains as being relatively large, with a highly folded neocortex, an elaborate visual system, and a reduced olfactory system. Le Gros Clark, one of the foremost experts on primate brains, wrote: "Undoubtedly the most distinctive trait of the Primates, wherein this order contrasts with all other mammalian orders in its evolutionary history, is the tendency towards the development of a brain which is large in proportion to the total body weight, and which is particularly characterized by a relatively extensive and often richly convoluted cerebral cortex .... In the Primates this expansion began earlier, proceeded more rapidly, and ultimately advanced much further [than in other mammals]" (1971). If we look at the brains of a variety of other mammals, however, we find that those features are by no means unique to primates.

Wirz (1966) has provided an important compilation of data on sizes of brain parts in most of the living orders of mammals which reveals that, with respect to relative brain size, prosimians are matched or exceeded by representatives of most of the other living orders of mammals, and anthropoids are matched by several orders of mammals, including carnivorans, ungulates, and cetaceans. Only one species of primate, *Homo sapiens*, has a brain that exceeds that of all other mammals in relative size. With respect to amount of neocortex, the same holds, with prosimians no more advanced than most other mammals, and anthropoids matched by several orders. The visual system is elaborated in several other mammals as diverse as cats, squirrels, ungulates, and marsupial phalangers (see e.g. Campbell 1972; Kaas et al. 1972), although in no other order of mammals are all members so specialized as are the primates. Olfactory bulbs

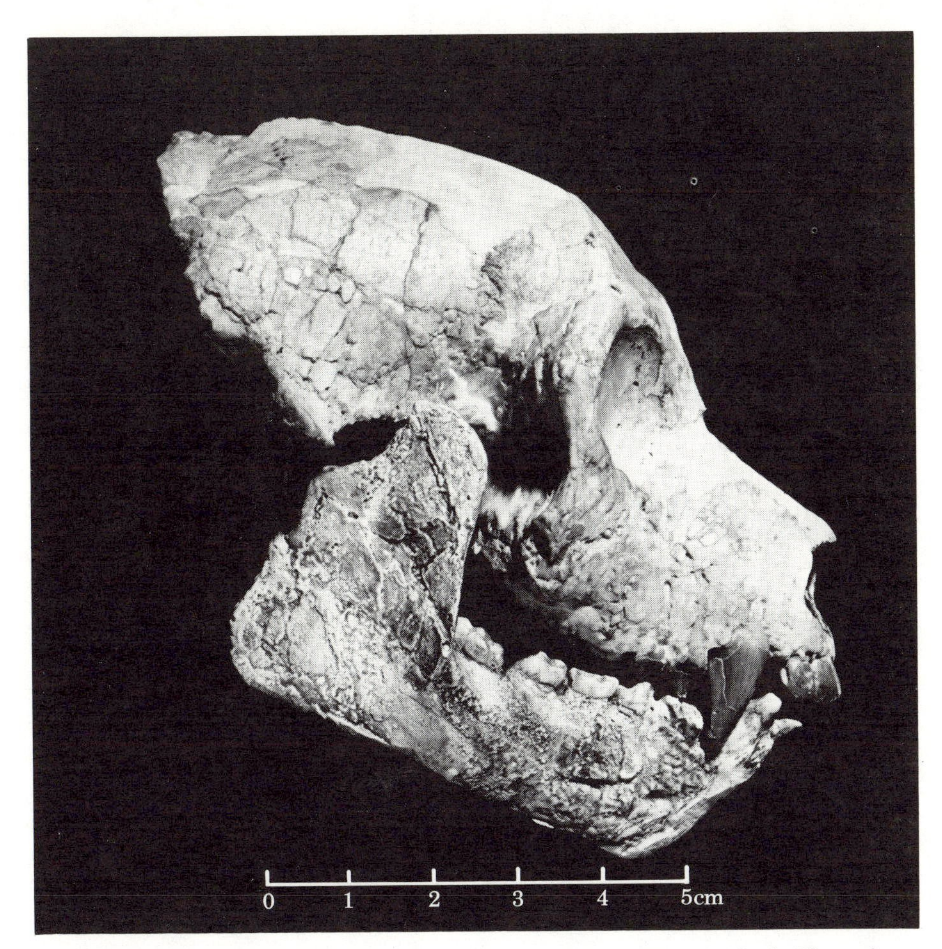

Figure 7. *Aegyptopithecus zeuxis*, a 27-million-year-old anthropoid from the Fayum region of Egypt. The skull and mandible are from two different individuals. The endocast of *Aegyptopithecus* is shown in Fig. 8. (Photo courtesy of Elwyn Simons.)

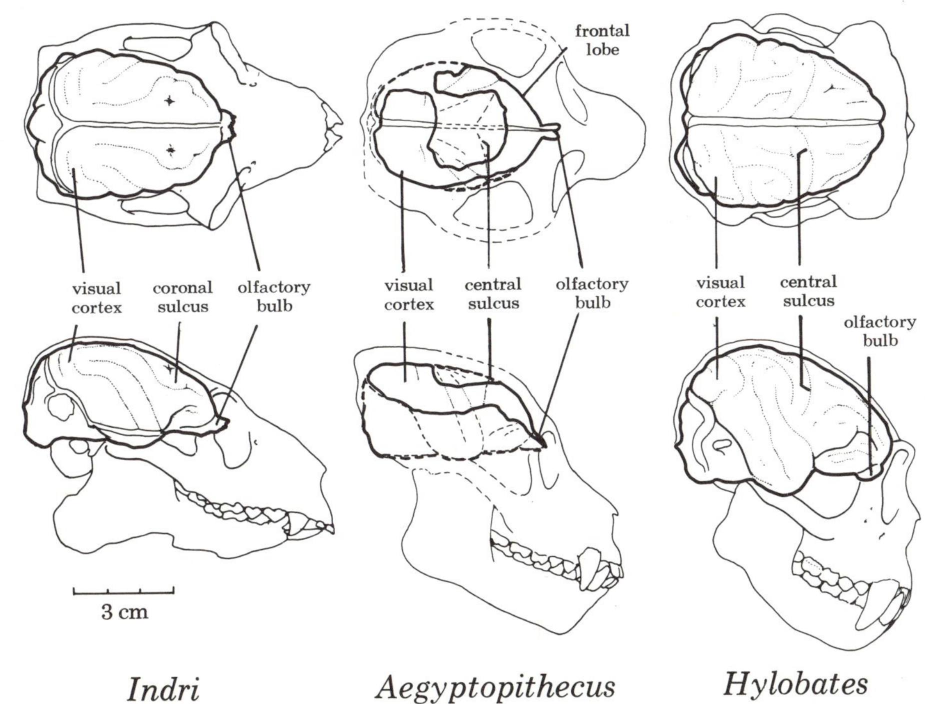

Figure 8. The endocast of *Aegyptopithecus zeuxis* is compared with endocasts of *Indri indri*, a living prosimian, and *Hylobates moloch*, a living anthropoid. All are drawn to the same scale. (Adapted from Radinsky 1974a.)

are reduced in many other mammals, particularly the aquatic forms, and in some, such as the toothed whales, the reduction has gone farther than in primates. Finally, the scanty evidence available on relative brain size in early mammals does not indicate that primate brains enlarged either earlier or more rapidly than those of all other mammals.

Humans have unique behavioral abilities, and because of this people have long sought to discover the neural features responsible for those abilities. The list of features implicated and believed to distinguish our brains from those of other mammals includes large size (both absolute and relative to body weight); relatively great amount of neocortex; relatively great amount of association cortex (in frontal, parietal, and/or temporal lobes); cortical speech areas (Broca's and Wernicke's areas); and asymmetry related to cerebral hemispheric dominance. Human brains are exceeded in absolute size by those of elephants and whales and in straight brain weight-body weight ratios by those of many small mammals. However, human brains are larger than would be expected for any other mammal—basal insectivoran, average living mammal, or even higher primate—of human body weight. In that respect, humans have the greatest relative brain size of any mammal, averaging about 3 to 3.5 times as large as would be expected even in a higher primate of human body weight.

Passingham (1973) reviewed the other features mentioned above and noted that if body weight and brain size are taken into account, few of those features can be shown to distinguish human brains from those of other primates. For example, while we have about 3 times as much neocortex as would be predicted for a primate of our body weight, we have no more neocortex than would be expected in an anthropoid brain the size of ours (see Fig. 3). Similarly, while we have more association cortex than any other primate, we have no more than would be expected in an anthropoid brain the size of ours. How areas such as cortical speech centers differ in humans as compared to other primates is still unknown.

Left–right symmetry, related to speech and other abilities, has been thought to distinguish human brains from those of other primates, but some recent observations (LeMay and Geschwind 1975) indicate that anatomical asymmetry exists in the brains of at least some other anthropoids.

Thus the only feature known at present to distinguish our brains from those of other mammals is great relative size. Other features—such as relative amounts of neocortex and association cortex, relative size of cerebellum, etc.—can be shown to be related to this large overall brain size. Of course, it should be kept in mind that the differences usually considered are gross anatomical ones, and it is possible that the features responsible for our unique mental abilities are on the cellular or molecular level.

The fossil record suggests that the increase in relative size that distinguishes our brains from those of other higher primates occurred relatively recently, beginning perhaps no more than 4 to 5 million years ago (australopithecine brains 2 to 3 million years ago were relatively larger than those of apes), and attaining its modern level within the past half-million years. One of the most fascinating unsolved problems in interpreting mammalian evolutionary history is discovering what selective pressures were responsible for that recent and unique increase in the relative brain size of our ancestors.

References

Bauchot, R., and H. Stephan. 1966. Donnes nouvelles sur l'encephalisation des insectivores et des prosimiens. *Mammalia* 30:160–96.

———. 1969. Encephalisation et niveau evolutif chez les simiens. *Mammalia* 33:225–75.

Campbell, C. B. G. 1972. Evolutionary patterns in mammalian diencephalic visual nuclei and their fiber connections. *Brain, Behav., Evol.* 6:218–36.

Clark, W. E. Le Gros. 1971. *The Antecedents of Man.* 3rd ed. Chicago: Quadrangle Books. 394 pp.

Connolly, C. J. 1950. *External Morphology of the Primate Brain.* Springfield, Ill.: C. C. Thomas. 378 pp.

Hershkovitz, P. 1970. Cerebral fissural patterns in platyrrhine monkeys. *Folia primat.* 13:213–40.

Hirsch, J. F., and W. S. Coxe. 1958. Representation of cutaneous tactile sensibility in cerebral cortex of *Cebus. J. Neurophysiol.* 21:481–98.

Holloway, R. 1972. New australopithecine endocast, SK 1585, from Swartkrans, South Africa. *Amer. J. Phys. Anthrop.* 37:173–86.

———. 1975. *The Role of Human Social Behavior in the Evolution of the Brain.* 43rd James Arthur Lecture. N. Y., Amer. Mus. Nat. Hist. 45 pp.

Kaas, J. H., W. C. Hall, and I. T. Diamond. 1972. Visual cortex of the grey squirrel (*Sciurus carolinensis*): Architectonic subdivisions and connections from the visual thalamus. *J. Comp. Neur.* 145:273–306.

LeMay, M., and N. Geschwind. 1975. Hemispheric differences in the brains of Great Apes. *Brain. Behav., Evol.* 11:48–52.

Passingham, R. E. 1973. Anatomical differences between the neocortex of man and other primates. *Brain, Behav., Evol.* 7:337–59.

———, and G. Ettlinger. 1974. A comparison of cortical functions in man and the other primates. *Int. Rev. Neurobiol.* 16:233–99.

Pilbeam, D., and S. J. Gould. 1974. Size and scaling in human evolution. *Science* 186:892–901.

Pubols, B. H., Jr., and L. M. Pubols. 1971. Somatotopic organization of spider monkey somatic sensory cerebral cortex. *J. Comp. Neur.* 141:63–76.

Radinsky, L. 1968. A new approach to mammalian cranial analysis, illustrated by examples of prosimian primates. *J. Morph.* 124:167–180.

———. 1970. The fossil evidence of prosimian brain evolution. In C. Noback and W. Montagna, eds., *The Primate Brain: Advances in Primatology,* Vol. 1. N.Y.: Appleton-Century-Crofts. pp. 209–24.

———. 1974a. The fossil evidence of anthropoid brain evolution. *Amer. J. Phys. Anthro.* 41:15–28.

———. 1974b. Prosimian brain morphology: Functional and phylogenetic implications. In R. D. Martin, G. A. Doyle, and A. C. Walker, eds., *Prosimian Biology.* London: Duckworth. pp. 781–98.

———. In press. Early primate brains: Facts and fiction. *J. Human Evol.*

Sanides, F. 1970. Functional architecture of motor and sensory cortices in primates in the light of a new concept of neocortex evolution. In Noback and Montagna, eds., *The Primate Brain,* pp. 137–208.

———, and A. Krishnamurti. 1967. Cytoarchitectonic subdivisions of sensorimotor and prefrontal regions and of bordering insular and limbic fields in slow loris (*Nycticebus coucang coucang*). *J. Hirnforsch.* 9:225–52.

Stephan, H. 1969. Quantitative investigations on visual structures in primate brains. *Proc. 2nd Int. Congr. Primat.* 3:34–42. Basel/N.Y.: Karger.

———, and O. J. Andy. 1969. Quantitative comparative neuroanatomy of primates: An attempt at a phylogenetic interpretation. *Ann. N.Y. Acad. Sci.* 167:370–87.

———, R. Bauchot, and O. J. Andy. 1970. Data on the size of the brain and of various brain parts in insectivores and primates. In Noback and Montagna, eds., *The Primate Brain,* pp. 289–97.

Verhaart, W. J. C. 1970. The pyramidal tract in the Primates. In Noback and Montagna, eds., *The Primate Brain,* pp. 83–108.

Welker, W. I., and G. B. Campos. 1963. Physiological significance of sulci in somatic sensory cerebral cortex in mammals of the family Procyonidae. *J. Comp. Neur.* 120:19–36.

Wirz, K. M. 1966. Cerebralisation und Ontogenesemodus bei Eutherien. *Acta anat.* 63:449–508.

"I'd say it was a male, 5 foot 3, 129 pounds. . ."

Ian Tattersall
Niles Eldredge

Fact, Theory, and Fantasy in Human Paleontology

Controversy in the study of human evolution reflects inadequacies in the formulation of hypotheses more than it does the supposed inadequacies of the fossil record

In the minds of many, paleoanthropology is more closely synonymous with controversy than almost any other branch of science. And whether or not this belief is in fact justified, it cannot be denied that the history of human evolutionary studies has been characterized by a great deal more dissension than agreement, a tendency that has in no way been ameliorated by the impressive recent augmentations of the known human fossil record.

Although paleoanthropologists have generally been content to focus on the arguments themselves, rather than to search for any underlying basis to them, it seems useful to ask, in general terms, whence comes this discussion. Is it inherent in the material of study itself? Is it simply that human evolution is, inevitably, an emotive issue? Or might it be that this lack of mutual understanding is related to the ways in which the study of human evolution has traditionally been approached? A dispassionate assessment would probably have to concede that all three of these factors have played their part in producing the multiplicity of conflicting hypotheses bearing on human evolution. And equally, it has to be recognized that, in an imperfect world, there is little that can be done, at least at the level of methodology, about the first two of them. We do believe, however, that much misunderstanding could be eliminated by attention not to the competing hypotheses per se but to the forms in which such hypotheses are customarily proposed.

The authors are Associate Curators in, respectively, the Departments of Anthropology and Fossil and Living Invertebrates of the American Museum of Natural History. Since obtaining his doctorate at Yale in 1971, Ian Tattersall has worked primarily on various aspects of the biology of the lemurs of Madagascar but maintains an active interest in problems of human evolution. Niles Eldredge received his Ph.D. from Columbia in 1969 and has been involved with the systematics and evolution of trilobites and the integration of evolutionary theory with the fossil record. *The authors would like to thank Eric Delson and Gareth Nelson for valuable discussion, and Marjorie Shepatin for help with the original figures.* *Address: American Museum of Natural History, Central Park West at 79th Street, New York, NY 10024.*

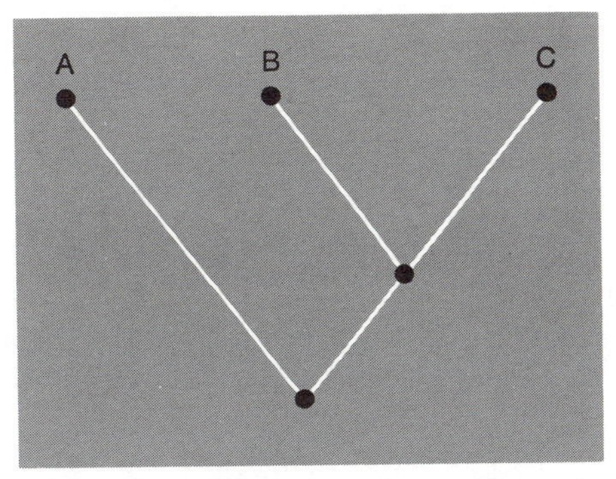

Figure 1. This is the simplest form of cladogram. The nesting of organisms in such schemes is based on the distributions of primitive and derived character states.

Types of evolutionary hypotheses

In essence, phylogenetic hypotheses can be formulated at three different levels of complexity, each successively further removed from the basic data available. The simplest is the *cladogram*—a branching diagram illustrating the pattern of distribution within a group of related organisms of derived characters, which one might describe as evolutionary novelties. Organisms are grouped together into nested sets on the basis of common possession of derived characters or, more properly, *character states*. The sharing of derived character states may be regarded as indicating an evolutionary relationship, whose exact nature remains unspecified at the level of the cladogram. The simplest type of cladogram is illustrated in Figure 1. Organisms A, B, and C all share derived character states that set them apart from others; B and C possess in common other derived character states not present in A. Beyond this, the cladogram carries absolutely no implications.

A more elaborate level of analysis is represented by the *phylogenetic tree*. At the minimum, the tree adds to the information contained in the cladogram by specifying the nature of the evolutionary relationships postulated. Such relationships may be of two types: that of the ancestor and its direct descendant (Fig. 2a) or that of daughter species each derived from a parental species as a result of splitting, or speciation, rather than linear evolution (Fig. 2b,c). Where stratigraphic information is available, the tree will obviously also specify the distributions in time of the taxa involved. Clearly, then, the tree is inherently a more complex statement than the cladogram (G. J. Nelson, MS), and from any one cladogram a variety of trees may be constructed. There are thus numerous ways in which the cladogram in Figure 3a can be transformed into trees, including those shown in Figures 3b–e.

More complex yet than the tree is the *scenario*—essentially a phylogenetic tree fleshed out with further types of information, most commonly having to do with adaptation and ecology. Just as a single cladogram can yield a variety of trees, a single tree can become the basis for a variety of scenarios, a reflection of the fact that the scenario is yet further removed from the basic "hard" data on which testable hypotheses can be based. An excellent example of a scenario is furnished by Robinson's elaborate vision (e.g. 1972), very inadequately summarized here, of the presence in Africa during the Plio-Pleistocene (ca. 2 m.y.) of two hominid genera. One of these, a large (150–200 lb.), robust, inefficient biped, somewhat "apelike" and "less advanced," was adapted to relatively wet conditions, probably did not make tools, and subsisted on an entirely herbivorous diet of fruit, nuts, roots, tender leaves, shoots, and so forth. This gave rise to, and subsequently lived sympatrically with, a smaller, lighter (40–60 lb.), more efficient biped: an omnivorous, toolmaking, hunting-gathering hominid adapted to drier conditions. The daughter outcompeted the parent, was thus responsible for the latter's eventual extinction, and ultimately gave rise to later species of *Homo*.

It will be apparent from this fairly typical example that the scenario is logically an extremely intricate type of proposition—if, indeed, many of them are logical at all. It is our belief that much of the current dearth of agreement between paleoanthropologists can be traced to the fact that paleoanthropological hypotheses are normally presented uniquely at the level of the scenario: precisely the type of formulation upon which agreement is least likely. It is not even evident that agreement on trees could be attained, since it is extremely difficult in practice to decide between the two possible types of evolutionary relationship even when, as is rarely the case, fossils are accompanied by firm dates. Extraordinarily difficult problems are encountered in recognizing both ancestors (Schaeffer, Hecht, and Eldredge 1972) and speciation events. It can be shown that two groups of organisms are closely, even extremely closely, related, but it cannot be *proved* that two taxa are both the daughters of a specified parent, known or unknown. The situation is exacerbated by the classical problem of negative evidence in the fossil record: we can never know how complete the known record is in its representation of extinct taxa.

Obviously, then, if there is to be agreement, we should expect it to come at the least complex of the levels of analysis we have discussed—the cladogram. This level of analysis is, like the tree, represented in the formulation of any scenario but is rarely separated from those other considerations unique to the tree or scenario. This effectively prevents the hypothesis of relationships—the only factor that can be evaluated in terms of the basic morphological data—from being presented in a testable form. And it is only on the basis of the rigorous testing and rejection of competing models, in terms of accepted criteria, that agreement will ever be reached. What all paleontologists require is a common approach, a set of ground rules; and as we have implied, these are almost by definition impossible to achieve at the level of the scenario, or even of the phylogenetic tree. Hypotheses of this order are simply too complex.

Figure 2. These simple phylogenetic trees show three possible patterns of evolutionary change. (*a*) A evolves directly into B in linear fashion. (*b*) As a result of speciation, B is derived from A while A persists. (*c*) In a less common type of speciation, A splits into two (or more) daughter species B and C.

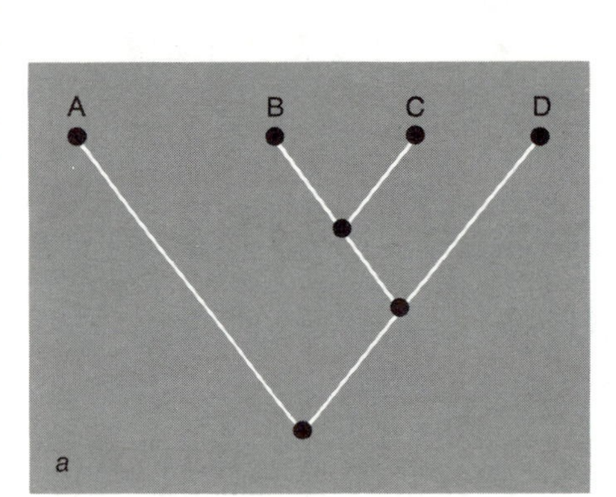

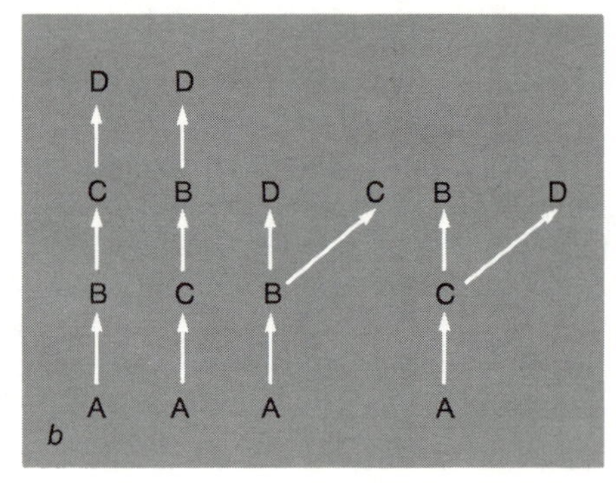

Figure 3. Many trees, such as those shown in *b*, can be derived from the cladogram *a*.

In saying this, we do not at all intend to imply that the devising of scenarios is a fruitless pursuit that should not be attempted. Indeed, scenarios are the most interesting and potentially the most informative of all paleoanthropological formulations. What we do claim, however, is that analysis should always proceed from the simple to the complex: that trees should always be based on cladograms and that scenarios should follow from trees. As things stand, the diverse components of scenarios are seldom separable, and much of the reasoning that goes into their construction is circular: the many elements involved feed back upon each other in an extremely intricate way. In particular, the evolutionary relationships expressed in scenarios are usually founded in ideas themselves based on considerations of adaptation, culture, environment, and so forth.

Formation of evolutionary hypotheses

How, then, should we proceed? A fundamental necessity is a change in the way we have tended to view morphological characters. Traditionally, only two kinds of similarity have been distinguished: characters of two taxa that are similar because of common descent and those that resemble each other because of parallel or convergent evolution (for our purposes the same thing, having to do with adaptation). "Primitive" and "specialized" characters have been recognized, but generally in a rather vague descriptive manner. General practice, in other words, makes no provision for the construction of cladograms where a basic requirement is the analysis of rigorously defined primitive vs. derived character states.

We have expressed elsewhere (Eldredge and Tattersall 1975) our belief that this lack of precision in approaching the evidence is due to the pervasive idea that phylogenies are a matter of discovery rather than of analysis. As long as we believe that the discovery of an adequate number of fossils will somehow reveal to us the course of evolution of the hominids or, for that matter, of any other group, there exists no necessity to approach the fossil and comparative evidence in any rigorous fashion. Clearly, the extensive recent discoveries of fossil hominids (many of the more interesting of them recently discussed in these pages by R. E. F. Leaky 1976) have done little to clarify the situation; indeed, the picture is becoming more complex than ever. We attempt here to indicate an alternative procedure which, at least at the simplest level of analysis, should help to minimize confusion.

In any branching scheme, the character state present in the ancestor is primitive; *any* change from this condition is derived. But it is well to remember that the ancestor itself may be derived in this character relative to its own ancestor. Thus, in Figure 4, D and E are both derived, albeit in different ways, relative to the condition possessed by C, which shows the primitive state for the D-C-E group; C, however, is in turn derived relative to A, which shows the primitive condition for the entire group A–E.

Thus defining the concepts of primitiveness and derivedness is easy; deciding in practice which character states in a given group of organisms are primitive or derived is often not. Nevertheless, a comparative procedure does allow the attempt to be made. The comparison of (presumably) homologous characters among a series of (presumably) related taxa inevitably reveals a great deal of variation between these taxa. Such variation is generally nonrandom—i.e. it falls into a more or less continuous spectrum. We have previously described (Eldredge and Tattersall 1975) how such spectra, known as *morphoclines,* result from the nature of the evolutionary process, whether the mechanism involved is splitting or gradual linear change. The primitive-derived sequence inherent in a simple morphocline has been termed its *polarity,* the determination of which lies at the core of the comparative process of phylogenetic reconstruction.

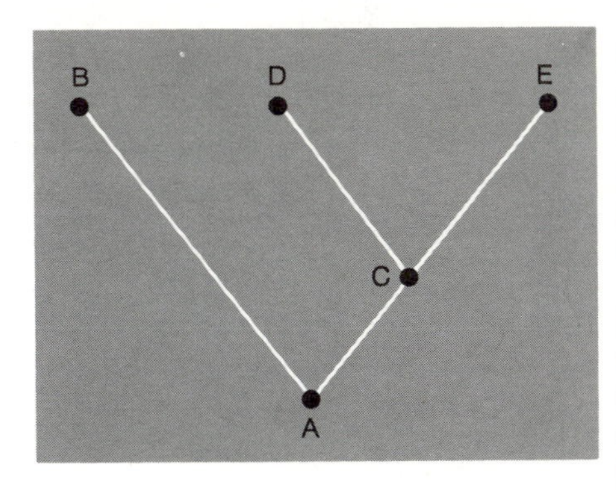

Figure 4. In this tree, D and E are both derived relative to C, which remains primitive for the C-D-E group. C, however, is derived relative to the common ancestor A, which is primitive for the entire assemblage.

Essentially, there are two approaches to the problem of the determination of morphocline polarity. The first lies in the communality of character states. If a given state is typical of a large number of the taxa under study, and particularly if this state is shared with taxa of similar rank closely related to the group under consideration, then it is reasonable to regard the state as primitive. This is why, for instance, we regard five digits as being the primitive complement of the mammalian forefoot; the evidence for this is purely comparative and does not depend on fossil data. An alternative way of assessing primitiveness vs. derivedness is provided by developmental evidence, even if ontogeny does not recapitulate phylogeny in any literal sense. But such evidence can never loom large in the assessment of fossils. It might also be argued that the stratigraphic age of a fossil provides yet a third way of evaluating the primitiveness or derivedness of its characteristics, for the appearance of the primitive necessarily precedes that of the derived in time. To use this criterion, however, it is necessary to assume that within a group older taxa must be equally or more primitive in *all* characters than those occurring later in the stratigraphic record. It is probably unwise to claim that such a situation has never arisen, but that it would be extremely unusual is undeniable. Organisms are almost invariably mosaics of primitive and derived characteristics, and the demonstration of primitiveness in twenty characters of a given form will never guarantee that the examination of a twenty-first will not show it to be derived. Although biostratigraphic evidence might possibly be useful as a last resort, it properly belongs to another level of analysis.

In practice it is hard, even impossible, to marshal a strong, logical argument for a given polarity for many characters in a given group. In these cases, however, one can at least recognize that such problems stem not from methodological inadequacies but from the limitations of the data available.

Once primitive-derived sequences have been established for single characters, analysis must proceed to the level of the organism, where complexes of characters are involved. Here other complications become apparent. Frequently a form will appear to be allied with certain relatives when one character is considered, but with others when a different character is used. Such situations stem from parallel evolution, a phenomenon that the comparative approach has revealed to be even more widespread than was apparent from studies using traditional methods. The final theory of relationships represented by the cladogram will thus not always be instantly apparent but must be arrived at by the assessment of the relative probabilities involved in cases of conflicting characters. In other words, a variety of hypotheses is generated; rival hypotheses can then be rejected until one remains, which will then stand until new data falsify it and suggest others. Of course, there will inevitably arise situations where this cannot be accomplished, but such situations merely demonstrate that the available biological data are insufficient to permit resolution of the problem—not that the procedure is inadequate.

This last is the crucial effective difference between the procedure we advocate and the rather formless approach that has traditionally been employed. Under the latter, it is rarely apparent whether disagreement lies at the level of interpretation or with the basic data themselves. In presenting theories of relationship in a potentially falsifiable form, the groundwork is laid not only for the possibility of agreement at this basic level but also for the recognition of the bases of disagreement if such should occur. In effect, a common language is established.

Proceeding from the cladogram to the phylogenetic tree, we begin treading more dangerous ground. To begin with, there arises the problem of recognizing species in the fossil record. At the level of the cladogram there is no necessity to define formal taxa, although, of course, there may be a temptation to do so for the convenience of reference. But when it comes to the specification of evolutionary relationships, the question of what may or may not be held to constitute a species becomes paramount. In some paleoanthropological quarters a certain fashionable reluctance currently exists to apply species names to some of the more recently discovered hominid fossils. Such reluctance is quite understandable, for to describe an assemblage of fossils as a species is to impute to it a wide variety of attributes that are undetectable from the basic data available.

Normal paleontological practice is to compare the ranges of morphological variation shown by fossil samples with those exhibited by closely related living species. In the case of hominid fossils this is rather difficult, because modern man is the only closely related species, and modern man is best regarded from this viewpoint as a " domestic animal," at least in the sense that the species evinces a great deal of variability which would, in the absence of the protective shield of material culture, have been eliminated by natural selection. Few would claim the vast range of variation in brain volume found in modern *Homo sapiens* is useful as a guide to the taxonomic significance of brain volume variations in early hominids.

And turning to the morphological variation seen in species of other living primates is not much more informative. The living great apes are highly unspeciose; and the patterns of within- and between-species variation typical of, say, lemurs, New World monkeys, and Old World monkeys are somewhat different. In general, one can fairly safely say that interspecific morphological differences between congeneric primate and other mammal species are pretty minor, to the extent that species may easily go unrecognized in the fossil record; closely related genera, on the other hand, are much more easily distinguished—about as easily, for instance, as one may distinguish Peking Man (*Homo erectus*) from modern *Homo sapiens*. The point here, however, is not to advocate a wholesale reclassification of fossil hominids on grounds that may or may not be relevant but to emphasize that in identifying a fossil assemblage as a species we are already working at a stage removed from the basic biological evidence.

The phylogenetic tree, as we have noted, additionally requires that the nature of the postulated evolutionary relationships be specified. In practice, it is relatively rare that an older fossil form can legitimately be regarded as ancestral to a younger one, or to a living species. For to be considered as such, a form should comply, in every detail of the characters used in the analysis, to the primitive condition specified for the appropriate ancestral node on the cladogram. Such an aggregation of primitive character states is often referred to as the "ancestral morphotype" of the group in question and is seldom exactly matched by a known fossil.

Discussion of the practical difficulties of forming trees could be continued ad nauseam but could not disguise the fact that we are unable to provide a clear-cut methodology for tree formation. At this level, however, one is sufficiently close to the cladogram that the reasoning behind the formation of a given tree should be apparent enough. But as vague as the methodology for constructing trees is, there is no methodology at all for the formulation of a scenario, with all its varied aspects of evolutionary relationship, time, adaptation, ecology, and so forth. In devising a scenario one is limited only by the bounds of one's imagination and by the credulity of one's audience—hence scenarios such as Sir Alister Hardy's aquatic theory of human origins, recently popularized and elaborated by Elaine Morgan in her book *The Descent of Woman* (1972). Infinitely too involved to do justice to here, this scenario, among many other things, ascribes the origin of human uprightness at least partly to the propensity of a remote ancestor to escape predators by fleeing into the water where, submerged to the neck, he (or, rather, she) would stand and wait for the predator, fearful of swimming, to leave in search of easier game.

Interpretation of the hominid fossil record

Reanalysis of the human fossil record in the way we have proposed is a monumental task. Despite the much-bemoaned inadequacy of this record, it is, in fact, extensive: hominid fossils are known from many sites widely scattered all over the Old World and are housed in numerous museums and other repositories. Moreover, since stratigraphic occurrence has so often conditioned views of the interrelationships of these fossils, insufficient attention has generally been paid in published studies to the morphological minutiae which, at the fine-grain level of relationships involved, play so important a part in the formulation of cladograms. Since we have seen only a fraction of the available material, the cladogram and trees we present are based strictly on our interpretation of the anatomical information as described and illustrated by others. It is no real comfort to realize that many of the existing statements concerning hominid phylogeny are similarly removed from the original data. As in any other problem of systematic biology, it is incumbent on the researcher to have a first-hand knowledge of the raw data; but where extensive comparative analyses are involved, this is difficult to achieve in human phylogenetic research.

In presenting this brief review we are further limited by the fact that much of the most interesting new material has as yet been described only in a preliminary manner, or has not been scientifically described at all. Immediately, this poses problems of phenon recognition—i.e. the ordering of fossils into basic sets (phena) for comparison (for a discussion of such problems see Delson, in press). We are unable to resolve these difficulties except by the somewhat arbitrary allocation of this material to various taxa. In order to facilitate discussion we have, with some misgivings, chosen to designate such arbitrary taxa as species.

Figure 5 provides a cladogram showing the hypothesized evolutionary relationships of the various taxa we recognize. We have elsewhere, on the basis of somewhat different cladograms, discussed in detail the distributions of the character states on which the theory of relationships is

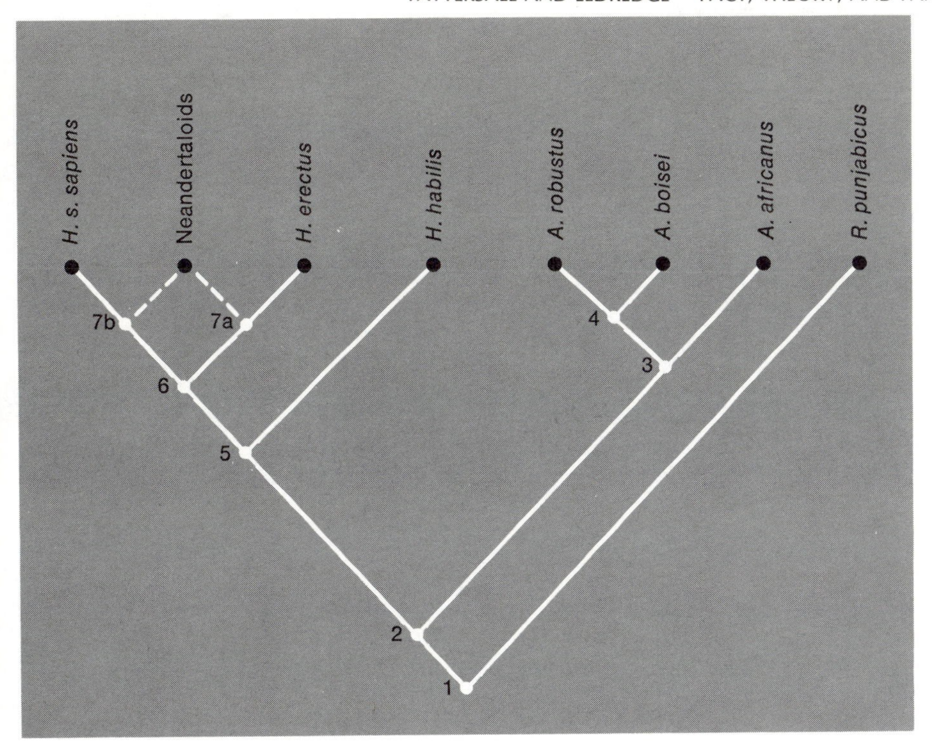

Figure 5. This cladogram shows a provisional theory of relationships among living and fossil hominids.

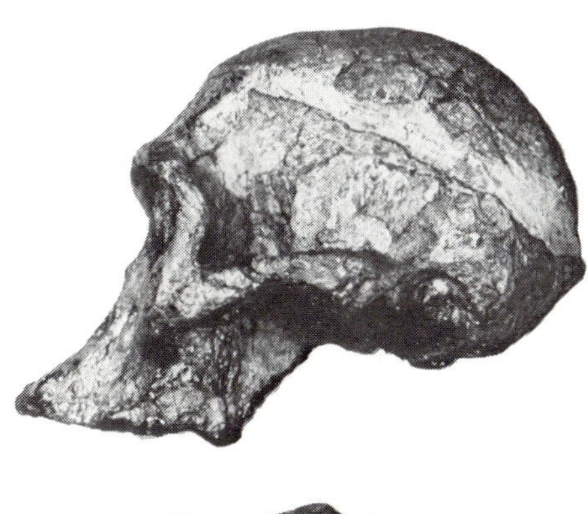

Figure 6. (*top*) The cranium of *Austrolopithecus africanus* (Sts. 5, a supposed female) from Sterkfontein, South Africa, represents a "perfect ancestor," conforming in every character used to the hypothetical ancestral morphotype of all non-*Ramapithecus* hominids. (*bottom*) The cranium of *Australopithecus robustus* (SK 48) from Swartkrans, South Africa, is highly derived relative to *A. africanus;* this is evident largely in size and morphology of the cheek dentition and concomitant changes in the supporting architecture of the face and braincase.

based, and characterized the ancestral morphotype (i.e. the collection of character states used to define an ancestral node on a cladogram) represented by each node (Eldredge and Tattersall 1975; Delson, Eldredge, and Tattersall, in press). It will probably be most useful here to note a few of the problems involved in forming the cladogram.

Possibly the most inherently interesting of all the morphotypes represented is that of the common ancestor (node 1 in Fig. 5) of all known hominids (we follow conventional usage in restricting Hominidae to those forms shown in Fig. 5, but some recent authors, e.g. Delson and Andrews 1975 have broadened the definition of the family to include the taxa generally classified in Pongidae). Unfortunately, available material of *Ramapithecus* (a middle-to-late Miocene fossil hominid known from Africa and Eurasia) is restricted to jaws and teeth, so comparisons including this form are limited to this anatomical region. In most features *Ramapithecus* is conservative and conforms to those characterizing node 1; but it is possible that it is uniquely derived in the shape of its dental arch. This is still open to interpretation, however.

Since all other known hominids share at least one derived character state not shared with *Ramapithecus,* they can together be regarded as constituting the "sister" of the latter. This much is clear, but in moving to the next few branching points we are approaching one of the most controversial areas of hominid taxonomy. This has not least to do with the identification of taxa. As we have said, this is an area with which we are equipped to deal only in a very arbitrary manner; assessment should properly be on the basis of adequately defined phena. For the purposes of discussion, we recognize two genera and four species of hominids in the African Pliocene and early Pleistocene: *Australopithecus africanus,* a gracile form known from South Africa and (probably) from some of the earlier East African sites; *Australopithecus robustus,* a more heavily built form from South Africa; *Australopithecus boisei,* a closely related but even more massive form from East Africa; and *Homo habilis,* a lightly built and larger-brained hominid also known from East Africa (Olduvai Gorge, East Rudolf, and elsewhere). We should emphasize, however, that there is nothing definitive about these taxa. Assessment of the most recently discovered material from Kenya, Tanzania, and Ethiopia, and reassessment of previously known fossils in its light, will doubtless show that our arrangement of taxa is oversimplified, and will enable the cladogram to be reformulated upon the basis of a better understanding of phena and a broader selection of characters. Our intent in this section is, then, primarily illustrative.

Apart from the problems of the recognition of taxa, perhaps the most substantial difficulty in assessing the evolutionary relationships of these Pliocene—early Pleistocene fossils is caused by the fact that, in its cranial and dental morphologies, *A. africanus* (Fig. 6) is, as far as we are able to determine, a "perfect ancestor." That is, it conforms in every character used to the hypothetical ancestral morphotype of all non-*Ramapithecus* hominids. Effectively, it is impossible to decide what the evolutionary relationships are of such perfectly primitive forms: there is no way to determine to which other taxa they are most closely related. We have tried to solve this problem by assessing the morphology of the postcranial skeleton of *A. africanus,* and in a previous paper expressed a belief that the apparent lengthening of the femur of this form constituted a derived characteristic shared with *Homo* (Delson, Eldredge, and Tattersall, in press). However, a reappraisal of details of pelvic and femoral morphology indicates that gracile and robust *Australopithecus,* despite minor differences, share a derived form of locomotor apparatus (reflected, for instance, in the elongation and reduction, respectively, of the femoral neck and head) not shared with species of *Homo.* Femoral lengthening would thus appear to be explained more plausibly as a parallel development in the two genera. It is for these reasons that we show all *Australopithecus* as constituting a sister taxon of genus *Homo.*

Within this assemblage, *A. robustus* (Fig. 6) and *A. boisei* form a clear-cut sister group, highly derived relative to *A. africanus* in ways having mostly to do with alterations in the size and morphology of the cheek dentition, and with the concomitant changes in the supporting architecture of the face and braincase. Distinctions between *A. robustus* and *A. boisei,* if, indeed, they are warranted, are largely of degree, the latter carrying to greater extremes the characters noted already.

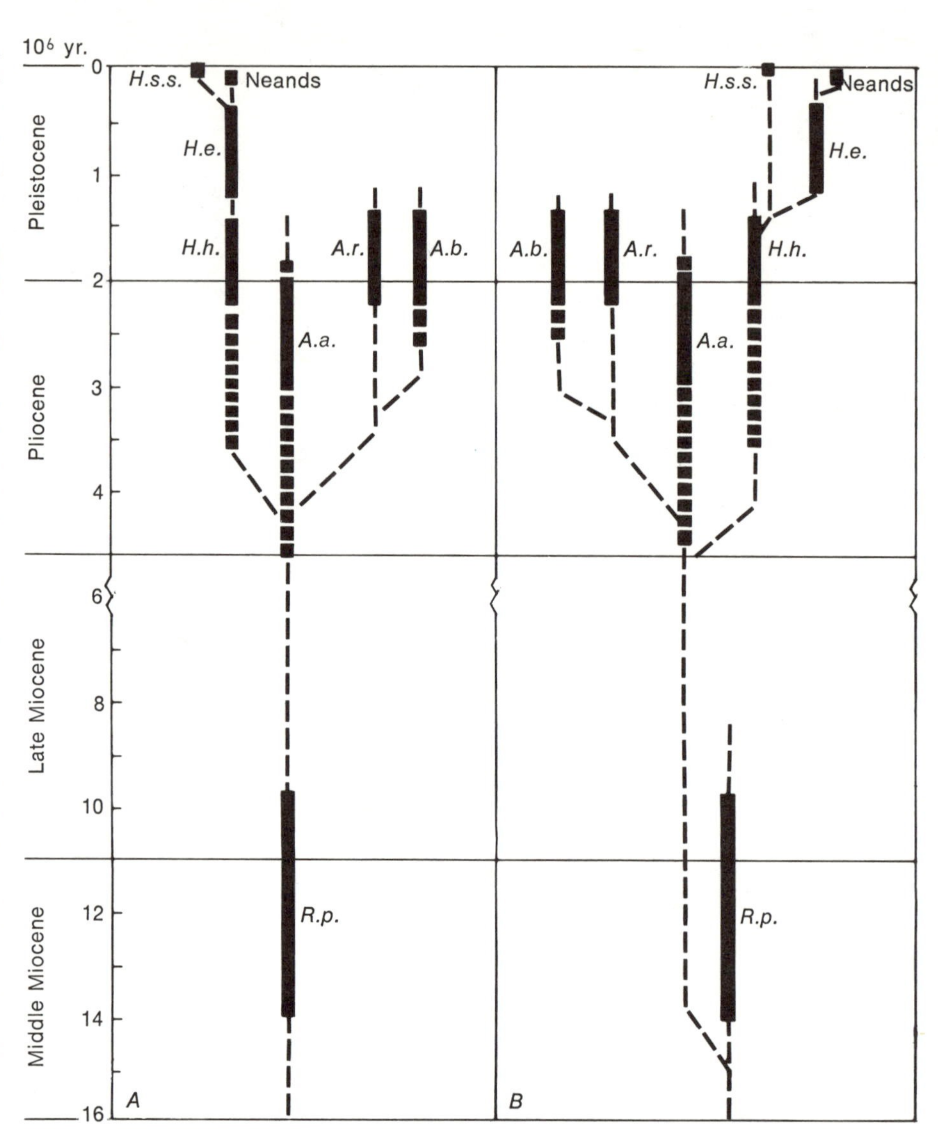

Figure 7. These are only two of the numerous phylogenetic trees compatible with the evolutionary relationships shown in Figure 5. (Abbreviations follow the names used in Fig. 5; broken heavy lines are used where fossils are only tentatively assigned to the species, and broken light lines represent inferential lines of descent not based on fossil evidence.)

The question of the content of *H. habilis* is a vexed one, and available material is not as easily interpreted as that of *Australopithecus.* It seems fairly clear, however, that at least during the early Pleistocene (and possibly even earlier) there existed a hominid with a body larger than that of *A. africanus* and a locomotor apparatus closer to that of later *Homo.* Its brain volume was substantially greater than that of *A. africanus,* although its teeth remained relatively large, and often substantially similar to those of *A. africanus.* These and other considerations justify establishment of node 5 (Fig. 5), although an insufficiency of characters available to us makes it hard to detect any uniquely derived cranial character states in *H. habilis.*

Yet further increased brain size provides the most obvious character for linking *H. erectus* with later *Homo,* and it is clear on a variety of grounds that *H. erectus* and *H. sapiens* (including, perhaps unwisely, the Neandertals and allied forms) constitute a sister group. The exact nature of the relationship involved, however, is another question. *H. erectus* possesses a number of characters that are plainly derived for this assemblage, but in which *H. sapiens* appear to remain primitive. This is most clearly seen in skull form. Node 5 (and, for that matter, node 2) is characterized by a high, rounded, lightly built cranial vault. This is equally true of modern *H. sapiens,* which would thus appear to show the primitive condition. Cranial shape is, however, conspicuously different in *H. erectus:* here the cranium is long, low, angled at the back, and formed of thick bone. Moreover, although *H. erectus* does share certain derived features of the dentition with *H. sapiens,* it does so little more than do certain fossils which we might best classify provisionally as *H. habilis.*

When the question is viewed in this light, it becomes easier to see why L. S. B. Leakey (1966) preferred to derive *H. sapiens* directly from *H. habilis,* without the interposition of a stage represented by *H. erectus.* In weighing the issue, the major problem becomes one of functional morphology. In proceeding to *H. sapiens* from a form possessing the primitive cranial conformation as exemplified by *A. africanus,* is it functionally necessary to pass through a stage showing the cranial characteristics of *H. erectus?* And even if not, is this what actually happened? Fossil evidence of an intermediate taxon not resembling *H. erectus* could decide the question one way; but more fossil evidence will not help us if the sequence *H. habilis* → *H. erectus* → *H. sapiens* is what actually occurred. Full understanding of the evolutionary relationships involved is dependent on an understanding of the functional significance of the various cranial types.

At the level of the cladogram such considerations are not of great importance once the sharing of other derived character states has established the position of node 6; this is unarguable whatever the precise nature of the evolutionary relationship involved might be. Not, that is, until the Neandertals and similar late Pleistocene forms enter the analysis. It could be reasoned that those features of the Neandertal skull that have vaguely been viewed as "primitive," such as the long, low braincase angled at the occiput, large brow ridges, and so forth, are derived characters aligning their possessors with *H. erectus* rather than with *H. sapiens.* In this case, node 7a would be appropriate to the expression of the relationships inferred. On the other hand, if such characters are in fact ancestral retentions, and the lightly built, rounded skull of *H. sapiens* is itself derived relative to this condition, then the evolutionary relationships involved are best expressed by node 7b. We are unable to decide at present between these two alternatives, and the cladogram reflects this ambiguity.

Pilbeam and Vaišnys (1975) have suggested that if paleoanthropological hypotheses are to be testable, they must first be converted to quantitative terms. Only then, they argue, is it possible to assess them in terms of probability theory. We emphasize, however, that the purely qualitative hypothesis of relationships that we have put forward is potentially falsifiable in any of three ways (apart from a rearrangement of phena). A different scheme may be shown to be preferable if it is based on a more accurate assessment of the distributions of primitive and derived character states. Alternatively, data on new characters may be added to the array and shown to be in conflict with the hypothesis as it stands. Or new data, in the form of new fossils, may require reassessment of the theory of relationships. In any event, a cladogram expresses all the *testable* conclusions that may be extracted from a particular data set. In moving to the more elaborate level of the phylogenetic tree, we are moving essentially into the realm of speculation, informed or otherwise.

An enormous variety of phylogenetic "trees" has been proposed to express diverse views of the process of human evolution. On examination, however, almost all such "trees" prove to be scenarios, since the data and assumptions on which they are based include all the elements that we have described as being appropriate to the scenario. In Figure 7 we present two alternative human phylogenetic trees, derived from the cladogram in Figure 5 simply by the addition of stratigraphic information and by the specification of the types of evolutionary relationship involved. These are only two of a large number of trees potentially derivable from the cladogram and compatible with it. Of the two, it may be that Figure 7b reflects reality in more respects than 7a, but there is no objective method of choosing between these alternatives or, indeed, among either of them and any of the numerous other possibilities. Because of the problems we have noted of identification of the hominid taxa existing during the Pliocene and early Pleistocene, and also because the dating of many of the fossils (e.g. the South African material) is equivocal, the stratigraphic ranges given in the trees are approximations only.

Either of the trees in Figure 7 could serve as the basis for dozens of scenarios. We cannot hope to suggest here even a modestly representative sample, but we can indicate a few of the elements which might reasonably be considered in devising a scenario. *Ramapithecus* possessed reduced anterior teeth and molars adapted for powerful chewing; by analogy with the functionally similar dental complex of the gelada baboon, it seems quite likely that this animal subsisted on a diet of small, abrasive foodstuffs gathered on the ground in open country. The fauna and geological settings of *Ramapithecus*—yielding sites in both Kenya and India are not incompatible with this suggestion.

An open-country environment is similarly indicated for *Australopithecus* species and *H. habilis,* although there is a possibility of some ecological distinction between robust and gracile types in East Africa (Behrensmeyer 1976). The dentitions of gracile *Australopithecus* and *H. habilis* remain relatively unspecialized, whereas *A. robustus* and, particularly, *A. boisei* show a marked shift toward a type of mastication in which the shearing of foodstuffs was sacrificed in favor of grinding. Whether this implies that the robust forms were specializing more in herbivory, while the gracile ones were omnivorous, is unclear. Robust and gracile *Australopithecus* shared a similar form of upright bipedal locomotion, although it has been argued that the latter may have been more efficient; *H. habilis,* also an upright biped, was closer in its locomotor apparatus to that of later *Homo.*

Identifiable tools do not appear in the record until about three million years ago, or perhaps slightly less. These artifacts cannot be associated with any particular hominid group but even at their first appearance exist in a number of recognizable types. Evidence of the butchering of large—even very large—mammals by hominids occurs very early in the archaeological record, although there is no way of knowing whether the butchers were also the killers or, again, which hominid(s) were involved. An excellent review of the achievements and directions of paleolithic archaeology in East Africa is given by Isaac (1976).

A certain continuity can be detected between some of these early industries, but the "Acheulean" stone industry occurs as a new development when it replaces "pebble tools" at Olduvai Gorge a little under 1.5 million years ago. The Acheulean is generally associated with *H. erectus,* a widespread species, but in eastern Asia tools either associated with, or apparently equivalent in age to, *H. erectus* seem generally relatively crude. The archaeological picture appears quite complex, however. There is little doubt that *H. erectus* was a full-fledged hunter (although it should be noted that among modern hunting-gathering peoples, meat accounts only for a fraction of the total diet), and evidence exists for the use of fire.

Later in the Pleistocene there appear various hominids usually regarded as representing archaic forms of *H. sapiens,* and sometimes referred to as "neandertaloids," an appellation that unfortunately tends to obscure their differences. "Solo Man" from Java is reminiscent of *H. erectus* from the same region; its dating is equivocal. "Rhodesian Man" is distinctly different from the Neandertals of Europe, and somewhat similar in comparable parts to *H. erectus* from Olduvai Gorge. The Neandertals of Western Europe appear to have been abruptly replaced by men of modern aspect; archaeological evidence for the evolution of the Neandertal "Mousterian" lithic industry into a more sophisticated stone technology of the type associated with fully modern *H. sapiens* is tenuous at best. We do not know at present where modern *H. sapiens* originated, or precisely when, although this must have occurred at some time between 100,000 and 50,000 years ago. By about 35,000 years ago, modern man had become the sole hominid species extant.

The preceding paragraphs represent only a tiny sampling of the sort of information that may be integrated into a scenario of human evolution. We hesitate to suggest a scenario ourselves, but perhaps the reader will be more imaginative.

References

Behrensmeyer, A. K. 1976. Taphonomy and paleoecology in the hominid fossil record. *Yrbk. Phys. Anthrop.* 19(1975):36–50.

Delson, E. In press. Models of early hominid phylogeny. In *Early African Hominids,* ed. C. J. Jolly. London: Duckworth.

Delson, E., and P. Andrews. 1975. Evolution and interrelationships of the Catarrhine primates. In *Phylogeny of the Primates,* ed. W. P. Luckett and F. S. Szalay, pp. 405–46. NY: Plenum.

Delson, E., N. Eldredge, and I. Tattersall. In press. Reconstruction of hominid phylogeny: A testable framework based on cladistic analysis. *J. Hum. Evol.*

Eldredge, N., and I. Tattersall. 1975. Evolutionary models, phylogenetic reconstruction, and another look at hominid phylogeny. In *Approaches to Primate Paleobiology,* ed. F. S. Szalay, pp. 218–42. Basel: Karger.

Isaac, G. Ll. 1976. Early hominids in action: A commentary on the contribution of archeology to understanding the fossil record in East Africa. *Yrbk. Phys. Anthrop.* 19(1975):19–35.

Leakey, L. S. B. 1966. *Homo habilis, Homo erectus* and the australopithecines. *Nature* 209:1279–81.

Leakey, R. E. F. 1976. Hominids in Africa. *Am. Sci.* 64:174–78.

Morgan, E. 1972. *The Descent of Woman.* NY: Stein and Day.

Robinson, J. T. 1972. *Early Hominid Posture and Locomotion.* Chicago: University of Chicago Press.

Pilbeam, D. R., and J. R. Vaisnys. 1975. Hypothesis testing in paleoanthropology. In *Paleoanthropology, Morphology and Paleoecology,* ed. R. H. Tuttle, pp. 3–13. The Hague: Mouton.

Schaeffer, B., M. Hecht, and N. Eldredge. 1972. Phylogeny and paleontology. In *Evolutionary Biology,* vol. 6, ed. T. Dobzhansky, M. Hecht, and W. C. Steere, pp. 31–46. NY: Appleton-Century-Crofts.

Richard E. Leakey

Views

Hominids in Africa

One of the world's foremost paleoanthropologists discusses recent finds from African sites that give evidence for very early differentiation among the Hominidae

The existing fossil evidence supports the contention that Africa was the crucible of human origins and development. African sites yield evidence illustrating many phases of the human story from the earliest stages. Fossils from early Miocene sites in Africa illustrate the very early differentiation of the Hominidae; from later strata there is evidence for more recent origin of the genus *Homo*. Asia and parts of Europe have also produced contemporary material of the earliest phases of development, but there is an absence of intermediate forms in Asia which could be held as possible evidence for migration of early *Homo* into Asia and the Far East. A growing interest in the field of prehistory and intensive work in Plio/Pleistocene sites (dating from 1 to 3 million years ago) of the Mediterranean region and Asia may yield data that will revise the present understanding. At present, however, the earliest stage of the human family is represented by fossils of *Ramapithecus* from Ft. Ternan in Kenya (*1, 2*), with a date of between 12.5 and 14 million years ago. Slightly younger examples are known from the Siwalik sites (*3*), dated by their contained faunas at about 9 to 12 million, and closely allied material recently reported from Hungary (*4*) is thought to be dated at perhaps 11 million.

Richard E. Leakey, Director of the National Museums of Kenya and well known for his fossil discoveries in East Africa, was born in 1944. He attended schools in Nairobi and received the education of experience by working with his parents in Olduvai Gorge and other parts of East Africa. Currently he is the leader of the East Turkana Research Project, a multinational, interdisciplinary investigation of the Plio/Pleistocene of Kenya's northern Rift Valley, where the group's discovery of many new fossil hominids has radically changed the thinking about human origins. Author of numerous publications, Mr. Leakey has received many honors for his research contributions and is Chairman of the Wildlife Clubs of Kenya Association and of the Foundation for Research into the Origin of Man. This article is based on an address given at the Sigma Xi National Meeting in November 1975. Address: National Museums of Kenya, P.O. Box 40658, Nairobi, Kenya.

The case for *Ramapithecus* as a hominid is not substantial, and the fragmentary material leaves many questions open. There is general agreement that the dental morphology is of the hominid pattern; the canines were "reduced," or proportionately small. Because we have no cranial elements other than maxillary and mandibular fragments and as the postcranial skeleton is unrepresented, the arguments for the hominid status of this genus are severely limited. The survival of this unspecialized hominid is not documented at present, partly because late Miocene and Pliocene localities are not well represented in Africa and also perhaps because known sites in Eurasia have not been investigated in great detail.

The better documented nonprimate vertebrates in the Old World provide many instances of middle Miocene forms surviving into the Pliocene. It is conceivable that *Ramapithecus* maintained itself on the forest fringes and savannas and might be found to occur alongside the more specialized forms in the late Pliocene. In this connection it has been stated that KNM-ER 1482 (*5*), a mandible from the Koobi Fora Formation, at Lake Turkana (formerly Rudolf), Kenya, shows certain traits that I believe distinguish it from the better documented *Australopithecus* and *Homo* of the same era, and perhaps it may be a *Ramapithecus*. Similarly, the recently reported finds from the Afar Valley in Ethiopia include a partial skeleton (called "Lucy" by its discoverers) that may be considered a late *Ramapithecus*. The study of this material and publication of detailed descriptions will be extremely important.

The African record between the middle Miocene and Pliocene is sparse. Two isolated molars are recorded from the Baringo Basin—one from the Ngorora Formation (*6*) at 9 to 12 million and the other at Lukeino (*7*) at about 6 to 7 million years. These sites are within the Kenya Rift Valley and could well yield further specimens. A mandibular fragment with a single molar in place is known from Lothagam (*8*), where a date of 5.5 million has been reported. A nearby site, Kanapoi (*9*), has yielded a fragment of humerus, and an age of about 4 million years has been proposed. Both of these specimens have been provisionally attributed to the genus *Australopithecus*, although there are few morphological features to support this. It is extremely difficult to identify mandibular fragments with any exactness, and the material might equally represent a very early example of *Homo* or a late *Ramapithecus*. Consequently, the citations for *Australopithecus* earlier than 3 million years have to be regarded with caution and some doubt at this time. It would perhaps be useful to place the two isolated molars, the man-

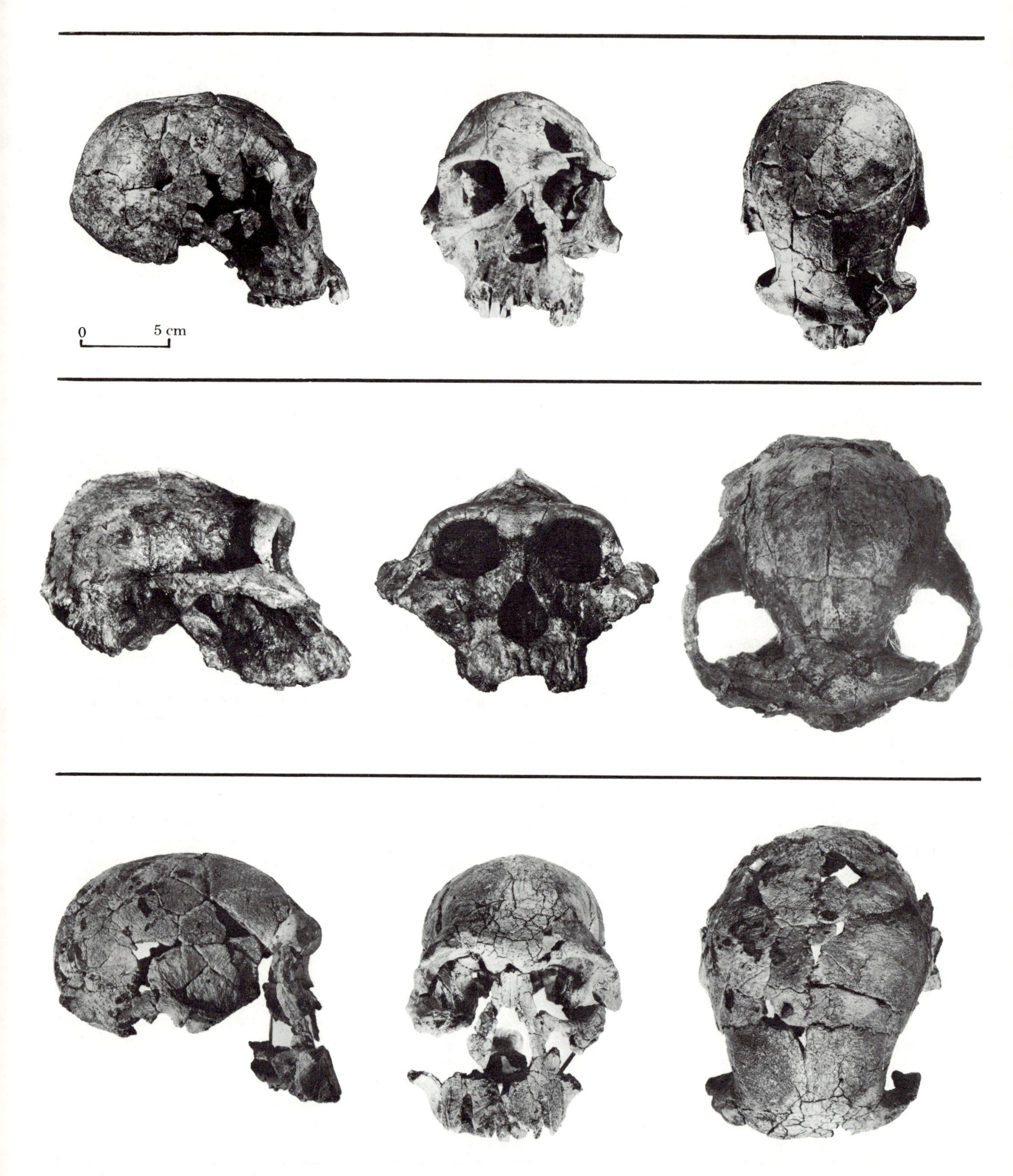

Figure 1. *Top:* Lateral, frontal, and superior views show a small-brained, small-toothed hominid (KNM-ER 1813) that is proposed as *Australopithecus* cf. *africanus.* The specimen is from the Koobi Fora Formation, an extensive Plio/Pleistocene site that lies to the east of Lake Turkana in Kenya's northern Rift Valley. *Middle:* Compare the same three views of another *Australopithecus* (KNM-ER 406) found in the same area. It is also small-brained but has large cheek teeth. The apparent considerable breadth in the frontal and superior views is somewhat misleading, because it is influenced by the broad zygomatic arches, which flare in order to make room for the massive chewing muscles. The photo of the left side of the cranium has been reversed to make comparison easier. *Bottom:* Similar views of a large-brained *Homo* sp. (KNM-ER 1470), from levels of the Koobi Fora Formation that date between 2 and 3 million years, are given for comparison.

dibular fragment, and the humerus in a "suspense" category as "Hominidae indet." until further evidence is available.

There is a considerable collection of fossil hominid material from late Pliocene times onward. In the past two years new field work has been going on at sites that appear, mainly on faunal grounds, to date between 3 and 4 million years. The Hadar in the Afar Valley in Ethiopia is being investigated by a multidisciplinary international team under the joint leadership of M. Taieb, C. Johanson, and Y. Coppens (*9*). The results since 1973 have been spectacular, with reports of mandibular remains, maxillae, cranial fragments, postcranial elements, and the famous partial skeleton "Lucy." The latest news is of a site in northern Tanzania, some 50 kilometers south of Olduvai, known as Laetolil, where isolated teeth and several mandibles have been recovered from alluvial sediments that have a reported radiometric age of between 3.3 and 3.7 million years. This project is being led by Mary Leakey, who is also continuing field work at the Olduvai Gorge.

The material from Laetolil and the Afar has not been described in detail, and popular announcements in the media are far from satisfactory. It is stated that the bulk of the specimens are best accommodated within the genus *Homo,* and this presumably relates to specific morphological characteristics. There is an inherent problem in hominid taxonomy caused by the present lack of any precise diagnosis for fossil forms. I am reluctant to anticipate further new discoveries, but I would expect that the genus *Homo* will eventually be traced into the Pliocene at an age of between 4 and 6 million years, together with *Australopithecus.* At present, however, this has not been firmly established, and it is very unlikely that mandibular or dental morphology alone will be sufficient for positive identification.

The extensive Plio/Pleistocene sites in the Lake Turkana basin are well documented, and the two principal sedimentary formations are the Shungura Formation in the Omo Valley at the northern end of the lake and the Koobi Fora Formation lying to the east of the lake. Since 1968, major projects have been underway, and a wealth of data has been collected. The formations span approximately the same period of time—bracketed between 1.3 and 3.2 million years. While some time breaks are manifest, the successions offer what can be considered a continuous record over the period in one extensive geographical basin. *Australopithecus* is represented by cranial and postcranial fossils in varying degrees of completeness, and *Homo* is also represented by quite substantial data. There is the possibility of a late *Ramapithecus,* as has been mentioned, but this cannot be taken as established until further material is recovered.

Australopithecus species

The problem of species definition within *Australopithecus* is far from settled, but I am of the opinion that evidence for two species of this genus can be established with some conviction in the Koobi Fora Formation. The most obvious, *Australopithecus boisei* (*10*), is very distinctive, being characterized by hyper-robust mandibles, large molars and premolars relative to the anterior dentition, cranial capacity values less than 550 cc, and sexual dimorphism manifested in superficial cranial characters such as sagittal and nuchal crests. An example (KNM-ER 406) is shown in Figure 1. The known postcranial elements, such as the femur, humerus, and talus, are also distinctive. This widespread species has been reported from other localities, such as Chesowanja, Peninj, and Olduvai Gorge in the southern Rift Valley of East Africa. *A. boisei* may require reconsideration as a full species and should instead perhaps be ranked on a subspecific basis, as a deme of the South African form of *A. robustus.* Additional data are needed if we are going to solve the problems that are always associated with such refined systematics in vertebrate paleontology. Consequently the retention of two allied but spatially separate robust species seems desirable for the moment.

The case for a gracile East African species of *Australopithecus* is less secure, but there seems to be too great a degree of variation if all the material is included as a single species. The best example of the gracile form from East Africa would be the specimen from the Koobi Fora Formation—KMN-ER 1813 (*11*) (Fig. 1, *top*). Various mandibles and some postcranial elements might also be included, keeping in mind the difficulty of classifying mandibles. In addition, certain specimens from Olduvai, such as OH 13 and OH 24, might be reclassified as "gracile" *Australopithecus.* No detailed proposal for such a classificatory scheme for the East African fossils has been put forward, but the typical characteristics would include gracile mandibles with small cheek teeth, cranial capacity values at 600 cc or less, and sagittal crests rare or nonexistent. The postcranial morphology appears to be similar to that seen in *A. boisei,* although at a smaller and less robust scale. In both species, one of the most distinctive features is the proximal region of the femur: a long femoral neck is compressed from front to back, and there is a small, subspherical head. There are other features, but very little is known about variation, and the sample is not impressively large at present.

I consider this species to be closely allied to the gracile *A. africanus* from Sterkfontein in South Africa; it may be a more northern deme of that species. The innominate bone is known for *A. africanus* and for *A. robustus* in South Africa, and slight differences have been noted between the two forms. No innominate remains are attributable to *Australopithecus* from East Africa, but *Homo* is represented by two specimens that are time equivalent, and they illustrate marked differences between the two genera that are greater than would reasonably be expected for a single, albeit spatially extended, species.

Differentiation in *Homo*

The presence of *Homo* in late Pliocene and early Pleistocene deposits is still somewhat controversial, but I believe that relatively complete cranial and postcranial elements such as the femur and innominate are diagnostic and conclusive. The best examples are from the Koobi Fora Formation, with specific examples such as the crania KNM-ER 1470 (*12*) (Fig. 1, *bottom*) and 1590 (*11*). The femur KNM-ER 1481 and the innominate KNM-ER 3228 (*13*) are also explicit. The specimens referred to are from levels that date between 2 and 3 million years and from localities that have also yielded material that can be confidently attributed to *Australopithecus*.

The principal characters upon which the early species of *Homo* might be defined are relatively large anterior teeth, with equally large, although not buccolingually expanded cheek teeth; moderately robust and externally buttressed mandibles that have everted basal margins; cranial capacity values exceeding 750 cc; and a high-vaulted skull with minimal postorbital waisting and no sagittal crest. The postcranial morphology is very similar to the modern human. It is proposed that certain specimens from East African sites fall within this category; such examples would include OH 7 from Olduvai, the type specimen for *Homo habilis* (*14*). I have commented earlier on the specimens from the Afar and Laetolil, and their final taxonomic position remains an open question in my opinion.

Nomenclature at the species level in *Homo* is not settled, but I do believe that *Homo habilis*, as represented by the type specimen from Olduvai, OH 7, is the same as the more complete and slightly earlier material, such as KNM-ER 1470. On this basis, *Homo habilis* has priority. However, there are going to be difficulties when considering the various stages of a single evolving lineage, and it may be appropriate to use a somewhat arbitrary definition based upon absolute dates. An alternative scheme, which I would favor, is to retain *Homo erectus* as the only species of *Homo* other than *Homo sapiens* and to classify grades or stages of evolutionary development. This would require agreement only on the diagnostic characteristics of the genus. Under this system, the earliest examples of the genus would be Stage 1, and so on. Of course, difficulties arise when new and earlier finds are made after the scheme has been presented in detail. A further variation upon this approach might be to consider only two species, *sapiens* and *presapiens*, and to identify fossil variations at a subspecific level. In this way we could consider the form *erectus* as distinct but related to *habilis*.

Major differences between KNM-ER 1470 and examples of *Homo erectus* (*15*) may be the result of significant time separation since, at present, intermediate forms are unknown. During 1975, a complete cranium was recovered from the Koobi Fora Formation which does show a number of features that are perhaps to be expected in a transitional stage. The skull is awaiting preparation and description, and thus final comments must be postponed.

The South African fossil evidence is complicated by the absence of reliable dates and the uncertainty of relative dates for much of the material collected from the limestone quarry waste. Various suggestions have been based upon vertebrate faunas, and the collections probably range from late Pliocene to the Pleistocene, with an age bracket of between 1 and 3 million years. At least two species of *Australopithecus*, the robust *A. robustus* and the gracile *A. africanus*, are represented. There do not seem to be any compelling reasons to assume ancestral-descendant relationship between these two species; the East African evidence would support this view.

Tools and habitations

The most impressive record of tools and habitation sites is from Olduvai Gorge (*16*), where numerous localities have been excavated over the past 30 years. The progression from simple "pebble tools" to intricate and perfect bifacial implements is well documented in this one area. There are also inferences to be drawn on the probable social organization, in the sense of community size and hunting preferences. At one locality, remains of a stone structure—perhaps the base of a circular hut—were uncovered; there is an excellent date of 1.8 million years for this. The threshold of technological ability is difficult to pinpoint exactly, and at best one could only suggest that it occurred during the Pliocene, perhaps in relation to the adaptive response embodied by the differentiation of *Homo*.

During the early Pleistocene, circa 1.6 million years ago, bifacial tools such as crude handaxes make their appearance. This development can be traced in situ at Olduvai and is supported by findings from other East African sites. The first record of stone implements in Asia and Europe is of handaxes, but, unfortunately, no absolute dates are known for them at the present time. In my opinion, the evidence available could suggest a migration of the "handaxe" people—perhaps *Homo erectus*—from Africa into Europe, Asia, and the Far East during the early Pleistocene, or a little earlier. The subsequent development of stone implements is complex, with impressive late Pleistocene and Holocene records from most of the world. It is not proven but can be postulated that the post-Acheulean, or handaxe, technologies can be related to the emergence of *Homo sapiens* and the subsequent success of this species. The association of stone implements with early hominid remains is rare, and many mid-Pleistocene and subsequent sites contain only one or two specimens, with certain impressive exceptions. The earliest record of *Homo sapiens* remains a problem because of the limitations imposed both by the small samples and by the dating techniques that can be employed.

It is clear that extraordinary advances have been made in recent years in our data records, and continuing investigations will presumably provide further evidence. There is now obvious evidence for considerable morphological diversity in the Plio/Pleistocene hominids of Africa, which has been interpreted as a consequence of a Pliocene radiation with different evolutionary experiments persisting into the early Pleistocene. The presence of at least three contemporary species in East Africa may be established on both cranial and postcranial material, and any review must incorporate the analysis of the entire fossil collection. The problem of whether two closely related and competing species could live side by side will be better understood when further studies are completed on paleoecological evidence, including palynology and micropaleontology under closely documented stratigraphy. It is known that several closely allied and morphologically similar cercopithecoid monkeys live alongside one another in African forests today, with no suggestion of mutual exclusion.

Darwin seems to be vindicated for his prophesy on the African origin for man. Many great scientists—Robert Broom, Raymond Dart, and Louis Leakey among them—were pioneers in the field which today has so captured the popular imagination. As the ancient record builds up with each new fossil find, we are reminded that extinction is a common phenomenon in vertebrate evolution—and modern man may be no exception.

References

1. L. S. B. Leakey. 1961. A new Lower Pliocene fossil primate from Kenya. *Ann. Mag. Nat. Hist.*, ser. 13, 4:689–96.
2. E. L. Simons. 1969. Late Miocene hominid from Fort Ternan, Kenya. *Nature,* Lond. 221:448–51.
3. E. L. Simons and D. R. Pilbeam. 1972. Hominoid paleoprimatology. In *The Functional and Evolutionary Biology of Primates,* ed., R. H. Tuttle, Chicago:Aldine, Atherton, pp. 36–62.
4. M. Kretzoi. 1975. New ramapithecines and *Pliopithecus* from the Lower Pliocene of Rudabanya in northeastern Hungary. *Nature,* Lond. 257:578–81.
5. R. E. F. Leakey. 1973. Further evidence of Lower Pleistocene hominids from East Rudolf, North Kenya, 1972. *Nature,* Lond. 242:170–73.
6. W. W. Bishop and G. R. Chapman. 1970. Early Pliocene sediments and fossils from the Northern Kenya Rift Valley. *Nature,* Lond. 226:914–18.
7. M. Pickford. 1975. Late Miocene sediments and fossils from the Northern Kenya Rift Valley. *Nature,* Lond. 256: 279–84.
8. B. Patterson, A. K. Behrensmeyer, and W. D. Sill. 1970. Geology and fauna of a new Pliocene locality in North Western Kenya. *Nature,* Lond. 226:918–21.
9. M. Taib, D. C. Johanson, Y. Coppens, R. Bonnefille, and J. Kalb. 1974. Decouverte d'hominides dans les series Plio-pleistocenes d'Hadar (Bassin de l'Awash; Afar, Ethipie). *C. R. Acad. Sc. Paris.,* Ser. D. 279:735–38.
10. P. V. Tobias. 1967. Olduvai Gorge, 2. *The Cranium and Maxillary Dentition of Australopithecus (Zinjanthropus) boisei.* Cambridge: Cambridge Univ. Press.
11. R. E. F. Leakey. 1974. Further evidence of Lower Pleistocene hominids from East Rudolf, North Kenya, 1973. *Nature,* Lond. 248:653–56.
12. R. E. F. Leakey. 1973. Evidence for an advanced Plio/Pleistocene hominid from East Rudolf, Kenya. *Nature,* Lond. 242: 447–50.
13. R. E. F. Leakey. In press. New hominid fossils from the Koobi Fora Formation, Northern Kenya. *Nature,* Lond.
14. L. S. B. Leakey, P. V. Tobias, and J. R. Napier. 1964. A new species of the genus *Homo* from Olduvai Gorge. *Nature,* Lond. 202:7–9.
15. F. Weidenreich. 1943. The skull of *Sinanthropus pekinensis:* A comparative study on a primitive hominid skull. *Paleont. Sinica,* New Series D. 10:1–298.
16. M. D. Leakey. 1971. *Olduvai Gorge, 3. Excavations in Beds I & II 1960–1963.* Cambridge: Cambridge Univ. Press.

Index